KU-699-316

The Harvard Lectures of Alfred North Whitehead, 1924–1925

The Edinburgh Critical Edition of the Complete Works of Alfred North Whitehead

Series editors: George R. Lucas, Jr., Notre Dame University and Brian G. Henning, Gonzaga University

The complete, collected critical edition of the unpublished and published manuscripts of British-American philosopher Alfred North Whitehead

This Critical Edition will bring together for the first time, in a series of critically edited volumes, the complete, collected published works and previously unpublished lectures, papers and correspondence of Alfred North Whitehead, one of the twentieth century's most original and significant philosophers.

Newly discovered materials, long thought lost or destroyed, including illustrations, equations and chalkboard diagrams, which Whitehead used in classroom settings, have never before been seen by contemporary scholars. A projected six volumes of unpublished material will illuminate many factors influencing the development of Whitehead's initial and later thought, as well as elucidate, in considerably greater detail than ever before, many of the principal concepts later set out in his body of published philosophical reflection.

New critical editions of his famous lecture series at Harvard University, such as *Science and the Modern World*, *Symbolism* and *The Function of Reason*, his world-renowned Gifford Lecture series at Edinburgh University in 1928, *Process and Reality*, together with new editions of justifiably famous collections of later essays and public lectures, such as *Adventures of Ideas*, *Modes of Thought* and *Essays on Science and Philosophy*, will round out this ambitious and authoritative Critical Edition.

Editorial Advisory Board

The editorial advisory board is composed of respected Whitehead scholars from the United States, Canada and Europe who will ensure the scholarly integrity and quality of the Critical Edition of Whitehead.

Editorial Assistants

John Becker and Richard Livingston.

The Harvard Lectures of Alfred North Whitehead, 1924–1925

Philosophical Presuppositions of Science

Edited by Paul A. Bogaard and Jason Bell

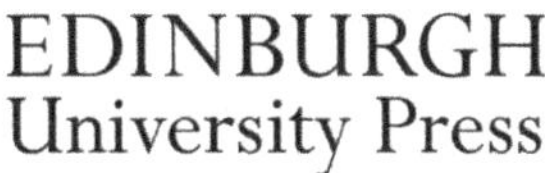

Edinburgh University Press is one of the leading university presses in the UK. We publish academic books and journals in our selected subject areas across the humanities and social sciences, combining cutting-edge scholarship with high editorial and production values to produce academic works of lasting importance. For more information visit our website: edinburghuniversitypress.com

Edinburgh University Press Ltd
The Tun – Holyrood Road, 12(2f) Jackson's Entry, Edinburgh EH8 8PJ

Typeset in Garamond and Cambria
by R. J. Footring Ltd, Derby, and
printed and bound in Great Britain by
CPI Group (UK) Ltd, Croydon CR0 4YY

A CIP record for this book is available from the British Library

ISBN 978 1 4744 0184 5 (hardback)
ISBN 978 1 4744 0185 2 (webready PDF)
ISBN 978 1 4744 0468 6 (epub)

Contents

General introduction

George R. Lucas, Jr, *General Editor*

Brian G. Henning, *Executive Editor*

Alfred North Whitehead, a British-born philosopher who attained widespread fame in America in the first half of the twentieth century, is perhaps best known among the wider public for his famous saying that the European philosophical tradition 'consists of a series of footnotes to Plato'.[1]

A student of mathematics at Trinity College, Cambridge, in the 1880s, Whitehead was elected a Fellow of the College in 1884, where he mentored the mathematical studies of such notable figures as John Maynard Keynes and Bertrand Russell, and was also inducted as a member of one of the most elite societies in the English-speaking university world at the time, the famed 'Cambridge Apostles'.

In 1910, Whitehead resigned his fellowship at Trinity College and moved to London, where (after a year spent in research and writing) he obtained a lectureship in mathematics at University College London. In 1914, he began lecturing at the newly organised Imperial College of Science and Technology (now Imperial College London) before being elected Dean of the Faculty of Science at the University of London in 1918 and Chairman of its Academic Council in 1920.[2]

While at the University of London, Whitehead successfully lobbied for a new history of science department, helped establish a bachelor of science degree in 1923, and made the school more accessible to less wealthy students. Finally, at the age of 63, Whitehead retired from his position at the University of London and sailed to America to accept a coveted chair at Harvard University, becoming a naturalised citizen and remaining an active and prolific scholar until his death on 30 December 1947 in Cambridge, Massachusetts.

At the beginning of the twentieth century, Whitehead began a collaboration with his student and protégé Lord Bertrand Russell that lasted well over a decade and resulted in the publication (in 1910, 1912 and 1913) of the three-volume *Principia Mathematica*, a monumental study intended to establish the foundations of mathematics in formal logic. Subsequently, in 1922, Whitehead himself concluded two decades of study of the geometry of space-time in *The Principle of Relativity*, a path-breaking work of theoretical cosmology explicitly intended to pose an alternative to Albert Einstein's

1. Alfred North Whitehead, *Process and Reality* (1929) (New York: Free Press, 1979), p. 39.
2. Both University College London and the Imperial College of Science and Technology were member colleges of the University of London.

General Theory. Other works spanning Whitehead's academic career in England include *A Treatise on Universal Algebra* (1898), 'On mathematical concepts of the material world' (1906), *An Introduction to Mathematics* (1911), *The Organization of Thought* (1917), *An Enquiry Concerning the Principles of Natural Knowledge* (1919) and a significant work on the philosophy of science, *The Concept of Nature* (1920). These works cemented Whitehead's reputation as one of the leading figures in the early development of analytic philosophy, alongside Russell, G. E. Moore and Ludwig Wittgenstein.

Upon coming to the United States, however, and with inspiration allegedly derived from the uniquely American philosophy of 'experience' (particularly that of William James and John Dewey), Whitehead undertook a series of works on far broader philosophical themes. His later work encompasses the history and philosophy of science, the sociological development of civilisation, the philosophy of history, philosophical reflections on education and, most famously, an ambitious attempt to develop a descriptive metaphysical system commensurate with early twentieth-century relativistic cosmology and quantum mechanics, which scholars now commonly label 'process philosophy'. Works during this American period of his career include *Science and the Modern World* (1925), *Religion in the Making* (1926), *Symbolism* (1927), *The Function of Reason* (1929) and his best-known work, *Process and Reality* (1929), based upon his Gifford lectures at the University of Edinburgh delivered in June 1928. In addition to subsequent books, such as *Adventures of Ideas* (1933) and *Modes of Thought* (1938), Whitehead also published a number of lectures and essays on diverse topics during this period, many of which he later published in collections such as *The Aims of Education* (1929) and *Essays in Science and Philosophy* (1948).

Whitehead's stature as a pivotal figure in American philosophy, his significant contributions to the rise of analytic philosophy, and the recent and worldwide renewal of interest in the constructive contribution of his seminal ideas to a number of important contemporary metaphysical and cosmological problems in philosophy argue for the appropriateness and timeliness of a complete critical edition of his works – an edition intended to rank alongside critical collected editions already produced or in progress on the works of his eminent predecessors and contemporaries, including William James, John Dewey, Josiah Royce, George Santayana and C. S. Peirce, as well as Einstein, Russell and Wittgenstein.

The need for a complete critical edition of the unpublished and published manuscripts of Whitehead is compelling. Beginning around 2010, an enormous body of unpublished material steadily began to come to light, thus helping to dispel the myth that all of the Whiteheadian *Nachlass* had been lost or destroyed.[1] The new discoveries include substantive correspondence

1. Whitehead's grandson, George Whitehead, confirms the story that Alfred North's heirs did destroy his personal effects at his request. There is ongoing disagreement among scholars regarding exactly what was destroyed. What have been discovered are items that were saved by friends, colleagues and students, as well as some of Whitehead's papers that were unknown to his family at the time of his death.

with students and leading intellectual and political figures of the period, along with copious and detailed lecture notes taken by both students and faculty colleagues attending Whitehead's classes. These lecture notes were recorded by philosophers who became quite accomplished in their own right, and some of whom (such as W. V. A. Quine) are characterised by an eminent historian of philosophy, Professor Jerome B. Schneewind, as numbering among the most significant philosophical figures of the twentieth century.[1]

Despite the fact that much of his work undertaken in America was written comparatively recently (mostly during the early part of the last century), scholars at present do not have a clear sense of what motivated Whitehead or what led to the development of his thought. Part of the reason for this has to do with the fact that his personal correspondence with colleagues – along with the lecture notes of his students which were not either discarded, donated or destroyed following his death – had been scattered in libraries and archives across the United States, Canada and the UK.

The editors and staff of the Whitehead Research Project have invested considerable time and effort locating, collecting and cataloguing a substantial body of these extant unpublished materials (many of which were mistakenly thought to have been lost to scholars forever). The newly discovered materials in particular shed considerable light on the factors influencing the development of Whitehead's later philosophy, and also often elucidate – in considerably greater detail than ever before – many of the principal concepts encompassed within that body of philosophical reflection (e.g. through the apparent use by Whitehead himself of illustrations, equations and blackboard diagrams in classroom settings which have never before been seen by contemporary scholars).

Whitehead's published works, meanwhile, remain in disarray: some are out of print, while others remain in print from multiple sources, replete with inconsistencies and extensive uncorrected typographical and textual errors. Many lack even a cursory index. Apart from a newly reprinted edition of Whitehead and Russell's *Principia Mathematica*, the sole exception was the re-issue (in 1978) of a so-called 'Corrected Edition' of *Process and Reality*. This proved to be a helpful interim work but it did not fully adhere to the rigorous protocols since established for scholarly editions generally.[2] Finally, while multiple copyrights for extant works are claimed by several publishers, in point of fact all of Whitehead's materials (published and unpublished) continue to remain legally in the sole possession of the Whitehead family estate, whose sole executor[3] has granted full permission for the future publication of both unpublished and published works to the Whitehead Research Project.

1. J. B. Schneewind, Chair, Department of Philosophy, Johns Hopkins University (letter of reference to the National Endowment for the Humanities, 30 June 2012).
2. See for example Mary-Jo Kline and Susan Holbrook Perdue, *A Guide to Documentary Editing* (3rd edition) (Charlottesville: University of Virginia Press, 2008).
3. George W. Whitehead (a grandson).

When finally completed, the *Edinburgh Critical Edition of the Complete Works of Alfred North Whitehead* will encompass the entire collected works of the author, published and unpublished. Furthermore, by organising, archiving, digitising, transcribing, editing and indexing all of his writings, the full breadth and significance of Whitehead's diverse work will, perhaps for the first time, come fully into view. Not only will this make available materials that have never been published, but it will also improve the scholarly understanding of previously published items. Perhaps the most useful and exciting aspect of this project to many scholars, however (beyond the publication of newly discovered material), will be the creation of a fully searchable electronic archive of these primary source materials.[1] In these ways, this *Critical Edition* should facilitate and energise the scholarly study of one of the twentieth century's most original and influential intellectual figures.

Outline and methodology of the Harvard lectures

The *Critical Edition* begins with a series of volumes devoted to the considerable body of recently discovered and almost entirely unpublished lecture notes, followed by two volumes of correspondence. Upon completion of the publication of all of the hitherto unpublished materials, the *Critical Edition* will turn, in chronological sequence, to the re-issuing of critically edited volumes of published works, from *Universal Algebra* (1898) through to the later American works, up to the date of Whitehead's death in 1947.

Whitehead mainly taught three courses during his Harvard career: Phil 3b ('Philosophy of Science'),[2] Phil 20h ('Seminary in Metaphysics') and Phil 20i ('Seminary in Logic'). This might imply a tedious and repetitive chronology of the same or similar topics over time, but this is not how Whitehead worked. Unlike the majority of academics – who most often teach a repertoire of courses for students in a major or graduate programme while attempting to reserve their own time for original and publishable scholarly research – Whitehead carried out his original research in the classroom and in public lecture halls, in dialogue with the undergraduate and graduate students in his core classes and seminars, and changed the course content based on the focus of his research at the time.

These changes are reflected to some degree in the shifting sub-titles for his Phil 3/3b, which during his 13-year teaching career at Harvard was variously called 'Philosophical Presuppositions of Science', 'Cosmologies Ancient and Modern' and 'The Function of Reason', among other titles. But in fact, topics changed more frequently than even changes in sub-titles suggest; every year was different. Louise R. Heath, who took Whitehead's Phil 3b two years in

1. Digital scans of many items can be reviewed on site at the Whitehead Research Project (WRP) in Claremont, California. Finding aid information for all items can be obtained by contacting the WRP staff. Visit http://whiteheadresearch.org for more information.
2. A course that Whitehead taught during both semesters, it was later split into two courses: Phil 3b (fall semester) and Phil 3 (spring semester).

a row, wrote at the top of her notes for the second year: 'Theoretically [this was the] same course as 1924, but I credited it because actually it was quite different'.[1] Whitehead did not tie himself to any particular set of topics from year to year, but simply taught his most current thought on whatever he was working on, often months before similar material would appear in print. Hence, the overriding significance of the materials included in these volumes of lectures is that they provide long-missing insight into the formation and development of Whitehead's thought during this crucial period.

It has long been believed, for example (as the record of his formally published work seems to suggest), that Whitehead first developed a provisional metaphysical synthesis of and foundation for modern science (in his first published American work, *Science and the Modern World*, in 1925) that differed substantively from his 'later' and quite distinct reformulation of these foundations in *Process and Reality*. Volume 1 of this *Critical Edition* demonstrates that this widespread interpretation is inaccurate, inasmuch as his later mathematical and geometrical foundations (especially Part IV of *Process and Reality*) are presented largely intact in the Emerson Hall lectures of 1924–5, while a more general, historical and publicly accessible account of these (same) foundations is offered in the Radcliffe College lectures (nominally under the same course heading) delivered at the same time (which formed the basis of *Science and the Modern World*). Having found during that first year that the technical account was somewhat daunting for his students – even for Harvard graduate students – Whitehead proceeded to alter the focus of the 'master course' significantly during the next two years, taking a more historical and descriptive approach to the materials presented in class as his thinking about these issues continued to evolve.

The lecture notes over the ensuing two years thus reflect the origins of at least some of the several divisions, or 'Parts', that would come to compose the formal publication of *Process and Reality* in 1929. These lectures themselves lead up to the delivery of the Gifford lectures in Edinburgh toward the end of the academic term in June 1928 (during which Whitehead reported encountering similar difficulties with audience comprehension and developing the proper 'pitch' for his lectures). What these notes together reveal is that the alleged 'transformation' of his thought from an 'early' (1925) to a quite distinctive 'later' version (1929) cannot be sustained on the basis of this new evidence. Instead, what we see is a more subtle 'evolution' of concepts and theoretical foundations, captured in a decided transformation in public dissemination and style. Indeed, the textual evidence suggests that Whitehead encountered a problem similar to that described by Wittgenstein in his own post-*Tractatus* lectures, in which the methodology involves abandoning a straightforward sequential-logical approach for a somewhat less systematic piecemeal criss-cross of the same subject-terrain from alternative perspectives,

1. Louise R. Heath, Philosophy 3b notes, Victor Lowe Papers, Ms 284 Box 2.9, Special Collections, Sheridan Libraries, Johns Hopkins University.

utilising varying modes of expression in order to gain insight into the complex whole of the subject more adequately.

In Whitehead's case, these lecture notes also demonstrate how he used his teaching and public presentations as the portrait or drawing board upon which to sketch and re-sketch his evolving thought, searching for a proper mode of presentation and delivery that would succeed in capturing the whole of a metaphysical vision that he himself had described to his son, North, in 1924 as 'the right way of looking at things', lacking only the opportunity that his Harvard years finally presented to set it all out systematically. This was not an easy task, of course, nor was it accomplished in a single year, or in a single course. Instead, we witness the true 'compositional history' of Whitehead's thought not in some hypothetical reconstruction, but in his very own words, during his own presentations, as recorded meticulously by his own students, all wrestling together with the problem of a proper reformulation of the metaphysical picture of nature as presented in the varieties of discoveries constituting modern science at the dawn of the twentieth century.

The annual succession of lecture notes in these volumes thus supplies the long-absent background for understanding the development and maturation of Whitehead's thought. For anyone interested in Whitehead, in process philosophy specifically, or in the evolution of early twentieth-century Anglo-American philosophy generally, these notes are a discovery of extraordinary significance.

Editorial principles

Lecture notes taken during Alfred North Whitehead's classes were recorded not only by his students but also by recent graduates and both junior and senior faculty colleagues. Some sets of notes were deposited in various library archives with which their owners were affiliated, while others were privately retained and discovered later by Whitehead's biographer, Victor Lowe, or by members of the Whitehead Research Project.[1]

In editing lecture notes taken during Whitehead's classes, the editors have operated on a policy of minimal interference with the text. Although we are dealing for the most part with notes on Whitehead's lectures taken by others, the text is the closest thing we have to Whitehead's words. This being the case, we correct what is clearly wrong (such as typographical errors and incorrect bibliographic information) but make no attempt to edit the text aggressively.

The following principles, listed alphabetically, have governed the way in which the text has been edited, standardised or silently corrected, across the *Critical Edition*. Elaborations and additions specific to each volume may be made as a matter of editorial discretion.

- *Angle brackets.* Angle brackets (⟨ and ⟩) have been used to indicate editorial intervention, since both parentheses and square brackets have been used by the original note-takers.
- *Capitalisation.* Capitalisation has been standardised without record according to the sixteenth edition of the *Chicago Manual of Style.*
- *Contractions/shorthand.*
    - If the shorthand or contraction is unambiguous, it has been silently expanded without record. This includes ditto marks.

1. For reflections on Whitehead's teaching style, see: A. H. Johnson, *Whitehead and His Philosophy* (Lanham: University Press of America, 1983) and 'Whitehead as teacher and philosopher', *Philosophy and Phenomenological Research*, vol. 29(3) (1969), pp. 351–76; William Ernest Hocking, 'Whitehead as I knew him', in George L. Kline (ed.), *Alfred North Whitehead: Essays on His Philosophy* (Englewood Cliffs: Prentice Hall, 1963), pp. 7–17 ; 'A. N. Whitehead in Cambridge, Mass.', which is chapter 5 of Dorothy Emmet's *Philosophers and Friends: Reminiscences of Seventy Years in Philosophy* (London: Macmillan, 1996); Victor Lowe's biography of Whitehead, *Alfred North Whitehead: The Man and His Work, Volume 1* and *Volume 2* (Baltimore: Johns Hopkins University Press, 1985, 1990); Lester S. King's letter to Lewis Ford, 3 May 1978 (LET322), in the Whitehead Research Project Archives, Center for Process Studies, Claremont School of Theology, Claremont, California, USA; also Lawrence Bragg's letter to Victor Lowe, 8 May 1969, in the papers of William Henry Bragg, Box 37, Folder B, Item 44, Royal Institution of Great Britain, London, UK; Harvey H. Potthoff's diary entry for 21 May 1936, diary held in the Harvey H. Potthoff Papers (Iliff 3-F), Series VII, Sub-Series G, Iliff School of Theology Archives, Denver, Colorado, USA; and George P. Conger's diary entry 'Notes on Whitehead's class lectures at Harvard University, 1926–1927', in Box 4 of the George P. Conger Papers (Mss 20), Literary Manuscripts Collection, University of Minnesota Libraries, Minneapolis, USA (that diary is to be published in a further volume of this Critical Edition); see also the Introduction to the present volume.

  - If a contraction or word(s) seems legible, but its meaning is ambiguous or unclear, a footnote has been added explaining the difficulty.
  - If the editors have a strong guess but cannot be sure about the reading of a word or a portion of a word, then they have placed angle brackets around those markings with a question mark at the end (e.g. ⟨example?⟩); or a footnote has been added.
  - If a word can be supplied through context, but is otherwise illegible, it has been placed within angle brackets with a question mark (e.g. 'Each age has its ⟨own?⟩ philosophy'); or a footnote has been added.
- *Dashes.* Dashes have all been standardised as en dashes.
- *Deletions.* Legible deletions have been retained and marked with strikethroughs (~~deletion~~). Insignificant accidental deletions have been silently removed without record. Insignificant accidental deletions include misspellings (~~freind~~ friend) and false starts (too~~m~~ much).
- *Diagrams/figures.* Whenever possible, diagrams and figures that would have appeared on the blackboard (and therefore were drawn by Whitehead) have been faithfully recreated or scanned and inserted in their original location relative to the rest of text.
- *Doodles.* Insignificant doodles have not been reproduced, but footnotes will indicate their existence.
- *Illegible words.* If a single word is completely illegible, the editors have substituted a question mark in angle brackets (⟨?⟩). If more than one word is illegible, the editors have used an ellipsis or series of ellipses of the approximate length of the illegible text within angle brackets and a question mark at the end (⟨……?⟩); or a footnote has been added.
- *Interlineations.* Authorial interlineations have been placed in their intended locations and marked with carets. Carets pointing upward (∧interlineation∧) indicate that the marked word has been moved to its present location from below, while carets pointing downward (∨interlineation∨) indicate that the marked word has been moved to its present location from above.
- *Italics.* All book titles and foreign words have been italicised. If a book title is underlined or enclosed in quotation marks, this has been silently removed and italics used instead.
- *Line and page breaks.* The finished editorial text will not retain the author's original line or page breaks, in most cases, and will not generally mark where line breaks occur, but for page breaks a record will be inserted between upright bars (|example|).
- *Line spacing.* Unusually large or small amounts of space between lines or paragraphs have not been retained in the finished text, in most cases, and no record has been made of them; but where deemed important to understanding the note-taker's intention some spacing may have been retained or a note added.
- *Marginal text.* Text sometimes appears in odd places on the page (e.g. vertical text, or text that has been circled and placed in the corner of the

page or in the margin). Placement of such text is a matter of editorial discretion, on a case-by-case basis. A footnote has been added explaining the original positioning of any text whose intended location within the larger text is ambiguous (e.g. 'This line originally appeared in the upper-left corner of the page').

- *Markings.* The retention of other marginal marks for emphasis, such as vertical lines, braces, arrows, or check marks, has been a matter of editorial discretion. See also 'Underlining'.
- *Markings not made by original author.* Any markings not made by the original author have been removed without record. These include archival markings, or any markings made by previous transcribers. If ambiguous, a footnote has been added.
- *Misspellings.* Misspellings have been corrected without record.
- *Notes.* The footnotes are of two types.
    - *Editorial notes/textual difficulties.* These notes explain textual treatment (e.g. difficulties with textual interpretation or describing odd text positioning).
    - *Substantive notes/contextual information.* These provide further information and context for the reader (e.g. information on a person or book that Whitehead mentions).
- *Page headers (including dates).* The note-takers' headers have been standardised to a uniform format. Dates marked on the second page or with page numbers may be retained at the editors' discretion if there are no other indicators of page breaks.
- *Paragraph indentation.* The indentation for paragraphs has been made uniform. However, for outline-style notes with a clear hierarchy the different levels of indentation have sometimes been preserved as a matter of editorial discretion.
- *Punctuation.* Punctuation may be changed without record for purposes of clarity; however, editors have been extremely conservative when making such changes, especially in relation to the addition of punctuation where the lecture notes are attempting to capture an oral presentation.
- *Separator lines.* Lines or other marks that the author has used to clearly separate one piece of text from another may be retained or silently removed as a matter of editorial discretion.
- *Supply, editorial.* Editors have inserted missing articles, prepositions or pronouns within soft angle brackets (⟨supply⟩).
- *Underlining.* Underlining (including double and wavy underlines) for emphasis has been retained, with the exception of underlines clearly not made by the original author and underlines of book titles, which have been italicised instead; the editors have not added underlining.

Acknowledgements

- Whitehead Research Project Executive Director Roland Faber, who supported this project from its initial stages.
- Claremont School of Theology, for its ongoing support of the Center for Process Studies and the Whitehead Research Project. Also graduate students from Claremont who have often served as editorial assistants, particularly Joseph Petek, who has made innumerable contributions to the Whitehead edition.
- Frederick and Nancy Marcus, for generous donations to fund editorial staff and the transcription process.
- Hocking-Cabot Fund for Systematic Philosophy, for supporting archival research at Harvard and Editorial Advisory Board meetings in Cambridge, MA.

Chronology for Alfred North Whitehead

1861 15 February. Birth as the last of four children.
1875 September. Whitehead leaves home to attend Sherborne School, a boarding school for boys (through to 23 June 1880).
1877 June. Whitehead's grandfather dies.
1880 October. Whitehead attends Trinity College, Cambridge, as a student (through to October 1884).
1884 17 May. Election to Cambridge Apostles.
1884 9 October. Whitehead elected a Fellow of Trinity College, Cambridge, and begins teaching mathematics there.
1888 First two papers published.
1888 Begins afternoon teaching at Girton College, Cambridge (then a women's college), probably for no pay, until 1913.
1890 16 December. Marriage to Evelyn Wade.
1891 31 December. Birth of T. North Whitehead.
1894 23 February. Birth of Jessie Whitehead.
1895 October. Bertrand Russell elected a Fellow of Trinity College.
1898 March. Whitehead's father dies.
1898 27 November. Birth of Eric Whitehead.
1899 Whitehead moves from Cambridge to Grantchester.
1901 Whitehead is appointed to the Council of Newnham College, Cambridge, although he never taught there.
1901 Bertrand and Alys Russell move in with the Whiteheads and live with them for about a year. Russell and Whitehead begin collaboration on the *Principia*.
1903 13 February. Whitehead is elected a Fellow of the Royal Society.
1907 January. Whitehead's brother-in-law, John Branch (married to his sister Shirley), kills himself. Whitehead visits to help put affairs in order.
1907 April. Whitehead moves back to Cambridge.
1907 May. Whitehead becomes Chairman of the Cambridge branch of the Men's League for Women's Suffrage.
1910 April. Whitehead resigns his Trinity lectureship and soon moves to Chelsea, London. He and his wife also buy a country home in Lockeridge, Wiltshire.
1910 December. First volume of the *Principia* is published.
1911 July. Whitehead is appointed to a teaching position at University College London.
1912 Whitehead is elected President for a year of both the South-Eastern Mathematical Association and the London branch of the Mathematical Association.
1914 10 July. Whitehead is appointed to a professorship at Imperial College London.
1914 15 July. Whitehead resigns from University College London.
1914 31 July. Gertrude Stein and Alice Toklas stay with the Whiteheads for six weeks.

1914 21 August. T. North Whitehead is enlisted to fight in World War I.
1914 1 September. Whitehead begins teaching at Imperial College London.
1915 Whitehead joins the Aristotelian Society, in which he is active through to his departure for America in 1924 (serves as President, 1922–3).
1915 Whitehead is elected President of the Mathematical Association, and serves for two years.
1916 11 July. Russell is dismissed from his Trinity lectureship due to his anti-war views.
1917 May. Eric Whitehead is enlisted in the Royal Flying Corps.
1917 The Whiteheads sell their country home in Lockeridge. They buy another in Oxted, Surrey.
1918 13 March. Eric Whitehead is killed in action.
1918 6 May. Russell is imprisoned for his anti-war views until 11 September of the same year.
1918 31 October. Whitehead is elected Dean of the Faculty of Science at the University of London and holds the post for four years.
1919 18 October. Whitehead delivers the first of his Tarner lectures (he gives seven in total, through to 29 November that year).
1920 Whitehead is appointed Chairman of the Academic (Leadership) Council of the Senate of the University of London and holds the post for four years.
1920 July. Whitehead is appointed Chairman of the Delegacy for Goldsmiths' College, London, and holds the post until he leaves for Harvard in 1924.
1920 3 November. T. North Whitehead marries Margaret Schuster.
1921 Eric Alfred North Whitehead is born, T. North's son and Whitehead's grandson.
1922 Whitehead accepts first invitation to cross the Atlantic to America in April for a conference at Bryn Mawr College in honour of mathematics professor Charlotte Angas Scott.
1924 2 February. Whitehead's mother dies.
1924 24 February. Whitehead accepts a five-year appointment at Harvard.
1924 1 April. Whitehead agrees to give the Lowell lectures, responding to the invitation of 18 March.
1924 16 August. Whitehead leaves England for Cambridge, Massachusetts.
1924 27 August. Whitehead arrives in Cambridge.
1924 25 September. Whitehead gives his first lectures at Harvard.
1924 4 October. Whitehead moves into a new flat in Radnor Hall (now part of the Memorial Drive Apartments Historic District), overlooking the Charles River.
1925 2 February. Whitehead delivers the first of his Lowell lectures, which he proceeds to give through to 2 March, every Monday and Thursday at 5 p.m., with the exception of 23 February (these would become *Science and the Modern World*).
1925 5 April. Whitehead delivers a lecture on 'Religion and science' at Phillips Brooks House at Harvard.
1925 12 April. Whitehead has lunch with John Dewey.
1925 14 April. Whitehead gives a lecture on 'Mathematics as an element in the history of thought' to the Mathematical Society at Brown University, Providence, Rhode Island.

1925 29 June. Jessie Whitehead arrives in Cambridge, Massachusetts, and begins working in the Harvard library on 20 July.
1925 30 December. Whitehead attends his first meeting of the American Philosophical Association and gives a paper on the subject of time.
1926 February. Whitehead delivers a second series of Lowell lectures (these would become *Religion in the Making*).
1926 February. The 'Committee of Four' begins meeting (Whitehead, L. J. Henderson, John Livingston Lowes, Charles P. Curtis). Together they choose four to eight PhD students per year for the Society of Fellows for a term of three years each with an option for three more (among these Junior Fellows was Quine, chosen in the first batch).
1926 During Harvard's spring break, Whitehead delivers lectures at McGill University (Montreal, Canada), the University of Michigan and the University of Illinois.
1926 September. Whitehead attends the Sixth International Congress of Philosophy at Harvard, and gives a paper entitled 'Time'.
1927 27 February. Whitehead accepts an invitation to deliver the Gifford lectures for 1928.
1927 April. Whitehead delivers the Barbour-Page lectures at the University of Virginia (these would become *Symbolism: Its Meaning and Effect*).
1928 1 June. Whitehead delivers the first of the Gifford lectures (which he proceeds to give through to 22 June).
1929 March. Whitehead delivers the Vanuxem lectures at Princeton University, New Jersey (these would become *The Function of Reason*).
1930 Whitehead attends the Seventh International Congress of Philosophy in Oxford.
1931 14 February. Whitehead attends a symposium for his seventieth birthday, arranged by James Woods.
1931 February. T. North Whitehead joins a human relations study group at Harvard Business School and moves to Cambridge, Massachusetts.
1931 29 December. Whitehead delivers his presidential address to the eastern division of the American Philosophical Association at New Haven.
1931 Whitehead is elected a Fellow of the British Academy.
1937 7 May. Whitehead delivers his final Harvard lecture.
1941 15 February. Whitehead attends a party for his eightieth birthday, organised by his Harvard colleagues.
1945 1 January. Whitehead is awarded the Order of Merit.
1947 30 December. Whitehead dies from a cerebral haemorrhage.

Published works of Alfred North Whitehead

1886 10 February. 'A celebrity at home. The clerk of the weather'. *Cambridge Review*, vol. 7, pp. 202–3.

1886 12 May. 'Davy Jones'. *Cambridge Review*, vol. 7, pp. 311–12.

1888 6 March. 'A visitation'. *Cambridge Fortnightly*, vol. 1(4), pp. 81–3.

1888 'On the motion of viscous incompressible fluids: a method of approximation'. *Quarterly Journal of Pure and Applied Mathematics*, vol. 23, pp. 78–93.

1888 'Second approximations to viscous fluid motion. A sphere moving steadily in a straight line'. *Quarterly Journal of Pure and Applied Mathematics*, vol. 23, pp. 143–52.

1891 20 February. 'The Fens as seen from skates'. *Cambridge Review*, vol. 12 (attribution to Whitehead not wholly certain).

1896 14 May. 'On ideals: with reference to the controversy concerning the admission of women to degrees in the University'. *Cambridge Review*, vol. 17, pp. 310–11.

1898 10 March. 'The geodesic geometry of surfaces in non-Euclidean space'. *Proceedings of the London Mathematical Society*, vol. 29, pp. 275–324.

1898 *A Treatise on Universal Algebra*. Cambridge University Press Warehouse.

1899 2 February. 'Sets of operations in relation to groups of finite order'. *Proceedings of the Royal Society of London*, vol. 64, pp. 319–20.

1901 April. 'Memoir on the algebra of symbolic logic'. *American Journal of Mathematics*, vol. 23(2), pp. 139–65.

1901 October. 'Memoir on the algebra of symbolic logic, part II'. *American Journal of Mathematics*, vol. 23(4), pp. 297–316.

1902 'On cardinal numbers'. *American Journal of Mathematics*, vol. 24, pp. 367–94.

1903 'The logic of relations, logical substitution groups, and cardinal numbers'. *American Journal of Mathematics*, vol. 25, pp. 157–78.

1903 14 May. 'The University library'. *Cambridge Review*, vol. 24, p. 295.

1904 'Theorems on cardinal numbers'. *American Journal of Mathematics*, vol. 26, pp. 31–2.

1905 'Note'. *Revue de Metaphysique et de Morale*, vol. 13, pp. 916–17.

1906 29 March. 'On mathematical concepts of the material world'. *Philosophical Transactions of the Royal Society of London*, vol. 77(517), pp. 290–1.

1906 Through to 1910. *The Axioms of Projective Geometry*. Cambridge University Press Warehouse.

1906 5 November. 'Liberty and the enfranchisement of women'. Cambridge Women's Suffrage Association, pamphlet.

1907 March. *The Axioms of Descriptive Geometry*. Cambridge University Press Warehouse.

1910 (With Bertrand Russell) 'Non-Euclidean geometry'. *Encyclopedia Britannica* (11th edition), vol. 11, pp. 724–30.

1910 'Axioms of geometry'. *Encyclopedia Britannica* (11th edition), vol. 11, pp. 730–6.
1910 October. 'The philosophy of mathematics'. *Science Progress in the Twentieth Century*, vol. 5, pp. 234–9.
1910 December (with Bertrand Russell) *Principia Mathematica, Volume 1.* Cambridge University Press.
1911 *An Introduction to Mathematics.* Williams and Norgate, Henry Holt.
1912 'The place of mathematics in a liberal education'. *Journal of the Association of Teachers of Mathematics for the Southeastern Part of England*, vol. 1(1).
1912 (With Bertrand Russell) *Principia Mathematica, Volume 2.* Cambridge University Press.
1913 (With Bertrand Russell) *Principia Mathematica, Volume 3.* Cambridge University Press.
1913 'The principles of mathematics in relation to elementary teaching'. *Proceedings of the Fifth International Congress of Mathematicians*, vol. 2, pp. 449–54. Also published in *L'Enseignment Mathematique.*
1913 March. 'Presidential address to the London branch of the Mathematical Association'. *Mathematical Gazette*, vol. 7, pp. 87–94.
1914 'Report of the Council of the Royal Society of London'. *Year-Book of the Royal Society*, pp. 177–87.
1915 'Report of the Council of the Royal Society of London'. *Year-Book of the Royal Society*, pp. 176–85.
1916 January. 'The Aims of education: a plea for reform'. *Mathematical Gazette*, vol. 8, pp. 191–203.
1916 'Space, time, and relativity'. *Proceedings of the Aristotelian Society*, vol. 16, pp. 104–29.
1916 'The organisation of thought'. *Report of the 86th Meeting of the British Association for the Advancement of Science*, pp. 355–65. Also published with slight differences in *Proceedings of the Aristotelian Society*, vol. 17, pp. 58–76.
1916 'La théorie relationniste de l'espace'. *Revue de Métaphysique et de la Morale*, vol. 23, pp. 423–54.
1917 *The Organisation of Thought.* Williams and Norgate, Greenwood Press.
1917 March. 'Technical education and its relation to science and literature'. *Mathematical Gazette*, vol. 9(128), pp. 20–33. Also published in *Technical Journal* vol. 10 (January), pp. 59–74.
1917 (With A. W. Siddons) 'Letter to the Editor'. *Mathematical Gazette*, vol. 9, p. 14.
1918 'Graphical solution for high-angle fire'. *Proceedings of the Royal Society*, Series A 94, pp. 301–7.
1919 'Fundamental principles of education'. *Report of the 87th Meeting of the British Association for the Advancement of Science*, p. 361.
1919 (With Oliver Lodge, J. W. Nicholson, Henry Head, Adrian Stephen and H. Wildon Carr) 'Symposium: time, space and material: are they, and if so in what sense, the ultimate data of science?' *Proceedings of the Aristotelian Society, Supplementary Volumes*, vol. 2, 'Problems of Science and Philosophy', pp. 44–108.

1919 *An Enquiry Concerning the Principles of Natural Knowledge*. Cambridge University Press.
1919 15 November. 'A revolution in science'. *The Nation*, vol. 26, pp. 232–3.
1919 'Address on founder's day [Stanley Technical Trade School, South Norwood, London]'. Coventry and Son.
1920 *The Concept of Nature*. Cambridge University Press.
1920 12 February. 'Einstein's theory: an alternative suggestion'. *Times Educational Supplement*.
1921 'Science in general education'. *Second Congress of the Universities of the Empire*, pp. 31–9.
1921 (With many others) *Report of the Committee Appointed by the Prime Minister to Inquire into the Position of Classics in the Educational System of the United Kingdom*. His Majesty's Stationary Office.
1922 20 February. (With H. W. Carr, T. P. Nunn and Dorothy Wrinch) 'Discussion: the idealistic interpretation of Einstein's theory'. *Proceedings of the Aristotelian Society, New Series*, vol. 22, pp. 123–38.
1922 18 April. 'Some principles of physical science', presentation at Bryn Mawr College conference put in print as a booklet and also published later that year as a chapter in *The Principle of Relativity*.
1922 16 July. 'The philosophical aspects of the principle of relativity'. *Proceedings of the Aristotelian Society New Series*, vol. 22, pp. 215–23.
1922 *The Principle of Relativity, with Applications to Physical Science*. Cambridge University Press.
1922 November. 'The rhythm of education'. *Bulletin of the American Association of University Professors*, vol. 9(7), pp. 17–19.
1922 6 November. 'Uniformity and contingency: the presidential address'. *Proceedings of the Aristotelian Society New Series*, vol. 23, pp. 1–18.
1923 (With H. W. Carr and R. A. Sampson) 'Symposium. The problem of simultaneity: is there a paradox in the principle of relativity in regard to the relation of time measured and time lived?' *Aristotelian Society*, Supplementary 3, 'Relativity, Logic and Mysticism', pp. 15–41.
1923 'The place of classics in education'. *Hibbert Journal*, vol. 21, pp. 248–61.
1923 'The rhythmic claims of freedom and discipline'. *Hibbert Journal*, vol. 21, pp. 657–68.
1923 'The first physical synthesis'. In *Science and Civilization*, pp. 161–78. Oxford University Press.
1923 17 February. 'Letter to the Editor: Reply to review of *The Principle of Relativity*'. *New Statesman*.
1925 August. 'Religion and science'. *Atlantic Monthly*, vol. 136, pp. 200–7.
1925 'The importance of friendly relations between England and the United States'. *Phillips Bulletin*, vol. 19, pp. 15–18.
1925 October. *Science and the Modern World*. Macmillan.
1926 September. *Religion in the Making*. Macmillan and Cambridge University Press.
1926 '*Principia Mathematica*: to the Editor of *Mind*'. *Mind*, vol. 35(137), p. 130.
1926 Autumn. 'The education of an Englishman'. *Atlantic Monthly*, vol. 138, pp. 192–8.

1927 'England and the narrow seas'. *Atlantic Monthly*, vol. 139, pp. 791–8.
1927 'Time'. In *Sixth International Congress of Philosophy*, pp. 59–64. Longmans, Green & Co.
1927 November. *Symbolism: Its Meaning and Effect*. Macmillan.
1928 'Universities and their function'. *Atlantic Monthly*, vol. 141, pp. 638–44.
1929 *Process and Reality*. Macmillan and Cambridge University Press.
1929 *The Aims of Education and Other Essays*. Macmillan, Williams & Norgate.
1929 *The Function of Reason*. Princeton University Press.
1930 January. 'An address delivered at the celebration of the fiftieth anniversary of the founding of Radcliffe College'. *Radcliffe Quarterly*, vol. 14, pp. 1–5.
1930 'Prefatory note'. In Susanne K. Langer, *The Practice of Philosophy*, p. vii. Henry Holt & Co.
1931 'On foresight'. Introduction, in Wallace Brett Donham, *Business Adrift*, pp. xi–xxix. McGraw-Hill.
1931 'Objects and subjects'. *Proceedings and Addresses of the American Philosophical Association*, vol. 5, pp. 130–46.
1932 'Symposium in honor of the seventieth birthday of Alfred North Whitehead', speech included in a booklet of symposium proceedings printed by Harvard University Press (pp. 22–9).
1933 *Adventures of Ideas*. Macmillan and Cambridge University Press.
1933 'The study of the past – its uses and its dangers'. *Harvard Business Review*, vol. 11(4), pp. 436–44.
1934 'Foreword'. In *The Farther Shore: An Anthology of World Opinion on the Immortality of the Soul*. Houghton Mifflin.
1934 *Nature and Life*. University of Chicago Press. (This book, which comprises two essays, was republished in its entirety as Part 3 of *Modes of Thought*, 1938.)
1934 'Philosophy of life'. In Dagobert D. Runes (ed.), *Twentieth Century Philosophy: Living Schools of Thought*, pp. 131–44. Philosophical Library.
1934 'Foreword'. In Willard van Orman Quine, *A System of Logistic*, pp. ix–x. Harvard University Press.
1934 July. 'Indication, classes, numbers, validation'. *Mind*, New Series, vol. 43(171), pp. 281–97.
1934 October. 'Corrigenda. Indication, classes, numbers, validation'. *Mind*, New Series, vol. 43(172), p. 543.
1935 'Minute on the life and services of Professor James Haughton Woods'. *Harvard University Gazette*, vol. 30, pp. 153–5.
1935 'The aim of philosophy'. *Harvard Alumni Bulletin*, vol. 38, pp. 234–5.
1935 'Eulogy of Bernard Bosanquet'. In *Bernard Bosanquet and His Friends: Letters Illustrating the Sources and the Development of His Philosophical Opinions*, p. 316. George Allen & Unwin.
1936 'Memories'. *Atlantic Monthly*, vol. 157, pp. 672–9.
1936 'Harvard: the future'. *Atlantic Monthly*, vol. 159, pp. 260–70.
1937 'Remarks'. *Philosophical Review*, vol. 46, pp. 178–86.
1938 *Modes of Thought*. Macmillan and Cambridge University Press.
1939 'An appeal to sanity'. *Atlantic Monthly*, vol. 163, pp. 309–20.

1939 'John Dewey and his influence'. In Paul Arthur Schilpp (ed.), *The Philosophy of John Dewey*, pp. 477–9. Northwestern University Press.

1940 'Aspects of freedom'. In Ruth Nanda Anshen (ed.), *Freedom: Its Meaning*, pp. 42–67. Harcourt, Brace.

1940 'The issue: freedom'. *Boston Daily Globe*, 24 December.

1941 'Autobiographical notes'. In Paul Arthur Schilpp (ed.), pp. 1–15. *The Philosophy of Alfred North Whitehead.* Northwestern University Press.

1941 'Mathematics and the good'. In Paul Arthur Schilpp (ed.), *The Philosophy of Alfred North Whitehead*, pp. 666–81. Northwestern University Press.

1941 'Immortality'. In Paul Arthur Schilpp (ed.), *The Philosophy of Alfred North Whitehead*, pp. 682–700. Northwestern University Press.

1942 February. 'The problem of reconstruction'. *Atlantic Monthly*, vol. 169, pp. 172–5.

1945 Preface to 'The organization of a story and a tale', by William Morgan. *Journal of American Folklore*, vol. 58(229), p. 169.

1942 October. 'Statesmanship and specialized learning'. *Proceedings of the American Academy of Arts and Sciences*, vol. 75(1), pp. 1–5.

1948 *Essays in Science and Philosophy*. Philosophical Library.

Introduction to *The Harvard Lectures of Alfred North Whitehead, 1924–1925*

Paul A. Bogaard, *Hart Massey Professor of Philosophy, Emeritus, Mount Allison University*

> On February 6, 1924, the President of Harvard University, A. Lawrence Lowell, wrote to Whitehead, asking whether he would accept an appointment as Professor of Philosophy for five years….[1]

Thus begins the 'Migration to Harvard' chapter of Victor Lowe's biography of Whitehead, detailing the steps leading to this formal offer and its aftermath. It marked the transition out of a career devoted to teaching mathematics and mathematical physics. It led into an extended career, well beyond five years, devoted to philosophy. This volume of lectures reflects the first step in that transition.

When Whitehead received this formal invitation he was days away from turning 63. A full career already lay behind him: 26 years as a Fellow at Trinity College, Cambridge, another 13 years teaching applied mathematics and serving in administration at the University of London, with over 50 books, articles and presentations to his credit.[2] Nevertheless, Whitehead was not yet ready to retire.

The Chairman of the Philosophy Department assured Whitehead that they would arrange plans within the department 'so that you could give your energy to the problems which interest you the most'. That might include a lecture course on the philosophy of science and one evening a week devoted to discussion of metaphysics and logic with graduate students.[3] When his wife asked him about this invitation to Harvard, Whitehead is said to have responded: 'I would rather do that than anything in the world'.[4]

They crossed from old England to New England in August and by 25 September Whitehead had plunged into the first course of philosophy lectures he had ever given. For these lectures he chose the title: 'The Philosophical Presuppositions of Science'. Within the departmental course listings it was

1. Victor Lowe, *Alfred North Whitehead: The Man and His Work, Volume 2* (Baltimore: Johns Hopkins University Press, 1990), p. 132.
2. See the chronology of Whitehead's life (pp. xvii–xix) and the chronological list of his publications (pp. xx–xxiv) provided in the present volume.
3. From the letter sent by Professor Woods quoted by Lowe, *Volume 2*, p. 135.
4. From the account of Lucien Price as quoted by Lowe, *Volume 2*, p. 135.

'Phil 3b' and it extended through both 'halves' of the year.[1] Arrangements were made for Whitehead to teach this course in Emerson Hall within Harvard Yard, and also at Radcliffe under the same course listing and title – likely in Fay House.[2] As Whitehead wrote to his son, North, on 9 November that year:[3]

> Tuesdays, Thursdays, Saturdays are my lecture mornings, viz. 9–10 at Radcliffe (repetition of Harvard course) and 12–1 at Harvard in Emerson Hall … they are 50 minute affairs.

He describes for North how he made his way through these teaching days and provides a glimpse of each class, first at Radcliffe and then in Emerson Hall:

> My class at Radcliffe has 9 or 10 members – 4 undergraduates, or postgraduates, taking it for 'credit' as the phrase is, and the others 'auditors' in the technical language used here. They are an intelligent set of young people.

One of these postgraduate students, Louise R. Heath, kept the notes she took from the course of lectures as delivered by Whitehead at Radcliffe. As for Emerson Hall:

> At my twelve o'clock lecture the audience is just 40 – mostly postgraduate and including about 10 members of the staff. I think that they are interested … and the audience has rather increased lately. At the end of the lecture men come up and ask questions.[4]

A senior colleague, William Ernest Hocking, attended many of these Emerson Hall lectures, beginning about a month into the semester, and kept his own notes. Both the Heath notes from Radcliffe and Hocking's from Emerson Hall have been known about for some time. What has been discovered only within recent years is that a younger colleague, Winthrop Pickard Bell, attended nearly all of the Emerson Hall lectures and his extensive notes have been preserved. Transcriptions of these three sets of notes form the major part of this volume.

There was also Whitehead's graduate seminar, Phil 20 h, 'Seminary in Metaphysics', during the first semester only:

> On Fridays there is the Seminary 7:30–9:30 in the evening. About 17 or 20 men, including some of the staff. I arrange for a student to read a paper, and we discuss it. I usually start the discussion, and intervene at irregular intervals. It is great fun. The men really discuss very well, with great urbanity and desire to get at the truth.

1. Although course numbers at Harvard often distinguished 'a' from 'b' depending on the semester in which a course was taught, Whitehead's course of lectures at both Radcliffe and in Emerson Hall was 'Phil 3b' and was offered through both semesters.
2. As we were told when enquiring at the Radcliffe Archives in March 2015.
3. Letters from Whitehead to his son, North, from August 1924 to August 1929 were transcribed and printed as Appendix B, to Lowe's *Volume 2*. That for 9 November 1924 appears on pp. 294–7.
4. Whitehead refers to 'men' at his Emerson Hall lectures, while we think it was women only at the Radcliffe lectures; although a formal list of graduate students (and colleagues) has not been found for the Friday evening 'Seminary', it seems to have included both men and women.

The following semester there was a 'Seminary in Logic' about which Victor Lowe managed to glean some information, although no notes have been found.[5] But for his Seminary in Metaphysics we do have notes from a few of these first-semester Friday evenings, written once again by Hocking. Notes for only a portion of the semester were found among Hocking's papers,[6] but those that have been found have been included in this volume.

So, we know quite explicitly what Whitehead was teaching at Harvard (including both at Emerson Hall and at Radcliffe), which matches what the department had asked of him at the beginning of the year. It was his first opportunity to explore his philosophical thoughts in the classroom. The lectures mark a major transition in his life and stand at a crucial juncture in his thinking. Especially since there are no remaining manuscripts of works that Whitehead published soon after this transition to New England, the two major ones being *Science and the Modern World* later in 1925 and *Process and Reality* in 1929, these lectures represent the only opportunity to gauge his ever-developing thoughts at this stage, and how these thoughts were growing beyond his earlier philosophical works: *An Enquiry Concerning the Principles of Natural Knowledge* (1919), *The Concept of Nature* (1920) and *The Principle of Relativity, with Applications to Physical Science* (1922). Assessing the importance of these lectures will now be possible, largely thanks to the efforts of his students and colleagues, and in remarkable detail. How has this come about?

Winthrop Pickard Bell[7]

Of the three people upon whom this volume depends for a window into Whitehead's teachings during his first academic year at Harvard, Winthrop Bell is the most extraordinary for the unexpected thoroughness of the notes he kept, and for the twists and turns that brought him to Whitehead's classroom in Emerson Hall. Both of these points require some explanation, and both present challenges.

Unexpected, because Bell's are not the typical notes one might expect to be taken to record university lectures. They are consistently thorough, to the point of providing an almost complete record of everything said that day – from beginning to end, including asides, jokes, reading and essay assignments – and a careful rendering of equations, diagrams and logical formalisms (in *Principia Mathematica* notation) copied from the blackboard.[8]

5. Lowe, *Volume 2*, p. 147.
6. Now held at Houghton Library, Harvard – see page li.
7. This section relies directly upon the 'Biography of Dr. Winthrop Pickard Bell' at the outset of the Winthrop Bell Virtual Exhibition located on the website of the Mount Allison University Archives, and the extensive work by Jason Bell (no relation) on W. P. Bell's years at Göttingen. See, for example, Jason Bell, 'Four originators of transatlantic phenomenology: Josiah Royce, Edmund Husserl, William Hocking, Winthrop Bell', in Kelly Parker and Jason Bell (eds), *The Relevance of Royce* (New York: Fordham University Press, 2014), pp. 47–68.
8. The Appendix to this volume provides sample pages scanned from the original handwritten notes kept by Bell, Hocking and Heath.

Was Whitehead's course of lectures the only example of Bell taking this amount of care? We can dispel that immediately by pointing out that during the same time, namely from October 1924 through to January 1925, Bell kept notes of the course of lectures for Phil 12c, 'The Philosophy of Aristotle', presented by Professor James H. Woods. There were no equations or logical notation, to be sure, but the same tiny handwriting similarly captured these lectures in full, replete with Greek in classical polytonic orthography. The best answer to when and how Bell developed this remarkable technique for taking notes is found early in his academic career.

However, Winthrop Bell's academic career was brief. After a promising beginning, with quite enviable opportunities, Bell nevertheless resigned from Harvard after teaching there only five years. His promising dissertation was never published (until recently rediscovered[1]) and his notes, papers and extensive correspondence were all taken with him back to his Nova Scotia home. And there they remained until a few years ago. How did it come to this?

Winthrop Pickard Bell (1884–1965) was born in Halifax to a prominent Canadian family: his father, Andrew, was a wealthy merchant and personal friend of the future Canadian Prime Minister Sir Robert Laird Borden (Borden would later request Winthrop Bell's services as his political interpreter at the Versailles treaty meetings); his maternal grandfather, Humphrey Pickard, was the founding President of what has become Mount Allison University; his brother Ralph served as Secretary of the Halifax Relief Commission after the Halifax Explosion of 1917 and served as Chancellor of Mount Allison University between 1960 and 1968.

Winthrop Bell's path into the academic world followed what was already a family tradition: a bachelor's degree in mathematics from Mount Allison in 1904, with honours, and a master's degree from Mount Allison in 1907 specialising in philosophy, history and German. Family wealth made possible his gaining a master's degree in philosophy from Harvard University in 1909; further graduate studies followed at Cambridge University, 1909–10, and in Germany at the University of Leipzig, 1910–11, and then at the University of Göttingen, 1911–14, although his PhD from Göttingen was delayed until 1922.

These years of delay were brought on by two unexpected twists: the declaration of war in 1914 found Bell not just unable to leave, but an 'enemy citizen' consigned to various German prisons for nearly the entire duration of the war, for the most part at Ruhleben. Even then, when peace was restored, Bell's papers reveal that he filed secret military reports on the political situation in Europe on behalf of the British and Canadian governments, activities for which he would, years later, be offered to become a Member of the Order of British Empire (an honour which he declined). Only thereafter did he return to North America, initially to his home in Nova Scotia; for the academic

1. It was Bell's complete doctoral dissertation that Jason Bell first realised was among the papers now kept in the Bell Fonds at Mount Allison University Archives. It is being published as: Husserliana Dokumente, Volume 5, *Winthrop P. Bell. Eine kritische Untersuchung der Erkenntnistheorie Josiah Royces. Dissertation 1914/22*, through the Husserl Archives at Leuven.

year 1921–2 he taught philosophy at the University of Toronto; and in 1922, with his PhD reinstated by Göttingen, he was appointed to the Department of Philosophy at Harvard University as 'Instructor and Senior Tutor'. By the end of his second year, his diary records that not only was he tutoring and teaching but he was also a reader for several PhD theses, including that of Charles Hartshorne.[1] During the 1924–5 academic year – the year Whitehead began his appointment – Bell shared his status in the department with Raphael Demos and Ralph Eaton. They each tutored and examined the students taking the basic Phil A course, for which J. H. Woods, R. B. Perry and W. E. Hocking provided the lectures. Each of them also taught a course of his own. It was Demos on whom Whitehead relied to do the examining of students taking Phil 3b,[2] the course on which Bell kept his extensive notes.

In 1927, to everyone's surprise, Bell resigned from Harvard and returned to Nova Scotia.[3] This curtailed any plans for completing a book already begun[4] but he did maintain contact with Harvard colleagues, had become good friends with Whitehead's son North, and maintained an extensive correspondence with Göttingen's phenomenological circle: Edmund Husserl, Edith Stein and many others. He was thereafter primarily engaged in business activities. During World War II, Bell lent his skills to the Canadian government to help constitute Canada's aircraft defences; he was elected Regent of Mount Allison in 1948; he was elected President of the Nova Scotia Historical Society in 1951; and he published a widely acclaimed work of local history, *The 'Foreign Protestants' and the Settlement of Nova Scotia*, in 1961. He died on 4 April 1965 in the province of his birth, in rural Chester, Nova Scotia, by the sea, at the age of 80.

Before we turn to the fate of Bell's books and papers, we must look a little more closely at his time in Göttingen. As a doctoral student between 1911 and 1914, during a peak period of the phenomenological movement, Bell attended several of Edmund Husserl's courses. In August 1912 Bell was invited to present an official seminar summary and Husserl was so impressed by the result that he asked to supervise Bell's dissertation. Here is an opportunity to shine light on the surprising level of detail that is found in Bell's transcriptions. Husserl (by Bell's own account to Herbert Spiegelberg, in 1955) initially asked Bell to put his pen down: he was taking too many notes and he should rather simply listen to see how phenomenology works.

1. Transcription of the 1924 diary by David Mawhinney, Bell Fonds, 8550/5/5.
2. Lowe, *Volume 2*, pp. 139–40.
3. Bell kept up correspondence with Demos for many years. A letter to Bell in 1930 from Zurich, where both Demos and Eaton were recuperating, reads: 'Well, Winthrop, you cleared out in time. I never thought the Harvard Phil Dept could run without us three – anyway, in the undergraduate part'. And a year later: 'It is a problem, I see, to be a philosopher and save your soul at the same time'. Bell Fonds, 8550/1/15, no. 4 and no. 6.
4. Decades later, Bell had apparently taken Husserl's defence of Royce to heart, writing: 'If I had ever finished the book I at one time had partly written, what is valuable in my dissertation would have been incorporated therein'. Bell to Hocking, 22 April 1957, William Ernest Hocking Collection, Houghton Library, Harvard University.

Yet for Bell, taking so many notes, and reviewing them later, was how he learned. Bell continues:

> It came to the last meeting but one. Now, said Husserl, we have had a good semester's work here. I think we should have for our last meeting a summary of the whole thing. 'Wer möchte das unternehmen? – Ach, Herr Bell! Sie haben immer so fleißig mitgescrieben. Ich glaube, Sie haben sich dazu bestimmt!' [Who would like to undertake this? Ah! Mr Bell! You have always so diligently written. I think you are the right one for the work!] ... I managed to please not only Husserl, but apparently the members of the seminar.... Then he said that if I wanted to do a *Doktorarbeit* under him I might come and see him at the beginning of the winter semester. It was an extraordinary piece of luck in every way – that he should have picked on me because of something he disapproved of, and then that the work of that semester's seminar had just been something that I should summarize with real satisfaction to myself and at the same time in a way that pleased him.[1]

Husserl's initial objection, that Bell wrote too much, points to Bell's skill as a recorder at both the early stages of phenomenology but also at the crucial transition towards Whitehead's process metaphysics. Likewise, Husserl's deeply positive impression with Bell's extensive notes mirrors our good fortune in 'hearing' Whitehead's lectures through Bell's extraordinary detail.

Bell's Göttingen studies occurred during a particularly formative time in the constitution of phenomenology. As Bell later recalled to Hocking:

> My experience was just at the time of some of [Husserl's] most intense absorption in his own work, in 1912–1913, when he was writing the first installment of his *Ideen* for the forthcoming *Jahrbuch*. He even, if I remember aright, had his meals served to him in his study, so that his trains of thought should not be interrupted by the conversation at the family table.[2]

Husserl was keen to build on Bell's familiarity with Josiah Royce[3]:

> Having sat under Royce at Harvard a couple of years earlier I was able to talk more or less enlighteningly on his name. Husserl asked to see some of his works ... [he] then would have nothing else than that I should do my Doktorarbeit on Josiah Royce.[4]

Despite an initial hesitation, Bell took up the task with his usual thoroughness, and Husserl termed Bell's dissertation an 'ausgezeichnet ...

1. Bell to Spiegelberg, 17 December 1955, Bell Fonds.
2. Bell to Hocking, 6 January 1962, Hocking Collection.
3. Josiah Royce (1855–1916) was born in California and spent nearly the entirety of his career as Professor of Philosophy at Harvard University. He defended a version of idealistic pragmatism, termed by him 'Absolute Pragmatism', against the comparatively individualistic pragmatism advanced by his personal friend and philosophical critic William James. Royce has been noted in recent scholarship as a progenitor of process and phenomenological philosophy. In addition to supervising Winthrop Bell's work on the master's degree, among Royce's notable students were George Santayana, T. S. Eliot, C. I. Lewis and William Hocking; Charles Hartshorne cites Royce as an early influence.
4. Bell to Spiegelberg, 25 September 1955, Bell Fonds. Correspondence from Bell to Hocking shows that Bell lent to Husserl Royce's collected works for a period of at least several years, and that he underlined key passages for Husserl. Bell later donated these books to Mount Allison University.

excellente Einleitung in die phänomenologische Erkenntnistheorie'[1] [a 'distinguished, excellent introduction in phenomenological epistemology'] which contained 'viel Schönes'.[2] Husserl sought to publish this work in the *Jahrbuch für Philosophie und phänomenologische Forschung*. It is finally appearing, over a century later, in the 'Husserliana Dokumente' series, along with Husserl's extensive commentary. Yet, why has it remained for so long unpublished? The answer to this may also help to shine some light on why Bell's other works, including his transcription of the Whitehead lectures, have so long remained unknown.

The conditions on 7 August 1914, the day when Husserl and the committee conducted Bell's oral examination, and promoted him with honours, were very far from ordinary. As Bell reports:

> I was caught in Germany by the outbreak of the First World War. In fact, my oral examination took place after the war had broken out, and under most unusual circumstances. I was in 'protective custody', having been hauled out of bed in the middle of the night when England declared war.[3]

Bell, as a Canadian citizen, was a British subject and hence an enemy alien in Germany. The university annulled Bell's doctorate and it was therefore quite impossible to publish the work at that time.

In 1922, Göttingen re-awarded Bell the doctorate, he took up his position at Harvard, and Husserl was finally in a position to offer to publish the dissertation in the *Jahrbuch*. Yet eight years after the completion of his work, Bell felt himself distanced from his criticisms of Royce, even as he remained in agreement with the essence of the work. Since in his later historical writings Bell showed himself to be something of a perfectionist, this might have been one element in his turning down Husserl's offer to publish the work. Husserl was certainly disappointed: 'es ist doch schade, dass Ihre Dissertation liegen bleiben soll. Hier dient sie mir gelegentlich als Lehrmittel'[4] ['It is a pity that your dissertation shall remain unpublished. It has served me here (in Freiburg) as a teaching aid'].

It is interesting to note that Husserl, the founder of the modern phenomenological movement, called Bell his 'weiser Mentor und Mithelfer'[5] – his wise friend and co-worker. As the first Anglophone doctoral student of Husserl's phenomenology, Bell introduced Husserlian phenomenology into the English language, as the first North American professor of phenomenology both in Canada (at Toronto), between 1921 and 1922,[6] and

1. Husserl's letter to Bell of 14 May 1922, Bell Fonds. Also published in Edmund Husserl, *Briefwechsel* (Husserliana: Edmund Husserl Dokumente 3/3, ed. Karl Schuhmann) (Dordrecht: Kluwer Academic, 1994).
2. Letter from Husserl to Bell, 30 September 1922, Bell Fonds.
3. Bell to Spiegelberg, 25 September 1955, Bell Fonds.
4. Husserl to Bell, 13 December 1922, Bell Fonds. Also published in Husserl, *Briefwechsel*.
5. Husserl to Bell, 7 December 1921, Bell Fonds. Also published in *Briefwechsel*, p. 30.
6. Upon this appointment, Husserl wrote to Bell (14 May 1922): 'Sie müssen doch ein famoser phänomenologischer Lehrer sein, in Ihrer übrigens echt englischen anschaulichen Sprache und präcisen Linienführung'. Bell Fonds. Also published in Husserl, *Briefwechsel*.

in the United States (at Harvard), between 1922 and 1927.[1] At Harvard, Bell taught the phenomenological method in graduate seminars that included such future Husserl students as Charles Hartshorne and Dorion Cairns, both of whom credited Bell for their introduction to Husserl. Indeed, Hartshorne would later be counted among those who were deeply influenced not only by Husserl but also by Whitehead.

William Ernest Hocking

Hocking's story is interwoven, if loosely, with Bell's. Almost 10 years before Bell's arrival in Göttingen Hocking had studied there, and may even have introduced to Husserl the idea of there being affinities between his recent *Logical Investigations* (1901) and aspects of the work of Josiah Royce, with whom Hocking had studied at Harvard. A curious thought, that the two colleagues each taking notes during the same lectures in 1924–5 had these features of their prior studies in common. More than that, Hocking had played a role in transmitting Husserl's recommendation for Bell to the Harvard department supporting Bell's appointment. One imagines them sitting side by side in Whitehead's lectures, responding each in his own way.

William Ernest Hocking (1873–1966) was born in Ohio and had studied engineering at Iowa State University before turning to philosophy. He studied under Royce at Harvard beginning in 1899 and completed his master's degree in 1901. The first American to study with Husserl at Göttingen, Hocking also studied in Berlin and Heidelberg before completing his PhD back at Harvard in 1904. His early years teaching were at Andover Theological Seminary, Berkeley and Yale before returning to Harvard in 1914.

Bell had also studied engineering for a time at McGill via Mount Allison and prior to his entering Harvard. But despite this sequence of similarities, the timing of these two careers was such that when Bell was studying at Harvard, Hocking was teaching elsewhere. And by the time Bell arrived in Göttingen, Hocking had returned to Harvard to join the staff. They were

1. Husserl successfully recommended Bell for the position at Harvard, writing to his former student Hocking, now a professor at Harvard, that Harvard would be well served by Bell's presence: 'Ein Wort wärmster Empfehlung. Eigentlich ist kein Wort stark genug, es zu seinen Gunsten auszusprechen. Ich rechne ihn mit Stolz zu meinen Freunden und danke dem Schicksal, dass es ihn mir zugeführt hat. Ich kenne ihn genau und stehe für diese Worte ein: Es ist eine der edelsten und bedeutendsten Persönlichkeiten, die mir in diesem Leben begegnet sind, einer der Menschen, die meinen Glauben an den Menschen aufrecht halten. Und nicht nur rein und bedeutend als Persönlichkeit, auch grundtüchtig, gediegen, vielversprechend als Philosoph. Schade dass seine Dissertation über Royce's Philosophie nicht zum Drucke kommen konnte, mit der er in Göttingen promovieren sollte. Die Fakultät hatte sie schon als 'valde laudabile' angenommen, das Examen rigorosum fand auch noch statt – in der Internierungsstätte (August 1914!), nachher wurde es aber als rechtsungiltig erklärt: und so hatte Bell das Göttinger Doctorat regelrecht gemacht und ist nun doch nicht Doctor! Würde sich Harvard seiner annehmen und ihm eine Stätte der Wirksamkeit bieten, so hätte es an ihm eine treffliche Kraft, die herrlich auf die Jugend wirken würde'. Original letter housed in the Hocking Collection. Also published in Husserl, *Briefwechsel*, pp. 164–5. Husserl's letter, by Hocking's report, had a good effect, as Hocking soon wrote back to Husserl with news of Bell's hire at Harvard, writing: 'I was rejoiced at what you said of Bell; and your word came at the right moment'. Husserl, *Briefwechsel*, p. 167.

both caught up in the Great War, nevertheless, if on quite different terms. While Bell was languishing in Ruhleben prison camp, allowed some study materials and correspondence but seriously restricted in what he could send out, Hocking was with the first detachment of American military engineers to reach the front in France. And while Bell was lingering in Germany reporting simultaneously for Reuters and British intelligence, Hocking was inspecting the courses used in army training camps.

Both men re-entered their academic careers soon after the war, but Hocking (more than 10 years Bell's senior) moved quickly into the senior ranks of the Philosophy Department at Harvard, and Bell began as a junior member of the department with his now reinstated PhD. Bell's personal diary[1] (quoted frequently throughout these lecture transcriptions) reveals that they worked together as colleagues and socialised as friends.

It is not quite clear whether each of these colleagues took on the task of keeping notes in Whitehead's first course of lectures simply out of personal interest, or whether there were arrangements within the department to ensure that such courses of lectures would be recorded. Each of them recorded these lectures, however, in a quite different manner: with only a few exceptions, Bell seems to have made an effort to record the whole of each lecture, and has preserved a complete run of all 85 lectures; Hocking's notes by comparison begin at the end of October, miss out here and there, and provide only a partial account of some lectures. On a few occasions Hocking seems to have arranged with another junior member of the department, Ralph Eaton, to cover for him. At the end of Eaton's typescript of notes for 16 December, he left a personal message for Hocking:

> Professor Hocking: I regret the meagerness of these notes, but find it impossible to listen and take a great many notes – The spirit of Whitehead is 'too wide a cosmos' to be got into notes. R.M.E.[2]

To some extent this must have been the reaction of anyone attempting to capture Whitehead's lectures in a page of notes.

Distinctive of Hocking's way of dealing with this challenge is his evident inclination to digest, organise and then record what are Whitehead's thoughts but also to arrange them in outline form in Hocking's own way. We can see this, of course, only because we now have Bell's notes with which to compare Hocking's. It is most heartening to find, therefore, that although Bell and Hocking often manage to capture material the other had failed to record, more often on key arguments, quoted phrases and pencilled copies of Whitehead's chalk drawings and so on, these two sets of notes support each other time after time, leaving us with the experience of reliving something quite close to Whitehead's original delivery.

1. Transcription of the 1924 diary by David Mawhinney, Bell Fonds, 8550/5/5.
2. See Hocking's notes for lecture 33 in this volume (page 147).

Louise R. Heath

Louise Heath, by comparison, was still a graduate student at Harvard, taking her courses at Radcliffe College, and we know of no one else recording these Radcliffe lectures. What she provides, thereby, is Whitehead presenting what was formally regarded as the 'same' course of lectures at Radcliffe as were recorded at Emerson Hall. Heath was earning credit for 'Phil 3b' under the same rubric – 'The Philosophical Presuppositions of Science' – as were the graduate student men taking the course on the same days. As Whitehead described for his son North, the Radcliffe lectures were 50-minute presentations at 9.00 a.m. each Tuesday, Thursday and Saturday, and the Emerson Hall presentations were each on the same three days beginning at noon. By comparing the notes taken at these two venues, we can determine, somewhat unexpectedly, that for much of the first semester each lecture given at Emerson Hall was then given at Radcliffe on the *next* class day (or even the one after), whereas for much of the second semester each lecture given at Radcliffe was the same (by and large) as given at Emerson Hall later that morning.

So, when Whitehead reports to North on 9 November that, after a 10-minute walk to Harvard Yard from Radcliffe, 'I then go to my room in the Widener Library – the University Library – and look over my notes for the 12 o'clock lecture and enlarge them',[1] he is not describing how he is reviewing and adjusting notes for a lecture he had just given an hour earlier, but how he is refreshing the notes intended for the next lecture, now to be given for the first time at Emerson Hall. The implication is that he has at the very least prepared notes for a number of lectures in advance, and that (through much of the first semester in particular) Radcliffe lectures were ones he had already presented a class or two previously. These are suppositions to which we shall return.

Louise Robinson Heath (1899–1988) has left behind somewhat less evidence of her career than the two male note takers, but it is a career devoted throughout to women's higher education. We think this is the same person who graduated from Newton Public High School in 1917 and then attended Mount Holyoke as a 'Mary Lyon Scholar' (named for Mount Holyoke's founder) in Philosophy and Biblical Literature, and earned her AB in 1921. She was at Radcliffe from 1921 or perhaps 1922 and completed her PhD there in 1927. In the preface to her one book, *The Concept of Time* (Chicago, 1936), Heath says:

> It is evident … that on this subject my thought has been very much influenced by Professor Whitehead. To him I owe thanks not only for sharing his thoughts on time but also for his time, both in class and in conference.

Her career from 1926 to 1947 was at Hood College, in Frederick, Maryland, while it was still an all-female institution; and then she was Dean at Keuka

1. Lowe, *Volume 2*, p. 295.

College, an all-female institution on the Finger Lakes in New York State, from 1947 to 1954.

Fortunately Heath preserved her notes from these 1924–5 lectures and at some point in her retirement (in the 1970s perhaps) passed them along to Whitehead's biographer, Victor Lowe. What Lowe found were student notes – the notes of a graduate student to be sure, but designed for her own use, with all the limitations of someone unfamiliar with much of what was being said concerning mathematical physics and complex logical notation. That being said, they capture a remarkable amount in a consistent manner and, given the notes available from the Emerson Hall lectures, make possible some substantive comparisons between the lectures presented in these two venues.

Three archival sources

In 1988 Victor Lowe's literary executor, J. B. Schneewind, was left with the unfinished second volume of *Alfred North Whitehead: The Man and His Work.*[1] Once Volume 2 was readied for publication in 1990, Schneewind declared in his Preface 'Thanks to [Lowe's] exhaustive investigations we know as much as it is possible to know about the life of a man who had the bulk of his papers destroyed'. Lowe himself, in the manuscript he had completed, mentions Louise Heath's notes from Radcliffe and also that there were some notes left by Hocking, but of Whitehead's own notes for these lectures Lowe explains:

> He did not keep those notes. I have not found anyone who attended the lectures at Harvard [i.e. Emerson Hall] in 1924–25, took detailed notes, and kept them.[2]

Fortunately, both men were wrong.

Even if Lowe had known about Winthrop Bell (there were those, after all, like Raphael Demos and Charles Hartshorne who had known him) it would have been challenging to track his papers to Nova Scotia. Bell had passed away there in 1965 and his papers had been sealed soon thereafter. They were destined to go to Bell's alma mater, Mount Allison University, the institution so well served by the Bell family. However, by the early 1970s only Bell's books and a few other items had actually been received. It was known at Mount Allison that the remainder of Bell's papers were coming but, for reasons known only to the family, there would be delays. Portions only were released to the Mount Allison Archives between 1986 and 2004. So, even if Lowe (or anyone else at that time) had tracked down the possibility of Bell's legacy, it would have been unavailable.

Whatever the reason for the delay of so many years – in large part Bell's work with British intelligence – the Archives staff at Mount Allison University were finally able to organise and index seven metres of material

1. Lowe's *Volume 1* of the Whitehead biography had been published in 1985.
2. Lowe, *Volume 2*, p. 144.

between 2003 and 2005, and by placing much of this description online[1] made it possible for scholars to discover, in the first case, the only complete copy of Bell's dissertation,[2] and soon thereafter the wealth of lecture notes he had retained both from his years in Göttingen and from the Whitehead and Woods courses of lectures at Harvard.

The notes on the Whitehead lectures were found in a rather innocent-looking folder containing about 90 pages tightly filled with handwriting in tiny script. Both sides of each page were used, occasionally even up the margins. Thanks to scanning (and then digitally enlarging and adjusting the contrast), and to the help and permission of the Archives, it has been possible to decipher this script and provide the transcriptions which set the baseline of information about these lectures – Whitehead's first ever lectures in philosophy.

Lowe may not have known about the Bell notes but he did point to Hocking's notes. 'On October 21 Professor Hocking began to attend the lectures. His notes from then to the end of the academic year may be found in Appendix 1 of Lewis Ford's *Emergence of Whitehead's Metaphysics*'.[3] As Lewis Ford himself acknowledged, what he brought to publication in 1985 was only partial; of those completely transcribed he seems not to have appreciated that at least three were missing, nor that the larger number which someone had summarised were, at best, summaries of summaries. We were able to locate the full set of notes that were among Hocking's uncatalogued papers, as organised and left to the Houghton Library at Harvard by his son, Richard Hocking. Arrangements were made to have them scanned and permission gained to transcribe them from these originals and present them here in full.

Lowe also mentions the Heath notes. At some point, perhaps in the 1970s, Lowe was able to contact her, and reports: 'Louise R. Heath kindly supplied me with her notes of Whitehead's lectures at Radcliffe that year'.[4] These have never been published and since they are now held by the Center for Process Studies at Claremont School of Theology, within the holdings of the Whitehead Research Library, we have been able to transcribe these in full, as well.

1. See http://www.mta.ca/wpbell/.
2. Jason Bell had noted the listing for Bell's dissertation in 2009 and, knowing this might be the only complete copy still existing, arranged for a postdoctorate position at Mount Allison beginning in the 2010/11 academic year.
3. Lowe, *Volume 2*, p. 144. At the outset of his Appendix 1, p. 262, Lewis Ford says: 'These notes of Whitehead's first course at Harvard were taken by one of his colleagues, William Ernest Hocking, and are here reproduced (in part) with the permission of his literary executer, Richard Hocking.... The lectures beginning with October 21, 1924 and ending March 28, 1925, have been summarized. The lectures beginning with March 31, 1925, and ending with May 26, 1925, because of their relevance to this study, have been put into publishable form without any major changes'. The beginning date of Hocking's notes, as it turns out, is 1 November. Only with Bell's notes by comparison (and a calendar) can one tell that Hocking misses three lectures in the set of notes Ford had transcribed in full.
4. Lowe, *Volume 2*, p. 144. In 1978, when Victor Lowe was at the Center for Process Studies (CPS) at the Claremont School of Theology as a visiting scholar, arrangements were made for a partial transcription of these notes with the assistance of Bruce Epperly, a graduate student employed by the CPS at the time. Our transcriptions have been completed directly from the originals, still held by the CPS.

The results are revealed in the table of contents of this volume: thanks to the notes taken and kept by these three individuals, we are now able to present the Harvard lectures of Alfred North Whitehead, his first course of lectures for the academic year 1924–5. These include the course of lectures presented at Emerson Hall as they were recorded by Winthrop Bell and, since William Ernest Hocking was recording a good many of these same lectures, we have presented them both for each of these lectures. Louise Heath's notes were from the lectures presented at Radcliffe, so we have set out our transcriptions of these separately, and as her notes do not consistently distinguish individual lectures (as Bell's and Hocking's invariably do) we have grouped her notes by semester, distinguishing individual lectures as best we can and linking them to each of the Bell–Hocking lecture transcriptions to enable cross-comparison.

Whitehead's only other teaching for the 1924–5 academic year were his semester-long 'seminaries' and since Hocking has provided us with notes for at least a few of these, from October, we have presented transcriptions of them as well. Lowe speaks of the first-semester and second-semester seminaries, but does not seem to have realised the extent, particularly in the first semester, to which there was an overlap of those attending nor that their discussions intersected at a few points.[1]

Finally, the Appendix presents single pages from each of these sources – taken from the same portion of the same lecture – to give a brief comparison of the three styles in their original form.

Whitehead's lecturing style

Lectures are by their very nature ephemeral; they are performances, and once over are gone. The score on which a conductor relies may persist, but that is not the same. By many accounts, Whitehead's performances were more like those of a conductor who only occasionally glanced at his score. There were no digital recordings captured by technicians, just handwritten notes taken by listeners; no archive of podcasts, just archival records documenting the occasion. Fortunately, we have some of these documents, but what we yearn for is to re-create the occasion. We have what a few listeners decided to write down and, insofar as it is possible, we can 'listen', through them, to Whitehead on the platform down front (Figures 1 and 2) and 'hear' him thinking through these issues and conveying his excitement and his conviction of their importance.

Lowe quotes Demos recalling 'a kind of oration' by the end of which 'the angels were singing'.[2] That was his response to Whitehead's inaugural lecture, on 25 September, about a month after the crossing from old England to New England. Bell records in his personal diary, 'At noon – Whitehead's

1. In fairness to Lowe, at one point (*Volume 2*, p. 143) he wonders if the topics considered may have been quite similar, but this can now be demonstrated.
2. Lowe, *Volume 2*, p. 142.

Figure 1. A drawing of Whitehead by William Ernest Hocking (found with Hocking's notes on lecture 74, 30 April 1925)

first lecture – Very interesting. – All the Grandees there'.[1] Lowe reports that not all the 'grandees' recall angel choirs, although dissenters were responding primarily to content. Whitehead was not known for a rich oratorical style but he certainly left many an audience with strong impressions. And one might suspect that the inaugural lecture was not going to be representative, since Whitehead will have wanted to prepare this one carefully; he probably read it out, or at the least kept closely to his score.

For almost 80 other performances, there are hints and suggestions that they were guided by notes drafted in advance, but not scripts read out to the class. Whitehead often commented at the end of one lecture where they would venture in the next, implying it had been prepared in advance; and he also paused occasionally, most often at the outset of another performance, to review the landmarks on the ground already covered. But there was no outline of the course handed out in advance, no syllabus in that sense, just the title and inaugural laying out of the kind of broad philosophical themes he would address.

We have quoted his letters to North explaining that between his two performances each Tuesday, Thursday and Saturday he would relax in his room in the main library and 'look over my notes for the 12 o'clock lecture and enlarge them'. As the earlier lecture at Radcliffe (during the first semester) was not the same one he was about to give at Emerson Hall, one can well imagine Whitehead looking over what was meant to come next, refreshing his intentions for the next 'movement'. Afterwards, 'We have Tea at about 4 to 4:30 in my study. (Callers don't want it.) After tea I read or write up lectures'.[2] So, the impression given is that his notes were worked up in advance, but

1. Bell's diary entry for 25 September 1924, Transcription of the 1924 diary by David Mawhinney, Bell Fonds, 8550/5/5.
2. Lowe, *Volume 2*, p. 295.

Figure 2. Whitehead's classroom in Emerson Hall (photograph by Ellen Usher, August 2008)

then reviewed, adjusted, enlarged and perhaps written up so as to incorporate what had been sorted out by thinking them through in and with his classes.

The number 80 was used above, because from the total of 85 lectures a few stand out as exceptions, and these exceptions might help us formulate the rule. Lowe was convinced that Whitehead was usually 'seated in a chair behind a table on the platform ... spoke to the class ... occasionally consulted what he had written down for his guidance'.[1] And with considerably more evidence for 1924–5 than Lowe had available, that still seems a good beginning. It is more difficult to discern in Hocking's consolidation of these lectures, although he sometimes marks direct quotes and notes Whitehead's asides; but in Bell's more complete scripting one can follow the flow, the pauses, the turnings of an oral presentation, and it leaves no imprint at all of being read out from a complete script.

One exception to that impression arises quite near the end, when Whitehead explains he wants to share with the class a paper on the history of mathematics he had presented the month before at Brown University.[2] This is the same paper he mentions in the Preface to *Science and the Modern World* that was added to the Lowell lectures, which form the bulk of that

1. Lowe, *Volume 2*, p. 144.
2. The paper, 'Mathematics as an element in the history of thought', required two lectures (81 and 82) to present, on 16 and 19 May, about a month after it had been delivered to the Mathematical Society at Brown University.

book, submitted for publication in July 1925. Hocking left no notes for the first of these two classes and only the briefest of notes for the second, but Bell's record of this paper being presented in Emerson Hall gives a quite different impression from most other lectures. The pace seems different – Bell has a hard time keeping up (or hearing) – and there are no diagrams on the blackboard. This one *does* seem like a paper being read out, which makes it the more evident that for most lectures one is not following a transcription that simply had been read to the class.

Another kind of exception are the occasions when it is clear that Whitehead is responding to questions. It is to questions from Demos that Whitehead says he is responding in the spring[1] and on two Saturdays in the autumn[2] when he says he is setting aside the lecture he had planned in order to deal with questions that had come up the previous evening – which must have been his Friday evening 'Seminary'. He had told North that those occasions were conducted quite differently from his course of lectures. 'I usually start the discussion, and intervene at irregular intervals'. The Saturday midday lectures that were set aside seem to be a kind of carry-over and may even have involved the same people. The implication is that the Friday evening seminaries were considering many of the same issues, so much so that it made sense to bring those discussions into the lecture classroom as a way to review and clarify what Whitehead had been trying to express. And thereafter, he would return to his solo performances, thinking aloud through the next set of issues.

The one other exception, surely the most marked exception and rather the 'elephant in the room' for this whole academic year, is the elephant known as the Lowell lectures. There is surprisingly little mention of them in his Tuesday, Thursday, Saturday scheduled lectures. They do, though, present a clear contrast in mode of presentation. The Lowells, there can be little doubt, were fully written out, and in delivery he must have simply read them aloud. Lowe details the circumstances of the invitation from President Lowell to deliver a series of eight lectures, in March 1924, which means after Whitehead had accepted the new appointment to Harvard but well in advance of his actual departure – time enough to try out some ideas. He proposed a preliminary title, 'Three centuries of natural philosophy', which suggests the chronology of the first six chapters we find in *Science and the Modern World* was already firmly in mind before he even crossed over to New England. He also consulted about the appropriateness of composing these lectures with the eventual book clearly in mind. Whitehead wrote to his son in December 1924 that he was bearing down on the task of writing out these lectures, and wanted them completed before the series began at the beginning of February. He later exclaimed it was like writing a whole book in two months.[3]

1. See lecture 74 (30 April).
2. See lectures 11 and 23 (18 October and 15 November).
3. The first comment was in his letter of 21 December 1924 (reproduced in Lowe, *Volume 2*, p. 301); and the second was from 15 March 1925 (reproduced in Lowe, *Volume 2*, p. 303).

The overall strategy of his course of Harvard lectures may also have gelled for Whitehead as had his Lowell lectures, well in advance. It is difficult to tell, but each lecture seems at most to have been sketched out and if written out, only afterwards. What we can see is that the chronology of the Lowells would not serve as the framework for his 'Philosophical Presuppositions of Science', a comparison we will consider in the next section, except to say here that it also seems clear that the course of lectures was not a preliminary run through for what would become *Science and the Modern World*. It certainly was the occasion for Whitehead to think carefully through those very issues, but that did not mean the overall outlines needed to be the same, and while the Lowells were written quite intentionally to appear in print and to be read out in advance as a preview for his Lowell Institute audience, his course of Harvard lectures was left open enough to become a genuine philosophical exploration.

Despite these two sets of lectures both dealing with the themes of most interest to Whitehead at this point in his philosophical development, he makes no mention of the Lowells in his course of Harvard lectures. He mentions the Lowell lectures of Russell and others but not his own.

There is one exception, however: a unique case that arises in the latter half of a lecture he delivered mid-April.[1] Hocking records a sequence of objections to 'subjectivism' but leaves no indication of Whitehead's mode of delivery, whereas Bell records the same objections but states quite clearly, in brackets, that Whitehead was *reading* at this point: '[Whitehead reads a lot, excellent, from Lowell Lectures.]' Once we are told – and it is Bell who tells us, not Whitehead – that these objections can readily be identified with chapter V of *Science and the Modern World*, on pp. 89–91. This is a most unusual case where we know Whitehead was reading, and from the same script he had already read out at the Lowell Institute, intended for print; it is a clear exception helping to prove the rule: speaking his thoughts aloud from a rough set of notes.

One final consideration that bears on classroom performance is the extensive use Whitehead made of the blackboard. This was a teaching tool Whitehead must have been using for the 40 years since he began teaching mathematics, and used extensively in applied mathematics as well. We now find Whitehead making frequent use of the blackboard during these philosophy lectures. (We need hardly point out that one does not use the blackboard sitting down, nor likely when reading from a fully written out script.) We find equations, symbols, logical notation, small doodles, examples from physics; even leaving mathematics and logic aside, there are 129 of these from Bell's notes. And lest one puzzle over whether some of these are Bell's own, or at all accurate, they match up favourably with most of the 104 recorded by Hocking, and are identifiably similar even to the 73 recorded by Heath from Radcliffe. Some of these chalk diagrams are quite detailed, quite

1. Lecture 71 (16 April).

complex and, we would argue, quite significant. Although a few diagrams are to be found in *Process and Reality* and Part III of *An Enquiry Concerning the Principles of Natural Knowledge*, those ones accompany his descriptions from physics. Many of the diagrams to be seen in this volume provide a vivid sense of how Whitehead himself pictured the relationships he argued were fundamental, and a spatial pattern depicting events and objects emerging through time.

While there are no self-referencing comments by Whitehead about his own lecture style, about his speaking voice or about sitting down (there are one or two references to his chair), there are many instances where chalk becomes the point of an example. The chalk seems always to be at hand. And there are many references to his reliance on diagrams. Many are the warnings that his audience should not take these as anything more than clumsy reminders, pointers. They are not to be taken as the truth; they are sometimes said to be 'silly'. But he does also defend their frequent use by saying, for instance (while chalking up a particularly complex example): 'Recourse to diagram rather than long verbal statement. Diagram is mathematicians' analogue to Plato's myth. (Whitehead can't make up myths!)'[1] We take that to be Whitehead's commentary on himself: I can't do myths, but as a life-long mathematician I can do diagrams!

Somehow this repeated stepping up to the blackboard seems emblematic of a lecture style in which Whitehead is thinking out loud, in front of the room, gesturing with a piece a chalk, demonstrating a young boy catching a cricket ball, and turning to the board to fill it with logical notation, mathematical equations, or remarkably adept Minkowski-like diagrams of the spread of space and the flow of time.

Content of Whitehead's lectures

Whitehead presented his class with no roadmap of where this course of lectures was going. The title and his inaugural lecture signalled the sort of issues he intended to address, however successfully those were absorbed by either students or colleagues, but there was no printed outline, no syllabus handed out. Nevertheless, it would be too hasty to conclude Whitehead was making it up as he went along. The topics he intended to consider were quite clearly those explored in *Principles of Natural Knowledge*, *The Concept of Nature* and *The Principle of Relativity*. He says so openly, referring to them often, assigning portions as readings and for essay assignments, along with frequent references to the work of colleagues, many of whom were the same as referenced in these former works, or thanked in his prefaces.

Whereas *Principles of Natural Knowledge* introduces his concept of 'extensive abstraction' in Part III, and *The Concept of Nature* picks up this

1. From lecture 39 (13 January). Notice that the note-taker, Bell in this case, cannot use the first-person pronoun, so we are left to struggle with transcriptions in which Whitehead oddly refers to himself in the third person.

method in chapter IV, within this 1924–5 course of lectures Whitehead holds this topic back until the beginning of the second semester, where it frames the development of his second semester of lectures, and in just this sense the first-semester lectures are devoted to the topics of *Principles of Natural Knowledge* Parts I and II, and the opening chapters of *The Concept of Nature*. Similarly, he introduces topics from Part I of *The Principle of Relativity* and he makes at least brief references to the more technical matters developed in Parts II and III.

This is not to suggest that the first semester is simply a review of the preliminary chapters of his previous work. And we are warned, as were his students,[1] that *Principles of Natural Knowledge* was 'obviously by someone immersed in mathematical physics'. In that work, he had 'let himself be fooled' and was often 'confused', 'did not bring that out well', or handled the matter too 'clumsily'. As he acknowledges throughout the year, he had often been 'in a muddle'. In *Concept of Nature* he was held back because he 'had not [a] full theory of Mentality then'. When he does turn to the method of extensive abstraction in the second semester, his introductory remarks warn: 'Not so concerned now in following systematic details, as in pointing out confusion in [all three] books … pointing out what lies <u>behind</u> ideas there'.

The topics addressed in this course of lectures were ones Whitehead had been mulling over for some years, but the lectures would be an opportunity for him to address earlier confusions and for him to introduce new insights. He had long been attracted to the philosophical challenges presented by presuppositions evident throughout modern science, but this course of lectures begins with a tighter focus on the scientific developments Whitehead had himself witnessed and the revolutionary new insights which he and colleagues he knew personally had uncovered. Maxwell's mathematical treatment of electromagnetism, Thomson's discovery of the electron, the rise of quantum theory and then relativity are quickly laid out for these unsuspecting students of philosophy. 'These lectures not philosophical', Whitehead explained, 'but to get Science into form in which one can expect Philosophers to understand it'.[2] Even so, it is quite remarkable that – as he reports to North – if anything, the numbers attending were growing.[3]

The inaugural lecture had introduced the alternative conceptualisations of a world characterised by continuity and that undergirded by atomism. His early lectures, presenting four decades of revolutionary science, had provided support for both of these and amplified the tension between them. Whitehead's growing conviction, openly shared, was that we are not confronted here by the choice between what seem to be radical alternatives,

1. The words and phrases quoted in this paragraph are respectively from lectures 80 (14 May), 58 (17 March), 19 (6 November), 41 (17 January), 38 (10 January), 43 (10 February), 59 (19 March), 8 (11 October), 37 (14 May), 74 (30 April) and 43 (10 February).
2. See lecture 3 (30 September).
3. Lowe, *Volume 2*, p. 298.

but by the challenge to unpack their underlying assumptions and articulate them within a metaphysical synthesis that enables a balance between them.

From his own experience with recent science Whitehead immediately turned to his other essential resource, the figures in the history of philosophy to whom he had recourse again and again. But here we are asked not only to consider Hume (as in his previous books) and the scepticism that left induction impossible to justify, but also to attend to Kant and Whitehead's conviction that the resolution of the former will require turning Kant on his head. He had suggested previously that the roots of Hume's problem are to be found already in the Greeks, but in addition to Plato these are now more narrowly identified with Aristotle's analysis of substance-attribute, which leads to the subject-adjective presumptions that become embedded in modern subjectivism. Even more striking is Whitehead's acknowledgement that the means for avoiding this philosophical dead end is *also* rooted in Aristotle – in his analysis of becoming as arising from potentiality. For all that we recall of the pedestal on which Whitehead places Plato, and the missteps he attributes to Aristotle, in these lectures he says, despite the 'unfortunate … bias he gave to Logic', Aristotle is 'the greatest of all Philosophers all the same'![1]

From these two resources Whitehead is prepared, by the beginning of November (about the time Hocking joins with Bell in attending and keeping notes), to begin a 'somewhat more Systematic consideration'. As he declares: 'you've got to see the Wood by means of the trees, you know'.[2] The wood he begins to articulate includes many of the implications he had already drawn from the needs of the new science, but his students would not have been in a position to appreciate just how new were his injection of not just 'events' and his own portrayal of 'objects', but a new focus on 'occasions' and the need for 'ingression', even the need for (not subject) but 'superject'. These concepts are known to us only because we recognise them as surviving into his later books.

We also find him musing over whether 'expansional' might better describe the core characteristic of his own metaphysics, rather than 'processional', and whether the insights he is determined to bring in from the biological sciences could better be articulated as a 'philosophy of evolution' than a 'philosophy of organism'. A particularly striking instance of trial runs that do not survive is Whitehead attempting to build off the Humean notion of 'impression' – he is, after all, determined to adhere to the empiricist assumptions he shares with Hume – where for a series of lectures he expresses his own insight as the 'impress' (the mode of interaction between events). Eventually he turns away from this trial run and turns instead to 'prehension'. By 23 November, Whitehead declares in a letter to North: 'I am gradually feeling my way into a metaphysical position which I feel sure is the right way of looking at things'.[3]

1. See lecture 72 (18 April).
2. See the beginning of lecture 17 (1 November).
3. Lowe, *Volume 2*, p. 298.

By December and carrying over to the final weeks of the first semester in January, Whitehead is no longer ploughing new ground, but thinking through an appropriate overview of what presuppositions have been unearthed and articulated. He explores the insights which physics should consider borrowing from biology. And, once again, he revisits the great alternative approaches of starting either from the immense permanences with which modern science has been enthralled – a 'Metaphysical fairy-tale', as he once called it[1] – or with the insights to be gained by beginning from change and generation. Curiously, he ends his first-semester lectures by looking back to the early modern assumptions of Galileo and Newton. At the same time (as he told North) he was composing his Lowell lectures, which become the core historical chapters of *Science and the Modern World* and which begin with Galileo and Newton and work their way century by century towards quantum theory and relativity. In contrast, in this 1924–5 course of lectures he proceeded the opposite way around. Clearly, the Phil 3b lectures were not simply a trying out of the new *Science and the Modern World* sequence, even while they were an exploration of the very themes and concepts *Science and the Modern World* and even *Process and Reality* will later deploy. We have the opportunity in these lectures of following along as he tries out some of the new concepts he will later adopt, and also try out some possibilities that he will decide to drop.

What then of the second semester? Before the teaching term began, on 10 February, he had already launched into his Lowell lectures each Monday and Thursday at 5 p.m., just off Commonwealth Avenue in Boston's Back Bay. These continued through the remainder of February, and his Radcliffe and Emerson Hall lectures during this same period shifted back into an introduction to the logic found in the first few sections of *Principia Mathematica.* According to Lowe's account of Whitehead's second-semester 'Seminary on Logic'[2] he was, perhaps even by student request, beginning with similar material there. It is tempting to suppose that while the Lowells were being presented, Whitehead had made his regular teaching duties a bit easier for himself.

The construction of symbolic propositions developed throughout February set the stage for a quite intense consideration of what the method of extensive abstraction (as first presented in *Principles of Natural Knowledge* and *Concept of Nature*) could achieve, and a comparison of the possibility of a geometry of space based on junctions, boundaries of volumes and the relation of 'extending over' as an alternative to the simpler pathway built on points. As with *The Principle of Relativity* we see Whitehead's openness to the implications of special relativity, and particularly the re-conceptualisation of 'simultaneity' that it requires, but also a deep resistance to time being folded into the fourth dimension of space.[3] Time remains, as it must for Whitehead, distinct; even

1. See lecture 30 (4 December) and also 32 (13 December) on 'permanences'.
2. Lowe, *Volume 2*, pp. 147–8.
3. By comparison, nothing of his alternative conceptualisation of gravity to Einstein's general theory of relativity is presented.

so, the Minkowski-like portrayal of events embedded in a time-cone is shown to hold out exciting possibilities for a Whiteheadian interpretation.

By early April Whitehead is articulating a 'temporal atomism' imposed partly by developments in science and partly because it makes possible his philosophy of organism. This has been remarked on before, most notably by Lewis Ford, who made the lecutre of 7 April[1] a focal point of his claims for a turning point in Whitehead's metaphysics.[2] Ford's Appendix 1 contains the only other publication of the Hocking notes from these lectures, as we have acknowledged, and since his full transcriptions begin at the end of March, that enabled him to capture the leading lines from 7 April. But we can only assume that without any Hocking notes before the beginning of November, Ford missed Whitehead's setting up of the challenge of finding an adequate way to interweave atomism with the continuity also required for any process philosophy.

'The real point', as Whitehead clarified on 9 April, 'is how to deal with this balance of Atomism and Continuity'.[3] With only summaries through February and March, Ford seems to have missed Whitehead's building up to an insight he had already anticipated. What was openly declared by Whitehead on 7 April was part of the overall trajectory of this course of lectures from its very beginning. Similarly, while Ford picked out Whitehead's acknowledgement of a definite 'muddle' enveloping these issues, full access to the Bell notes (or even fuller access to the Hocking notes) might have warned him that Whitehead was professing a muddled state on many matters, leading all the way back into his previous publications.

This Introduction may not be the appropriate place for a serious analysis of Lewis Ford's *The Emergence of Whitehead's Metaphysics: 1925–1929* (1984), but he has left us with a misapprehension of how this course of lectures was organised, one we feel compelled to dispel. The tunnel vision induced by reliance on Hocking's notes shows itself as well when Ford turns to the chapter on abstraction in *Science and the Modern World*. 'According to W. E. Hocking's notes, the term "actual occasion" was first used in Whitehead's lectures for April 30, 1925'.[4] A correlative concept with that of 'eternal object' which (as Ford correctly notes) was the central issue of the chapter on abstraction, Ford notices in the Hocking notes the careful attention to this topic in, apparently, the lecture of 22 May. However, while Hocking had marked his set of notes at 22 May, a calendar plus Bell's notes and diary – Heath's notes do not help, since they are not consistently dated, and match the Emerson Hall lectures only roughly at this point – make it clear that Hocking actually took these notes on 21 May. Hocking was absent for the class on 23 May or Ford might have noticed that these two lectures,

1. See lecture 67 (7 April).
2. Lewis Ford, *The Emergence of Whitehead's Metaphysics: 1925–1929* (New York: SUNY, 1984), pp. 51ff.
3. See lecture 68 (9 April). This same point does not get recorded in Hocking's notes for the same lecture.
4. Ford, *Emergence*, p. 67.

together,[1] are clearly an anticipation of what becomes chapter X in *Science and the Modern World*, 'Abstraction'. That is what Whitehead says these lectures are going to address, or, more pointedly, they are 'considering things from the point of view of Eternal Objects rather than from that of actualities'. More importantly, what Ford could not have realised was that both 'eternal objects' and 'actual occasions' had been introduced by Whitehead in his October lectures, and were being treated specifically in these terms already in the lectures of early November.[2]

In all these cases, the lectures were so organised that Whitehead had already introduced most of his more forward-looking concepts by the end of October and refined them more systematically during November. Then as the second semester constructed a propositional framework not only for systematising geometry but also for undergirding the metaphysical vision Whitehead was still developing, many of these key concepts were bought back into his lectures as he sought to demonstrate their collective coherence. It misconstrues the trajectory of these lectures to focus on any one of these key concepts as coming into play for the first time this late in the academic year.

It is relevant to point out, and perhaps even deeply important to appreciate that Whitehead paused in the midst of the concentrated work of the March lectures to consider methodologically the role of 'hypothesis'. This seems to have been not just a procedural step essential within the sciences, but equally typical in metaphysics, where one needs to articulate, Whitehead suggests, the vague apprehensions we develop into hypotheses. Here we have a critical indication that there is no hard and fast boundary between where scientific theory, natural philosophy (or, as Whitehead sometimes termed it, cosmology or the natural order) and metaphysics either begins or ends.

Whitehead's own hypotheses led him finally to what he had warned was still missing in *Concept of Nature*: a theory of mentality. This provides a particularly clear example of the pattern we have suggested can be seen across this course of lectures. At several points in his lectures through November into December he pointed ahead to the possibility and need for cognition. But not until the last two lectures in April does he return to it, saying: 'To avoid [yet another] muddle now need a point kept in background so far'. Cognitive experience was the point he had not yet developed: 'how Cognition looks sticking close to point of view developed so far in this course'. No one familiar with Whitehead's later publications will be surprised by what one finds in these two lectures.[3] The mind must be recognised as 'imaginal', rather than images mental. But in the context of this course of lectures it is interesting that he waits to develop 'how Cognition will look', till this later

1. See lectures 83 and 84 in this volume.
2. The term 'occasion' is introduced by 21 October (lecture 12) and as 'actual occasion' from 4 November (lecture 18), whereas he is considering the role of the 'eternal' by 30 October (lecture 16), 'eternal existent' on 8 November (lecture 20) and 'eternal object' – and consistently thereafter – by 13 November (lecture 22).
3. See lectures 73 and 74 (28 and 30 April).

point of the course, not as something new, but long recognised by him as the precursor he needed to take the next and final step.

The 'Test of self-consistency of an ontology is its possibility of giving place to Epistemology'.[1] Whitehead openly acknowledged that his was not a widely held conviction, that one should not, philosophically, begin with epistemology. He would have agreed that the issues philosophers confront in epistemology are tremendously important, but the way to begin is with metaphysics:

> Whitehead [read this in the first-person!] objects to point of view that first step in Philosophy is an investigation into powers of human mind (How Knowledge is possible). You can't express yourself until you have a certain minimum of ontological doctrine.[2]

As he had argued already in the autumn, he 'repudiates Epistemology as the one foundation of metaphysics. But it's the first source and Critic of the Metaphysics'.[3]

So, while metaphysical hypotheses need to be developed first, the first test of such hypotheses will lie in the epistemology they make possible. It stands as a remarkable feature of this whole course of lectures, which began with Whitehead teasing out the presuppositions exposed by the revolutionary changes in the science of his day, then working through those possibilities against the greatest philosophical minds he could resurrect, followed by an attempt at systematising what he had uncovered into a coherent system, and then, nearing the end of the course in early May, the *testing* of his system against the epistemology his ontology could provide. However one judges the results of this testing, it was a remarkably self-conscious, self-critical assessment of where his philosophical work stood, tested directly in front of his colleagues and students.

Whitehead's final lecture seems a particularly apt opportunity for a final comment. Ford had noticed the rather oddly timed return to considerations of Newton and Einstein, but even in Hocking's notes he might have noticed that this return is prompted by a discussion of the experimental results of work by Miller on the top of Mt Wilson. (It is not surprising that there is no mention of this in Lowe, since the Heath notes from Radcliffe do not include this last lecture at all.) Dayton Miller was an American physicist known for holding onto the aether theory and resisting Einstein's theory of relativity. Miller worked with Edward Morley to improve the sensitivity of the Michelson–Morley apparatus, and continued to refine these techniques for many years, claiming he could obtain results indicating aether drift. In 1926 Miller published results in the *Physical Review* of measurements he had made at the top of Mt Wilson during 1925 and which he had already announced that spring.

1. The opening statement of lecture 76 (5 May).
2. See lecture 76 (5 May).
3. See lecture 29 (2 December).

Whitehead saw this announcement as an irresistible opportunity to share with his class an example of science in the making.[1] Here was the kind of scientific work which exposes underlying assumptions and therefore the occasion to review one last time the kind of presuppositions his whole course of lectures was organised to explore.

Editorial handling of the lectures

A full list of the editorial principles for the entire *Critical Edition* is given on pages xiii–xv. These have guided the work of this volume and are intended to guide the editorial effort behind what will become several volumes of Whitehead's Harvard lectures. In addition, there are considerations which apply specifically to this volume and to the handwritten material we confronted.

- *Lecture numbers.* The 85 lectures Whitehead gave at Emerson Hall during the 1924–5 academic year have been numbered consecutively, as this proved to be the only reliable and unambiguous means of identifying each. None of the persons who took and kept notes for these 1924–5 lectures attended every lecture, and though most lectures were dated this was not always accurately done. It was possible to establish how many lectures there were only by checking across all three sources; and it does seem, fortunately, that from all these sources we have at least some record of each lecture. Bell also kept a personal diary, which has proven to be an essential additional source of information, and where appropriate has been quoted in the editors' notes to the lectures. From institutional records we only ever found general listings of the courses offered in the Philosophy Department[2] for that year and little else.
- *Page numbers of originals.* Since no attempt has been made to retain original page length, and since each of the handwritten pages of notes had been numbered – through the full year for Bell's notes, through each semester for Heath's notes, and dated-numbered by Hocking – we have inserted these numbers (for example |23|) to provide a clear indicator for whenever one might wish to check the transcript against the handwritten originals.
- *Retained look, layout and line breaks.* Bell's handwriting is so small, and fills most of his pages, that it has not been possible to maintain much indication of the layout of his transcription in this printed version, but we have endeavoured to give a rough impression at least of the layout of both Hocking's and Heath's notes. Thus, for example, where Bell breaks for a few lines into two columns this has been retained, and for Hocking

1. It is difficult to know for sure, but lecture 85 (26 May) almost seems like an extra added on. It was not even given at Radcliffe. And the opportunity to examine the implications of Miller's claims seems a more likely explanation than Whitehead having long planned to end the course on this note.
2. Course listings for the Philosophy Department can be found, after the fact, in the Reports of the President and the Treasurer of Harvard College, which are reproduced at http://pds.lib.harvard.edu/pds/view/2574586?n=6828; and for Radcliffe at http://pds.lib.harvard.edu/pds/view/34299933?n=2547.

we have sought to maintain the outline style he created. Similarly, the use of arrows, lines and odd spacing has been retained insofar as it has been feasible, because it seems an underlying feature of how each attempted to capture what was an oral presentation. Marginal marks have similarly been included.

- *Lack of punctuation.* Following directly upon the last point, we have introduced more standard punctuation only where confusion might arise, but have otherwise retained lack of punctuation, excessive use of dashes, extra spaces and so on, all in an effort to not lose the pace and flow of Whitehead speaking.
- *Placement of diagrams.* In a similar way, we have not only worked to retain the sketches and diagrams each attempted to copy from the blackboard, but also to retain the relation on the page with its most closely associated comments, since they are often Whitehead's oral clarification of what his chalk drawing was meant to convey.
- *Rendering of non-text items.* The mathematical equations, symbols, logical formulae (in *Principia Mathematica* notation) and Greek with accents are all typeset reproductions of the handwritten original, whereas the sketched diagrams are digital scans taken from the original page, enhanced to improve contrast, erase overlapping text and decrease the interference from text leaking through from the opposite side of the page, and finally replaced on the now transcribed page.
- *Abbreviations.* Contractions and abbreviations were encountered in all three sets of notes, and are particularly rampant in Bell's notes, according to a shorthand of his own devising. As stated in the list of general editorial principles, our intention has been to expand all abbreviations and most contractions, silently, and to comment on them in notes (as with other indecipherable terms) only when all else failed.
- *Angle brackets.* As each original course of notes makes use of both parentheses and brackets, we have signalled our own editorial insertions using ⟨angle brackets⟩. We have attempted to keep this 'noise' to a minimum, opting to use footnotes instead.
- *Footnotes.* In addition to footnotes being used to clarify editorial issues with the text, they have also been used to share information prompted by items that arise in the content of these lectures. Most often these are people, publications, poems or events familiar a century ago. Our objective has been to clarify and identify but not to engage in scholarly interpretation. Thanks to Lowe's biography, the *Stanford Encyclopedia* and online search tools, this has been possible, but we have stopped short of tying these references even to other of Whitehead's publications let alone secondary literature. They are intended to invite further research, not to forestall it.

Acknowledgements

In addition to the support and sources listed in the General Editor's Acknowledgements for the Critical Edition (page xvi), this volume would not have been possible without access to the three major sources on which it has drawn:

- The original pages of notes by Winthrop Pickard Bell have been transcribed and published here with permission. Winthrop Bell's original Alfred North Whitehead lecture notes for Philosophy 3b at Harvard University, 1924–1925, were found in the Winthrop Pickard Bell Fonds – 6501/11/2/8 and were supplied compliments of the Mount Allison University Archives – Sackville, New Brunswick, Canada.
- The original pages of notes by William Ernest Hocking are used here with the permission of Houghton Library, Harvard University, Cambridge, MA. The folder of original notes were found in the uncatalogued collection 92M-71, Box 3, Houghton Library, Harvard University.
- Due diligence did not reveal any heirs of Louise Heath, whose original pages of notes are owned by and published with the permission of the Center for Process Studies, Claremont School of Theology, Claremont, CA. The original artifacts are archived as Whitehead Research Project, Center for Process Studies, STU004.

The Heath notes had already come to the Center for Process Studies when this overall project began, and while it had been known that Hocking, too, kept notes (and these had been published in part as Appendix 1 to Ford's *Emergence*) the editorial team sought out and found the folder of notes kept by Hocking and deposited by his son, Richard Hocking, in the Houghton Library.

This was prompted by Jason Bell (no relation) discovering in 2010–11 that Winthrop Bell had also taken and retained notes from these lectures. They were located amongst Winthrop Bell's papers that had only recently been organised, described and placed online (see http://www.mta.ca/wpbell). This had been accomplished by the Mount Allison Archivist, David Mawhinney, who has continued to support this project through the intervening years. No one knows Bell's biography better, and his transcriptions of Bell's personal diary, for example, have enriched the backdrop to this story.

An extra note of gratitude is owed to Jason Bell for recognising the unusual quality of Bell's notes, for getting the ball rolling, and early draft transcriptions of challenging handwriting, and who has served as co-Associate Editor. Through Jason Bell support was provided by the Social Science and Humanities Research Council of Canada during the early stages of recovering Bell's handwritten notes. A special word of thanks is also owed to Joseph Petek, the master of the materials now held in the Whitehead Research Library, whose tireless efforts to draft and recheck transcriptions of Hocking's and Heath's handwriting, tease out administrative mysteries and directly assist in salvaging hastily sketched diagrams (despite the intervening continent) have proven invaluable every step of the way.

Technical challenges lay in store that ranged from logical notation, relying upon a symbolism no longer actively utilised, to mathematical equations utilised only within physics, and Greek terms carrying classical accents. These were all dissolved with expert assistance from Nick Griffin (for whom *Principia Mathematica* is still ever present), Ronny Desmet (with expertise in applied mathematics and a keen interest in the Whitehead years leading up to Harvard) and my colleague Bruce Robertson, who not only corrected my corrupted Greek but also solved some linguistic mysteries. Indeed, each of these generously offered assistance and advice well beyond my plea for help.

It was George Lucas and Brian Henning who immediately recognised how unusually informative were the notes and sketches left to us by Bell, which along with those from Hocking and Heath could make for so valuable a volume of material from the first of Whitehead's Harvard years. They have encouraged, guided and supported this first volume throughout.

Finally, a personal moment to acknowledge how wife and family have watched over the retired professor, still at work when it was time to be out at play.

'I am gradually feeling my way into a metaphysical position which I feel sure is the right way of looking at things. I am endeavouring to get it across ... in lectures.'

A. N. Whitehead, letter to his son, North, 23 November 1924

Emerson Hall lectures, Harvard Yard, 1924–1925

Notes taken by W. P. Bell and W. E. Hocking on Phil 3b, 'Philosophical Presuppositions of Science', delivered by Alfred North Whitehead

First semester

Lecture 1

Thursday, 25 September 1924[1]

Bell's notes[2]

|1|[3]

Every Philosophy dominated by some type of difficulty. – Some problem of fundamental kind lying at root of it.

What are peculiar difficulties of daily life which "this" Philosophy is calculated to solve? is the question

So (1) Sort of difficulty which systematized "Science" presents.

Then 2nd lecture:– question of unifying to coherent whole the presuppositions of Science.[4]

(1) Elucidation of:– What Science is in itself. Might take our start from question of Motives (in human nature) from which Science arises. But not human psychology as starting point here. Rather:– What is there in the Nature of Things which leads that there is or can be any "Science" to be what in outline it is? Certain relations between Theology, Metaphysics, and Science. Theology warned off scientific field in ca 1600, Metaphysics about 1700.

There must be complete freedom for scientific hypotheses.– This is 1st presupposition (and extends to influence from one Science to another). The complexity of things is beyond our power to cope with. New ideas always look ragged and crude and somewhat silly.

But must come back to fundamental fact that it's a rational synthesis we're seeking. The Antinomies are only means to this end. There are no completely autonomous entities in World. Philosophy and even Theology are capable of rendering services to Science. Perhaps even Medieval Theology rendered service in fostering Scientific Spirit. The modern corps of devoted Scientists.

---

1. Each lecture will provide a transcription of notes taken by W. R. Bell, as explained in the Introduction, and a transcription of notes taken by W. E. Hocking – or someone substituting for him – although the Hocking notes begin only in late October (lecture 14 and then from lecture 17 and thereafter). Also, for each Emerson Hall lecture, the equivalent set of notes by Heath will be cross-referenced. The Radcliffe version of lecture 1, from notes taken by L. R. Heath on 27 September, begins on |1| of her notes, page 411.
2. In his diary for this date Bell notes: 'At noon – Whitehead's first lecture – Very interesting. – All the Grandees there'. Wherever we find comments in Bell's diary that seem relevant to these lectures, we will include them in a note; the transcriptions are due to David Mawhinney, Mount Allison Archivist. Courtesy of Mount Allison University Archives, Winthrop Bell Fonds, 8550/5, E.
3. Atypically, there is no heading on Bell's notes for this lecture.
4. Notice that the title Whitehead had given to this first course of lectures was 'Philosophical Presuppositions of Science'.

How does Romance pass into exact Investigation and thus eventuate in Science? Belief back of it that there is a simpler order back of the appearances with the very rough regularities there. The Easterner, when he wonders, retires to a cave and continues to wonder. In two centuries 1500–1700 in Europe more done for Science than in the 1500 years of peace and prosperity following death of Aristotle.

Why didn't observant and careful Chinese do it in their Millenniums. Why should there be the search for corresponding Spectrum of Elements? We believe, as ultimate motive for scientific speculation, that there's something in fundamental nature of things that makes regularities – rationality there. Weakness of Hume's Philosophy and its modern derivatives ∧in opposite here∧. Intimate belief in inherent rationality of things. "I assume that there's a fundamental decency of things". Here's where Theology comes in. Men of Western Europe in 16th and 17th centuries inherited Theology which insisted on "decency" of God; Whereas the other peoples had Gods too capricious or the like. It all came from belief in rational God who attended personally to the detail of things. Kant:– We never know the real nature of things, but a suitably expurgated edition.

Knowing (Cognition) is just one of the groups of relations between Every Universe by reason of its relational Essence --- ? ? ?[1]
Kant, in effect, worked out this theory for particular case of relation of cognizance.

Whitehead asks:[2]– How is any entity possible, allowing for relations which it presupposes? This comes down to problem of Inductive Logic. Why are these processes of generalization valid? [Not what are these processes, etc. (Mill, Bacon[3], Pearson etc.)] Why do they carry any probability? Metaphysical account of relations of things must allow for Inductive Logic give us any probable Knowledge of Universe at all.

If A has no power of giving information of B, then A′, A″ ... A^{1000} has none. Then Bacon's, Mill's, Karl Pearson's analyses all assume what we're after.

How can one fact be relevant to another fact which is not contained in it?
Science collapses if you assume an independent atomicity of facts.

2ndly: you don't get out of the difficulty by introducing "probabilities"

3rdly: you don't get out of the difficulty by saying that "nobody doubts it".

4thly: no help to be got by basing your trust on past Experience:– You can thus (Hume) account for your habit of belief; but your belief in the rationality should be weakened by this ∧analysis∧. And Hume assumes a Causation of Association to account for the assumption of Causation.

1. Both here and in the line above, Bell seems to lose the end of Whitehead's sentences.
2. Bell's notes refer to Whitehead in the third person although of course Whitehead would in his lectures have referred to himself in the first person.
3. Bell's handwriting gives 'Bain' here but 'Bacon' four lines below.

Becoming is a series of modes of explication of essential togetherness of things.

|2|

Togetherness of things is fundamental. "Organisation" (in Biology) exists only in Relationships. There can't be any one thing (a horse, e.g.) except in certain circumstances. Its happiness too, a further dependence.

Every individual entity, Whitehead calls "abstract", because it supposes this sort of togetherness, in order that it may be at all. Each individual one must have definite relations to all things (which may vary, or be rather indefinite). No individual is living indefinitely in any environment. Brilliant and deserved success of Aristotelian system of classification obscured his followers' attention to this togetherness of things. Half the difficulties of philosophy come from too exclusive attention on the isolated individual. The predicates belong to some part of Environment as well as to the thing. (Examples) Predication is a one-sided way of seeing the togetherness. This Togetherness of Things takes form of a flowing process of becoming – of realisation. This was so obvious that overlooked. And we get a world interpreted as "appearance" and therewith divorcing of Philosophy and Science (which deals with this appearing world).

For Whitehead there is nothing behind this veil of becoming. The metaphor of "veil" is wrong. There is nothing but this process of becoming real. – Ideal of objective togetherness of things is becoming realised – in varying degrees – Thus there are varying degrees of reality.

A may be real for B. B for A. But it's nonsense to say that either A or B is absolutely real. B's becoming real for A = B's inherent, individual qualitative character becoming significant for A.

Thus realisation is a matter of reactive Significance – of Valuation. But there are Stages and degrees of Reality. – according to how completely B acts on A and A receives action. Every objective Existence much somehow be envisaged from standpoint of Reality. An existent ⟨blank⟩[1]
unless the realized Valuation is reciprocal.

Is a Doctrine of complete relativity of Reality. – Standpoint which finds Spinoza the most suggestive of Modern Philosophers.

Transition of ⟨blank⟩ --- into ⟨blank⟩

Metaphysical Philosophy, however, not the mere handmaiden of Science.

Metaphysical Philosophy stands as near to Poetry as to Science, and needs them both: has to deal, too, with the Individual in its complete concretion. Concerned with what can be and what will never be, too.

1. In this case, and hereafter, we will insert ⟨blank⟩ to indicate Bell, in his own notes, left the remainder of sentence blank. Presumably, he lost the train of what Whitehead was saying. Where he uses dashes (often a series of short dashes or hyphens) or ? ? we will retain those. Our own insertions make use of angle brackets, to distinguish from the square brackets used by Bell.

Lecture 2

Saturday, 27 September 1924[1]

Bell's notes

|3|

Whitehead: 27, Sept. ~~1924~~, ~~Friday~~ Saturday[2]

Consideration of Process. Fundamental and underlying fact in every detail of Experience. Consider it as if no other Philosopher had ever considered it. Then come back and show relations with the older philosophers.

Consideration of Process throws light on two contrasting ideas:– Continuity. Not ordinary mathematical form of Continuity (as in mathematical books) though closely allied with it. True contrasted idea is Atomicity.

The continuous always has divisibility in it somewhere. Continuity and Atomicity have maintained a very equal duel through history of human thought. Whitehead thinks[3] neither can be missed. Through slight mishandling can exhibit Atomicity as Discontinuity, and this won't do. Process exhibits atomic character imposed upon a continuous field. Science illustrates this.

Now highly speculative lecture on the Structure of Energy ("Energy" < !!). In what direction to look for fundamental assumptions of Physics[4] when framed with reference to Continuity. In other direction misunderstanding of Aristotelian "Substance" led to concept of "Matter"– as if always under everything a passive Being which bears all characters, etc. Science has been wholly developed by men bound down to notion that that was necessarily fundamental view. (Process.) Process as change to something else.

Process as realization of Values by transition. But in addition to change you have to have Retention of Value Value a reactive significance between entities. Retention comes from ⟨blank⟩

Whitehead should expect to find at basis of physical field change and recurrence – a structure of "Vibrations"

Now for Conclusions of Science itself. If 40 years ago, should have come down on ideas of "Material" of[5] one sort ~~of~~ or another. – The old Victorian

1. The Radcliffe version of this lecture, from notes taken by L. R. Heath on 30 September, begins on |8| of her notes, page 414.
2. In his diary for this date Bell notes: 'Another interesting lecture by Whitehead'.
3. We will point out, as we did in lecture 1 (and here for the last time) that where Whitehead likely referred to himself in the first person, Bell uses the third person.
4. In slightly different terms, Whitehead seems here to be echoing the title and central objective of these lectures.
5. Bell has 'or' but it should, clearly, have been 'of'.

"Ether" e.g. – Outcome of a diseased metaphysical craving. They found something going on, and had to subsume something to put it in. – Doesn't do much harm, but wholly beside the point.

What is going on? Taking modern electromagnetic theories.

(1) the goings-on of the electro-magnetic field. That represented (Clerk Maxwell All modern theory just a modification of Maxwell) a sort of continuity without slightest hint of Atomicity. Then arose discovery (20 years ago) of the Electron. – Requires you to conceive superposed upon your electro-magnetic field a certain Atomicity. – An unexpected but[1] not inconsistent thing. Then later another nasty bump in the Quantum Theory. Suggests that there's an essential discontinuity somewhere – and doesn't quite say where. Has been suggested that Space-Time is discontinuous.

Up to Whitehead to show that idea of Atomicity as fundamental is sufficient --- [?] ⟨blank⟩

No reason to be compelled by Quantum Theory to accept discontinuity of Space and Time.

Outline of most general sort of thing Maxwell was talking about. – Maxwell reduced 8 equations upon which whole theory of Electricity is built. Maxwell took 3[2]

1. Charge of Electricity at a point in a field – ρ

2. Electric Force at a point in a field. ∧(Empty Space)∧ Resolved into 3 components (f, g, h)[3]

3. Magnetic Force at a point in a field. ∧(Empty Space)∧ Resolved into 3 components (α, β, γ)

Electric currents:–

4. Something only in a "Conductor" or the like – (i_1, i_2, i_3)

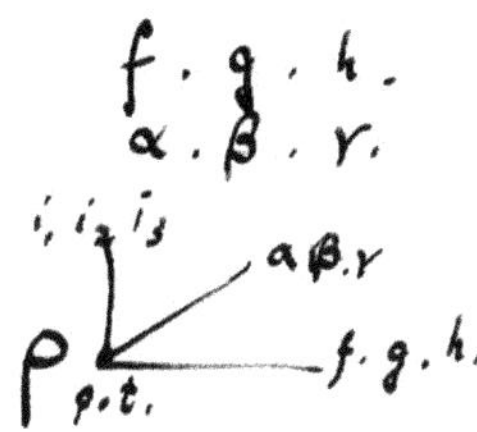

Point p at time t

Now as move from p in any direction can try to estimate rates of variation in the 3 components as you keep time constant. – Spatial rate of Variation of the 3 Vectors. – The S V of ρ

But you may also sit down at point p and watch what goes on at p while time goes on. Temporal rate of Variation...etc. The T V of ρ

What Maxwell did was to connect in certain way certain aspects or properties of S V of (f, g, h) with ρ [Details don't matter for Philosophy but quite essential that you should grasp the sort of thing that's being done quite clearly]. Also S V of (α, β, γ) with r.

(The "Divergence" of (α, β, γ) etc.)

1. Bell has 'by' but it should, clearly, have been 'but'.
2. Bell's notes definitely record '3' here but perhaps it should have been '4'.
3. The convention, as we understand it, is to present these sets of three components with parentheses, which in all three cases we have added.

|4|
The equations are exactly the same! – only (α, β, γ) substituted for (f, g, h). But there's no exact analogy in theory of Magnetism with this r.
In second group of Equations Maxwell has three equations. [You're apt to get equations in threes of course in directed material – expressing some one fact in mathematical physics.] – relating in a definite way S V of (α, β, γ) with T V of (f, g, h) and with (i_1, i_2, i_3) the "Current"
Third group is just got by substitution again (f, g, h) for (α, β, γ) and vice versa in 2nd group.

But when you come to (i_1, i_2, i_3) again you have no strictly magnetic analogue to Electrical Current. – So you just leave it out.

S V of (f, g, h) with T V of (α, β, γ) then.

——— This is type of idea which lies at basis of all Mathematical Physics. Maxwell very vague as to what he means by a "Current." When Electrons and Protons came along people said Current is when your Electron (little charge of electricity) moves [Whitehead doesn't believe in this view.] –But this motion of Electrons simply dims a more fundamental view of Current.

[?? – I may have Whitehead just *verkehrt* here.][1]

Lorentz[2] here. Current zero everywhere where there's not an electric charge.

All this very nice until modern investigations about interior of an atom. – Couldn't make it work. Now view that Maxwell's equations work everywhere except inside a nucleus (Electron or Proton). But it's only inside these that there's any charge. Then the Equations hold everywhere except where they apply!! Whitehead searching for an interpretation of i, i_2, i_3. which would allow Maxwell equations to hold everywhere.

In crude form – but explains the Quantum Relation (the purpose of it) Maxwell's equations presuppose a general continuity. Quantum theory has to do with some idea of an atomic structure of energy and action. Kinetic Energy as ½ mv^2 In 19th Century Kinetic Energy regarded as the real thing – all others somewhat suspected.

Now your m can only be withdrawn or increased molecularly – i.e., jumpwise. But v can vary continuously. Energy therefore essentially capable of continuous variation. Quantum theory throws doubt on this and Physics wasn't prepared for it at all.

But the theory "works" in so many respects that it must be on right road in some way or other. ~~~~~~ [Time]

Reflections:– Assumption that concepts that work in one context in Natural Science must work in others, is erroneous. (½mv^2 not inside Molecules etc.).
A different set of conditions will bring up entirely different set of actions.

---

1. This seems to be Bell's own comment. The German *verkerht* could be translated as 'upside down'.
2. Hendrik Antoon Lorentz (1853–1928) was a Dutch physicist known, among other things, for deriving the transformation equations subsequently used by Albert Einstein to describe space and time.

But this doesn't show earlier forms of theory wrong – but modification by increasing richness. Not that old idea was wrong but inadequate and not rich enough.

Lecture 3

Tuesday, 30 September 1924[1]

Bell's notes

|5|[2]
Whitehead: 30 September 1924 (Tuesday) –

Quantum Theory–Energy can only come out in definite lumps. – Hence discontinuity of Space-Time. Whitehead disagrees. – "Nonsense". Now to envisage more accurately and definitely at is main basis of Quantum Theory – General character (only can get hold of it if taken in some perfectly definite embodiment.) – (These lectures not philosophical, but to get Science into form in which one can expect Philosophers to understand it.)

Substance of Quantum Theory.  Nucleus and vibrating elements. Each mode of Vibration has its own "frequency." 9/10 of properties of Atoms as known to us comes from these (blank)
In an atom each mode has its definite frequency. This is certain. Each mode of Vibration means a certain store of average Energy. You may coax Atom to give up the Energy of that Vibration. When it does this it start radiant Energy (light etc) of that same period. Now it is found that quantum of Energy from a mode of Vibration is always s. h .n.

(ν = nu here (Greek letter) Tν = 1)

S is an integer (1, 2, 3 ...) and h always has same value, whatever the atom or the circumstances. So you can get an amount of Energy hν or 2hν or 3 hν etc. So atom either gives 1.h.ν etc. or none at all. It won't give up $1^{1}/_{10}$ h.ν. from a little more coaxing.

This law of all or none is familiar enough. In American currency, can't spend less than 1 cent.

But this is puzzling in Energies. Idea that your v (in ½mv^2) must go in jumps because Space and Time discontinuous. – Whitehead holds this absurd.

It looks as though Energy × Time (= "Action" in language of Mathematical Physics) was divided up in lumps $E = Sh\nu$

$$ET = Sh$$

Fundamental fact we have to deal with, is Process. Essence of Process is this:– Obviously there's a passing on, – a transition to something different.

Obvious that idea of instantaneous Present as resting ultimate fact may be wrong.

---

1. The Radcliffe version of this lecture, from notes taken by L. R. Heath on 2 October, begins on |13| of her notes, page 416.
2. In between lectures, on Sunday 28 September, Bell noted in his diary: 'Reading in Whitehead's *Principles of Natural Knowledge* etc.'

If process is fundamental, then idea of static distribution of things now must be, instead of fundamental, a highly artificial (and useful) working up from the consequences[1] of process. So instead of attempting to define process in terms of a succession of instantaneous nows we'll attempt to define ⟨blank⟩

We fundamentally live in Stretches of Time and not in Instants of Time. Here it is that Whitehead feels very sympathetic with Bergson.

But departs entirely from Bergson in his Anti-intellectual point of view. We live in Volumes (Time-Space volumes) so to speak.
I.e., distinction between Presents is between Volumes (Psychologist's "Specious Present"[2]). Distinctions between Present and Past and Future then very much blurred. "Logical constructions" [is "Fake"!] Process contains within itself Transition. But if there were mere Transition, then because there's no sharp present you'd never have anything realized. So you must allow for Retention too.

In very particular case of Physical field you'll find this provided for:– e.g. through Recurrence. Also intuitively requires Sameness – Identity. – "There it is again". What can be identical through process, Whitehead calls an Object.[3]

– All this required in Process.
Whitehead surprised that his use of term "Object" has puzzled people. It's, so far as Whitehead can see, exact correlative of Spinoza's "objective".

Whitehead therefore looks on the physical field as being ultimately made up of Vibrations and Atoms as having identical Sameness through process – by being a Structure of occurrence – thus enclosing in itself still process – Vibrations.

Vibrations, then, fundamental.

Whitehead also conceives Reality as made up of "higher Values" feeding on the "lower" ones. Higher ones unrealizable unless
|6|
starting with aboriginal simplicity of Structure. Higher ones using lower as their bricks so to speak. Structure upon Structure.
Another feature exemplified in physical field.

Every entity has a significance which requires a corresponding patience in everything else for its entering into being. Whitehead would expect each of these Structures to express itself throughout everything else. Faraday's paper in 1847.[4]

---

1. Abbreviation looks like 'consc[ss]' and could be 'consciousness' but more likely is 'consequences'.
2. In *Principles of Psychology, Volume I* (H. Holt, 1890), p. 609, William James credited E. R. Clay with this phrase, and added: 'the practically cognized present is no knife-edge, but a saddle-back, with a certain breath of its own on which we sit perched, and from which we look in two directions into time. The unit of composition of our perception of time is a *duration*…'.
3. This is the first mention of a concept that will take up much of Whitehead's attention in these lectures.
4. This may be a reference to Faraday's paper in *Philosophical Magazine*, also mentioned by Whitehead in *Concept of Nature* (p. 146), although the year is puzzling. Or it might have been an 1844 paper in *Philosophical Magazine* (vol. 2, pp. 284ff.), 'A speculation touching electric conduction and the nature of matter'.

Another point with regard to Occurrence. – Various stages of Occurrence definitely called, in Physics, the "Phase". You get peculiarly important results with question of Phase adjustments where Periods the same (in Sound:– whole theory of Resonance depends upon Sameness of Period.)[1]

Next great advance in Physics to be made in realm of Biology. – There you get matter in entirely different circumstances. – Just as Vacuum Tube revealed electrons – so living organisms will ⟨blank⟩

– A peculiarly delicate adjustment of periods and the question of timing (adjustment of phases) is secret of extreme delicacy of the organism.

Forms in which Electro-magnetic field (fundamental fact in all Physics today) manifests itself as electro-magnetic waves:– (1.) Light waves; the radio-waves (∧those on which we do our∧ "listening in") All have this peculiarity.

Mean firstly certain vibration of Electric Force and of Magnetic Force. May have various frequencies. Forces – Electric forces and Magnetic forces – all tangential to Wave front. These involve no current whatever and no density of charge at any point.

– Essential character of all these waves of radiant energy. Next way in which Electro-magnetic field manifests itself is in electrons and protons – little lumps of Electricity – In current theory ∧(5–10 years ago)∧ each of these are little static charges of Electricity. This of course false because Electron would blow itself to pieces at once.

With regard to matter it has been assumed Atom is higher structure of these very elemental structures. 4thly [allied to Quantum Theory] Modern investigation of light emanations from Bodies has revived interest in Newtonian corpuscular theory of light. – Looked on 40 years ago as utterly dead and buried. Wanted for some things and won't work for others. William Gray [?]: "State of science now just as though on M.W.F. we had to use Wave theory of light and on Tu, Th, Sat had to use the Corpuscular."[2] Utterly contradictory and yet each explains one enormous mass of phenomena, which the other doesn't; and vice versa.

There's a set of scientific Bolshevists that is inclined to acquiesce in that view. Whitehead is too much of a rationalist to acquiesce in that. Let us suppose physical field consists ultimately not only of Waves but also "primates" – an atomic structure of Vibrations with a definite frequency. What kind of Vibrations are open to us – These primates (same amount of Energy in each and with same frequency) are the "corpuscles"[3] of light etc. Idea of a structure of vibrations superimposed upon waves of electro-magnetic field. Got to guide yourself with sort of conditions to be satisfied. When you have an Electron at rest gives you nothing magnetic

1. Seems like the correct place for missing end parenthesis.
2. This oft-quoted remark is usually attributed to Sir William Henry Bragg. It occurred in his Robert Boyle lecture at Oxford for the year 1921: 'Electrons and ether waves'. Perhaps that is why Bell questioned whether he heard the last name correctly.
3. Seems the appropriate place for these missing end quotes.

whatever; merely electric. When a primate at rest in space then (α, β, γ) simply doesn't exist. So there's to be vibration without any magnetic effect.– Knocks out at once those waves in which Electric and Magnetic effects are combined. Now this sort of possibility hasn't been considered yet. So the next thing to do would be to go to Maxwell's Equations and put the magnetic force = 0 and assume that the: one definite frequency ν $(f,g,h) = (\bar{f},\bar{g},\bar{h})e^{\nu^{T}}$ [1] – where the last represents the vibratory factor

He'll find:–

(1) He's got here not waves which move away – But "stationary Vibration."
(2) Electric force instead of tangential is normal to the wave-face.
(3) That he can't get rid of his "Current" anywhere. – And he must get physical interpretation for this. (4) There must be some region in which ρ (electronic density) is not = zero. This shows point where strict reasoning of mathematical physicist comes in (Experimental physical can't have it all its own way!) Then he will see (5) when he looks at his Equations:– that he has absolutely nothing to tell him how the ρ rises and falls. – He's one equation short. Then he'll bethink him: What's the right kind of Equation to bring in? Natural one would be to try whether ρ doesn't obey there the wave equation.

This lecture has been on "Scientific Method", To do this in sheer generalities is sheer "gas".

1. It is not clear whether the 'e' is the base of an exponential function or (as above) energy.

Lecture 4

Thursday, 2 October 1924[1]

Bell's notes

|7|
Philosophy 3b.– Whitehead.– 2. Oct. 1924 (Thurs.)

Essay (Oct. 16th) (Thurs.):– "Why are the generalizations of Inductive Logic sensible procedures?" "Science collapses if you once admit an independent atomicity of facts." Discuss and criticise this last statement.

On view deriving from Hume and shared (or professed) by 9/10 of the Scientists today all justification of Science collapses (because justification of Induction vanishes).

References:–
Hume: *Enquiry* §§3,4,5 (1st part), 7.
J.S. Mill, Huxley, – etc. – birk[2] the difficulty in same way as H.
Far ablest man living who'd practically accept Hume's principles *Analysis of Mind* – Lecture 5 (particularly).
Whitehead's Presidential Address of Aristotelian Society 1922-3 "On Uniformity and Contingency."
2000 words

Now finish off the Physics:– Conclusion last time:– granting assumptions (e.g. Maxwell's equations, idea of Vibration, of "Primate" [as fundamental structure out of which all other structures built] etc.) that if atom at rest no magnetic phenomena.

Would expect a structure having certain essential character of identity, of change into something else, and yet retaining Value which is really controlling etc.–Whitehead would expect something rhythmical. It would be an eminently rational thing on part of Nature if fundamental structure were rhythmical.

So assuming everything vibratory for a Primate in one frequency ν. As result must have a region in which there's definite occurrence of magnetic energy [?] rather [= electric density??]. Electric force normal to these surfaces

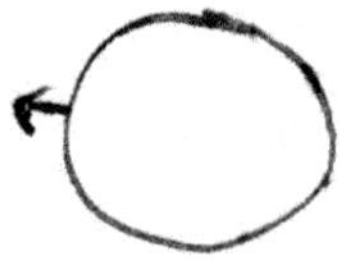

Also pointed out that you have to bring in further conditions to show how ρ comports itself inside spheres where it isn't zero.

1. The Radcliffe version of this lecture, from notes taken by L. R. Heath on 4 October begins on |17| of her notes, page 417.
2. Although 'birk' is what Bell has written, which makes no sense but, given the following lines and title, this is surely 'Bertie', who was, of course, Bertrand Russell.

Whitehead chooses for this the "Wave-equation."[1] Note how in framing scientific hypotheses there's a mixture of Empiricism – Pragmatism – and added conditions which, for some reason or other, you think may be likely [as e.g. assumption of Rhythm, above]. Finally Physical Science starts with a certain number of absolute blank assumptions ∧only justified pragmatically because they work.∧ Physical text-books slide over this and try and make you feel that they're in some way necessary or the only things a rational man could do.

"You can feel that they're consonant – in natural harmony with – what you already know." But this isn't scientific. Demand for Simplicity a dangerous guide, but always one. Unless it's simple to a certain degree, it beats you. But planets around earth in circles because simple!!!

The region around Primate divided into a series of ridges.

– Nodes as in theory of sounds – and "Loops" – Nodal surfaces and Loops here, too. – Wave vibration in three dimensions.

– Electrical density falling off according to quite complicated law.

When you get outside,– natural to suppose that this ends up at a nodal or loop surface. Here again a blank assumption. Suppose it a node surface, no force there, then you satisfy your demands if everything beyond were nothing – no effect beyond.– So this won't do. So natural to look on outer limit as a loop-surface where you get maximum energy (all this continuity inside the "primate")[2]

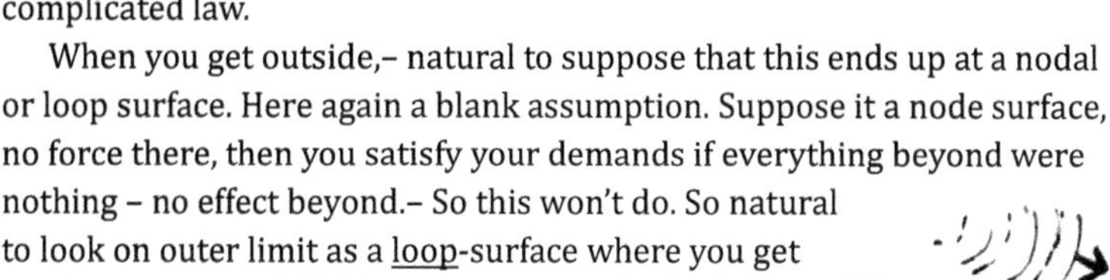

Electrical force outside then vibratory, varying as – $[\propto] \frac{1}{r^2}$

But inside, much more complicated formulae.

In modern physics, we're finding that inside proton all sorts of things happening not accountable on simplicity of $\frac{1}{r^2}$

Suppose whole radius S.a (or σ.a) ∧–[Integral]∧ where both S and σ are integers. (Centre must be a node – Energy zero – otherwise why any direction?)[3] When you come to see how these conditions can be fulfilled (must have a certain continuity of transition in force from centre to circumference. – How you choose the conditions of continuity is always just a little arbitrary. – Science shot through and through with these

1. To provide some historical context: the 'wave equation' to which Whitehead refers was first devised by eighteenth-century mathematicians and then further developed by Hamilton and others in the nineteenth century. Of course, Whitehead taught this material throughout his career, his own dissertation on Maxwell would have involved his electromagnetic wave equations, and remained the basis of Whitehead's own interpretation of developments in relativity. These equations were the background for Schrödinger's famous quantum wave equation, published in early 1926 based on work he had concluded during 1925. Immediately beforehand, de Broglie had proposed that matter behaves like a wave in his 1924 PhD dissertation, having conceived the first basic ideas (he said of himself) in 1923–4. While Whitehead makes no mention of either, it is clear that he was continuing to read about such developments into the early 1920s (as evidenced by his quoting Bragg on the wave/particle duality – see lecture 3 – from 1921).
2. This parenthetical phrase was placed up the left margin and then directed to this location with an arrow.
3. Seems a likely place for this otherwise missing end parenthesis.

prejudices and arbitrary assumptions of Simplicity – just like Greek argument for planets in circles around Earth.)

Now there's this wretched Quantum theory to be considered. But first 2 ways in which you can satisfy all these conditions (∧of∧ making up this structure)

1. the Electronic S.a
2. the Corpuscular σ.a

Can this give us any physical conception as to meaning of ρ. – Man of Science

|8|

always trying to get a picture of what's going on. [Physicist therefore viewing Mathematician ~~wit~~ a little askance]. Assume it as a sort of Energy [Well known Energy-equations].

We can work out total average Energy of the System, and Energy at any point etc. What happens to the Energy in its vibratory alterations? You can trace this. But if you're outside you find this entirely accounted for by flow of Energy in dimensions

i_1, i_2, i_3 – a Vector characteristic of fact that there's a flow of Energy.

(a reciprocal one backward and forward)

Then inside – if ρ not zero – then flow of energy won't account for change of energy inside; but a flux – a reciprocal ∧production and disappearance,∧ creation and destruction of Energy.

This only exists if there is ρ and depends on flow of ρ

Focal region is that in which there's a decay[1] in the production or destruction of Energy. "Energy" – term used because ordinary Energy Equations. What it means beyond these, Whitehead doesn't make any professions about.

Another point – Velocity of Light in vacuo – turns up till one's sick of it – "c". You'll find our a connected with the frequency ? ? ⟨blank⟩

by formula $2\,a\,\nu = c$

When you have primates of different frequencies their a's will be different.

Now what's their average energy? This important because we now know that Mass is simply average energy --- ⟨blank⟩

Mass and Energy have come together today.

How. Just like Amplitude of Swing of Pendulum $\frac{3/4 E^2}{Sc}\nu$ is Average Energy of an Electronic Primate.

Mass is this divided by c^2 or $\frac{3/4 E^2}{Sc^3}\nu$

With Corpuscular Primate it's different. Its average energy takes this form

$\frac{\sigma\pi^2 E^2_\nu}{16c}$ Mass ∴ $\frac{\sigma\pi^2 E^2_\nu}{16c^3}$ ($\frac{\pi^2}{16}$ is awfully near $3/5$) so write it $\frac{3/5\,\sigma E^2}{c^3}\nu$

1. This term might be 'delay' or even 'relay'.

Mathematician or Scientist asks himself it this reminds him of anything. Scientist is man that what[1] comes into whose mind is relevant.

Reminded Whitehead of the σ. h. ν of Quantum Theory. Suppose h, then, $= \frac{3E^2}{5c}$ [Another blank assumption]. Then you'd say you'd have explained Planck's constant. When you know Energy out of Atom in "lumps" you've disturbed the organization so that one of these primates dissolves out into Electro-Magnetic waves. What you prove is that there's nothing in the Quantum Theory that makes it necessary for you to assume any discontinuity whatsoever.

Now of course hundred and one other questions to come up here. This especially neat example of a "Structure" superimposed on a field which in essence is Continuity. Example of a structure which in essence extends everywhere etc. etc.

Grave difficulties here still. In physical theory while you're feeling your way you mustn't worry too much about things you can't explain. E.g. Wave-theory of light when first suggested by Huygens it couldn't explain (most obvious thing about light) that it goes in straight lines and casts shadows. That's why Newton gave corpuscular theory. Another difficulty: Because its vibratory how can you explain steady attraction (Answer through perfect timing).[2]

1. Although the abbreviation here is 'wh.' and this is standard for either 'what' or 'which' (determined by context), in this case might be 'whatever'.
2. Bell ran out of space on his second page of notes and added this final sentence up the left-hand margin.

Lecture 5

Saturday, 4 October 1924[1]

Bell's notes

|9|
Whitehead. – 4. October 1924 – Saturday[2]

Recapitulation without Mathematics the sort of ideas Whitehead has been wanting to bring out. – re Scientific Methodology – ~~Point Whitehead was~~ (1) Enormous influence of unconscious metaphysical assumptions. In 19th century Scientists would almost all have told you with disgust that very idea of being influenced by Metaphysics was absurd. But this only meant freedom from some traditional Metaphysics. Whitehead:– Science of 19th and 20th centuries greatly influenced by Substance, Quality, Category. [Aristotle's explicit formulation]. Men of Science have always looked for some passive stuff – something with definite permanent quality underlying everything. Newton – Mass and Inertia. – Some Stuff, in some vague sort of way, looked on as passive and permanent. First as "Matter" then (19th Century) as "Ether" ("a sort of distilled Matter, you know!").

Whitehead contrasts --- ⟨blank⟩

Whitehead simply wants to ask which of these more fundamental. Whitehead thinks the less "abstract" idea is that of "Process" – This leads us to search for some more active or dynamical idea as fundamental rather than stuff. Idea of Stuff doesn't carry with it necessarily idea of duration. Can realise itself completely at an instant. Carries as fundamental idea Confirmation at an instant. Then you provide for lapse of time by succession of configurations at successive instants with their mathematical type of continuity.– Cinematographic view of universe. [Differential Calculus – "infinitely small increment" etc.] Whitehead thinks this unduly "abstract" way of putting things. Whitehead says what we immediately know about is living through a period. And totality of Event in that period is important thing. And analysis of this into ∞ number of instants etc. gives you a high abstraction. Period through which, and which as a whole is there for your knowledge, and which contains within itself material for your knowledge. Something happening within "Slab of Reality presented for your knowledge" – and when, for simplicity, you've analysed it down to concept of what's happening at an instant, you must never let the dynamic character lapse. – Must never become passive. Change at an instant must

1. The Radcliffe version of this lecture, from notes taken by L. R. Heath on 7 October, begins on |25| of her notes, page 421.
2. Bell notes in his diary on 4 October, 'fine lecture by Whitehead'.

inevitably be (rather than configuration) – or Rate of Change – must be the fundamental fact. Abstraction to let yourself come to an instant at all. [Why not both since unless you have the other you haven't got any change from anything into anything.][1] Bergson right: Thought has to spatialize otherwise things too complex. But wrong that thought comes down to mere configuration here. When you go to Mathematics for definition of rate of change and analyze its definition you find it's really a way of relating the instant to its environment – Supreme merit of conception of mathematical ideas underlying differential calculus. Leibniz was here less philosophic in this than Newton the mathematical physicist. Because Leibniz had idea there was something real as an "infinitesimal".[2] Weierstrass ("that great German Mathematician"[3]):– establishing a property at an instant which expresses its relation to a whole neighborhood. – And one of indeterminate length. – therefore can look on it as small as you like. Thus can get perfectly definite idea of a rate of change. Therefore all advantage of reduction to point of space and time without getting rid of dynamical idea of process.

Come thus to conception of happenings in the physical field. You've then got to account for the steadiness you find in things. The Statical point of view rather presents to the imagination change as founded on steadiness.

Enormous effect of Metaphysics as reflected in imaginative means of Science. Use of Metaphysics is to make Science imaginative. (Scientists ought to read Poetry, Metaphysics and Novels).

Dynamical point of view on the other hand – You've got to embroider on the change a steadiness of things somewhere. Long periods in Science where fundamental imaginativeness not very important.– When some new concept of high degree of self-consistency introduced with great train of consequences. E.g., after Newton's *Principia*. Immense number of deductions to be made here, and so generations occupied in working out results and tests etc. etc. – Imagination for Mathematics, Experiments, etc; but not to reconstruct fundamental ideas. So: rise of mere Specialist etc.[4]

|10|

How are we to embroider Steadiness on the Dynamical view. You embroider it on atomic structure of change – Change having focal region somewhere. We took, then, one example of a simple type of Atomic structure. – And one accounting for Quantum Theory etc. We took an Electric field – Idea of vibrating changing Charge of Electricity inside a region and vibrating Force outside. That showed certain types

1. This bracketed comment completed in a right-hand column and the next sentence in a left-hand column. Thereafter the notes continue without columns.
2. Although Gottfried Wilhelm Leibniz (1646–1716) is sometimes credited with introducing the term 'infinitesimal', he remained sceptical that such a thing could be objectively real. And this is clearly related to his basing his use of the calculus on differentials. It was Newton, as Whitehead observes, who was willing to make use of instants of time, whereas Leibniz's 'monads' were not divisible into instants.
3. The mathematician Karl Theodor Wilhelm Weierstrass (1815–97) was a major contributor to modern analysis during the nineteenth century, concerned (among other issues) with the soundness of calculus.
4. Whitehead's description of such 'long periods of science' anticipates what Thomas Kuhn (1922–96) would later call 'normal science'.

of permanence:– (1) Its own type of structure. (2) Its individuality. Whitehead distinguishes between the Individuality and the Type. All primates would be one existent from point of view of Type. But each is different – gets its Individuality from (Minkowski) its "Worm" of Space-Time[1]∧ – A certain mathematical curve∧ – Way we always do test Individuality in existence [In physical world only]. Man of Science in addition wants some very simple idea of ⟨blank⟩

Wants certain unity of Conception. – Not a lot of elemental varying things. – Calls this "finding a Sound Physical Basis". "Let us look on foundation of Activity as being really the changing Structure of Energy" (whatever you mean by "Energy").

How are we to correlate all these ideas; – to reason from them. Man of Science takes those formulated laws empirically found in past to be useful. So in recreating his notions he has recourse to most general formulations he can find, which have justified themselves.

ρ
$(\alpha.\ \beta.\ \gamma)$ represent these most general ideas – no
(f. g. h.) matter what they are
(i_1, i_2, i_3) Clerk Maxwell's well tested equations

Then there's the well-established way in which what's called the "Energy" of the field is calculated from these Adventures of Energy in Space-Time. Then you want $(\alpha.\ \beta.\ \gamma)$ and (f. g. h.) to give you Energy as a quantity. One form in which Energy has adventured you want Electro-magnetic phenomena – Light waves

"Poynting's Formulae"[2] – for transference of Energy at "this" point Electro-Magnetic Waves gives you no Steadiness however. You find that i_1, i_2, i_3 defines ∨as leading factor∨ direction and flow of Energy. You find that outside, that flow of Energy accounts for all that happens:– inside not so. This is an example of "trying to get a sound physical idea of things".

This may be well or badly done in this case. But at any rate it's an example of the sort of method used. – For discovering types of adventures of your fundamental factor.

(1.) With regard to that atomic Structure – what has a focal region – This (Whitehead) is essential character that Atomism must take if you take Process as fundamental. Whitehead said this in his last book ("on "Relativity"). "Tower of London" an expression for --- ⟨blank⟩ etc. etc. [Spinozistic]

(2.) Beside preservation of type you've got in higher structures some Steadiness of Scale that's impressed upon you. Certain scale of Amplitude,

---

1. The term 'space-time worm' can still be found, occasionally (see http://plato.stanford.edu/entries/time). It arises with Minkowski's re-presentation in 1907–8 of the Maxwell equations in four dimensions, in which space and time are combined, followed by his 1908–9 work in which each point in Minkowski space corresponds to an event in space-time.
2. In 1884 the British physicist John Henry Poynting presented his partial differential equation for the rate of energy transfer in an electromagnetic field.

e.g. Might treat this as preservation of ? ? But "Scale" is better [= Units in which Scale made?]

Amplitude as Space scale	e.g. in Pendular swing	leading toward configuration
? ? as Time scale		idea of activity

In complicated Structures you may have more than 2 scales. – Chances are you've only One True scale. E, v as the two Scales Whitehead used last day ⟨blank⟩ scale.

Recapitulation: (1) You preserve good empirically tested generalizations.

(2) These are the more satisfying the greater the generality of them. Evils of Maxwell's Equations from present point of view because perhaps they don't hold inside the nucleus of an Atom. – If some lack in generality, like that, it's the beginning of the end for that "*Auffassung*".[1] Seeking for Spinoza's "Infinite Attribute." One of the merits of the very crude idea Whitehead put before us that it would hold inside as well as outside.
(2)[2] Utilization of more fragmentary hints from our general knowledge. Electron at rest without magnetic field. Whitehead takes this as a hint to be utilized [missed this --- ⟨blank⟩][3]

3) Search for sound physical interpretation.

4) There's the use of the idea of Simplicity.– Simplicity a very bad Master (Mustn't say a thing true because it's Simple). But for one thing only a Simple Law that's interesting or important. You can't argue from complicated laws – If too many variables can make almost anything conform to it.[4]

5) ∧Observe the very∧ Complicated interweaving there was between deductive reasoning and things drawn from Inductive Process.– in ∧Process of shaping Scientific Concepts.∧ But not true that you can use Induction by itself.

Two uses of mathematics: 1) The "Engineer's" use, so to speak.
2) Suggestive and Critical use when you're in a creative mood, etc. – Utilitarian vs creative and critical use.

Then, when you've got hold of your Concept you'll find some things it doesn't explain – rough edges. Depends on how rough as to what you do.– Quantum vs. Wave Theory on face of them quite contradictory. But each explaining so much that you can't believe either quite wrong. – People burned each other through ages[5] because of these rough edges in morals etc. – They're in Science too. –

---

1. Bell here (or is it Whitehead?) uses German for 'conception'.
2. Not clear if this intended as continuation of (2) or a mis-numbering.
3. Clearly this bracketed comment is Bell's, and one of the few times he explicitly acknowledges missing something.
4. In subsequent decades, Karl Popper became well known for making much of this methodological point.
5. From this point on, Bell completes his notes on this lecture by writing them up the left-hand margin.

6) Some use of Common sense about judging your rough edges. Then (7) an enormous importance of very trivial facts. – Example in discovery of Argon (slight difference of atomic weights in isolating Nitrogen[1]). "Never repeat a successful scientific experiment". Minute Clues. – Example again Newton's gravitation explained everything except one silly little shifting of long axis of paths of plan etc.[2]

1. Although argon was suspected of being a component of the atmosphere, it was not isolated and identified until the 1880s and 1890s. This made it the first of the 'noble gases' to be identified. The name means 'inactive' and, like all the inert gases in its column of the Periodic Table, it was difficult to isolate or characterise chemically.
2. For example, Uranus takes decades to complete one revolution. After that planet was discovered in 1781, it therefore took some time for astronomers to be able to observe the small perturbations in its orbit. Utilising Newton's laws of motion they could calculate that this effect was the result of an additional planet in an even wider orbit, and this led to the observation of Neptune in 1846.

Lecture 6

Tuesday, 7 October 1924[1]

Bell's notes

|11|
Whitehead – 7. October. 1924 –

Dominant Question here now:– What is it that we observe --- In what sense, if any, do we observe the "physical field" – the scientific machinery of atoms, electrons, light-vibrations, etc.? Answer is fundamental for Philosophy of Science.

Quotation from [advance copy] *Space, Time, Motion – An Historical Introduction to General Theory of Relativity* by Prof. Vasiliev (translated Lucas and Sanger) (Introduction Bertrand Russell) <Excellent Book[2] Whitehead doesn't quite agree with Russell's way of saying the Introduction [Russell's statement re "Realism" of Scientific theorizing as ca[3] professed Phenomenalism. Observable interpreted as Wider than the observed, etc. Phenomenalism as ideal unattainable. Opposite (perhaps merely Pragmatic) principle – that Science should be maintained]. Something wrong about a Principle which can never be applied. Very bad tradition set by Hume – holding something you can't possibly believe or apply. Sign that there must be something wrong in Premises or Logic. Hume's conclusion should have been:– I see my principles quite clearly but somehow I can't have got them placed quite correctly.

Whitehead very much doubts whether Einstein's Relativity can be represented as an advance towards Phenomenalism! Einstein's idea of "Force" controlling the "physical field." Control of our bodies etc., etc. is something we do observe. – Carrying a portmanteau and feeling we're holding it up against something. But take Einstein's 16 equations and tell the poor experimental physicist that that's what he observes and he must get rid of "push" "pull" etc., etc. is "rather tall".

But Whitehead quite agrees with fundamental postulate of Empiricism that you can't and mustn't lug in anything that you can't and don't observe. "Purism" as a scientific principle X X[4]

1. The Radcliffe version of this lecture, from notes taken by L. R. Heath on 9 October, begins on |28| of her notes, page 423.
2. Alexander Vasilievitch Vasiliev (1853–1929), who studied under Weierstrass in Berlin, was with the Mathematical Institute in Moscow when Knopf published his *Space, Time, Motion: An Historical Introduction to the General Theory of Relativity* in 1924. It was translated from the Russian by H. M. Lucas and C. P. Sanger and had an introduction by Bertrand Russell.
3. Bell's abbreviation here should perhaps have been 'cf.'
4. It is uncertain what Bell intends by the 'X's, but likely to indicate he missed something.

"Phenomenalism" means really adoption of Hume's point of view that only observable entities are sense-data and spatio-temporal relations between them.

But ought these to be the only entities we bring in for science [for Philosophy of Science?]? What we want to knock out are Entities of a purely hypothetical type where nothing of this type ever observable. But must allow for variations of type. We all agree earth made of molecules. But no direct evidence of billions at centre of earth. We don't know whether they're of gold or what not – perhaps some element not yet discovered.

So wide interpretation of "Observability" needed.

Now: What is it we do observe – at the very beginning? Science started from ordinary entities, which in ordinary life we "observe"– Sounds and Sights, bars of iron and green trees and birds, bodily feelings of wide gradation, watches etc. Lost his watch and didn't ask attendant if he'd found a bunch of sense-data – something round, hard, shiny, etc., etc.

As it exists now Science presumes Electrons, Protons, and perhaps more fundamental Energy-Quanta, as ultimate elements. Ultimate presuppositions expressed in Equations expressing Spatio-temporal adventures of these Energies. The entities are nothing but Spatio-temporal adventures.– A form or structure of the energy which has such adventures. "Bodies" are merely comparatively close packed aggregations of these fundamental things. "Attributes" of these are "secondary qualities."

Whitehead disagrees with view of secondary qualities of "Bodies" as merely Sensory in existence. Current view however – secondary qualities are reactions to stages of spatio-temporal adventures. These are views of "naive" – very naïve – Scientific people. [Cantab.[1] rather than Harvard] Secondary qualities interest man of Science because thereby he can obtain information re spatio-temporal determinations of said "physical field". ∧Slightest scrutiny of that scheme of thought shows no rest there. What's become of ordinary "bodies" – stones, trees, watches, etc.? Do we observe them and how? What are "we" who are doing the observing? – Relation of us to the sense-data or to the Electrons which condition the observation in some way.∧[2]

Because of this, it is, that B. Russell says that you can't rest on Phenomenalism alone.

We have here as result three types of entity to play with:–

(1.) Observers (2.) Things observed (3.) Fundamental Entities (whose Spatio-temporal relations are sufficient conditions of that observed).[3]

|12|

(2.) is the Field of Display.

1. 'Cantab.' is the contraction derived from the Latin 'cantabrigia' and from the Anglo-Saxon name 'Cantebrigge' referring to Cambridge, England, and often by the University to refer to itself.
2. Bell uses an arrow to indicate where the beginning of this passage needs to be moved to, but does not make clear how many lines to shift. The remainder of the paragraph has been moved.
3. An end parenthesis has been supplied here.

(3.) is the Physical Field – ("Field of Control")

But this not satisfactory, either. Observers given a very thin role there. We don't feel mere Observers – we feel ourselves taking a hand in the game, too – in various ways.

Permanent Unity of such Entities seems to lie below mere observing level – touches something like groups (3.)

Now (2.) – what are the things observed? Sense-data and spatio-temporal relations alone (as with Hume)? Hardly.

Now at (3.) what about that very sound fundamental principle of Empiricism? These Fundamental Entities not observed. Brought in because of "some morbid metaphysical craving."

First thing to learn:– that these very neat abstractions

[the merely observing, the merely displayed, etc. entity]

only got at by high degree of abstraction. Great errors of Philosophy come largely from spotting quite a clear idea but taking it as more concrete than it is. The fallacy of over-completeness always worrying Philosophy. (1.) (2.) & (3.) run into one another.

If your keep them concrete you can never get your Entities pure.

[?] Now how to come down on Entities of 3rd group. – Can't get at them by analogical reasoning. No analogy to adventures of Electrons.

Accordingly has been held that they're merely a useful myth. The quaint, rather naïve use of the "Mythical" theory Whitehead can't agree with – a fairy tale about things that aren't there – how can it explain things? Can't!

Another answer possible:– Various formal analogies between mythical entities, having a reality of their own, and the observed; and some advantage of simplicity etc. in former. Easier to deal with by mind etc.

Then there are really some Fundamental Entities actual give you handier picture etc.[1]

So long as more concrete entity has some formal relation – then turn whole thing into observable terms at end again – Analogous to writing instead of hearing.

1. The correct sequence of these three phrases is not clear.

Lecture 7

Thursday, 9 October 1924[1]

Bell's notes

|13|
Whitehead. – 9. October. 1924 (Thursday)

Remarks about the "mythical" theory.[2] – Viewing it in its naïve form, sheer and absolute nonsense.– First: "must be so", says Whitehead. But we can give it a sense by stating that there are various formal analogies with regard to interrelations of mythical energies and those of the observed entities; and therefore so long as you confine your reasoning and thinking to those relations they will equally be true of the actual entities; and it may be more convenient, psychologically, to reason about the mythical entities. ∧Give you an easier picture∧ – The "mythical" then must also be actual entities. You must know them as having this structure. Leaves you with problem of finding the real entities in Nature. – the things you're really talking about. It's like ~~with~~ dealing with National Statistics (unemployment etc.) by Graphs, and dealing with the latter. But if there were no such thing as Unemployment in the Nation, all the graphs in the world wouldn't give you any useful information. Mythical theory thus interpreted doesn't at all relieve you from asking: What is an Electron? etc.

One of the most poisonously fruitful errors is "Fallacy of Misplaced Concreteness". Our habit of thinking Electron as much more concrete than it really is. Whitehead sees a "little spot of mist" etc. Now if he says this is the Electron he's making this fallacy. All you really want of the Electron is a sort of structural or imposed character with a focal centre in the sphere we call the Electron. That type of vibration with that focal location. Electrons would be, then, highly abstract in relation to concrete fact of Electro-magnetic field.

Question of what is degree of Abstraction of Entities you're talking about is import. Many of differences between Philosophers and their schools depends on differences as to what types of things are really Concrete and which not. Very difficult question – because we habitually reason and think with surrogates – you use words or images as counters. Proper placing of Concreteness is one of fundamental questions in Metaphysics.

1. The Radcliffe version of this lecture, from notes taken by L. R. Heath on 11 October, begins on |33| of her notes, page 425.
2. In lecture 6, Whitehead considered whether fundamental entities like electrons might simply be a useful myth – a position with which Whitehead warned he cannot agree.

But if you have made the Electrons very highly abstract Entities, you will in fact raise again in acute form Hume's problem of Induction. Making it, e.g., a mere class of sense-data.

Another type of Entities which control or condition Entities of Science is independent type --- X X[1]

Class Theory of these Entities abolishes them as a distinct type – Another theory says that great distinction between (2) and (3) doesn't hold.

[The 3 types again:– (1) Observers (2) Display (sense data) (3) Conditioning Entities (Electrons etc.)][2] (3) are "observed" entities – but with a particular function with regard to the entities of group (2): while for Science you've to consider them in the two different functions. And then, as to (1), – how do we know anything about them unless somebody observes them? You only get the distinction as an absolute one at a very high degree of abstraction. Fallacy of misplaced Concreteness comes in here in thinking you've got here the more concrete instead of the more abstract Entities.

Whitehead's Analysis of Problem of Physical Science:– It has the field of display – intrinsic nature of various ~~field~~ families of sense-data and then the (3) – the "Physical field." Now when you consider how you define an Electron – in terms of how it affects future position, motion, vibration of other Electrons or Protons etc. – or light or other radiant Energy. So Electron is merely a way in which other Electrons in future are going to be influenced. You go round and round.

The "character-distribution" is simply really a statement of abstract conditions. And you don't get complete concreteness "until you wed the condition with the display" – and vice versa (no sense-data running wild, so to speak).

In your analysis you have taken two sides of the concrete:– you have these sense-data as they are here and now, no matter what --- ⟨blank⟩ But with Electrons you can't define characters you're assigning to them without reference to the Future and the Past. – But there's also a "Value now" –something which breaks in upon you and affects you now.

The Display is what it is now; but is definitely indicated by adventures of spatio-temporal character of the Conditioning Series.

[The display is always for something else. For an Observer[3] "might be itself a piece of chalk" – i.e., "Appearance" to a certain point.]

Display for A (observer) completely determined by immediate special spatio-temporal adventures of bundle of Electrons in (3).

1. It is not clear what Bell intends here. It is not obvious that something is missing, but also not clear whether it follows this sentence or underscores the previous sentence with XX appearing under 'sense-data'.
2. This bracketed reminder (of the three-part distinction from lecture 6) has been inserted from the left.
3. A confusing use of brackets, here, has been resolved into what seems to make sense.

|14|

Therefore ∧the∧ State of the Conditioning Entities definitely determines the display for A. But when you again ask how we go about it ~~have~~ to ask what state is of all these conditioning entities – you find it can only be expressed in terms of how these conditioning entities condition the future. "Theory of the retarded Potential" [? ?] Display for the future will be that display subject to Condition laid down in the present.

One gap in the Argument:– When there is the ∨immediate∨ display for A does that display include concrete knowledge of Entities displayed as ? ? ⟨blank⟩

Whitehead:– In ~~most~~ condition of greatest Concreteness every Concrete Fact is both observer and Display and Control Entity (Physical field).

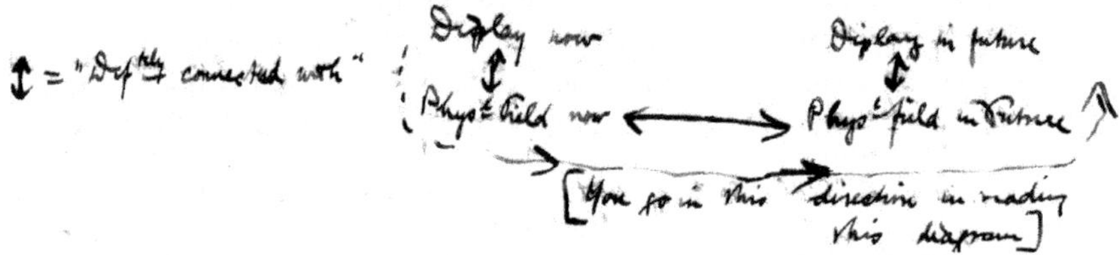

This is the case actually presupposed by every scientific work put before you.

Then when you analyze what you mean by the physical field now you can't do it without reference to Past and Future and other portions of Space. $\frac{\partial}{\partial t} \cdot \frac{\partial}{\partial t}$

Whitehead defies you to define temporal change without reference to ? ? ?[1] ⟨blank⟩

Always assumed that as soon as you've got Mathematical formulas you've settled all future history. But only if you had sufficient and complete rather than just conditioning presupposition. Now question whether in diagram above "Physical Field now" oughtn't to be – isn't – included in "Display now" (sense data). Is this total itself a display. Whitehead's view:– The perceptual object (lump of chalk "and all that") is simply and solely our apprehension of those sense data.

The object is a permanent group of character of Conditioning with steady spatio-temporal relations. If mankind does in any real sense observe such objects then what Science does is to achieve a more precise rendering of the perceptual objects which are somewhat vaguely defined and observed. – and in doing this produces the Electro-magnetic field, the Electron, etc.

Other theories:– Lump of chalk is merely classes or groups of sense data [Whitehead poked Russell up to re-adopt this after his excellent little

1. While Bell was unable to complete this sentence, the previous sentence makes clear that Whitehead was likely saying 'and to space'.

Problems of Philosophy.[1]] Whitehead's only objection to this theory is that he doesn't think it true.

(1.) There is no determinate class of Sense-data – these things perpetually altering as you go along. You need some Control Entity to define your Class of Sense data – as basis of Classification of Sense-data itself [X X X][2] Particularly difficult with lapse of time – How far back.

(2.) ~~Analogy~~ Illustrations in terms of Cricket. – Catching ball that stings hands. It's the object exhibiting the color that interests you – not the display of sense-data.

How account for vividness and universality of these Classifications?

1. Russell published this 'little' work in 1912. In it he cites Whitehead's *Introduction to Mathematics*, which had just been published in 1911.
2. It remains unclear what Bell intends by such 'X's. They could be another way to signal he has not caught what Whitehead has said, as at one point he combines '--- X X'. But they also seem to mark important passages, perhaps ones he wants either to clarify with Whitehead or to come back to himself and reconsider. They are not commonly used in the notes for later lectures.

Lecture 8

Saturday, 11 October 1924[1]

Bell's notes

|15|
Whitehead: 11. October 1924; (Saturday)

Discussing the status of a perceptual object. One of those critical points where philosophies diverge.– Kant's conceptual machinery becomes particularly relevant here. Objects are outcome of conceptual activity. Whitehead holds that all philosophic world has got to go to Kant. But when you leave school you forget learned details but retain a certain precious impress – how to go about things. There's got to be a certain active machinery to explain what is outcome of Experience. But Whitehead puts Center of Gravity of that machinery not in the Subject of that Experience but in the Cognitor.

First deliverances of cognition are that Cognitor in process of realization. (Whitehead believes himself here fundamentally with Hegel, and with Aristotle) [?][2]

Whitehead puts the activity in the process. There is a process of realisation and included in this process is what I know as that process of Cognition in which there's a cognizant on one hand and things cognized on the other.

Then asks:– What is true analysis of this process of realisation? Expects to find there the activity which Kant places in a fragment of this process – namely in cognisance as centered in cognizor. Whitehead asks whether we can't define this process of realization in terms of something which we can discern in process of cognisance and consciousness: but which may exist at level far below the conceptual.

Then Consciousness might be defined as highest achievement of process of concrete realisation. Whitehead's machinery is Behaviorism in which End to be attained is already there conditioning its own attainment. Analysis of Existence and Possibility as something conditioning Reality. The End to be attained has already an Existence so as to condition Behavior through which the End comes to realisation. But the Togetherness of End and Behavior not primarily because of Cognisance of it. But Cognisance is an extreme case of this fact in which some portion of this process (Body etc. ??) has attained utmost level of concreteness which any portion can attain.

1. The Radcliffe version of this lecture, from notes taken by L. R. Heath on what was likely 14 October, begins on |39| of her notes, page 428.
2. These bracketed single and sometimes multiple question marks seem to be Bell's.

Activity we find is, then, what makes Knowledge possible, but Whitehead holds it only makes Knowledge possible because it's what makes process of realisation possible. Process of knowing is only an exceptional example embedded in total process of realisation.

In order to substantiate this analysis we must discover among our procession of Cognitors a function of conditioning --- ? ? ? ⟨blank⟩ Entities discerned as setting the conditions satisfied by concrete procession of reality itself. Whitehead discovers these where Kant does:- in perceptual objects. Strains of self-conditioning process. First thing you insistently observe is the Objects – insistently associated with definite regions of Space-Time.

This all compatible with considerable lack of precision. But consider insistence on certain unity underlying strain of conditioning (of control)

Possibilities of Error, too (Illusionists and Mirrors etc.). (seeing a very distant star, e.g.). We associate light-twinkle we see with an immediate strain of control – localized.

Sounds – Either stray Sensa or to be associated with particular strains of control. There is a generally insistent apprehension of control an insistent localization of control. But also partly sheer error in this procedure of apprehension.

Now discussion of alternative theory:- Perceptual objects are mere classifications of sense data, and there are no other underlying entities. Argued brilliantly by B. Russell in Lowell Lectures and in *Analysis of Mind*.[1] – A theory to which Whitehead susceptible himself. Gains its plausibility from surreptitious slipping in reference of strain of control.

Perceptual object is defined by the class. Therefore mustn't use ~~this~~ object to explain how you form the class. Mustn't explain Classification by perceptual object; because perceptual object is to issue from the classification.

Whitehead maintains classification theory entirely too vague. It partly falls under criticism of Berkeley –

|16|

Sense data differ widely for different observers.– (Crimson cloud for one man is black mist for another, etc.) And even for same observer sense-data differ wildly and are subject to Breaks and without possibility of tracing strain of control there is no way of telling what you're talking about.

You get extremely indefinite class – "the observable." You don't know what you are talking about. Whitehead in *Principles of Natural Knowledge*[2] is under of influence of Class theory – continuously expressing himself in ways, though, that don't jibe with this. – Was in a muddle.

---

1. Bertrand Russell delivered his Lowell lectures at Harvard in March–April 1914 and later that year they were published as *Our Knowledge of the External World: As a field for Scientific Method in Philosophy* by Open Court, and his *Analysis of Mind* was published by George Allen & Unwin in 1921.
2. For the first time in these lectures, Whitehead mentions his own *An Enquiry Concerning the Principles of Natural Knowledge* (Cambridge University Press, 1919) and one week later Bell is recording in his diary for 19 October: 'Read Whitehead's *Principles of Natural Knowledge* until pretty late'.

In *Concept of Nature* and 2nd chapter of *Principles of Relativity*[1] was very much more clearly coming down on Control theory.

Very acute discussion of whole subject in Broad's book "Cause p. 266–283.[2]

Broad himself comes down on side of control theory – but doesn't handle his arguments rightly. – Doesn't quite extricate Self from Berkeley, and whole point of Control Theory is to enable you to do this. Theory of multiple relations would have helped him here to do it better. Real fact is much more complex than Broad allows for.

"Ingression"[3] is Whitehead's term for this complexity of situation. Broad does refer to a multiple relation theory of Dawes Hicks and George Moore[4] -On right lines but have oversimplified things and therefore get into all sorts of difficulties. Whitehead's example of cricket with vivid experience:

What the school boy is emphatically not doing is classifying the phenomena. Not anticipating sensations etc. etc. But is thinking of the entity itself – the ball. If he were classifying he would be wavering etc. etc. When you try really to see what you're apprehending in moment of stress – the classification of display is not what's interesting you.

Alternative theory which also gives you a unit entity underlying the "display":– Theory of substantial Matter of which display is the qualities. Whitehead doesn't[5] this can for a moment stand against Berkeley's criticisms. As soon as you've admitted idea of secondary qualities everything perceptual goes into it – predicates of your own mentality then. – All you want to make qualities of the object become qualities of your mind. Only to be got over by control theory – enabling you to get on with much more complex situation.

Substance-Quality arrangement simplifies situation much more than legitimate [Yes! – Appearances come in !!][6] Whitehead then holds that transition from this vaguely but persistently perceived world of sensible objects to the rather conjectural world of the scientific objects (Electrons etc.) is effort to proceed from a type of aspect which is perceived.

You then substitute for these something of same generic character as perceptual object but having specific differences.

You represent generic character of what you observe but with a conjectural greater definiteness.

---

1. Whitehead's *The Concept of Nature* (Cambridge University Press, 1920) was based on his Tarner lectures delivered at Trinity College in November 1919. *The Principle of Relativity with Applications to Physical Science* was published two years later (Cambridge University Press, 1922).
2. It does not seem that Bell managed to get this reference down completely. The book is probably C. D. Broad's *Scientific Thought*, just published in 1923. A section fits the page numbers given in which Broad considers the role of cause in perception.
3. This is the first occasion in these lectures in which Whitehead uses this, which becomes an important technical term for him.
4. Broad and Russell were both members of the Aristotelian Society along with Whitehead, as were Hicks and Moore (and each served in turn as President).
5. Bell seems to have missed a word, here, perhaps 'think'.
6. This seems to be a rather uncharacteristic interjection by Bell, but still carefully bracketed.

Lecture 9

Tuesday, 14 October 1924[1]

Bell's notes[2]

|17|
14. October: Whitehead.[3]

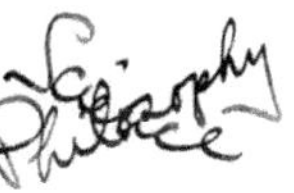

Whitehead's Metaphysical Standpoint (Eliciting this from consideration of what's put up to us by all Science – particularly Physics). – Pushing these ideas back we arrive at certain metaphysical standpoint.

Aesthetic factor involved in very essence of Realisation. Taking account of – being an interest for – apart from this no aesthetic value – and no Realisation. Aesthetic value arises when we analyze what we mean by Realisation.

Procedure of Thought is always procedure of abstraction.

Confining ourselves to mere achieved Value of the Display (considering the Red Vas existingV for us alone e.g.) is an abstraction – because abstracting from ~~the~~ existent and its envisagement. Envisagement of Possibility rooted in immediately present Past – Cause of realization from behind. Envisaged Existent as Value for achievement is Final Cause – lies in front of you.

So concrete fact embodies in one Unity both Efficient and Final Cause. But former is the more concrete. Final is the more abstract – involves analysis of efficient cause into --- cognitor and ⟨blank⟩

"Process of Reality viewed as including its own Conditions." "Fact" is both process and Conditions of Realisation – These only separable by abstraction.

Whitehead accepts basis that Phenomenalism lays down:– we must include 0 that's not observable. But converse principle ⟨blank⟩
Process of Cognisance My envisagement of a future fact as an efficient cause with counterpart in certain arrangement of molecules of my body. – My state of envisagement being evolved in Reality. But final cause is taking it not from point of view of process. – Abstracted from this is idea of lunch by itself – something fitted for realization "Myself envisaging lunch" as concrete thing.

---

1. The Radcliffe version of this lecture, from notes taken by L. R. Heath on what was likely 16 October, begins on |42| of her notes, page 429.
2. At the top of this page of notes appears the name 'Bartlett' (inside a drawn circle), for which no connections have been found. Also, it is difficult to imagine how the crossed words 'Science x Philosophy' could have been enunciated orally, so the assumption is that Whitehead drew this up on the board. Hocking was not yet attending these lectures and so cannot confirm this conjecture.
3. In his diary entry for 14 October, Bell notes: 'read Bergson's "Metaphysics" over again'. By this date Whitehead had already mentioned Bergson in at least two lectures.

You can't get complete Concreteness unless you take in everything. therefore you never get [I.e., in theoretical description] to it, but only to relatively more complete Concreteness. This is looking on it from highly developed state of reflective sophistication.

What each isolated entity is viewed for itself becomes ∧a modification –∧ a Value – an interest – an efficient agent – of another entity.

But that modification under conditions spatio-temporal – of reaction. There's no realization except with limitations. ∧To be realized is∧ Mutual reactive synthesis under specific temporal limitations.

(Spinozistic) Character of "Mode" of Realization (Whitehead – the relational essence of the entity viewed as a mere existent)

"Greenness" of object. Whitehead looking at.

Relational essence {prior to/always in} realization is Space-time – Theatre of Realisation – "Nothing but --- - [? ?]"

Way in which Space-Time enters into relational ∨essence∨ of Existents are very different for different existents.

Way sense-data enter into the limitation in respect to their relational essence are extremely complicated. That's what Berkeley pointed out re Crimson cloud.

You can never look on sense data as merely occupying a spatio-temporal region. Whitehead's conclusion other than Berkeley's.

One strand of ∨the∨ concrete reality is the molecular structure of the eye and body turned to see the green. Green is involved – an "ingression" into the concrete event in which it's being realized. Insofar as I'm looking at it its grade of realization it is increased. But green has other reactive significance too. If we all went out of room Green would lose in reality – because would lose the high degree of complex reactive significance that we are capable of.

Controlling Conditions have obviously a much simpler relation to ∨relative∨ essence of Space-Time than the Green. The control entity is there and has considerable reactive significance (collection of molecules or what not) whether we're here or not.

[Primate as structure of Vibrations throughout all of Space and Time but focused somewhere. When you say anything here or there in Space-Time there are many possible meanings.][1]

|18|

[Whitehead isn't distinguishing between the green ~~and the~~ perceive and the appearance of it.][2]

But some of these stand out. Tendency among Philosophers to muddle these up. – Don't separate ideas that seem to Whitehead clearly and obviously different. Or if you make them – get confused by multiplicity

1. Missing end bracket provided.
2. This bracketed comment appears at an angle in the upper left-hand corner of page 18 of Bell's notes and might be Bell's own comment.

of them. Whitehead thinks under significance of Quality – Substance categories you get an over-simplification.

Realisation is a question essentially of degree for Whitehead. We live through Periods and in Volumes – So the concrete fact is an Event. So you can't get on there by talking about Instants. When we do that we've reached a very high degree of Abstraction. [Discussed in books:– "Extensive Abstraction"][1]

Since there's no such thing as "At an instant", – realization involves in some way Retention through a period. And Retention is closely allied to Extension.

E.g., if Green realized for me as "there" there must be a retention of Green through Time. (Discussion of Motion in Display[2] for later. A different point to Transition. – Example of steadily altering color etc.)

Essential Transition with Continuity in Process. – with this one seems to be up against all psychological questions of minima, etc. etc.

P of v. [3] No retention of Value for me at higher frequencies than those involved in ^physical^ constitution of body. (Nothing to do with Spatial-Temporal structures as such). There may be being in same Temporal process whose whole span of a very varied life might be comprised in least of my perceptual minima. The thing is relative to my body.

Last time[4] concerned with perceptual and scientific objects.

Now take up question "Beyond" and how it arises in our immediate observation. Things going to happen beyond this lecture – we're not quite certain what they are. Also something on other side of that Wall – we don't know what for sure. There's an inside to drawers of the table that we don't know. Space behind moon, etc. Now how can we know anything about it, if we've never observed anything about it. (on basis of Phenomenalistic axiom).

B. Russell:– we can't get on in science without accepting the merely observable. The observable must be observed in some very broad sense. Apparently we know that the observable have a relevance to our scientific ideas. Why should the observable be more relevant than the (utterly irrelevant) Tale of Aladdin. –

Whitehead answers:– Because in a broader sense these "observables" are observed. Suppose we take this immediate display of sense-data. There's something in Observation besides this "immediate Fact".

1. Whitehead is referring especially to his *Principles of Natural Knowledge*, part III; and *Concept of Nature*, chapter IV.
2. 'Display' is the second in the three-part distinction Whitehead discusses, especially in lectures 6 and 7.
3. Bell includes an arrow from a point just below this symbol to the end of the paragraph. Presumably he wants this marker for 'point of view' to be extended to whole paragraph.
4. It is not certain whether the horizontal line indicates a break in the lecture, although 'last time' seems to refer to previous lecture.

Lecture 10

Thursday, 16 October 1924[1]

Bell's notes

|19|
16. October 1924 Whitehead.[2] 

Another paper; on "The perceptual object: its Relation to Sense Data and its General Status." [Thurs. Nov 6, Three weeks][3]

Literature: Broad C.D.– *Scientific Thought* 266–283
Moore G.E. – *Philosophical Studies* Chapters V–VII
Russell, Bert. – *Our Knowledge of the External World* – relevant parts
Laird, J – *Study of Realism* – Ch. II.

Discussion might be taken up from entirely different point of view – a Kantian one, if one wishes – Alternatively – or from point of view of modern American school of Critical Realism (Santayana etc.)

Questions on Sat.[4]

Now:–

Analysis of what we mean by "Beyond". What observational basis has it? Beyond what? Simplify this "what" as much as possible. – Take immediate display of Sense data as the what (avoiding difficulty of perceptual object). Then something beyond. We remember past, anticipate future, etc. etc. There is ⟨blank⟩

[Husserl's "intentional background" etc. etc.][5] It's not merely other.

Mere otherness tells you nothing about anything [You can identify it when it comes as ∨the∨ right ∨one∨ or not].

Must be a positive fact. There aren't any negative facts for you to know.

You meet no "not's" on the street. Must be some [definite] relation to the present experienced.

The term "beyond" is frequently used where a phrase expressing mere otherness is logically required (so in Math). But although Logic may sometimes be satisfied with this (or with mere identity), it's always an abstraction founded in some specific distinction of mode of being in a relationship.

1. The Radcliffe version of this lecture, from notes taken by L. R. Heath on 18 October, begins on |49| of her notes, page 431.
2. In his diary for this date Bell notes: 'Whitehead nearly unintelligible'. It is not known what Bell intends by the symbol he has placed to the right.
3. The deadline is actually written in, at an angle, in the right margin.
4. In addition to the written assignment, Whitehead is reminding the class that the next lecture (on Saturday, 18 October) will take the form of questions and answers.
5. This is most likely a comment from Bell, who had studied with Husserl.

"Modal" here in Spinoza's sense. Always some reference to Being involved in a mutual relationship in distinct modes.
Value to you of disjunction in distinct ? ? ? –

= Contrast Value of their Essences.
or Identity Value of their Essences.

Fundamental relational diversity. Relation establishes Entities as mutually relevant and their stations in that relationship distinguish them and yet keep them together as having this contrast or identity-value.

But the relation is always definite. Foundation of any observation of identity or diversity requires a more fundamental diversity of instances – Instances joined together as stationed in a relationship.

Two distinct meanings to diversity. Pelican and Hippopotamus. Relational Essences of the two – no distinction. Both occupy Space Time etc. Diversity wholly in individual essences of the two. Green diverse from Red but ⟨blank⟩

But Green also diverse from a blind pig. There the relational essences are diverse and here fundamental difference. Owing to diversity in their relational essences we're precluded from any comparison of them apart from their relational essences.

Stations in Relatedness is homogeneous when Relations filling them are essentially of same type – i.e., have same relational essences.

"Brown" not in this room, occupying space in same sense as chairs or you and I. – but in much more elaborate sense. Scientific Object occupies it in a somewhat different sense from that in which Perceptual object does.

Since relation of Beyondness allows same ? ? to appear as Relator in different stations it is fairly obvious that the brown patch doesn't get its particularity from the brownness alone. Qua brown the patches are indistinguishable. But qua patches are otherwise. The station gives you the more concrete individuality.

Spatio-Temporal system of relationship marked out by these objects extends beyond them. – Systems indicate and carry with them modes of relation with regard to events not marked out to us by any specific apprehension of what's going on there.

We don't know individual essences of the "Beyonds" but we know them as related to this or that in apprehension – finding modal realisation in --- ⟨blank⟩

|20|

Events outside Room we don't know by individual essences by[1] only by relational schema to things inside.

Idea of "Beyond" Two types of knowledge. – (1) Full Perception – of Individual Essences and their spatio-temporal relations ^– "Cognisance by adjective"^

1. While in Bell's handwriting this term is clearly 'by', from the grammar of the sentence it needs to be 'but'.

(2) A minor type is simply Cognisance by Relationship – an extremely imperfect type.

Cognisance of some unobserved individual essence as in process of some definite spatio-temporal realization.

Cognisance by Relatedness leaves individual essence undefined; but it's not Cognisance of Indefiniteness. But it's ignorance of what is definite. Based on Ignorance or deliberate ignoring of the definiteness. Owing to that fact is that you can hold fact that there is an Entity in certain relations (definite) etc. etc. is basis of "logical variable" (Idea of any and some – Very perplexing – you never meet Mr. "Any man.")[1]

Something you can place to a certain extent, without grasping full individual essence.

You're definitely able to place essence with some knowledge. while etc. etc. (Very necessary for Philosophy to be able to account for Ignorance).

"Modal" used is limited to a specific ∨way, referring to specific∨ structure in Space-Time. "Modal Exhibition" is ⟨blank⟩

Space Time then is field both of modal limitation and [Whitehead here becomes incoherent for n^{th} time][2]

Geometric Structure of Space-Time results in Plurality of Being not collapsing into Monism of ---

So far not sensibly diverged from Kantian view of Space-Time. But now divergent.

Quotation from Caird [Kant not consistent]

Philosophy of Kant.[3] – "Whenever therefore we can make any universal assertions as to objects presented through sense; whenever we can say of such objects that they must be so and so, we can say they are based on (our own nature) and not on that of objects affecting us."

1. Missing end parenthesis provided.
2. Clearly, this is Bell's comment.
3. Edward Caird (1835–1908) published *A Critical Account of the Philosophy of Kant, with an Historical Introduction*, in 1877 (James Maclehose and Sons), which was superseded by *The Critical Philosophy of Immanuel Kant* in 1889 (published by the same press).

Lecture 11

Saturday, 18 October 1924[1]

Bell's notes

|21|
Whitehead - Sat. 18. Oct. 1924[2]

Question:- To justify Induction we want something, beyond, what is specifically included in Hume's premises. But we all have these beliefs. Whitehead's additional premise: practically the sheer assumption that Events are correlated as assumed in induction.

Or:- Can anything less than this full assumption be justified. Assumption of relevance from present to Future is precisely the assumption of whole case.

Whitehead's answer:- In exactly same spirit as Kant (but Anti-Kantian himself really) It's not a question of first observing a fully determinate fact and then plastering it over with assumptions from without. The process of Reality carries with it all that you can infer about it.

Concrete fact observed is not a bloodless dance of Sense-data, any more than it is, as Bradley says:- "a bloodless dance of Hegelian categories."[3]

Any analysis of concrete fact therefore exhibits comparatively abstract result of Being is fusion of ∨the∨ two factors - it's the fusion which is realized. In abstraction it appears as controlled as to its ingression into realisation.

Realisation is a process of Value both achieving and in achievement. The Existents itself are without value. The Analysis shows this process of creative fusion shows on its achieving side a throwing forward from this fluid fusion --- ⟨blank⟩
into the static possibility of such modes. - In every relation which their relational essence ~~show~~ allows for ⟨blank⟩

This throwing-forward is the protension of present into Future.

---

1. There do not seem to be notes taken by L. R. Heath at Radcliffe for this lecture. It may be that Whitehead did not attempt to repeat this class at Radcliffe, perhaps because he is here responding to an audience that was very likely all or very nearly all male, unlike at Radcliffe. In fact, in the notes taken by Hocking of the Friday evening Seminary (see page 531) there is some evidence that Suzanne Langer was attending at Emerson Hall, and these discussion sessions may therefore have included some women students.
2. For this date Bell notes in his diary: 'interesting discussion in Whitehead's course'. But in the notes found among Hocking's papers (identified as coming from the Friday evening 'Seminary in Metaphysics'), it is recorded that these were questions raised in the Seminary. There was already some overlap in those attending the lectures and the Seminary, and it may be all were invited to participate in this Saturday discussion. This carry-over happens again, a month later, on Saturday, 15 November.
3. This oft-quote phrase seems to be a variation on what F. H. Bradley (1846–1924), a leading British idealist, first coined in his *Principles of Logic* (Oxford University Press, 1883), p. 533.

Theoretical statement of which Induction is in itself is an abstraction from immediate observation of what concrete fact is in itself. (Rather more general than mere Spatio-Temporality is this "Beyond.") – Conjecture in Science arises because of our partial knowledge of present fact. If we knew present fact fully our knowledge of physical field wouldn't be conjectural. Kant centers all this ∧(this fusion)∧ in the Cognisant and doesn't give it the aesthetic turn.

Question:[1] ⟨blank⟩

Answer:– Induction can only apply to ⟨blank⟩
You might turn [?] into another line of Experience altogether – no protensions from present. Dream life, etc. May be just as vivid as reality. (Examples.) Whitehead's only reason for believing it a dream is that no time and no place at which Whitehead can fit it on to anything in relational schema of present experience.

– Only in the individual memory of its being between sleeping and waking etc.

Relations between Imaginary and Real are quite complex.

Question re Memories from Dreams etc. etc.

Whitehead's denial that we experience an instantaneous present. Series of instants is a "dense" series [Math.] You have present, immediately, before you:– Transition. And of the resting thing you have the Experience of the Endurance of what's before you.

The permanents are All the Universals by which you'd characterize the door [object] It's "World path" [Mod. Physics] – Its "path of endurance" is the individuality of the door. Whitehead argues that this "path of endurance" is too highly abstract.

Path of controlling endurance. – That because of which the Universals are characterizing the door for me. – Asking how? reveals a very complex story. To say "It's a quality of the door" is a little too simple.

Time as an abstraction from this immediate Process. The universal is irrelevant to time.

Shimer[2] talking not of his Experience of this table but of Tableness ∧[Cf. for point of view]∧ Making fallacy of misplaced Concreteness.[3]

Point of View [The permanent entities upon the process][4]

1. While Bell leaves us to puzzle over the question being considered, in the notes left by Hocking for the Seminary, for Saturday, 18 October, those notes do provide the questions as well.
2. This comment on Shimer appears on the right of the page. Shimer may be the student who used this example of a table in the discussion. In May 1925 Bell notes in his diary that someone named Shimer defended his PhD thesis, 'Relativity'. He is also named in the notes for the Seminary left by Hocking.
3. Sequence of the last three of these four sentences is ambiguous in the handwritten notes.
4. This missing end bracket provided.

[1] Physical field not <u>more real</u> than what Common Sense sees and posits [with its secondary qualities etc. etc.]

Common Sense is much more <u>Concrete</u> than Physics – But physics tells you something <u>more</u>

Contrast between Wordsworth and Shelley in attitude to science.

1. The Hocking notes for the Seminary participation in this Saturday discussion seems to suggest that at roughly this point it was Bell who introduced a question.

Lecture 12

Tuesday, 21 October 1924[1]

Bell's notes

|22|
Whitehead Tues. 21. Oct. 1924

Kant himself re the "Caird" position.[2] Cf. Kant in Preface vs. what Hume says he is worrying about. "Experience tells us what is; but not that it necessarily is etc. " – " therefore never gives us any truly general truths. – Reason rather aroused than satisfied by it etc.

Character of information there in Experience and our Apprehension of it to be distinguished. Characters may be there for awareness and analysis may err.

Now for relation Kant to Hume:– "Let me be once fully persuaded of these two principles etc."

(1) – Nothing in any object ∧considered in itself∧ which gives ground to draw conclusions beyond it etc., etc. and 2nd showing that same principle applies to a collection of objects.

(2) Then Hume's claim we must find impression for idea of necessity if we maintain we have the idea. Kant and Hume seem to agree on the 1st of these two (with a little explanation or clearing up of meaning of "object" or "experience") and differ with him sharply on the 2nd.

Whitehead agrees with Hume's 2nd, and disagrees with both ⟨on⟩ the first – unless "any object" is construed in a way not meant by Hume or Kant. But Whitehead believes Aristotle would agree with him. (Whitehead) [Whitehead means what Aristotle has conveyed to his (Whitehead's) mind].

Whitehead's own account of what he's endeavoring to do is to rewrite Aristotelian line of thought – hopelessly entangled in actual positive scientific situation and thought of his own day. – To put that line of thought exactly as Whitehead thinks it suits present situation.

"Non-intellectual teleology in Nature" Whitehead has been hinting. Many persons find this in Aristotle. Whitehead thinks very important there.

Hume's way of getting out of Skepticism is appeal to practice. Bertrand Russell follows him there so far. Whitehead doesn't believe this line open to a philosopher. William James taking fundamental line there – if you appeal

---

1. The Radcliffe version of this lecture, from notes taken by L. R. Heath on 23 October, begins on |52| of her notes, page 432. Despite a listing for 21–25 October, Hocking's notes under this listing do not seem to pertain to lecture 12. See footnote 1, page 50.
2. Caird had been quoted at the end of lecture 10, on 16 October. See footnote 3, page 38.

to practice you must bring practice into fundamental meaning of truth itself. In being tried, this point of view gave very valuable contributions to thought.

William James did what Bertrand Russell and Hume ought to have done.

Kant's way out. Then (3)[1] people who seem to agree with both the Hume's

Points and then make "postulates". Whitehead can't see any sense or basis in this, if we really don't know anything there.

You can throw over both, or (with Whitehead) agree with second and disagree with first. Whitehead. – commenting on every thought of sentence about: if we have any general truth --- it must be due to nature of our Sensibility etc. etc. from Kant's intro. "Whenever therefore we can make any universal assertions as to objects presented through sense..."

What do we mean by "Universal Assertions" and by "Objects presented through sense"? Aren't there perhaps some types of "Universal Assertions" we can make on basis of experience, and perhaps other types of "Universal Assertions" we can't make on basis of experience

First as to "Objects presented through sense" – Take "that particular green patch there and now." Caird-Kant guilty of "Fallacy of Misplaced Concreteness". Analysing this into greenness and patchiness – provided by the mind. Then says all general assertions here concern the patchiness. Relatively concrete Cognition there is more concrete than either Greenness or Patchiness. [These are products of conceptual activity of mind. – Known by abstraction neglecting feature equally necessary for concreteness.] Greenness essentially incapable of realization. Greenness is not a sense-datum. The immediate sense datum is the green (green patchiness).

Greenness = individual essence of this

Patchiness = relational essence of this

~~Immediate deliverance of perception~~ Caird-Kant seem to consider Greenness as "naked datum" [= standing up before Cognisant as ʌobjectiveʌ otherness]; Whitehead regards it as green patchiness

But the actual objective otherness is not a universal "Green patchiness" but a particular Modiness of this – one for you; other for you. Involving e.g. superficial shape (triangularity etc.) or a vague Voluminousness of color etc.[2]

|23|

This is the most Concrete Universal of pure Sense datum type. This level of abstraction is simply separating this one from other such of same grade of Concreteness – from other Sense-data of same general concreteness. It touches its "limitation" and not its "conception" (Spinoza. Limitation is by something of own kind). ("Beyondness" as stations for relator of same type of relational essence)

1. Bell uses '3rdly' here, but seems to be in sequence with '(1)' and '(2)'.
2. These two sentences are preceded by an arrow pointing up to the bracketed comment above.

Green (in sense of green patchiness) is higher abstraction isolated from lower abstraction from that Modiness. Greenness is higher still and correlative with the abstraction. Patchiness is the Space-Time.
Even here Whitehead has "made much too rash a jump." Has omitted a relatively essential to whole primary procedure of abstraction. We started from a definite instance or "occasion"[1] of realization and the particular "modiness" has its Modality (its relational essence as expressed) in relation to definite particularity of that occasion. The particularity of that modiness owe then to the beyond of that occasion.

Thus Space-Time finally isolated in our abstraction is Spatio-temporality relative to that occasion. We can go higher to extensive manifoldness divorced from any relation to definite occasion.

At any rate we mustn't confuse stages of our Abstraction or are vague on it we're apt to go wrong on adjustments of Metaphysics to Science.

This Principle of "abstractive relativity" is of great importance. "Beyond"is not mere Beyondness in vacuo, but from some definite occasion of realization. Abstraction in its most lowly – primary – form proceeds from particular anchorage in a definite occasion; and when you divorce from this, you get "Abstraction in 2nd degree"

Abstraction in 1st degree = "occasional abstraction."
Abstraction in 2nd degree = "praeter-occasional abstraction" – referring to no occasion of realisation.

We can make no universal assertions in respect to --- ⟨blank⟩ when you've got outside of occasional abstraction.

Logical use of "~~Manifold~~ Universe of Discourse." More fundamental type of Universe:– A Universe of percepts – "Manifold of modal realizations included in

occasional beyondness which is derived from
some definite occasion of realization."

Occasional abstraction bears in itself relational --- to [Context]
"Indications" by "Relatedness"

Memory is cognition of earlier happening as filling definite Station in Beyondness of later happening and vice versa. And it's only in this double immediate mutual fitting on of different experiences [only sure when close together] that you have any ⟨blank⟩

Everything must bear within it impress of its own possibilities of relationship.

[Then seems to imply this relating only matter of Spatio-temporality.]
~~My~~ Relation of to-day and yesterday to dream of last night.

1. This seems to be the first instance of Whitehead introducing the term 'occasion', which he develops through the next lectures, especially lecture 17 and thereafter.

Lecture 13

Thursday, 23 October 1924[1]

Bell's notes

|Continuing on 23|
Whitehead Thursday 23. Oct. 1924:–

Was in middle of asserting certain relevant distinction to be kept in view re Abstraction. "Occasional" vs. "Praeter-occasional" Abstraction. The lowest type of Abstraction is relation to a particular occasion. Lecture on "Beyondness" dealt with same phase of thought. Then going on to "Universe of Perception." Manifold of modal ... derived from the "Beyondness" of an "occasional" Abstraction.

Occasional Abstraction always has reference to a particular occasion. Hume's dictum then comes from not recognizing distinction between the
P of v. lowest Universality got in this occasional Abstraction and the higher kind
[and Thesis] where you ---
[Here ⟨blank⟩
|24|
Whitehead thinks the sharp distinction here really justified. [False simplicity
necessary to presentation – critics come down on this!]

Whitehead thinks "occasional Abstraction" not sufficiently emphasized as yet. Differences because one man has the idea obscurely in mind, other not. But all of us recognize it in flashes now and again.

Now back to the quotation from Caird:– "Whenever therefore we can make any universal assertions ∧as to objects presented through sense∧ ... etc. ...etc."[2]

Whitehead says (with Hume) that we can make no "praeter-occasional" assertions whatever. But perhaps this restriction is meaning of "presented through sense." Whitehead would read it so:– "Whenever therefore we can make any universal assertions which are occasional, these assertions being with respect to objects given through sense....
... must be based upon the occasion from which the universe of perception is derived". This occasion includes our own cognition and bodily life.

The chain of Abstractions again:– (1.) This modiness of green patchiness

1. The Radcliffe version of this lecture, from notes taken by L. R. Heath on what was likely 25 October, begins on |60| of her notes, page 435. There may be notes by Hocking on this lecture, but they seem to have been incorporated into those from Saturday, 25 October (lecture 14, page 50). It is immediately apparent that, unlike Bell's notes, Hocking's notes record whatever he has captured from Whitehead's lectures but organised into his own outline. Indeed, Hocking's first set of notes already demonstrate his intention to organise Whitehead's content in his own way.
2. First quoted at the end of lecture 10, 16 October.

(2.) Green patchiness (3.) Greenness. Of course dozens of such transitions from one starting ∨point possible, e.g.∨ from this same no. (1) to (2) This modiness of color Patchiness and so to "Coloration" merely.

There is, however, thus, an *Infima Species* – a lowest member, where the this assigns "occasional" perceptual world and particular relations and specifications – "*Infima Existens*." – Involving greatest particularity (and concreteness) which thought can compass.

Relations of this to Space-Time ("Structural Limitations"). Generality within this Spatio-temporality is all I get here. – Generalising into Beyondness of this occasion without which it cannot be understood. Existens infima is only an Entity as expressing limitation within this particular Space-Time. ∧A limitation in respect to the Universality of the "Beyond."∧ (Bergson's emphasis on importance of Limitation as being inherent in reality) (Spinoza too) (That what's happening is best grasped

⟨blank⟩

A certain difference between Space and Time. Modern Relativity strikingly original in its assimilation of Space and Time. But after all there is a difference ("I put it to you as sensible people.") Somewhere or other we've got to unravel as to how far we can push the Identity of Space and Time. – Idea at present run to death. – Overstatement (Typical). [Good side-remarks. Scientists of any time are almost infallible as to what's worth doing. But their mode of expressing that infallibility just is to ~~claim~~ assert in an unqualified way the truth of this special enunciation; and there ⟶ they're equally infallibly wrong.]

Limitation involved in the different stages of Abstraction again.

The Higher abstractions only exist as in the lower Existents. [Of course] –Greenness only as in some definite modiness of green patchiness.

Existence is Relevance of mere Being to Realisation.

Aristotle here (contra Plato). Whitehead really taking a middle position. – Lowest Existents are pure forms. – They do transcend their immediate realizations.

This is not Aristotelian. But Whitehead doesn't hold Aristotle's idea of Matter. – Agrees with Aristotle that we have to have a theory of Matter – ὕλη.

Lowest Existence is an εἶδος or μορφή – a form.

εἶδος and ὕλη is shape and wood!!

For Whitehead the Aristotelian Matter is Value [! !] and that the process of Realisation is the Becoming of Value – namely as the "reactive significance of forms". – As informed Value.

Realisation is Valuation.

Lloyd Morgan – *Emergent Evolution* – and Alexander – *Space, Time and Deity*[1] – process of Emergent entities out of Space-Time. Whitehead has a little quarrel with both (thinks Alexander the better man but perhaps

1. A zoologist and student of T. H. Huxley, C. Lloyd Morgan's *Emergent Evolution* was published by Henry Holt & Co. in 1923. Influenced by earlier work of Morgan, Samuel Alexander presented the Gifford

Lloyd Morgan got idea first). Alexander a bit tends to look on emergence of entities as chance or incidental. Whitehead goes one step further:- The Space-Time process is the emergence of entities; the entity that emerges is Value – is reactive significance.

Here again where Whitehead differs from Aristotle – Aristotle's metaphor of Wood wrong. It's the shape you've got first. – which you know as in thought. In so far as "anything" can be put into priority it is the Form and what emerges is
|25|
the Shaped-Mattered. Whitehead regards it as a Superject rather than Subject (i.e., ὕλη) Pliny and Virgil use superjacio in this very sense –"lying over". This sort of metaphor isn't essential to any philosophy whatever – but is to the emphasis and the emphasis governs the development.

Perception is aware of the actuality of Form-value – Of some ingredient Forms which are shaping or informing that particular case of Form-value.

"Occasional particularity" is station in the ⟨blank⟩

"There is the conceptual analysis and there is the occasional particularity." Realisation is the formation of Value and Value has no being apart from its information in Realisation. But the lowest existents transcend this particularity. – Have a significance which can't be expressed entirely in detail of their actual ingression into process of reality.

lectures at Glasgow, 1916–18, and these were later published by Macmillan & Co. as *Space, Time, and Deity* (1920).

Lecture 14

Saturday, 25 October 1924[1]

Bell's notes

|Continuing on 25|
Whitehead–Saturday, 25th October, 1924:–

Start with passage from Ross's book on Aristotle. (few months ago) p. 65–66.[2] Passage about Change and becoming.
2 kinds. –
(1) That which becomes, persists.
(2) That which becomes, doesn't persist.
But always what happens is that
A, not-B becomes A B.
Always two elements, a substratum and a form.
A third element always presupposed:– Privation of Form.
Three Suppositions of change: therefore is Matter, Form, Privation.

Necessity of getting out of Provinciality of Times we live in. Specialized physical methods and their effect. Our modern ~~prov~~ intellectual provinciality. Must soak ourselves in great minds of the past with different outlooks. They lack some of our finesse, but they have something we haven't often. Never mind distorting them somewhat in translation. Aristotle provincial too; hadn't read Darwin.[3]
Real Defense for a broad Humanism

No becoming from bare Privation but only from a Substratum, and from it not as Being *simpliciter*, but etc etc. Difficulty removed by recognizing grades of Being and difference between Actuality and Potentiality.

Passage [Aristotelian i.e.] combines two sides of Aristotle's thinking. Starts with view of Predicate qualifying something. A-not qualified by B is succeeded by A-qualified by B.

Transition is excluded by this point of view. – Cinematographic point of view – jump from one bit of film to the next. There are no "middle terms." So rest of paragraph is devoted to explaining away the Subject-Predicate point of view and replacing it by a more elastic schema. But Subject-Predicate point of view remains absolutely fundamental for Aristotle. – So there's a difficulty and confusion somewhere here.

Now psychologically how understand this. – Instead of Subject-Predicate you began to talk about Matter and Form. (More elastic – not so worried by "Excluded Middle" law. – That law's what's in the way of Transition). Finally in last clause of quotation we pluck up courage and use

1. The Radcliffe version of this lecture, from notes taken by L. R. Heath on what was likely 28 October, begins on |64| of her notes, page 437.
2. Sir William David Ross (1877–1971), *Aristotle's Metaphysics* (Oxford University Press, 1924).
3. The arrow in Bell's handwritten notes suggests the last two lines here be shifted above last sentence in this column, as they have been.

Potentiality. You get around Excluded Middle without verbally violating Substance is Quality Categories (or Point of View). This Point of view is just a cloak for more fundamental point of view.

Whitehead. – second half of paragraph is absolute nonsense if first half to be taken quite seriously. Whitehead thinks second part very good sense. But can only be rescued from nonsense by admission that Subject-Predicate relation is only a cloak – not an analysis – just enough to indicate. We must then explain what you mean by "potentially but not actually."

Again--- 176–177– Aristotle is discussing various senses of δύναμισ. One use is Power, second is Potentiality for Change. And Aristotle says second sense is one he's using – Potentiality in a single thing of passing from one thing to another. Aristotle saw that Potentiality indefinable. Eleatics had denied this [Megareans].

Point is that Change is not Catastrophic – It's not true that A that wasn't B suddenly became B. Whitehead agrees with Ross–Aristotle here; but sympathizes with any Megarics who lived long enough to read Aristotle. Thinks Aristotle accepts Megaric premises and refuses their logical conclusion!

Too bad if you can't say A-not qualified by B without involving A in a sort of potential B-ness. (Fraudulent bankruptcy e.g.)

Processional View of Reality, on other hand, as conditioned, lets you conceive it. Conditions

|26|

when Abstracted from Process is a certain Envisagement (taking account of) aggregate of relevant existents.

This Aggregate is in a way the conditioning basis of that processional character of reality. In being a conditional basis it is not a static basis from without. --- Existents enter into Process as conditioning Forms. Formation the ὕλη by the Ingression of these Existents into the particular stages (or parts of Stages) of this ultimate processional reality that is there before us.

The Greek term for this unshaped [An abstraction obviously. The mere unshaped can't exist, obviously] is Wood. Hume says ultimate fact is "Impression." I.e. impress – form – upon something.

Unjustified use of words > The ultimate matter better conceived not as Wood, but as Value. – No unshaped Values possible but inasmuch as there's a procession of Shaped Value. And this is ultimate character of Realisation. The conditioning static form --- ⟨blank⟩

"Togetherness of their individual essences breaking in upon one another." ὕλη being "informed" – but not as Subject but as Superject. It's the outcome which is the shaped value.

Spinoza's Mode as privation, limitation in relation to the one Substance Modal Limitation. There's an emergence of Value by that peculiar way of Ingression of B into A.

A not qualified by B	A qualified by B
A not taking account of B	A involving necessarily modal limitations and in transition to transcending this limitation

This is to some extent a contradiction. Second is *ipso facto* a certain type of qualification.

In a certain sense you never get rid of being qualified by B.

Our modes of thought are absurdly – childishly – simple. Actual facts of the case is complex. We always speak, then, in childish simplifications that won't hold water for a moment.

Now reads last half of the Ross–Aristotle over again. That: Implication of Privation of Form. Whitehead points out that if it's mere privation it doesn't help you at all. But if it's "Envisagement" without any further realization --- etc. Use of "Incidental" ("Simpliciter", too). ("bare privation" vs "privation in a substratum").

Whitehead – An Existent is ? ? And defines Existent as having relevance to processional realization. Being as relevant to Reality is an "Existent"; and totality of Existents, viewed as static elements for human thought – just there is static grounds which condition the process.

Hocking's notes[1]

Whitehead, October 21–25[2]

The Aristotelian description of change

Describing change in terms of substance and predicate in which S-not-P becomes S-P and P and not P have no alternative, i.e., the law of excluded middle holds we get into the difficulties which the Megarians showed

Describe it as Potency passing into realization, and ~~Not~~ P is already present as a possibility before it appears: Not-P does not exclude the possibility of P, (but rather indicates it)

In this sense we may say that in perceiving change we actually Perceive possibility and its realization.

(i.e., we perceive the particular possibility which the situation S-not-P contains)

1. Presented here are the notes by Hocking which are labelled: 'Whitehead, October 21–25'. However, they do not seem to include content from Whitehead's lecture from 21 October. They may include content from 23 October but not as Whitehead presented it. Instead, it seems to be worked into the outline as organised by Hocking on the page presented, as here. The editors thought it best to keep this page of notes intact, and let the reader decide from which lectures they have been drawn, as evidenced in Bell's closer presentation of each lecture.

   The reader will find that, beginning with lecture 17, and quite consistently through the remainder of the series, Hocking invariably organises Whitehead's content into an outline of his own making.
2. While most of Hocking's notes remain in handwritten form, this page is preserved in typescript. In other instances where typescript is preserved, there are indications the notes were taken by someone else in the class and provided for Hocking in this form.

(And perhaps we perceive also the necessity by which the passage is made; i.e., we perceive the passage as something more than a mere fact, as having a trend, and a requirement in it, a potency as well as a geometrical possibility, and perhaps a value-trend a nisus toward the P-becoming of S whence we may say that

In the process of Realization, Value becomes 'informed'

To have value is to have reality
Realization is the ingression of value into form

Reactive significance is the relevance of any existing state to another state that may fit

The element of structure, I should say, in all perception, as the (Kantian) understood element

The fundamental proposition here is that we can perceive the term and its relation without perceiving the other term, the relatum.

It is this inherent connectedness of things that make induction possible

Lecture 15

Tuesday, 28 October 1924[1]

Bell's notes

|Continuing on 26|
Whitehead – Tues. Oct. 28th, 1924:–

Two ideas running concurrently in this course – (1) To explain sort of ways in which Whitehead makes plain to himself philosophical problems and their rise. (2) To direct attention to what it is that has to be explained: How it is that Philosophy arises. Difference between Philosophy and Natural Science:– [It's "this" I'm going to put in the melting pot – no matter what it is we are talking about.] But getting in state of mind to see the big puzzle of Universe before you is the philosophical difficulty. Reason for living lecturer rather than book – a sort of *habeas corpus*. Greatest way: "Problematic"[2] getting into touch with great philosophical minds of the past and see how they saw it. They were all seeing same great fact, but emphasized different sides of it.

Last time a dose of Aristotle. Now further:– Aristotle's idea of the prime mover, etc. Aristotle's two ways of taking things – The Subject–Predicate one and the ὕλη - εἴδοσ (μορφή) way (freer and deeper). The Ross paragraph is excellent example of the Aristotelian tradition. – It's in this tradition that you find fallacy of misplaced concreteness. Aristotle himself (in de Gen.)[3] is better.

[Question out of Quantum theory (raised by student). – Number of states (physical) limited.]

|27|

You've got to allow for facts before you bring transitional reality. – Got to allow for change. – Also that there's some form or other of ground of condition, which is therefore not itself changing:– That side of fact by which change is what it is. You may reduce it to the ghost of this – mere contiguity in Time and Space and similarity (Hume). Or you may make it the fundamental thing – God – like Aristotle. Some way or other you've got to have what's in some sense a ground. Aristotle puts it without, as God – prime mover, itself unmoved etc. Hume looks on it as a ghost of a general statement which he can take from within.

---

1. The Radcliffe version of this lecture, from notes taken by L. R. Heath on 30 October, begins on |67| of her notes, page 438. There do not seem to be any Hocking notes pertaining to this lecture.
2. This is added at an angle in left margin and would seem to be Bell's comment.
3. The full title of this Aristotelian treatise has long been referred to in Latin as *De generatione et corruptione*, or more recently in English as *On Coming to Be and Passing Away*.

Something general (blank)

Beauty of Aristotle is all inclusiveness of his philosophy. He doesn't turn things out because they don't fit into a system. Whitehead thinks Spinoza gives best shot at solution, but now Aristotle – p. 93–94 Ross [reads] – *Primum mobile* eternal (out of Space and Time) [W. seems to muddle Aristotle's God and *primum mobile*][1] Whole of modern physics is a study of recurrence [Aristotle's "circular motion"] – Aristotle has got the essence of the thing – transition and recurrence which retains itself; and a condition previous and logically prior to the sheer movement. Assumption that movement must start from centre or circle [? Whitehead can't see this][2] Idea that impulse must be maximum at initiation [Last word of Relativity is that there is a maximum velocity which governs all else.] Certain haunting of relevant ideas through Aristotle. Aristotle fixed star sphere as fastest ?? – Now "supposedly observed fact" that movement of light is the fastest. Then the idea of transmission of movement throughout universe in Aristotle. "Transmission" is what makes it impossible for us to look on colour as quality there in the desk. Modern idea – Action at a distance impossible ? ?

Ross's queries at end of *Physics* (2) How can incorporeal being initiate push or pull? Answer in the *Metaphysics*. (1) How can prime mover be put at the circumference.

As to push and pull idea – 19th century view. (Lord Kelvin[3] would have told you the same.) Now it's lost the delightful obviousness of the billiard balls and greater difficulty – more abstruse and metaphysical idea of fundamental field of force.

In *Metaphysics* first mover causes motion as an object of desire – not a physical agent at all. But this raises equally great difficulties (p. 175 – Ross) Things of different genera have only analogically "same" causes as one another. Individual is cause of individuals – No general cause etc. (P. 199 – more of Aristotle's Theology):– Adaption ascribed to unconscious Teleology of Nature rather than to divine agency. Change and Time is unperishable. Only continuous change is change of place; and only continuous change is circular, etc., etc. – Reason for eternal Substance. Its essence must be not power but activity. Must be immaterial Substance.

Platonic Forms out of time (Whitehead's "Existents") but comes down then on their "essential isolatedness". "Existents":– beings conditioning this real world. Processional side doesn't come from these forms as beings; but from a side of fact which is other than the forms. Whitehead:– this other aspect of the essential transition is conditioned by relevance of Platonic Forms to it. Platonic Forms are part of its structure. – It's because

1. The 'W.' within this bracketed comment might be 'Whitehead', which suggests it is a comment of Bell's (although the muddle may be his own). But it could also be 'which' and refer to Ross.
2. This missing closing bracket has been provided.
3. William Thompson (1824–1907) was made the 1st Baron Kelvin in 1892, the first ever British scientist to be raised to the House of Lords.

of them that it can be and without them there could be no transition. But it's realisation through limitation.

Whitehead would point out that it is the condition of the activity. It's not a 1st cause starting it but it's an immanent condition there all the time. Realisation is essentially only itself under modal limitations. Otherwise you'd have simply Monism.

Common world of envisagement is just Space-Time viewed --- ? ? Within the common environment there's modal limitation – Structural realization – which as limited has its differentiations. But as limited essentially also transcends itself and goes on to something else. Reality is limitation and always has to go on.

What Realization is? – Being something for each other. Hume calls it "Impression." What is Realised are not the Platonic Forms – They're never real. – But the ὕλη – The interest – the taking account of – the impression [Leibnizian]. The intrinsic fact of what one thing is, being something for the others.

|28|

Whitehead "not a 'modern' philosopher."

There's no such thing as Impression in the abstract. Whitehead drops Hume's "modern" Subjectivism. Impression or taking account of is always specific. It's always the ὕλη as formed –shaped. A Form never becomes Real – a Form is envisaged as in Reality. Possibility clings to them not in that they are possible; but that a specific realization is possible.

Transitional achievement of mutual interest involves impression of 'black'. – And my interest, too. – I am being shaped by black.

"Envisagement" is a "mild form" of realisation.

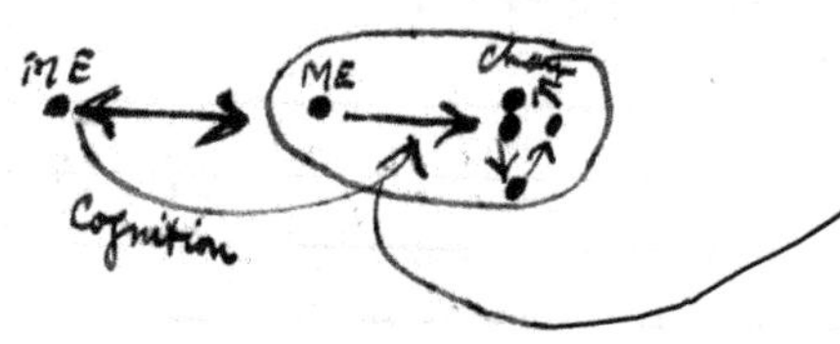

Chair itself as an event is very complex – a process of mutual taking account of, there (at chair). Then, that reactive significance – Realisation of chair for me and me for the chair; and then there is my reflective consciousness of all this.

My image of the chair:– Image not inside me but I'm inside the image. But the chair is simply a set of reactive significances. Chair as a set of Images. The Forms which shape the chair cut vividly into my envisagement. – My direct perception of yellow or green etc. etc. Envisaged

! → by thought as mere possibility, I get a fainter one, that's all. Whitehead agrees with Hume in this point.

Lecture 16

Thursday, 30 October 1924[1]

Bell's notes

|Continuing on 28|
Whitehead – Thurs. 30. Oct , 1924:–[2]

[need of term Problematic][3]

Mill, interesting, but unsatisfactory because never really got hold of what there was to discuss. One reason why Whitehead goes back to Aristotle; because of his seeing what really is the problems of Metaphysics.

We started with problem of Induction. Why is it valid? Man of Science sheerly assumes it and nothing in Science itself to make it anything but a blank assumption. Then we took up perceptual objects and they brought us up against something extremely vague (Vagueness of Class) or a sheer χ (τόδι τι).

Then to idea of Becomingness, Transition, Change. – What is there in it? What is it that goes on? Why isn't it nothing? What is it that becomes (question of Reality – Problem of ὕλη and εἶδος) and in what shape does it become?

Any answer must arise from consideration of some ground of things by virtue of which they are what they are. Is there any eternal principle which carries with it the solution (from within! [✓]) of these questions. Something self-explanatory including these (Hegel would say "hear! hear!") Something to mitigate the "Blankness" of the Assumption. If we can't find something such we're forced back to Hume's position (His virtue in showing this clearly) – something which we can't justify but there it is – sheer irrational practice – which you can't justify nor forecast its success. Bertrand Russell, too.

Haunting idea of an underlying principle – in West individualized and personalized; in Eastern thought more wisely left impersonal. In Western Thought only three dominant lines of Thought: (1) Theology with dominant influence from Semitic sources [Aristotle too] (2) by Physical Science which is of Greek origin. Ionian Philosophers essentially Scientists. (3) A modern mode of approach, largely psychological – starts as Hume starts. – Gives us the enormous subjective bias of modern thought. Each has its danger of warping the truth. (1) Western Theology permeated by idea of the best. Object is to define that essential aspect of world that defines the best, and find its status within the whole. But philosophy seeks

1. The Radcliffe version of this lecture, from notes taken by L. R. Heath on 1 November, begins on |71| of her notes, page 440. There do not seem to be any Hocking notes pertaining to this lecture.
2. In his diary for this date, Bell notes: 'Wonderful Lecture from Whitehead'.
3. It is not clear just what term Bell (or Whitehead?) is referring to.

Theism in Point of View

the whole. – The metaphysical Whole is equally with the robber as with the saint. Doesn't take sides.

(2) is too abstract – Leaves out Spiritual and Moral side and aesthetic side. Loses beauty of rainbow. The Scientist doesn't want his heart to "leap up" when he beholds the rainbow – gets out his spectroscope. Science commences with Abstractions and ends therefore with them. Ends up with a set of problems which are such because of initial omissions.

|29|

(3) apt to commence by leaving out whole problem. Studies particular Organism in relation to an environment. Always assumes just that underlying ground of things which is interest of Metaphysics. Each of the three then takes problems up to gates of Metaphysics and leaves them there. Metaphysics seeks that Self-explanatory ground – which must be present, then, in every bit of reality we find. A non-passing, self-explanatory side of things.

The problem so elusive that each age has a provincial way of approaching this topic. – Even granted the adequate mentality. Each age shut up to its own emphasis. Not only each land but[1] each age has its own provinciality. But the problem itself has always haunted each age with its own particularity. Examples: from Aristotle last day. Modern Psychology way muddles up a lot of statements about way organism behaves etc. ? ? ? Perhaps some of Easterns have made most[2] persevering attempt to speak of this without particularity. But they only tell you what it's not – Too elusive.

Various great moments (movements) of thought – Greek, Medieval (with Semitic infusion) etc. Each has its culmination. Can't underestimate any movement of civilized thought (art or what not as well as philosophy) without the form – moments, stages –

There's the period of preparation ("the primitives")[3] – genius with detached achievements – brooding surprise and haunting pathetic failure. It's got hold of something it can't do. – That's secret of its charm to later ages when seeking for fresh inspiration for genius.

Second moment – Culmination – Achievement – found; how to put the thing across the footlights. Stands up as a beacon for all future achievement – That's beginning of the ∧history of the∧ thing – Something is found out. But also a limitation *ipso facto* at same time. Then long period of carrying on, improvement, carrying out, etc. etc. – Stage has been set.

Third, finally the thing gets worked out. – overdone. Then a (perhaps self-conscious) effort to be primitive again. – Barbarians come back, or "Revivals" – anything to give a new beginning.

Philosophy of the Greek civilization culminated in Plato and Aristotle – dual side to the culmination. Plato is the culmination with the primitive

1. Bell definitely has 'by' here, but context suggests 'but'.
2. Occasionally Bell will split into two columns, but in this case seems not to have used right-hand column.
3. Missing closing parenthesis provided.

element still there. That's the charm of Plato. Aristotle might have been one of his own followers. (The difference between them). They do sum up and express fundamental elements of any philosophy founded upon a naïve objectivism. And this is implicit philosophy of Scientist when he's really doing science. Read Huxley:– impression that Hume was the scientific philosopher. But reason for this (Hume and Scientists both making it uncomfortable for Theology). (But for Science too – not seen).

Aristotle, then, particularly interesting for very masterly way in which he includes in some way, all the elements which he had there.

Then long history of this movement. Concluded by Spinoza – who horrified the world by the last edition of Aristotle's God.

At same time Descartes started modern subjectivist trend. Both the *cogito* and *sum* are in the first person singular. – There's the battle flag of subjectivism.

Whitehead's view that none of these points of view wholly right and none wholly wrong. Eastern idea of Brahma [?] too vague. We've got to combine somehow something which acknowledges Subjective element but links it up with and gets beyond *Cognoscitor ergo Cognita.* [?]

Hume and Kant stand in middle stage of modern philosophy. – Culmination. The two of them set the stage within which modern philosophical drama has to be played. But which is the hero and which the villain.

With Hume:– Very first few words of Kant's introduction and Descartes etc. etc. – You
|30|
find this keynote. – Hume's division into Impressions and Ideas – it's all on the mind. But he's not got rid even there, of the Eternal. Wherever you get anything general you've got something that lies beyond particular occasion. You're essentially presupposing the Eternal. He begins with an All.

But its underlying Eternal confesses itself very soon as not self-explanatory. – At end of *Treatise* and even shorter *Enquiry*[1] – gives up. – Practice – But this may be good enough. Bertrand Russell also has grown a little happier with practice.

Kant has same subjective standpoint and shows that within this you have more that is self-explanatory. But you start with particular sensibility and conceptuality. Kant too gives us *Critique of Practical Reason* to get beyond the Solipsism of the First one.[2]

("Ghost of the old world wailing over corpse of the new." – a black practicality about it.)

---

1. Single-term references are to Hume's *A Treatise of Human Nature* (1739–40) and *An Enquiry Concerning Human Understanding* (1748).
2. The 'First one' presumably refers to Kant's *Critique of Pure Reason* of 1781, followed by the *Critique of Practical Reason* in 1788.

The First thing we've got to do is really to consider the "Static Basis" which gives the character to the process of Realisation. And in considering that basis (which is found in every particular occasion) – we don't necessarily go to the *a priori*. We must entirely depend on *cognosco* (first clause of Whitehead's phrase)

"I am apprehending an occasion". "We can know nothing which is not embodied in the occasion that we are apprehending". – We can't get beyond that but we can see in the occasion anything which[1] is general and both in making that occasion what it is and must turn up in any other occasion what is relevant to that one. That one tells us that there are other ones relevant to it. The Eternal is that which has that necessary relevance throughout all these [interlocked ---]

You can't know that there aren't other lives which to us are as nonentites. Our lives may be part of other lives.

Whitehead's metaphysics is an endeavor to {discern/express} that ∧which is Eternal in∧ "occasion Eternal" relevant to the being of that occasion which in any moment I am apprehending.

Must be equally necessarily the Eternal of any other occasion bound to present one by any relevance whatever.

Relevance not synonymous with implication. If the one is in any way connected with the other then the Eternal of the one must be Eternal of the other. Whitehead talks of "occasion" as full concrete occasion when you go on to something which really limits it in Spinoza's sense.

Spinoza's fundamental philosophical definition of "limitation". Only by something of same adequacy of concreteness. Dream does not limit the two occasions of the day before and day after – but does limit the occasion of the night between.

1. The contraction 'wh.' is used three times in this sentence, and a good illustration of our judgement that the first should be expanded as 'which' and the other two as 'what'.

Lecture 17

Saturday, 1 November 1924[1]

Bell's notes

|31|
Whitehead – Sat.; 1st Nov., 1924:–

Somewhat more Systematic consideration. Difficulty with Metaphysics. So vague – about everything. But "you've got to see the Wood by means of the trees, you know." Point of view that we've got to start from the: *Cognovi* and go to *Cognitor* [?] ["The plural in *Cogniter*" ??]
Fact = Factuality. That general sense of what is over against us in Cognition – including ourselves. Split it up into a few general ideas.
Fact = Factuality = Determinate, Processional, Realisation.

This word = a battle-flag of accuracy.– That there are details and they can be determined.
We mustn't think always in technical terms.– Things are always getting along.

"Realisation" – "Some how or another there's a being real" – You discern in this something less than Real, otherwise you'd not have the idea of this becoming realized. You're always comparing everything to a definite ideal of Reality.– There is a Realisation. (Whitehead says he wants to "fatten this out.")

Talking last time about The Eternal. How this has haunted the imaginations of men you feel that nothing too pretentious to be said about it. When we think of us sitting down to do it:– nothing too homely to be said about it (Hume vs. the great Mystics)– They're both right – but Whitehead feels in the homely mood to-day. Generalities apt to look a bit grand. But only in this way.

Contrasting it, first, with Endurance – not the always enduring but something out of Space and Time is what Whitehead is trying to get hold of.

Phrases, no one of which will do, but each suggests a phase.
The Eternal = The Determination = The Ground exhibited in every Occasion of Realisation = that which is true respecting (Accent, "high light", here) every occasion because it is an Occasion (Processional character in itself).

= The Substance of which the Occasion is a limited affection or the Substance which requires all occasions. (But Substance looked on passively. Not so here:– it determined the procession)

1. The Radcliffe version of this lecture, from notes taken by L. R. Heath on what was likely 6 November, begins on |76| of her notes, and continues on 8 November, on |78|, pages 442 and 443.

= The Ground of Generality

= Fact as self-contained with its own reason.

You never solve one of the great philosophical questions by showing that they're really something else that causes them. – What you want to do is to show it in such detail and such generality (same thing [?]) so as to see that everything must be so ∧(in general type, at least)∧ because things are what they are.

Relation between what is Eternal and what is Contingent is the prime puzzle really.

→ You're always pushing the Contingent back and back but never get quite rid of it. If there's no Eternal in this sense there's nothing to be said – no Metaphysics, no Science, etc. etc. It's because there's something eternal that we keep going at all. Even Hume:– sat down to write treatise on general things.– Hume's generality in opening sentence! Hume made this one sentence dominate all his own Philosophy – and for 200 years (∨people∨ declaring that he had influenced them etc.)

Every sentence with slightest generality in it ---

But even mere fact of indication – ☞[1] "That!" – I've got hold of a general procedure. – Implies faith in the Eternal, in some way.

Spinoza really conceives task, where Hume too "homely." But Spinoza talks rather as though the Divine went on its own and other beings on their own – The one a model for the other. Whitehead thinks Spinoza wouldn't have stumbled thus had he been brought up in the East.

Comparison of distinct occasions appears to be appeal to Eternal for what is exemplified in each.

Similarly:– All reference to Possibility is a reference to some ground which is in the Occasion, but which is not occasional. – Same holds for Impossibility. – There's something which is General, which the actual refers to: but *ipso facto* doesn't embody.

"Actuality" (using many terms because each one becomes a neat little symbol meaning nothing)

|32|

"Actuality" is Being an Occasion of Realisation [This is uncommonly like what Whitehead, in books, has called an Event] Difficulty with Eternal is to get into grips with the particular occasion. – Spinoza's mistake. Indian Philosophy at fault here.

– The East lets the definite occasion slip.

– The West lets itself get swallowed by the definite occasion.

The "how" of limitation and also the transcendence of limitation [is the occasionality] ∝[2]

Every limitation essentially transcends itself – limitation as self-transcending – The how of limitation is the Eternal. How fact does come before us in a plurally monadic form.

1. Bell actually draws in this little hand, pointing. It is unique.
2. It is not clear what this symbol is intended to convey.

"It's" The organic unity in the pluralistic diversity. Finite entities which are diverse aren't what is in same sense in which the "Fact" is a Unity.

Question of Attributes of the Eternal

In its function of Conditioning the processional Realisation. To condition is to impose limitation. You don't lose limitation.

Having regard to what the occasions of realization are in themselves (Having regard to the Actualities the Eternal wouldn't be what it is without its attributes being what they are) the Eternal as thus conceived is just the determining of processional realization.

The Limitation means essentially the reference to alternatives inherent in that particular occasion of Realisation. Cites Spinoza – What is finite in *suo generi* can be bounded by something of its own kind – the "Alternative."

Now difficulty:– Thus every occasion of realization bears impress of the Eternal.

(A) generally, by reason of its general shaping by the Eternal (Expressible in general terms).

(B) occasionally, by reason of particular limitation in which it exhibits the eternal attributes.

The Eternal is generally relevant by --- and particularly relevant by ---
The fact that there are particular occasions is part of the nature of the Eternal.

Here again Spinoza breaks down. Part I of *Ethics* is great. But Part II etc. Spinoza never explains why his Eternal should have any affections, at all. It's there you get that haunting contingence. Why are we all sitting here in this room? – Back and back explaining but always something equally contingent.

Now coming to the second word in our title:–[1]
"Processional" The passing limited occasion passing on always to something other than itself. – It is otherness in transition. A becomingness of Content. The Content is always omission. There's something left out. This is side which Bergson emphasizes.
[But Values !!] There's a two-fold omission. – You drop the past and omit all the alternative possibilities that don't get realized. There's also the reaching forward – the putting in the Actual. The Occasion is therefore a limitation within Procession. The Occasion has an internal processional ˅nature itself.˅

Always going on to Otherness, than itself. This processional nature doesn't imply [uniqueness] in processional direction. [This comes in in connection with Paradoxes of Relativity]

There's not one neat little row of occasions.

There's not one neat little line of dissolving views. Very fundamental implication of "Simultaneity" for Whitehead. – Genuine modern Relativists haven't this. – A real sense of Simultaneity between what's happening

1. He could be referring to *Philosophical Presuppositions of Science*, and then the next word introduced, 'Processional', would be a leading example.

here and in Sirius or in London. But take succession of Steps of the Simultaneous.

The whole point of modern view, with which Whitehead thoroughly agrees, is that this hasn't got simplicity assumed. – Alternative Stratifactions going on "like that" →

You need the whole bundle of these. Unless it's just poetical mathematicians giving poetical meaning to their symbols.

You've got to drop the instantaneousness and to have a conception of reality which allows reality to come on the stage in a lot of different ways. Procession partly a growth in intensity. In many ways "Expansional" would be a better term than "Processional." In our own experience we do live processionally.

Hocking's notes

Whitehead – ~~Oct~~ Nov. 1

Actuality = Being an occasion of Realization –
Like an event.
Fact = Factuality = Determinate Processional Realization
The Eternal = Contrasted with Enduring –
= Something out of Space-Time. In back of it all Determination
"Nothing can be too homely & nothing too grandiose
Feel in a homely mood today –

Ground Exhibited in every occasion of Realization.
That which is true respecting every occasion because it is an occasion
The processional character in itself –
The substance of which the occasion is a limited affection
The substance which requires all occasions whose whole being is that there are occasions
The substance determining the processional occasions
The ground of generality
Fact as self-contained with, or exhibiting, its own reason

You're not looking out for causes – you're merely exhibiting – how things must be so because things are what they are.

Never seems to get rid of a certain contingency –
Keep pushing it back – the relation is the puzzle.

Unless there is the eternal in this sense there is no science, no metaphysics, –

Unless something self-explanatory nothing to be said.

The fact that Hume sits down to write a metaphysics at all.

Every statement with a touch of generality in it is meaningless without the eternal.

Even every indication, pointing, which I expect you to understand.

Nov. 1 – 2[1]

Expansional would be
in many ways better than Processional

The Comparison of distinct occasions is an appeal to the eternal exemplified in each

All reference to possibility is a reference to something in the occasion which is not occasional.

Likewise all reference to impossibility – something which is determining.

How to get the Eternal into grips with Reality

The how of limitation & of the transcendence thereof

Shows how limitation essentially transcends itself.

How all this plurality gets together.

Attributes of the Eternal – What can you say of it reflects the plurality in unity of processional realization.

The Eternal could not be itself without these attributes.

Limitation refers to alternatives

Quotes Spinoza on Finite – bounded by something else of the same kind. In becoming, something is left out

Realization is omission. Bergson

Every occasion bears impress of eternal in two ways

1. General Shaping
2. Occasional relation, the particular Shaping or limitation.

The fact that there are particular occasions is an eternal fact – one of the attributes.

1. Whereas we have supplied page numbers |#| for each of the pages of Bell's and Heath's handwritten notes, Hocking includes his own dates and page numbers, as here: Nov. 1 – 2 indicates the second page of notes he took on 1 November. We retain his own indicators and supply page numbers only in the few instances where they are missing.

Why are we here in this room. Explain by something previous – but always the haunting contingency.

A twofold omission – in reference to actual, & possibility
Dropping what we were doing.
and omitting what we might have done.

There is also a reaching forward –
The occasion is a limitation within procession always reaching forward – otherness – never get away from this.

--- But a whole bundle –
Another remark. Not a neat little row of occasions

Insist on fundamental simultaneity – a definite meaning. There is a total now

There are alternative stratification

Lecture 18

Tuesday, 4 November 1924[1]

Bell's notes

|33|
Whitehead – Tuesday 4· November; 1924:–

Conception of single temporal line of advance as not really adequate way of describing Reality. (Psalm XIX)[2] What we immediately and obviously have in immediate Experience is a passing on. People have deplored it or welcomed it; but can't get away from it. In addition is in own experience things happen one after the other; but not [simply] "quite"[3] true.

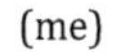

(me) (Whitehead)

"Expansion." This theory has its Endurance; that one its. Seem to have common Simultaneity with respect to our cognition. This Simultaneity is a real and dominant fact. But is it unique? Until recently such speculations would have been looked on as fantastic. But in modern physics found that if you drop this uniqueness you get practical applications for a doctrine which recognizes:– Alternative types of relations expressing "expansive" processional character of things.

Unique Time Series – accepted too easily as obvious by Philosophy ? ? Isn't there evidence for a more baffling and complex view? The processional aspect of things is an "expansive" development. There's a certain interest if not in one unique direction. The development takes form of an emergence into things. What emerges, when you try to get hold of it, is imbedded in an immense (incomprehensible totality) "beyondness." What emerges is always a detail, an occasion.

The occasion itself is processional. This lecture isn't just there –Passing on within occasion (within detail) and then from occasion to occasion. Temptation to make it static – to grasp not Transition but Configuration (Bergson:–"Spatializing") – This done (Whitehead) by high degree of abstraction.

Seeing "Greenness" within modal transitions of greens enduring.– This getting at something out of transition is essential characteristic of Perception. But you've got there something that stands outside the Transition. It can be in other Transitions.

1. The Radcliffe version of this lecture, from notes taken by L. R. Heath on what was likely still 8 November, begins on |80| of her notes, page 444.
2. In the opening line of Psalm 19, 'a psalm of David', 'The heavens declare the glory of God; and the firmament sheweth his handywork' (King James Bible). The heavenly bodies, it would seem, bear witness through the continuity of their movements.
3. This technique has not been used previously, but it seems to indicate the bracketed term is Bell's own, while the 'quite' is Whitehead's.

It's part of the Conditioning ? ? ? But the realization is limitation of this. Now want to fix attention on the Transition and consider some of the difficulties that arise.

Take momentary Spaces – "Spreads" [*Ausgebreitetsein* vs. *Ausdehnung*][1] How to correlate these ΛdifferentΛ spreads? – Because this thing has a where in the one and the other. Dominating idea to correlate them ? ?

You see something tumbling. How do you correlate the positions in order to see it. This correlation which takes us over from instantaneous and minute.[2] Spaces into the one Space as meant by the Physicist.

The classical view doesn't give you any grasp of any inherent relation of these spaces one to ⟨the⟩ other that would account for correlation. You can say we do so , as blunt fact. Only alternative way to-day is psychological:– It's the carry-on of the Mind. Then you're getting a subjectivist point of view. – [Doesn't solve problems by denying – explaining away etc.]

Space of the Physicist is "conceptual". "Conceptual" as a modern "Mesopotamia" comfort.[3] "Of course physical Space is conceptual; so is everything we think about." But want to know relation between this and what is immediately there before us in Cognitum.

Realist holds that in immediate awareness object imposes Concept on Thought. Concept arises because of Cognitum. Activity arises because of Cognitum.

Kantian holds that Activity of Thought imposes character on Cognitum.

Kantian must hold that first delivery of Consciousness is priority of Cognisant.

Realist must hold that ⟨blank⟩

Primary fact is that an entity S is there in complex togetherness of the Cognitum. S realizing its own Cognitum.– "Secondary Impress".– In this – accompanying and forming it, – you have impressed Thoughts, impressed Concepts. The concept is imposed on S by the facts of the case. And the mere S is abstracted from its Environment. To find what S is in this Environment, must think of it as having Concepts impressed on it by its Environment. Vice versa – same for Environment – the "O." You can only explain what either S or O is by reference to larger togetherness. Whitehead would protest equally against Realism that would look on "O"
|34|
as unaffected by S-O relation; and Transcend^{m} wh.[4] ⟨blank⟩

When we say that Space of physical Science is conceptual there's ambiguity:–

---

1. It is not clear whether these German terms are Bell's or Whitehead's. The translation might be 'being spread out vs. extension'.
2. These two terms are uncertain. In Bell's handwriting the first seems like a contraction (instant^{s}), which makes it 'instantaneous', but the second is not a contraction and looks like 'minud' ('mind' or 'minute'?).
3. This is a puzzling insertion of 'Mesopotamia'. Might it refer, again, to Psalm 19 and its precursors?
4. With the remainder of sentence left blank, it is hard to say how 'wh.' should be expanded, and 'transcendentalism' seems unlikely for a previous contraction, but it is not clear what else it could be.

(a) Object is object of merely abstract thought. "Objective Thought" [*objectum*] is then only an indirect ingredient of actual occasion of this thinking process. Qua indirect ingredient is merely an ingredient via Envisagement in lowest sense. – May then never have been direct ingredient in any Procession. If that's what you mean then Space would have no more importance for actual Experience than Mother Hubbard story.
(b) Actual occasions of Sense awareness impose this concept on the Cognisant. So analysis of primary Consciousness must disclose this feature.

Your objective thought would be a real fact in the objective world. Our assent to general statements often gained in sense of second alternative and admission then used in sense of former alternative. One impulse to that was given when discovery was made that our metrical system was to some extent not necessarily unique. But ⟨that⟩ there are alternative procedures was taken as meaning that these alternatives had no reference to what's imposed by facts of case but are (Poincaré) "pure conventions."[1] But it's perfectly obvious that unless Convention has some relevance to what is there; there might have been other sides etc. etc. Complete analysis never imposed on us. ^But^ that we have some imposition from fact seems fundamental. Statement that physical Space is Conceptual:– want to ask at once: what is there in the facts that imposes this?

No ground for Correlation of one instantaneous Space to the next one. Perceptual Space.– Summed up in Art. 4b.4: *Principles of Natural Knowledge.*[2]

All paradoxes of Relativity are explained by realizing that persons may be talking ~~of~~ relevantly to instantaneous Space of different time systems.

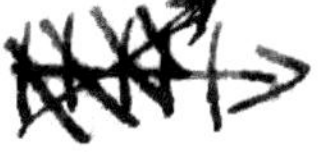

[Examples ? ?]

Geometry:– Nothing inherent in Classical Theory. Why should instantaneous experience organize itself in such Volumes – Geometrical relations etc. Structural Limitation becomes a blank fact for traditional view. You can see it as arising out of larger fact in this new way. – Get nearer to a self-explanatory whole.

The Time-extension is now given, too, the same manifoldness as Space-extension. You get a "Spread". Temporal Expansion requires for its full development a "Spread." – Our Experience will be with reference to one dimension. This is at bottom of Bergson's protest against looking at Time as simply divisible?

You can split it up under any one of (innumerable) its aspects; but itself you can't. Questions again of Times and Spaces of dreams and day-dreams.

1. See, for example, Poincaré's 1902 *La science et l'hypothèse* (Flammarion; translated as *Science and Hypothesis*, 1905).
2. Reference is made to Whitehead's own *An Enquiry Concerning the Principles of Natural Knowledge* (1919).

Chap. III. P. 66-70 in *Concept of Nature.* – Passage of Mind distinguishable from Passage of Nature and closely allied to it. Mind not in Time in same sense in which Nature is in its passage. Mind in Time and in Space in a sense peculiarly its own etc.

There are realizations of Value whose modal limitations are of the Spatio-temporal type, which nevertheless cannot be adequately comprehended within the --- of --- occasions. (Mind & brain-states etc.)

Abandoning lineal theory of processional realisation. Next question is the Question of Individuality – Limiting off an Occasion.

Qua realization of Value limitation is essential

Hocking's notes

Whitehead, Nov. 4. Eaton's acct.[1]

1. Conception of a single line of advance is not an adequate description of reality

What we have in immediate knowledge is an expansion a multiplex of directions

What emerges is a detail, a limitation, from an immense totality, an attempt to freeze your detail

2. Realization is essentially limitation.

There is a correlation of instantaneous spaces in the one physical space of the physicist

Can't be attributed to MIND: subjectivistic

Can't be disposed of as CONCEPTUAL

3. The realist holds that the cognitum imposes concepts on thought

The Kantian holds that thought imposes concepts on the cognitum, the cognizant is prior

The primary impress is of S on O

Secondary impress of S on (S on O)

Both S and O can be found reflected in the other

You can only explain S and O by making them part of a larger togetherness. One must protest against realism "digging out O" and transcendentalism "digging out S"

(The interest is in an indifferent One, – Schelleing[2])

1. Apparently this typescript of notes was taken by Ralph Eaton, who was an instructor and tutor in the Harvard Department of Philosophy, like Bell and also Raphael Demos. They all tutored students in Phil A, to whom Woods, Perry and Hocking lectured. Demos and Eaton taught a course together: Phil 19, 'Problems of the Great Philosophers'. Bell taught one on his own in the second semester: Phil 25b, 'Philosophy of Values'.
2. Presumably a misspelling of F. W. J. Schelling (1775–1854), who did speak of indifference.

Lecture 19

Thursday, 6 November 1924[1]

Bell's notes

|35|
Whitehead – Thurs.; 6th Nov., 1924:–

Next Essay (3 weeks):– Change, Motion, Time. Literature – Relevant Parts from Ross on Aristotle
[or: read Aristotle *De Generatione*];
Whitehead: *Principles of Natural Knowledge* Ch.VI,VII
(frightfully confused chapters);
Concept of Nature Ch. III
(better, but one gets out of a confused state by dropping the difficulty, so in some sense this is not so good);
Alexander. – *Space, Time and Deity* Ch. I & VI
(gets in muddle in some parts by trying to do a lot of math without being a mathematician);
Broad: *Scientific Thought* Ch. II;
Bergson: *Creative Evolution* – Ch. IV

Analysing initial Entry we were to make in Metaphysical Discussion. Under three headings:– (1) there's a definite determinate subject for knowledge – a definite occasion with determinate relations; (2) the processional character; (3) character as Realisation – Things becoming real.

At point at which every philosopher becomes confused. (There always is some point of confusion for every philosopher.) – (Whitehead's here). Things we know as being realised are the relations to ourselves of other things – That's what immediately being realised. These relations named by Hume as "perceptions of the human mind" and classified by him as "Impressions" and "Ideas" [Quotes initial sentence of *Treatise* again, as significant for all that follows].[2] Whitehead is to talk of "Impressions" today. Can't talk of all at once – Though ought to (6 tongues all going at once) – Character of things not sequential but expansive. [again].

Whitehead differs from Hume. (1) holds that what we know as immediately real is the relation. This is the cognitum. Accordingly, the cognition is a further relation requiring separate consideration. That

---

1. The Radcliffe version of this lecture, from notes taken by L. R. Heath on 11 November, begins on |82| of her notes, page 445.
2. Book I, Part I, Section I: 'All the perceptions of the human mind resolve themselves into two distinct kinds, which I shall call IMPRESSIONS and IDEAS' (from the Oxford University Press reprint of the original edition).

thing which is being real is an "Impress" (rather than Impression [obvious reasons]). Whitehead wants to single this out. Hume didn't do so enough.

An Impress is a togetherness of Myself and something diverse from myself. ("The mind is in the idea [the "impress"] and not the idea in the mind") (On this line, Whitehead thinks, it is that you get the escape from Solipsism).

(2) A further difference from Hume. – who is adversely affected by assumption that Subject-Predicate form of proposition is the only type which expresses a fundamental type of relation. Until lately, no corresponding reform of Logic ∧to correspond∧ with recognition of inadequacy of this. Hume and others appear to treat his Impressions and Ideas as Predicates of the Mind in a way going beyond conveniences of verbal Expression. An Impress is a relation on equal terms between two entities expressible equally well as predicate of either relation (if you want to do so. – You're then employing a convenience of elliptical phraseology in language).

(3) An Impress has then no essential connection either with humanity or mentality of the human type or with cognition. An Impress is cognized in human knowledge. But the impress is apart from and independent of cognition and doesn't require this. Cognition of an Impress requires ⟨blank⟩

Relatedness of Cogniser in Impress is independent of his further relation in the Cognition. – He is not primarily impressed as Cogniser.

An impress is cognized when one of the implicated Relata happens to be cognizant of it. But in itself Impress is most concrete of things. – Immediately real. An Impress in itself is transitory and limited; and these characters are connected. It's limited because transitory and vice versa. This passage or transition is of the very nature of the Impress. There is an essential transition of an Impress to an enlargement of itself – or to other impresses. An Impress is internally transitory or externally transitory – But it's the same transitoriness. The thing limited is itself transitory ---

An Impress is thus finite in exact sense defined by Spinoza (second definition)[1].

There is thus in Impress, in its general character, nothing which marks it off in any Atomic sense. As within ocean no intrinsic marking of bucketfuls etc. To that extent the limitation of an Impress is arbitrary. But we do find individual entities within all this. There are limitations coming from very character of the content.

|36|

Impress as Passage – Transition of Relations. But a Transition of what? Impress is breaking down of isolation of mere general individual Essences of things.

1. Spinoza's second definition: 'A thing is called finite after its kind, when it can be limited by another thing of the same nature; for instance, a body is called finite because we always conceive another greater body. So, also, a thought is limited by another thought, but a body is not limited by thought, nor a though by body.' (From R. H. M. Elwes' 1883 translation online at http://www.yesselman.com/e1elwes.htm)

P of v. X Realisation (the becoming of the Impress) is something beyond an Essence being for itself. – and being for other things. – Breaking down of isolation. [But Essences have relations – need distinctions ^between intrinsic and empirical etc.]^ I'm enjoying the impress of the Watch and me. – It's a value for me good or bad. Alexander's "compresence" but ---

The Watchiness of the Watch is for me there etc.

What becomes is of the basic stuff of Value. The whole of what there is in realization is things becoming of Value to one another.

This use of "Value" for the emergent entity which both is the Impress and emerges from it presupposes that Value can be fitly applied to that general ground of Becoming which Whitehead has called the Eternal.

Value is the ὕλη – the wood which is realised. But mere ὕλη is an abstraction and furthermore Value as characterized by what is generally true of every occasion. The general how of Becoming is "the Eternal". But value as shaped and particularized upon any finite occasion is the Eternal captured by the finite. But that's all the Eternal can do:– to become so captured.

The Eternal as captured by and in the "Impress" – the Impress is then also a substance:– the Eternal particularized in this finite mode.

Point where you get confused (Alexander becomes confused there). It's an invariable fact of Experience that Limitation is of the Essence of Value. Without Limitation no Value. But since always there we have no proper words to express this.

In respect to the higher Forms of Value we may be able to see the sort of things. – E.g. the aesthetic Values of Art. The first commonplace that Art is selection. [But aesthetic value not coterminous with Works of Art].

Every work of art depends on its content being disengaged from any irrelevance.

Realisation is just as much a process of omission as commission. (Bergson here emphatic). If Irrelevances made the work loses. If you put in more into a work of Art you get less.

Greeks:– The half better than whole:– If the other half is irrelevant namely.

Especially good illustration in Chinese Artist having made his point with exquisite rhythm of line etc. then ruthlessly suppresses everything else that would have been there in the screen.

Even in Italian Renaissance (more full-blooded) (not necessarily more exquisite). Riot of life and color on the canvasses. – But brought into a limited organic relation by which just that[1] life and those colors are adjusted to produce just their values without ? ? Vitality of a work of art depends on this:– That every detail bears its weight and serves its purpose.

But you're also charmed by delicious bits of pure irresponsibility – the cherubs and landscape. It's there to relieve some tension of formalism. [Like wave breaking into foam to relieve itself ---]

1. There are extra words at this point in Bell's handwriting that are puzzling and seem unnecessary.

But Art doesn't start with Selection. – First comes Content – Envisagement of the Value to be realized – But procedure is Selection.

Another Example in that Architecture which tries to realize Volume. (Plenty of vSpacev in the Sky but you feel the real vitality of Volume in domed structures – Byzantine and Renaissance domed Structures etc., etc. It's the walls and dome which gives the Volume its Value for you. You don't get it in an open courtyard.) You don't get your Impress of Volume from mere size – you must have a proportion between Volumes. – finite ones balancing themselves against ⟨each⟩ other. Finally all relative to proportions of human body.

Albert Hall is vulgar and unimpressive – a lot of obstructive detail which defines nothing. Compare St. Paul's. Its limitation has achieved that

(View pt.) ultimate Reality only expressible by "I am that I am"

– Persistent Individuality of the work of Art.

Inmost character of realisation is this atomic character of the individuality – achieved by structural limitation and definition expressing itself within this unbounded transition.

The recapture of the Absolute by something which is relative. The Eternal ground

|37|

of things revealing itself in Individuality. – The transient having a ground in that which does not pass. In a sense, the particular occasion has captured the whole. That which depends on relation to all other things achieves an individuality which makes something true for itself alone. Intense value of this recapture of the Eternal by some passing circumstance. You get this in poetry – Permanence of Passion where occasions for it are all past. Story of Newman reading Thucydides about revolution in Corcyra and his sleepless night.[1] Sense of Intrinsic permanence of the Value that has been achieved. Keats, again, last stanza of Eve of St. Agnes.[2]

The irrevocable Past as yet the relevant past. And yet somehow an enrichment of that Eternal that never changes. – The Supreme Paradox. And yet because of which Eternal all the future will be what it is – and yet that only because of what was in the past.

1. J. H. Newman wrote a poem called 'Corcyra' when travelling in 1833, but we have not located this story of sleepless nights.
2. And they are gone: aye, ages long ago
These lovers fled away into the storm.
That night the Baron dreamt of many a woe,
And all his warrior-guests, with shade and form
Of witch, and demon, and large coffin-worm,
Were long be-nightmar'd. Angela the old
Died palsy-twitch'd, with meagre face deform;
The Beadsman, after thousand aves told,
For aye unsought for slept among his ashes cold.
(From *The Poetical Works* of John Keats, 1884, online at http://www.bartleby.com/126/39.html)

Hocking's notes

November 6.[1] Limitation necessary for realization of value.

What is immediately being realized is a relation to ourselves: the relation is the cognitum and is the real, the "IMPRESS", not the impression of Hume

A togetherness of myself with something diverse from self. The mind is in the idea, not the idea in the mind: the mind is IN the impress, – the escape from solipsism.

An impress has no essential connection with cognition. An impress is the most concrete of real occasions and happens to be cognized in some cases by one of its members

If you are thinking about what is actually being realized, and the cognitum is imposing itself upon you, space is conceptual in this sense, i.e., it is one side of the fact.

|Nov 6 – 2|
5.[2] Space must correlate the different systems of space-time
Has a "SPREAD" see *Principles of Natural Knowledge*

6. Time and space of dreams and day dreams
There is a passage of mind which is distinguishable
From the passage of nature
Concept of Nature ch iii

7. Impresses are limited and transitive in their very essence
Limited as bounded by others of the same kind Spinoza
We do find individual...

8. The basis stuff of VALUE, – each become as a value for the other. Value both is the impress and that which emerges from it. Value is the eternal, the general ground of becoming, the hyle which is realized

The eternal as CAPTURED in the finite, as particularized in the impress, eternal substance particularized in a finite mode

9. Illustration of ART, art is selective
Realization, like art, is omission, if you put in more you get less
"I am that I am"
The capture of the Absolute, achieved a self-sufficiency as individuality
Individual is the reflection of the whole
The irrevocable past is yet the relevant past
The past is eternal in the future.

1. This page of notes is preserved as typescript, and continues from the typed notes credited to Eaton, who seems to have taken notes in Hocking's absence. Cf. notes for 4 November, and see note 1, page 68, lecture 18.
2. The numbers 5–9 are added in pencil to the final five typescript paragraphs (the previous four paragraphs were left unnumbered). It is possible that Hocking added these to Eaton's typescript. It does look like his handwriting.

Lecture 20

Saturday, 8 November 1924[1]

Bell's notes

|Continuing on 37|
Whitehead – Sat. 8. Nov. 1924:–

Opposite difficulties in trying to shape a Metaphysical problem. Either Clear-headed by leaving out half the facts. Or you're adequate and muddled. Everybody steers uneasy course between desire to be clear and desire to be adequate. So everyone is partly superficial and partly muddled. Clearness here now (and of last lecture) practically impossible. Having started with the flux – no definite individual entities which can be taken out of it. – Now fact that there are such; and only because you can limit off things and get to this or that, that any discourse or rationality is possible at all. The very inmost character of Realisation (this fluent togetherness) is after all this atomic individuality which is achieved somehow or other by this limitation of structural definition. A measure of ~~concreteness is~~ attained.

Many-sidedness of "Impress". – It is (1) a ⟨blank⟩

(2) It's individuality shown as a definite enrichment of the Eternal. All the "Static" has to take account of what "happens". The "Eternal" is particularised in particular occasions. Enters into occasions with access of Value owing to definite station of that occurrence within the whole. That condition of things which is enriched by whole Past. The past transitory emerges into a permanent Value – part of Static condition of all subsequent occasions.

Every definite occasion points beyond itself by enriching the Eternal – the very ground of Becoming. – Here the essential distinction between Past and Future. The Past is relevant to us by how the Eternal has been enriched by those occasions with reference to us here. So the Individuality of an occasion is in one sense eternal. Unless you've got that in your Mind, there's no possible ground for Memory. While you're remembering, that's an immediate fact – here and now. There can only be any immediate evidence for "past" in Memory with this ground for it. – This variation of the Static condition for all Becoming. Otherwise your Metaphysic's only alternative is to hold that the memory is a pure delusion – an illusion.

Enrichment of realm of "Existents" (= "Platonic Ideas"). Essential attribute of Eternal. It holds in envisagement before itself – the whole realm of ideas. But as the Eternal is now for us the total nature of the past belongs to it. Realms of ideas itself alters in its reference to the particular

1. The Radcliffe version of this lecture, from notes taken by L. R. Heath, begins on |86| of her notes, page 446.

occasions as they flow by. How it gets altered:– the shadow of truth falls on it. Viewed in the pale apartness from any reference to what happens, there's nothing true or false – There can be logical and illogical ideas. But Truth – what has to do fundamentally with question of Realisation. And shadow of this falls on Realm of Ideas and affects it. Its that shadow of truth that we comprehend in Memory. Individual Memories are simply one share in this Enrichment. Each fact of realisation takes its place as an eternal truth. (Neoplatonic ideas. – First limitation of the Eternal into Eternal-as-enriched etc. etc.) From that point of view:– Very laws of Nature are relevant to the past. Therefore every law of Nature ΛisΛ [has] to that degree subject to expansion of realisation.

Coming to another idea re emergent individuality of Impress. – Endurance – different from Eternality which lies below Time and is its ground. Endurance is retention of individualized outline within time. We're inclined to view endurance over
|38|
unlimited time as Eternality. Wrong there!

It is the individual Value which is an individual substantial thing emergent from the Eternal and preserving Spatial and Temporal ---

Distinction between Temporal transition and Spacial Extension. Both have characters of Transition and Extension but --- ? ? The thing demands as whole just that total Space [Paper-clip].[1] But in Time I can take a fractional bit and get same thing.

Modern Relativists tell you that distinction between Space and Time now abolished. Whitehead thinks you can't abolish it unless you abolish distinction between Sense and Nonsense. There's in some sense a Spatiality that's different from Temporality – lies on face of things and anyone who tries to deny. –

There may be and are phases VaspectsV (e.g. Extensity) which they have in common. Endurance has in some way or another

The paper-clip is, really, something on its own. Any Metaphysics that denies this must have got into a false simplicity. At whatever cost of muddle, we must satisfy in some way that primordial craving of Common Sense.

It may be derived substance etc. but for all its reference, its something in itself. So it not merely endures in Space and Time, with definite Spatial and Temporal limitations of Structure. A structure is an impress which has reverberations to all points, Λextends to all Space,Λ and yet have a home.

Idea (common) of saying of Nucleus:– "There it is", and talking of rest as "Influence"! Whitehead slipped in the word. – Nucleus harbors a centre of Activity. This is where Whitehead's difficulties begin – because the Emergent is in some sense an Emergence of something that's over, and yet remains limited, etc.

1. Bell's bracketed 'paperclip' likely presented as an aside, as an example. Notice it appears again below.

P of v. X In a sense the occasion can't pull itself together into one until it's realised. It's not there until it's over, so to speak.

The "Superject". – That entity bears within itself somehow a further limitation of the Eternal. – It is in itself an envisaging Eternal.
--- "Limited derivative Substantiality"
Notion of the individual realised Entity enduring – an achievement – a unity of different things. Eternal realised by just that --- ? ?
"The eternal is nothing but the Unity of its limitations."
Spinoza's "Unity" of eternal substance is nothing but --- ⟨blank⟩
The fact of there being a plurality is additional feature of its very essence.
Eternal in absolute ? ? also "Substance" [1]

The Eternal underlies and the Superject emerges. – Unity of Eternal Existents and emergent of Plurality in Superjects.

Whitehead has merely generalized two problems:– (1) what do you mean by a perceptual object; and (2) what do you mean by a Mind. –

Joined:– what do you mean by the limited enduring Individualities that emerge from ocean of change.

Nothing is easier than to deal with the paper-clip if you bring in mind. – What holds the fleeting phenomena together? The mind, which holds them together, Mind as flow of its perceptions. But what holds these together? Final great metaphysical problem of getting within the plurality of things some one eternal ground, and within latter some ~~indivty~~ principle of individuality of plural
~~essen~~ ? ? ?

Hocking's notes[2]

Whitehead. Sat. Nov. 8. 1924

1. Dilemma of Metaphysics
   Clear & leave out
   Adequate & Muddles –
   Come to a point where clearness is impossible

2. Individuality of every great work of art.

   Having started with the fact of togetherness & flow, brought up with the fact of presence of set off entities, without which no rationality is possible

   The inmost character of realization is after all This atomic individuality achieved by structural definition

3. The impress gets itself into something by virtue of this individuality
   A definite enrichment of the eternal because that thing has happened.
   The eternal, which is general, is particularized on particular occasions.

1. These sentences are hindered by so many question marks and blanks.
2. From this lecture on, Hocking's notes remain in handwritten form, and run parallel to Bell's.

Emerges into an individuality of value which is in a sense eternal, entering into the structure of all subsequent occasions

4. Without this, no possible ground for memory

No assurance of truth of past unless there is something eternal in things. No immediate assurance of the past.

Unless your metaphysics hold that, your memory is a pure delusion. The past is present as the underlying condition of things.

The realm of ideas in its pale apartness from what happens is neither true nor false. Truth has to do with the act of realization. The shadow of truth falls on the ideas, (makes them particular – in enriched envisagement), & as such are taken up into memory.

11/8 – 2[1]

5. Idea of Endurance. distinct from Eternality which lies below time. End. is the retention of individualized outcome within time.

(Thinks he uses ~~Spinoza~~ ˅eternal˅ as Spinoza does.)

Endurance differs from Spatial Extension –
If I divide the paper-clip in space, I have no paper clip but fragments –
But if I divide the Endurance, the whole paper clip is there –
To say that the distinction between space & time is abolished is to abolish the distinction between sense & nonsense.

---

6 – Individuality expressed in spatial-temporal terms – is something in itself

Any metaphysics, at whatever risk of muddling, must recognize that.

Can be in all parts of time & space & yet be somewhere. Like the "primates"

Cant cut off the "influences" from the "nucleus."

7 – Nucleus houses an activity –

Difficulties begin here –

It is itself an envisaging entity –

It is not merely an achievement – in being itself, it contains the quality of the eternal, a derivative substantiality, of being the ground of what happens.

---

8 – Enduring achievement is the enduring fusion of individual entities, but also the emergence of envisagement (?)

"The eternal is nothing but the how of its limitations" The unity of the eternal is the organic unity of the limitations in one whole. The plurality is of the very essence of the eternal.

Substance is the eternal in its generality.

1. Hocking records his dates in different styles, but since we are relying on what he provides, we will not try to standardise them.

In its particularity it is a "superject"

"The concrete fact is eternal substance emerging in a plurality of superjects"

<u>11/8 – 3</u>

9 – Has merely generalized two problems

1. What do you mean by a sensual object
2. By a mind.

One problem –

1. What do you mean by the individualities that emerge in the stream of change?

The mind is nothing but the flow of its perceptions with something to pull it together as one mind –

Lecture 21

Tuesday, 11 November 1924[1]

Bell's notes

|39|
Whitehead – 11. Nov. 1924:–[2]

Last two lectures – considering "Impress" – Emergence of substantial individuality in realisation. From point of view of conceiving a static ground of things (in relation to particular occasion). And this ground as emerging in realisation as outcome of a particular limitation.

Next (not yet) to other point – relation between "Impression" and "Idea". – Idea:– Thought:– the Envisagement, then try and bring the two together.

First –Whitehead's own criticism of something. – already there in Broad in Introduction: to *Scientific Thought*. W.K. Clifford's idea: "The still small voice that whispers 'Fiddlesticks'."[3] Philosophy gets into trouble by disregarding this.

Broad: p. 18–25 – two branches of philosophy (Analysis + Definition of fundamental concepts) (∧clear∧ statement and criticism of fundamental beliefs)

"Critical Philosophy". Concepts of Substance, Cause etc. – Philosophy pretends to analyze these. Not specific substances. etc. – All scientific conclusions depend on certain assumptions which Philosophy wants to criticize. Philosophy concerned not mainly with conclusions but with {starting-points/bases} ✓

Over against this is "Speculative Philosophy." [But Broad isn't right, I feel, as to his account, as read by Whitehead, as to what this other philosophy is]. Hoernlé's idea: "Object of philosophy is the synoptic vision."[4] Broad hints that this is not quite sure of the good scientific morality of this.

Has Whitehead been deserting Critical for Speculative here recently? Whitehead feels that Broad's book and others of the kind sometimes fail to elucidate the very points they're considering by refusing to see the real

1. The Radcliffe version of this lecture, from notes taken by L. R. Heath, begins on |90| of her notes, page 448.
2. In his diary for this date Bell notes: 'Excellent Lecture from Whitehead'.
3. William Kingdon Clifford (1845–79) preceded Whitehead at both Trinity College, Cambridge, and thereafter University College London. As often as this quotation is attributed to Clifford, its source remains unknown.
4. Alfred Hoernlé was at Harvard from 1914 to 1920, and Whitehead's quote is from *Matter, Life, Mind and God: Five Lectures on Contemporary Tendencies of Thought* (Harcourt, Brace, 1922). Hoernlé's idea was also an objective announced already in Whitehead's first lecture.

breadth of the problem they're considering. You've got to make up your mind whether there is or isn't a taking account, in course of Nature, of what was, will be, may be, etc, etc. If there isn't, then all these manners of concluding etc. are mere baseless habits of our minds.

If there's no relevance to the future in the present, Induction has no basis; however much you talk.

Information as to Eternal Basis of Process of Occasions

There can be only three general lines of procedure for theory – (a) some theory of a priori ideas, which mind postulates in dealing with Nature, etc. (unless information from immediate ⟨blank⟩) (b) Kantian theory of subjective activity, giving form to Experience. (c) Naïve Realism. – We discern in each Experience elements reaching beyond the occasion and there just as anything else in it.

(Example of Whitehead getting and reading here letter bringing Knowledge of occasions of experience in England. – Knowledge of this – [at least in general] present in the occasion of reading the letter.) In the particular occasion must be discernible some basis of possibility of all occasions.

We can't go a step beyond Experience; but we must go all the steps which Experience imposes upon us. English Empiricism follows Hume unfortunately in stopping Critical Philosophy right in the middle of its stride.

Prejudices. In Chemistry, you get back to alchemist in black cap and wand. But don't say, we must abandon chemistry. Astronomy and the successful experiment of the dancing savages to save the Moon in Eclipse. We never arrive at such success that we can afford to cease being critical. We mustn't give up anything because Broad thinks it a bit dangerous. Isn't Broad being a bit provincial; – taking habits of mind among friends at Cambridge and saying: these mark the limits of sanity. Some Provincialisms get a swagger and status about them, but really just as bad as the others. Broad does guard himself.

Broad evidently dislikes taking in results of ethical and religious experiences. Of course, these have to be subjected to critical examination as any other experience of human mind. But one distinction between this and physical science – There's a certain obviousness of your subject matter in physical science that there isn't in other case. – Can point to them. We can define exactly the Subject matter of discourse by mere induction before we've got much intellectual grasp of them.

But in other case you've got a closer reliance on critical philosophy necessary. If you neglect critical philosophy here you can't get ahead.

That's why Psychology lodged in Emerson Hall[1] and allied to philosophical division.

---

1. Emerson Hall, named after Ralph Waldo Emerson, was constructed in 1906 for the Department of Philosophy. At various times it made room for Psychology and Sociology. It is of course where Whitehead was lecturing.

Science can seem, only, to get away from philosophy. – Longer periods in which Science can go on without renewed reference to Philosophy.

Another fallacy of Broad – seems to think a sort of fund of fact, like old wage fund theory. – That this occasion empties itself of all significance before those of us who are experiencing it [or: experienced in it?]. This is wrong. We all of us
|40|
feel our appreciation of Nature greater and deeper because of poetry, etc. We perceive significances and relevances because of some haunting lines that discipline our mind.

Every occasion is more to us because of our critical philosophy. It's not true that we can see only what the untutored rustic sees. Nor does he see only what Hume sees. He sees more – only he thinks wrongly about it. He discerns only (being undisciplined) vague and monstrous shapes of thought.

The first thing the human race did when it began to think was to think wrong. We are criticizing those persistent discernments which always have been in the Race. We're criticizing and shaping them. We can't escape them; but we may civilize them.

Broad's objection to putting religious and ethical experience with the others. Why not? Who is Broad and who are we to put Experience into these compartments (really abstractions)? There are surely ideas and relevant factors of situation to be found by cutting across the lines: ∧– by the synoptic view∧. You can't even keep the various sciences apart – chemistry running into physics. You can't keep mentality and physical world apart. Philosophy ought to cut across lines and get new points of view – gets things together in a new way and thus get a new and illuminating abstraction. (We'll always be thinking in abstractions, of course.)

Reverting to the three general types of theory from earlier in the lecture:

(1) A priori postulates – (2) Subjective Activity Forming experience (3) Or:– have to discern in any Experience elements reaching beyond the individual occasion and there --- etc.

(2) is obviously framed to meet obvious objection to (1) – i.e.: Why postulate? What good unless something there to correspond? (2) says postulated is simply what has to be there for the thing to be experienced. Both start from or lead to Modern Subjectivism. – Experience as in the Subject. (3) says Subject in Experience. (1) and (2) conceive Experience as qualifying the mind. (3) views the mind as adventuring amid objective experiences and --- ⟨blank⟩

Whitehead regards Realism that's not quite naïve as ∧in∧ a quite impossible position – Then a public world becomes a dream inside a private one – once you've got the private world. Whitehead thinking of Bertrand Russell here. Whitehead thinks Russell makes a concession here that bowls him over. If you once admit private images [Depends on how!?][1] you can never get beyond this.

Realism and the justification of Common Sense is bound up with view that Mind is inside its Images and Impresses and not vice versa. Not analyzing mind and thinking of finding impresses, (X) etc., in it, but vice versa. Not Images in mind, but finding Mind as "ingredient" in Images. That's just what we do say. "I'm immersed in such a topic." Whitehead says that's the literal truth. We think of ourselves as actors in Scenes. We don't think ourselves as actors with the scenes inside of us. It looks like a horrid paradox, but Whitehead's really a plain man on top of a bus, etc.

Modern Subjectivist tradition entirely inverts the immediate delivery of common sense.

Capture of Substantial Individuality by emergent final Value.

The finite as an abstraction with essential to subordinate transitions within itself and so much beyond itself.

This relational essence seems to dissolve entity itself into relational complex.

Point is that this final pulling-together into an emergent entity something made finite by being disengagement from other Values and this makes something definite and unit.

Hocking's notes

Whitehead. Nov. 11, 1924

1. Arrested by a doubt which affects own mind and no doubt your own.

Printed by Broad in Introduction to *Scientific Thought* "that chill atmosphere when you think you have been getting ahead rather" pp. 18 & 25.

2. Critical philosophy as analysis of concepts (as substance∧& cause∧)

Speculative philosophy – taking over the results of the sciences, religion, ethics, reflecting on the whole in the hope of reaching general conclusions

The "synoptic vision" of Hoernlé as in point.

3. Defence

Either there is or is not in the nature of things a "taking account of" the past & future.

[How are memory & induction possible?]

Only 3 possibilities.

a. Theory of a priori ideas or postulates which are also a priori

1. This bracketed comment seems to be Bell's.

b. Kantian theory of subjective activity

c. Naive realism – We discern in each element of experience elements that reach beyond – objectivity there on some level with other

Receiving a letter from abroad.

"In each particular occasion there is to be discerned some eternal basis for all occasions" "We cannot go a step beyond experience, but we must go all the steps that experience imposes on us."

"but Broad rather provincial – Cambridge habits of mind, extraordinarily sane, giving the very limits of sanity"

4. What Broad objects to is (1) considering the results rather than the basis of science (2) taking in religious + ethical experience

---

5. Have to be considered same as other Experience.

There is a certain obviousness about the physical experience.

|Nov 11 – 2|

6. Another fallacy in Broad.

Seems to think there is a certain definite observational wages-fund.

Observation of nature infinitely more delicate because of poetry – perceive associations and relevancies.

The occasion is always more to us because of the significances and relevancies. Nor is it true that we see only what the untutored rustic sees – nor does he see only what Hume sees. Savage sees the devil moving things. Human race begins by thinking wrong. We are now criticising and shaping discernments that have always been in the race.

7. Broad objects to getting the different lines of thought together – religion & ethics with the rest. Why not?

Experience does not come in compartments. Can't keep sciences apart – physical & mental – Almost childish.

Cut across in new ways – get your religion into your physics & physics into aesthetics

8 Revert to 3.[1]

Kantian procedure intended to meet objection to past. Why postulate?

Both Kant & the other lead to subjectivism.

The naive realist conceives mentality as adventuring amid realities.

Realism not naive is in an indefensible half way house. B. Russell's private world is but a private dream of a public world. admit private images & that is all you are dealing with.

1. This seems to point back to the three possibilities listed above under item 3.

|Nov 11 – 3|
Realism is bound up with this view
Mind is inside its images, not its images inside the mind.
Is inside its ingresses, not vice versa.

From the standpoint of common sense that is what we do.
"I am immersed in the topic, in mathematics" not the reverse
We are actors in scenes – you & I in this scene –
Not the scene inside of us.
Not "a horrid paradox" but "a plain man on top of a bus"

Capture of individuality by the emergence of finite value –
Disengagement from other values –

Lecture 22

Thursday, 13 November 1924[1]

Bell's notes

|41|
Whitehead, 13, November, 1924:–

How "Form" is related to "Matter"? Individual Essences and Eternal Existents to Emergent Hyle. In what way they determine the determinate process?

Plato has recourse to Myth.

Santayana has recourse to paragraph after paragraph illuminated by a magnificent style – not inferior to Plato's.

Modern Tradition in German philosophy has recourse to magnificent apparatus of technical terms that rather overwhelm one.

Whitehead has series of scraps dealing with it. Final conclusion first and then go back on our tracks. 11 short statements:– (Redolent of Spinoza's use of "Object")

Whitehead calls an object anything that stands up before us in perception or thought as disengaged or transcendent from the particular occasion. – Something therefore in itself eternal. It is these objects,[2] which give their meaning to "Sameness", "Identity", "Analogy" – the latter is always reducible to an identity of ingredient objects in some sense. An object, in the same way as any particular entity, has a dual aspect. It is what it is in itself and how it is relatable to what's beyond it. Of course, the second is part of the first. The finiteness, definiteness of an object etc. depends on the fact that you can make an abstraction, in which whole of object in its accidental particularity can be abstracted from experience in so far as concerns its systematic relation to the object which is before you.

Whole "Togetherness" can lose its accidental immensity and object can stand before you simply as systematically related to its Beyond. This is "Relational Essence" of an Object.

That aspect of systematic relation is, after all, simply the housing/∧generality∧ of every particular relation.

Therefore Whitehead lays down as:–

(1.) That an object pervades its whole relational essence. Viewed as an abstract object it is equally everywhere and at all times. – Relational Essences under guise of Space-Time:– In so far as object can be realized anywhere and at any time – viewed in abstraction. Qua mere object it pervades its whole relational essence. But:–

1. There do not seem to be any notes taken by L. R. Heath at Radcliffe on this lecture.
2. Just to be clear, these are hereafter and quite consistently referred to as 'eternal objects' (as opposed to sense objects, etc.).

(2.) Community of Relational Essences. – It wants another Relatum – at other end – of course. The Space-Time of one object in abstraction is exactly the same as that of another. Because Space-Time simply says how they're all mutually relatedness. This "Togetherness," in a sense, always there; in another, never there.

(3.) Being real is more than this formal externality (Space-Time as showing how the things are formally to one another). It's more than just you sitting there and me here. That might be known to God. That 3 comes after 2 is nothing for either 3 or 2 themselves. Something overwhelms the mere formal relationship – This is the realization – That things are something particular in relation to each other. It is within their general relationship but it comes about by selection out of the eternal fact of the possibility in which the general relationship stands for us.

Within the community of possibility the limitation within this is the Realization.

"What is real is a fusion of what they are in Themselves". – "The Becoming of the basic stuff of Value". – and the Value is limited by the fact that it's just the Selection – the limitative aspect – which makes possible and gives you the Realization.

? Realization is "the extension of the possible and what remains thereby [sic!!!] becomes the actual." Bergson, Aristotle (can't explain process of Becoming without reference to "Potentiality").

For P. of V. Ch. V ??[1] (4.) What you get emergent is Value. (Within limitation. There is no value apart from limitation.) There isn't an unshaped Value, which then gets an accidental shape. But it is the general process of Realization which is the emergence – the obtaining of Value. Aristotle had tendency to limit his Forms within the process of Realization (They were non-transcendent) but to see his ὕλη as there for Realization

[?? Whitehead queries this itself.] Whitehead:– It is the ὕλη which is non-transcendent and the Form which is transcendent. Whitehead more Platonic or Neoplatonic than Aristotelian. Providential that there were both. If ~~they~~ hadn't ⟨been⟩ different no one would have ever dared to express an opinion again.

What lies behind the ground is what's beyond all occasions because in all:– The eternal determinateness of the How it's all done. That How is conditioned and shaped by the eternal essence of this How of all possibilities now being realized in this Actuality.

But the eternal How must be an eternal taking account of Possibility. Eternal How therefore is nothing but consideration of that general community of Relational Essences and taking account of Individual essences which can emerge into Fusion of Value. – The putting together of all those as an eternal possibility.

---

1. The only book of Whitehead's that contains a chapter V at all relevant is his *Principles of Natural Knowledge*: 'Events and objects'. Or, it may be from another author.

|42|

The process of Realization is a process of Garnering in. You have "The Shadow of Truth". Therefore the How is always a particularized How – having relevance to the particular – The irrevocable Past is there and what becomes has to become, in relation to what is actual. So you get this constant modification of eternal ground of Becoming. – Very Laws of Nature themselves have a quality of passingness in them. So Emergent Value is not merely general; but is a particularized Value – Emergent Value has Endurance – I.e. this Value is qualifying. (The individual notes of a tune – passing along, each – But only with the retention do you have a growing of Value and get the Tune)[1]

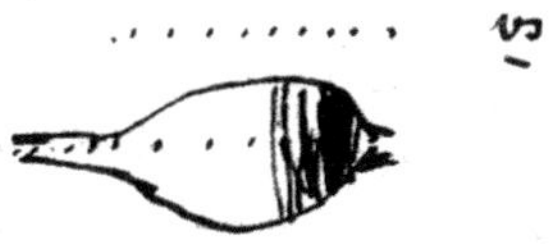

The sheer passage of Time-Space relationship is a mocking-fiend like that in Macbeth. When you say: What is this Space-Time region? It falls to pieces in the very act of giving itself to you as a whole.

The fusion of Realization it is which retains Content which pulls together the whole. This is where Time differs from Space. – Because e.g. the piece of Chalk is not in every one of its Spatial parts; but it is in every Second of Space.

Emergent Value is[2] "The Capture of the Eternal" – What has been is an eternal fact that nothing can blot out. – an eternal reality held and captured there and now. This is a myth too. – drawn on black board.[3]

Emergent Value because it's the "Capture of the Eternal" has, first, endurance. And it is a "Superject" because within its endurance it embodies within itself every quality of the Eternal.

The determining side of the Eternal } (Psalm XIX)[4]
and the embodying side of the Eternal } –
are embodied in the endurance.

"Superject" because it's what comes out of the process. Its growth, retention and passage.

(7.)[5] "That" same shape of Superject might have been real in other ways. But the "Shape" of that is the realization of a Social object.

Concrete form of Fusion of these realizations is what Whitehead calls an Impress. Mistake to think of Realization as built up out of its simpler elements.

1. Closing parenthesis provided here.
2. This 'Emergent Value is' seems have been added in the left margin.
3. Whitehead is referring to his chalk, to the blackboard and now to this 'myth' drawn on the blackboard (a comment captured by both Bell and Hocking); this is direct evidence that this and other diagrams are, in fact, blackboard sketches. Less obvious cases, like mathematical equations, logical formalisms and the like (not easily provided orally) may also have been chalked up on the board.
4. Whitehead referred to Psalm 19 at the beginning of lecture 18; see note 2, p. 65.
5. Bell seems to have missed opportunities to list points 5 and 6, whereas Hocking in his notes interrupts his own numbering to attach Whitehead's points 4, 5, 6 and 7.

There are grades of Realization:– You get less vivid realizations sometimes.

"Superject" which comes out of Fusion of Possibilities of "all these".

"An Impress is the fusion of Superjects into higher Values" and therefore it is enduring Values becoming related to something beyond themselves.

Our knowledge is always of ourselves as being more than ourselves. Our Experience gives us our experience as something more than we are – in which we are ingredient. (Religious Experience, e.g. – Expression of losing yourself to gain something more, higher.)

Hocking's notes

Whitehead, Nov. 13

1. Where Plato has recourse to a myth – Santayana to a –

Modern German philosophy to a magnificent apparatus of technical terms.

Have here only scraps of different ways of looking at it.

2. Object is something which stands up Vfor conception & thoughtV as beyond particular occasion – as in some sense eternal. The object –

Gives meaning to sameness & identity –

3. The object like any particular entity has a real aspect –

1. Is what it is in itself &

2. It is how it is related to what is beyond itself –

Nor entirely separable – but distinguishable.

The whole of what is beyond the object can be abstracted from except as to its systematic relation to the object before you

This is the "relational essence" of an object – you can never get away from it

It is the generality or home of all particulars.

4. The first point:

An object pervades its whole relational essence

It is everywhere at all times – In so far as it can be anywhere at any time

5. The second point

Community of relational essences

Wants a relatum at other end.

Space-time of one is same as sp-time of other

6. Third

Being real is more than this formal externality of relation.

Something more descends upon & overwhelms this – the fact that these are something for each other – within this general relation –

13 – 2

Selection from the totality of possibility – a limited realization. Taking account of what they are in themselves.

That which is realized is value

What is potential is always in background

4th Point

What you get emergent is value in limitation

There is no value apart from limitation

"Process of realization is the attaining of value"

Has a suspicion that Aristotle had a tendency to limit his forms within the process of realization. They were non-transcendent.

Whitehead thinks the hyle[1] or value is non-transcendent which the forms are. Here more Platonic than Aristotelian.

5. "The eternal determination of how it's done" beyond all occasions –

The how of all possibilities

Must be an eternal taking account of possibility. It is how things are possible –

– The How is always a particularized how a particularized eternal

Emergent value is a particularized value

6. has Endurance. Has Retentiveness.

As in the notes of a tune   a growing of value

a myth drawn on the blackboard

Space time as relational essence is the mocking fiend – divide it up. What is it – falls to pieces as you put it together

---

And the whole of that value is in every part even the smallest of its endurance.

11/13 – 3

7th Point.

This is a social object underlying that concrete form of fusion which I call an "impress"[2]

Not built up of simpler elements

Superject that comes out of the relations of all these.

Impress is the fusion of superjects into higher values

Our knowledge is always of ourselves as being more than ourselves

There is something in which we are ingredient & which is more than we are.

In religious experience people are endeavoring to state something which they feel to be intensely true.

1. The usual way to transliterate what in Bell's notes remained in Greek: ὕλη.
2. While Hocking and Bell usually record very similar diagrams (drawn we suppose from what Whitehead is doing in chalk at the blackboard) this time they seem puzzlingly different.

Losing self to gain something higher.

Obvious way in which it has struck mankind from Buddhist and ~~to~~ gospels.

In Asia the limitation of the higher superject is apt to be dropped.

Lecture 23

Saturday, 15 November 1924[1]

Bell's notes

|43|
Whitehead – Sat. 15th November 1924:–

Lecture not the one Whitehead had prepared. Result of discussion in Seminar last night.[2] Main point not made clear. Essence – to get main outlook clear. The difficulty of Metaphysics is to be sufficiently concrete. Then when you've been sufficiently concrete, you've also got to be sufficiently general, too. Utmost generality needed to be got into most complete concreteness.

What about the finite, real world in which we find ourselves? The idea of static plurality of self-sufficient reals shows itself as ~~fal~~ nonsense. The object is always embedded in its environment.

Second point is that reality is becoming of realization – It's passing before one. This ought to come in Book 1, Chapter 1. Not at the end.

Third, what is it that is being realized? – A structural Togetherness of things! A system of realisation of Structure – Its Structure that's being realized. What Whitehead means by "Structure"? – Essentially a limitation. Structure is just that which it is, in an ocean of possibilities. A Structure is a definite limitation amid an unbounded possibility – It's a mode. Refers to a definite How. It's becoming in this particular and definite way!

Now for: "Togetherness":– Not merely formal fact that you're there and I'm here – It makes a difference who you are and who I am, etc., etc. ~~The I~~ There's a mutual imposition, fusion, going on. And there's emerging an Entity which is that which is being realized in this room. Individual Essences are under limitations in fusions – mutualities – taking account of – within all the vague and fluctuating that is "this room" or "this lecture". When I know of the room as real, I'm knowing myself as ingredient in a reality beyond myself – in which I'm being fused.

I am there for the chair; as well as the chair for me.

An entity which is the outcome of mutual interest – of value – it's emergent value.

1. There do not seem to be any notes taken by L. R. Heath at Radcliffe on this lecture.
2. In addition to lecturing at Radcliffe and Harvard each Tuesday, Thursday and Saturday, Whitehead conducted a seminar every Friday evening primarily for graduates – a 'Seminary in Metaphysics' in the first term and a 'Seminary in Logic' in the second term. It seems that some of the same graduate students and colleagues attended both the lectures and the Seminary. This seems to have been the case with Bell, Hocking and others, like Rafael Demos. This sort of carry-over of discussion had happened previously, on 18 October, and the fact that it happened more than once underscores the likelihood that Whitehead was considering many of the same issues in both of these (indeed, in all of his) courses.

The Unity is more or less vague, in original definite. Two sides to the unity. – Can say this room is a certain 3-dimensional volume etc. – It lacks pulling together, so it all falls into parts. Space alone is only a form both of holding together and apart. What makes it definitely this realisation is the mutual imposition of things upon one another in this room.

Then you get an atomic individuality. – That particular outcome, e.g. It has a certain endurance. Every outcome remains an eternal truth for always.

It's this emergent resultant as an Entity within total reality which ---

In Physics you get these emergent entities – Electrons, Molecules, Crystals, definite conscious characters – but as yet have no independence and is to be looked on as entity of derivative character and requiring its relationships.

Why do I say that table is real? It has its own internal realisation of becomingness. Mutual imposition of subordinate entities. It is a relation becoming a Substance almost. It's the attaining of Individuality via Relationship. And if I didn't think that Table the outcome of any mutuality of interest, I'd say I had an illusion – I'd be doing it all. Mutual imposition of Ingredients issuing in a Unity. (Ingredients not all "Qualities". Nor all completely inside this unity.)[1]

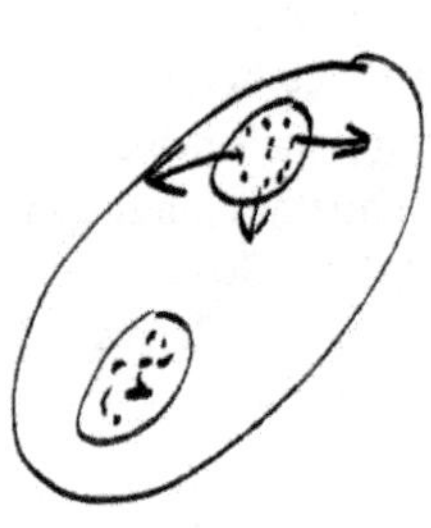

"Emergent into substantiality of value:– individuality of value". [Everything always inside the bigger] Is there one Entity that is the totality of Reality? Whitehead doesn't think so.

A structure of mutual imposition attaining individual value. Whitehead thinks concept of all Reality as one entity is Nonsense – "Entity" concept essentially a finite one. [Reference to Whitehead's own book on "Relativity"] – Any one "Entity" is an abstraction – You've left something out, to get to it.

This Becomingness not managed from Without by a stage-manager.

We've got to make what we are talking about all-inclusive/self-explanatory. – It must include its own "motive" ("nicely ambiguous between what moves and why it does the moving"). Embodies in itself its own conditions; and this means that we have got, so far as any Metaphysics is open to us, in the passing occasion, to see that which does not pass but is the self-conditioned motive which can therefore be conceived as the reason for the passing, and its "How".

|44|

Because of that essential inclusion ⟨blank⟩ That common ground of Passing has got to be found as in its embodiment in every particular achieved entity. – Each as embodying outcome of a definite system of mutual impositions. This system insofar as ~~an~~ entity --- ? ? ?

1. Closing parenthesis provided here.

The conditions must be expressed as again in terms of that which does not pass. But is there in the real world anything so discoverable? Whitehead says we find it in everything that remains self-identical etc. You have the same colors, shapes etc. These are the "objects" which stand up before us. (The process of omission again. – omission of a potentiality.)[1] Potentiality means something, after all. All objects are to be looked on as standing in the static condition of things. For some things they have the form which might be there but is not.

(Almost quoting Ross-Aristotle here. Aristotle's analysis only vitiated by Aristotle's attempt to get it all under the Subject–Predicate form.) Thus there's this general ground which expresses itself in a taking account of (call them "Platonic Ideas", or what you like – But with no flavor of "high mentality"!! – that's why Whitehead calls them "objects".)

This "general ground" doesn't stand outside and the Becoming gets added on to it. But it's just the other side of the Becoming. This ∧Emergent of Becomingness∧ in each measure expresses its own grounds. Each emerges in respect to its own grounds. It's this adaptation of what is general, to the particular fact that's what is hard to get hold of.

It's here that we get in each fact a taking account of past, future, possible, etc., etc. This expressed also in the physical world. – Each Entity takes account of surrounding field and vice versa (Molecule in Electro-magnetic field, e.g.). The Entity then isn't the Entity in itself alone but in interaction with Environment.

In thinking of "Physical field" you're thinking of about the most abstract thing possible [?] You get stages of Entities. You can't have a vicious ⊙ downward.[2] Must be prepared to produce a prime entity somehow. And must show how you'd treat Cognition, Mentality, as coming into this. In your basic conception of the Real you mustn't leave out anything. In what's real there must be just that trace of everything that was and will be and can be.

Why treat the mutual Togetherness under title of a "Fusion"? This is result of direct experience of others imposing on one and stream of my life as without that a thinner thing. I merge myself in a larger ⟨blank⟩

– Take this as simply as a fact of observation. But not a watertight all or none conception? – No – But there are all sorts of ways of "taking account of".

I have an apprehension of Table as mutual taking account of the parts.[3]

A stream of detail imposing on one another and bringing about an entity. Not "Subject":– Not underlying any detail but arising out of this.

1. Closing parenthesis provided here.
2. Although the sense of Whitehead's remark seems to be 'spiral downward', Bell has drawn a circle with centre dot.
3. This term is not a contraction in Bell's handwriting, but the third letter might be 'r' or it might be 's', in which case this should be 'pasts'.

Hocking's notes

Whitehead. Nov. 15, 1924

~~1~~ Not going to give the lecture he had meant to Seminary discussion showed the "main outlook" was not clear.

1. The notion of a static world of independent entities, without relation to other things is baseless.

2. Reality is becoming, is passing before one – "a remark too obvious to make" – "can't catch a moment by the scruff of the neck – it's gone you know".

3. What is being realized? A togetherness of things
   It is structure that is being realized – Structural togetherness

4. What do I mean by structure? It is essentially limitation. The table has just the shape it has. In an ocean of possibilities. The "mode" – or definite how

5. "Togetherness" – not a mere formal abstraction
   It makes a difference who & what the elements are. Each is something for the other, chairs are something for the users.
   Something is emerging, an entity having mutuality, entities in fusion. Taking account of. When I regard myself as in a real room, it is as in something beyond myself. There is an emergent value. It is a unit.

6. Unity has two sides –
   Spatio-temporal volume. Holding apart & also together.
   Mutual imposition of things on each other in this room. Taking account of.

Nov. 15 – 2

7. Emergent entities (Alexander's) in physical world – molecular crystals – Why do I say the table is real?
   Not merely because it is with other things in the room
But because there is a becomingness of structure, almost a relation becoming a substance.
   Some mutuality of interest

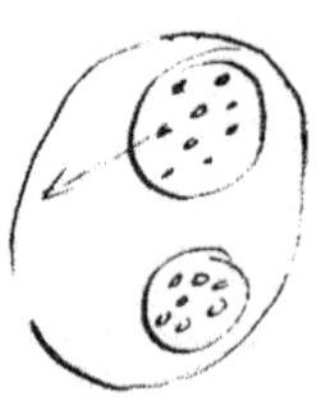

Dr. Johnson's foot stamping meant to him that the paving-stone was lower than his body (and was being affected?) Mutual imposition of ingredients issuing in a unity. Won't speak of dots as qualities. Each little dot is all over the shop. Then there may be one entity which is the totality. But rather thinks not (?). An entity is always an abstraction. All reality as one entity is nonsense.

Essential relativity – within & without

---

8. Becoming is not managed from without.

No stage manager. We've got to make what we are talking about all inclusive.

Its own explanation. Includes its own nature. "Motive being nicely ambiguous between" what moves and what is moving (?)

There is a common ground of passing –

---

9. Conditions not in themselves passing

Must be expressed in terms of that which does not pass.

To be found in everything so far as identical Matching colors. Finds then the same color stands up before the mind.

Objects. (Not Platonic Ideas)[1]

Possibilities. Not everything happens in a high mentality

<u>11/15 – 3</u>

10. This general ground is not fitted on to the becoming from outside.

11. Stages of entities.

Must be prepared to produce a prime entity –

Can't say a is real because b is real forever –

12. How does cognition come in?

Can't leave out anything.

Whatever the prime entity, it must have the trace of everything that has been or can be.

Self is felt real because self is felt as result or outcome of a taking-account-of. I <u>emerge</u> from my details as an entity. I do not lie behind them. knock me on the head –

---

1. This line, without parentheses, was written up the left margin.

Lecture 24

Tuesday, 18 November 1924[1]

Bell's notes

|45|
Whitehead, Tues. 18th November 1924

The fundamental question is relation of "objects"[2] (those elements in past occasions which are not passing – to which concepts of identity apply) (and which we can think about – which form the content of possibility – potentiality) – how that element is related to the passing real occasions – how this to be expressed.

For any "Realist" that's obviously the fundamental question. Not really less pressing question for any other type of philosopher? (Lindsay's new presidential Aristotelian Soc. Lecture: "What does the mind create?"[3] Whitehead would answer "Nothing." Lindsay –Glasgow Idealist).

Whitehead's name for this assemblage of elements ("eternal" – i.e. out of passing show) is "Objects." That which stands up against the Mind, in its own rights, so to speak.

For Point of View | Divides objects into "primary" and "emergent" Objects. The primary one belongs to Eternal principles as such. Any perfectly general statement containing any content not merely logical, must be expressed in terms of primary objects. Greenness is a primary object.

Any particular relevance of any primary object to any particular occasion is incidental to the eternal primary object. The general potentiality gets its character because of the being of primary objects. General relevance of process of realization to primary objects. Primary objects perfectly neutral to all times and places and occasions. Primary objects are, in their potentiality, in a community of relational essences. – There is the general potentiality of "External relatedness" about them. This general potentiality represented for us in some reference or other to Spatio-Temporal relatedness There's a general ⟨blank⟩

Any primary object pervades all Space-Time. – All Space-Time exhibits that potentiality of relatedness. Potentiality, in a sense, nothing from point of view of a particular occasion. It is a definite fact, however, – one which stands in very nature of the eternal principle:– that there are these objects in every possible relatedness as "in envisagement". Part of the eternal

1. The Radcliffe version of this lecture, from notes taken by L. R. Heath on 20 November, begins on |94| of her notes, page 450.
2. Later, Whitehead will refer to these, consistently, as 'eternal objects'.
3. Alexander Dunlop Lindsay, 1st Baron Lindsay of Birker, was President of the Aristotelian Society, 1924–5. Whitehead had been President in 1922–3.

principle that there is for it this eternal universal pervasive relatedness which is there for realisation.

Every particular real occasion has its own particular adjustment to this manifold of primary objects.

Every particular real occasion is, as matter of fact, in some sense the realisation of a selected structural relatedness out of whole manifold of primary objects which has the " accent of reality", and that the particular occasion is furthermore in ∧its own∧ definite relation to every possible structural relatedness. This definite relation receives its focus in the particular limited structure which is receiving in that occasion the accent of realisation. There's a definite relatedness to every particular occasion.

To explain this further we must divide primary objects into two kinds:– "Pure" objects and "Social" objects or "Situations". "Pure"; or "Simple"?? A bit afraid of "Simple". A pure object is purely an object. Its only relevance to any occasion is either that it's ingredient in it or has a relation of potentiality. You can't look on it as the concept of an occasion. It's never the whole concept of the occasion. Thus it doesn't achieve any further ~~concept~~ status than that of being discerned in an occasion; whereas a "Social" Object is a concept of an occasion – is the structural relation between primary objects. Primary objects allow such and such relational structures between themselves. This external (to their individual essences – Same e.g for red or green or brown as to their status in it) structural relation stands in potentiality of their relational essences; but viewed merely as a Social object there's no reality about it. Only trace of reality is as it stands in this envisagement of the Eternal principle. It's equally "potential".

But has this difference from a pure object that you can ask if this is being realized. "Pure" object – Nonsense to ask whether it alone is being realized. Realisation is a togetherness – a mutuality. A social object is the concept of a togetherness – even if only as standing within the potentiality of the relational essence. It's the concept of a realized togetherness – out of Space, out of Time. Social object presupposes therefore "pure" objects. "Situations" can be realized.

The generality of the relational essences is nothing else that the community of any possible (silly or beautiful etc. etc.) social objects – is pervaded by all possibilities of social objects.

But in talking of Social objects qua mere Social objects, pure objects aren't related except via their relational essences.

For Point of View Contra mere Symbolism ⟩

How is it that we can describe a social object. Wittgenstein's *Tractatus Logico-philosophicus* [1]

(1.) You must describe its mere formal structural relations. How this? Any account where

1. Wittgenstein's *Tractatus Logico-philosophicus* was published in German in 1921, and then along with an English translation and with an Introduction by B. Russell in 1922.

|46|
individual essences don't come in, must consist in some method of pointing at some occasion of realisation of some social structure but with other individual essences. This, when you can't directly point at the object in question. This occasion must be in process of realisation. Ways of pointing at it may be very devious. Also, you've got to point at the pure objects which ~~your~~ really turn up in your structures. Only way is to point at another set of occasions and say: They're the ones realized here and there and then – You've got to point at your pure objects somewhere.

Process of description always only half carried out and then you hope the mind will jump at it. Showing structural model and pure objects in various occasions and you've got to put the second in the first. You produce an alternative picture painted in wrong colors and then produce samples of the right colors. Every description must have essentially that character.

Hume recognizes differences between "pure" and "social" ~~phil~~ objects. Hume's "simple and complex" perceptions (covering both "impressions" and "ideas"). Says you can't have a new "simple" impression – Although with candid exception. Whitehead should think that in modern dying industry. Women with keen color sense – Wanting certain color I've never seen just to go there. Hume puts this aside as a trivial exception – where really for him no exception can be trivial.

Now the "emergent" objects. Whole essence of an emergent object is:– it emerges as being actual. The fact of Actuality in Realisation is the emergent object. Emergent objects are of various grades. The grade is reckoned according to grade of other objects ingredient in it. You've got to start from a basic object or an emergent object of zero grade – I.e. an emergent object whose ingredients are pure objects. The very essence of an emergent object is that it is actual. You've got the stuff of reality in it. [Initial Concreta etc.] It's a fairly comprehensible job to think of fusion of subordinate objects in higher ones.

Molecules of body to Organism. Nothing can be explained – all we can do is to set the things forth in fundamental generality. When you come to basic objects, (Otherwise infinite regress. Whitehead thinks this regress would be vicious) you must start with one in which all ingredients are primary objects. A "basic" object is essentially – has no accent of reality in any of its ingredients. So we're up against difficulty – (pure initial[1] difficulty) of what we mean by realisation.

The almost insuperable difficulty of that first start – always getting cloaked in our minds because we start further on – Reality emerging into something more than itself. But transition from Platonic ideas to reality – that's the problem. Whitehead can't get more than a ghost of a glimmer of light here.

[But Concreta also *in essentia* as well as *in existentia*] Any really thorough-going Realism must have faced that question. That, Whitehead

1. The first three letters of this term are not clear.

supposes, why Aristotle and Plato classed as once and thorough-going Realists; because they did face that question. "Modern" Realists too tinged with Subjectivism, so only face question of Becoming of Realisation halfheartedly.

The first aspect of Reality is the fact of it as limitation. Amidst the wealth of potentiality there's one definite limited selection. Equally, the first or second aspect is the transition in selection. There's a selective transition going on. But the transition itself is an aspect of the selection. The selective structure has an aspect of display (endurance) and one of transition (it goes on into other structures) :· ⟶ :

Diagram or Myth:–

e in future always "imbedded" then in past

... a″ b″ c′ d e, f, g‴ ... —becomes→ ... a b‴ c″ d′ e f, g″ h‴ ...

→ ... c‴ d″ c′ f g, h″ i‴ .

Also more than it contemporaneously – as well as when you take temporal transition. The realisation of basic objects.

Whitehead doesn't think that "basic" object is the one first turning up in time. Logically prior. No temporal priority. Whole stuff of reality comes on together.

Its vigour of realisation may differ from time to time

Hocking's notes

Whitehead 11/18 –

1. Lindsay's Aristotelian Society paper
   What does the mind create.
   I would say nothing.

2. Objects same for all entities that are eternal –
   Stands up against the mind in its own rights – Identical

3. Primary & Emergent Objects
   a. Primary Object belongs to the eternal principle as such.
   All perfectly general statement applying equally to all occasions must be primary.
   Greenness would be primary object
   Perfectly neutral to all times & occasions capable of being so viewed.
   They are in a community of relational essences. General potentiality of external relatedness. Always with some reference to spatio-temporal relatedness.
   It is there for realization.
   Every particular occasion has its own adjustment to this manifold of primary objects. And is a realization of a particular structural relatedness

of primary object. And is in definite relation to every possible structural relatedness, some of which receive the accent of reality.

Primary objects are of two kinds

Pure objects (afraid of 'simple')

Social objects

Pure object is purely an object. not a concept of an occasion never the whole concept of the occasion

Doesn't achieve any further status than that of being discerned in an occasion

Social object is also the concept of an occasion

Structural relation. concept of a togetherness

Has no ~~relation~~ reality about it

11/18 – 2

Social object presupposes pure objects

Pure objects are not related except via their relational essences

How can we describe a social object

Wittgenstein *Tractatus Logico Philosophicus.*

You must describe its structure

By pointing at some occasion which is the realization of some other social object. "That has the same structure"

[How then not real]

Then you have to point to the pure objects in another set of occasion. "Realized there & there"

Never carried out – half carried out & you hope the mind will jump at it.

Hume recognizes this distinction

"There is another division of our perception ~~impressions~~"

Simple & Complex –

Thinking of new simple ideas –

H.[1] thinks it the commonest thing going

Women want a dye color never seen.

11/18 – 3

b. Emergent Objects

Has to do with the definite limitation of eternal principle by fact of actuality.

Of various grades

According to the objects ingredient

Must start somewhere if avoid ∞ regress.

Has no temporal priority } Basic object, is of zero grade

one whose ingredients are pure objects –

Fusion of molecule into organism is fairly comprehensible. Nothing is explained, only set out.

--- Up against the initial difficulty of what you mean by realization

1. In Hocking's handwriting it is not clear whether this is an 'H.' for Hume or a 'W.' for Whitehead.

Can only get "the ghost of a glimmer" –

Any thoroughgoing realism must face that question – Aristotle & Plato did face it. Modern realists half-hearted, all tinged with subjectivism.

What is realization

The first accent of reality the fact that it has limitation.

Selection

The second (?) Transition in selection

An aspect of the selection

An aspect of display

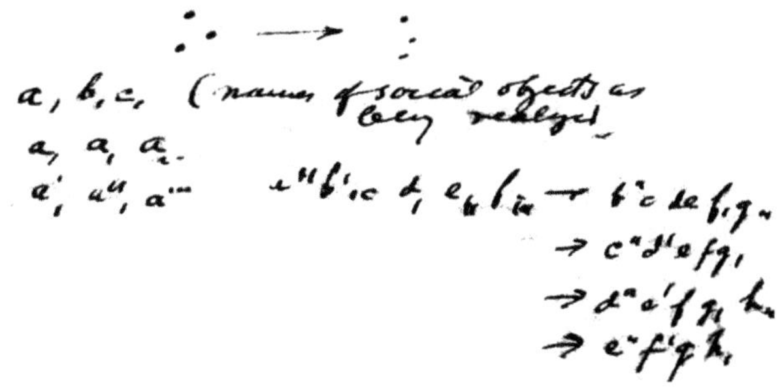

Lecture 25

Thursday, 20 November 1924[1]

Bell's notes

|47|
Whitehead – 20, November, 1924:–

Basic objects. General description of it must be founded on observation [?] – most general statement of what we do find in things observed. There are four incidents of knowledge. In perception, the perceiving is here; object there. You perceive an aspect of the object – dependent on structural relations between "here" & "there". There's a double reference here.

In physical field, talk of Electron "here". But describe it? – only in terms of what it does "there"! Our whole Knowledge shot through and through with memory. But the past There is here for Recollection – There's the immediate recollection here and now of the past. (of the immediate past, not vitiated by any thoughts of indirect way of arriving at knowledge of it). Recollection only justified as knowledge of the aspect of the past in present occasion. In Anticipation, "which includes ground of Induction", Present is known as having aspect in the future – from here to there.

Duplicity here seems oftenest spatial, sometimes temporal. Certain assimilation of Space and Time in Modern Thought. Duplicity of Reference to Region – of Spatio-temporal reference. Conclusion is that ^relevance of^ a pure object to Space-Time in any occasion of realization is of a highly complex character. Idea that:– It is "there", isn't really stating facts of the case. It's assuming that there is a meaning of "being there", *simpliciter.* There are senses in which the object is everywhere – only it's everywhere in varying senses. The discussion of the perceptual object in Broads *Scientific Thought.* Whitehead's own opinion is that it's not worth the paper it's written on – doesn't have regard to Complexity of relation of pure object to Space-Time. When we consider that this is the first thing that ought to strike us. A pure object is "all over the shop".

Russell (following Leibniz) got nearest – only doesn't take up in sufficient depth – in Lowell Lectures[2] – brings in idea of "perspectives", which is obviously getting near it. – How an ^emergent^ object, really, is made up. Whitehead:– the four essentials of – categories which apply

---

1. The Radcliffe version of this lecture, from notes taken by L. R. Heath on what was likely 22 November, begins near the end of |94| of her notes, page 450.
2. Bell has Whitehead referring here to Bertrand Russell's Lowell lectures, whereas Hocking refers to *Knowledge of the External World.* Russell's lecture series was given at Harvard in 1914, and the book based on these lectures was published by Open Court in the same year as *Our Knowledge of the External World as a Field for Scientific Method in Philosophy.*

to – an emergent (basic) object. – Actuality, Selection, Pattern (System), Unification.

(1) The Actuality means that in the emergent object we find a delegation or limited embodiment or exhibition of the "eternal underlying principle". It (this ground of becoming) is itself only through this embodiment in emergent objects. Whitehead thinks the very foundations of Time are to be found in "Actuality". Take "Subsequent" in broad sense. – Means that the Eternal becomes stamped with ∧"a definite limitation which is that object"∧ the impress of that object – takes account of that Object – "The Shadow of Truth". The object as past is a living past. By word "Subsequence" of an object means that ∧Spatio-temporal∧ region throughout which the Eternal is so stamped.

General transition has an Expansiveness not completely accounted for ---

By the subsequence of Being means that region within expansive --- of becoming. A Basic object is the fundamental example of this principle of Subsequence.

(2) It exhibits itself as a Systematic Selection with a pattern throughout its whole Subsequence. This Selection is systematic or patterned and is throughout the Subsequence in two senses. (a.) every part of Subsequence receives a character by reason of its place with reference to the basic object; (b.) That the qualification thus received, from very nature of the Basic object. Now (α.) What is it which is selected? (β.) What is the unit object which issues from it?

The selection is a question of omission. Indefinite range of manifold of potentiality and you omit from this. Then this selected range obtains its individual colour just from the finite definiteness which it has.

(4)[1] What about the Unification? Firstly it is Structural. But it's more than that. It's distributed throughout the subsequent region. In every portion of the Subsequence there is a fusion of pure object each of which carries with it the character ∧[?]∧ that it's also otherwhere. Thus the ingression of a pure object into a basic object – In one sense the pure object is in certain regions of the subsequence and in quite another sense in others.

Where you have the fusion of various objects that's the peculiar function of the Subsequence. The pure objects are not there, merely, simpliciter – but as pure objects moded and limited by their separate existence elsewhere here and there in the Structure.

(Trying to put in most general terms what turns up everywhere)

|48|

Ingression of a pure object into Space-Time has always complex character ("from here to there" etc.) "Aspects" here or there. In so far as we find an impressed Unity it's not a single

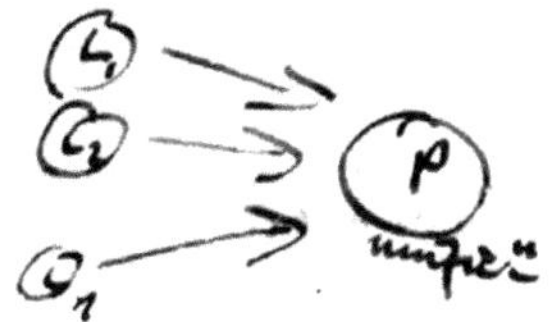

1. It is not clear what, if anything, happened to point (3). Hocking and Heath number these sections somewhat differently.

detached element or region of Space-Time we're thinking of; but the whole Subsequence (Widener Library – you don't think of it simply as seen from one particular point in the yard. But the whole as it impresses itself on the whole of its Subsequence). P as a "Subregion". – might be bigger or smaller. L's as "Standpoints" for P.

From P' we'd have a slightly different set of objects: O′, O_1′, O_2′, etc. and slightly different locations. Now the whole logical pattern in which all these L's are located to all these Subregions is --- ⟨blank⟩

The "Structure" – highly complex and highly related. That's why you get logical incompatibility. Relations perfectly definite and obvious as outcome of harmony of plan.

Now, what is it that we're talking about at P.? Various cases:– the Electron which is what it is simply somewhere else. Then there's what we are discerning – inside our heads and --- ⟨blank⟩

what is the general character

One can say what Categories apply to it.

(i) Quantitative cumulation

(ii) Qualitative cumulation

(iii) Qualitative contrast

(i) illustrated by intensive quantity and vividness etc. Essential point is fusion of like upon like. Cumulation takes place with reference to identity in the likeness. There are diverse modalities of that which is alike.

(ii) has reference to actual diversity of the individual essences of O_1 and O_2. But even within their differences there is a certain congruence – a certain unity of Nature. The Unification has its basis in that Unity of Nature – Congruence of diverse essences. (A tune heard is best example. "P") Fusion of Colors, etc.

Quality of a note from its harmonies, a color from mixing colors. Note influence in modality in difference between Chord and Tune.

Unification achieved etc.

This means they're all different examples of some higher abstraction.

(iii) has its rise in very opposite. A unit relation arising from diversity of essences. Kinship here in background. In (ii) kinship in foreground and diversity in background.

Illustrations here are discord – Contrast in sensa between sound and color. In (ii) Relation is likeness through identity in higher essence.

It's always a togetherness which is being realized. Unification within standpoint. P. will exemplify all three.

Hocking's notes

Whitehead. Nov. 20 – 1

Basic Objects

1. Description must be founded on observation-

Most general statement of what we do find

Four incidents of Knowledge: viz

2. In perception –

1. Relation of here – to there –

2. In physical field another incident

We talk of electrons – active there.

3. Our whole knowledge is shot through & through with memory but the past there is a recollection here – Immediate recollection of past –

4 Anticipation – present is known as having its aspect in the future –

Always this duplicity – The relation of the pure object to the occasion is complex. And to say that it is simply there is not stating the facts of the case. Broad's discussion of perception not worth the paper it is written on because of ignoring this complexity. A pure object is "all over the shop" – Russell following Leibniz comes nearest in *Knowledge of External World* takes up "perspectives" – One wants to go further.

3. The four essentials of an object – categories

Actuality

Selection – or Finiteness

Pattern or System

Unification

11/20- 2

4. Actuality

Limited exhibition of the eternal underlying principle. It stands in the essence of the principle – which is only through embodiment in emergent object.

Time depends on actuality.

Whatever is "subsequent" to an emergent object ---

Eternal principle as stamped with a definite limitation.

The subsequence of an object VbV means the spatio-temporal region within the expansive transition of becoming during which the eternal principle is subject to b.

5. Systematic Selection

Is throughout the subsequence in two senses.

These questions

1. What is it that is selected

2. What is this unification
3. What is the unit object wh. issues

1. Selection is omission –
2. Unification is structural –

It is distributed throughout the subsequent region. Fusion of pure objects each of which carries with it the structural limitation that it is also elsewhere. Ingression (a term urged by him years ago.) of pure object into a basic object, is in a particular part & in another sense in other parts.

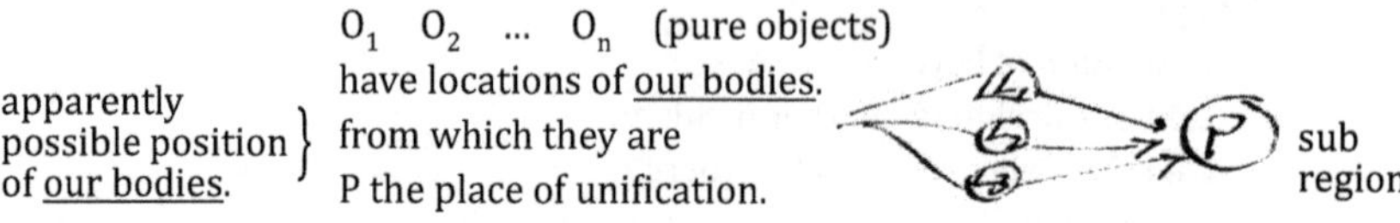

The library is the pattern of how it impresses itself on the whole of its subsequences. i.e. as one takes up Different views of P

11-20-3

As one changes position Ⓛ "standpoints"
you don't have in Ⓟ
the same pure objects Ⓞ

O_1'
O_2'
O_n' Ⓟ'

Structure of object is P, P′, – "patternness"
Systematic & definite character though complex.

---

3. What is this unification at P ⓘ?

Think of electron, & what we discern , etc.

Being of the very nature of realization; cant be explained in the sense of being pulled to pieces but only what categories apply to it.

a. Category of quantitative cumulation
b Category of qualitative cumulation
c. Category of qualitative contrast

a. Intensive quality & vividness extent of patch – adding same to same takes place w. respect to the identity

b. Refers to diversity of O_1 & O_2, but with a certain congruence in wh. the unification takes place. (as of a tune at P) Means that they are generally akin – All instances of some higher abstraction

c. Just the opposite. A unit relation arising from diversity. Kinship in the background q. foreground. Discord vs. harmony. The contrast comes out as a single fact from the discord; but it also is a sort of togetherness.

"Process is after all not explained in the slightest degree"

Lecture 26

Saturday, 22 November 1924[1]

Bell's notes

|49|
Whitehead – Sat. 22 November: 1924 –[2]

Basic objects. – Fundamental type of object of which everything is built. Various characters in them – but a touch of Woodenness in them. Whitehead's general view of Cosmology is rather like structure of American civilization (with its good and its bad side). Everything based on mass production. Real trend of Evolution – Mass production of inferior types which are absolutely necessary conditions for production of superior type. Basic objects are very final ultimate groundwork of everything.

Getting general outline of how conceived. Now reiteration. Whole nature of things based on a fundamental duality of aspects of realization. –

(1) Aspect of Monism. – Monistic totality of realization, which you can't tear apart. Founded in something, which stands behind and within process. – a unity.

(2) Aspect of Pluralism – having its basis in Existents – within the world of Potentiality. – Neither is itself without the other. You can't tear the concrete fact apart really. Machinery which gets the two together. Within the Monism there's the passing into the Pluralism via Selection and by very ground of Being having a stamp on it which expresses finite actuality of what has become (is becoming) actual. The togetherness of --- ⟨blank⟩ realization.

Pluralism finding itself in ingression into the Monism. No emergent entity is fundamentally independent. Only an emergent from the essential Monism; and at same time from point of view of Being itself as an actuality --- ? ?

Question then as to how a pure object enters into the monistic realization within Space–Time. Whitehead's point is that that Substance is of a complex character and cannot be expressed under the simple guise of a two-term relation (Quality – Subject, etc.). Even when you give it a false simplicity you get it as a many-term relation [Hence term "Ingression", "Ingredience" as a general term for many types of relation]. – Whitehead also pointing out that where you had a basic object (emergent) B in process of realisation; that again is really a complex fact. Whitehead's four incidents merely only describing the Subsequence (comet's tail). In

1. The Radcliffe version of this lecture, or a portion of it, may be found within from notes taken by L. R. Heath on what may have been 24 November, on |106| of her notes, page 455.
2. Bell's diary for this date says: 'Good lecture, on the whole, from Whitehead'.

any Spatio-temporal region (P) in Subsequence; for that region there's an aspect of B for P. B is there in a sense:– There's the P-aspect of B. Then there's the actual location of B. B is anywhere throughout its Subsequence in aspects.

Where is B? It's pervading its Subsequence in one respect. But it's in its actual location in quite another sense. Electron – Modification of Electro-magnetic-field at P through B. Can look on it under name of Causality, too. The whole of these "senses" can be kept together as "systematic" pattern Structure. The outline of this it is which Whitehead wants, really, to unfold. (Carrying out what rushed over at last time).

All the perplexing questions which come up when you deal with philosophical presuppositions of Science[1] boil down to one. – When you see how Basic Objects handled in Science – it's not simple idea of something (matter) simply there in some one sense. All the difficulties arise out of this.

This position not new. Whitehead read out to us Ross's summary of Aristotle. Pointing out that he based his very notion of becoming on fact that you can't say A is not-B simpliciter. "Potentiality" of B hanging about. Whitehead says B is not simply there simpliciter. Whitehead only going one step further. Whitehead criticizing Aristotle's worse self from point of view of his better self. The false simplicity gets enshrined in the Logic. An object is really everywhere. If you simply take electrons and atoms in "billiard-ball" view. --- The electron is its field and nothing else (there's a nucleus of course) but "location" of Electron rather loses its sense – you can't disjoin it from its field. Whole question of Perception too:– Can't get out of the difficulties of Subjective Idealism by running a more complex view. Causality and Induction require this, too.

Whitehead trying to give definite description in most general terms of what we actually find. Not a mystical fairy-tale. Not only Sciences requiring this, but Poetry. Which brings as a flash before you some more Concrete fact. Shelly's first Stanza of lines from *Mt. Blanc*.[2] (Written after he'd been

1. Note that Whitehead is utilising his title for this, his first course of lectures at Harvard.
2. The everlasting universe of things
Flows through the mind, and rolls its rapid waves,
Now dark – now glittering – now reflecting gloom –
Now lending splendour, where from secret springs
The source of human thought its tribute brings
Of waters – with a sound but half its own,
Such as a feeble brook will oft assume,
In the wild woods, among the mountains lone,
Where waterfalls around it leap for ever,
Where woods and winds contend, and a vast river
Over its rocks ceaselessly bursts and raves.
(From http://www.poetryfoundation.org/poem/174397).

reading Kant??). How extraordinarily right Shelly is if you're sharp in apprehension.

[Leibniz's every monad mirroring everything else, etc.][1]

Now questions of ingredient pure objects in emergent object B. What is happening at P includes among other things the P-aspect of B. Asking what this is??

|50|

Green

| | |

At P you have the P objects (O_1, O_2, ... O_n) – Selection. O_1, O_2 etc. not there simpliciter, but modally. Mode in which object O_1 is there is as located (for P) in L_1.

Mode = whole complex of Spatio-temporal relations. O_1 is only there in that particular mode. It's modally in a Standpoint (P). It's not simply there in L_1 – but it's there for P standpoint. For P′ the pure object O_1 may not be present ("there") at all.

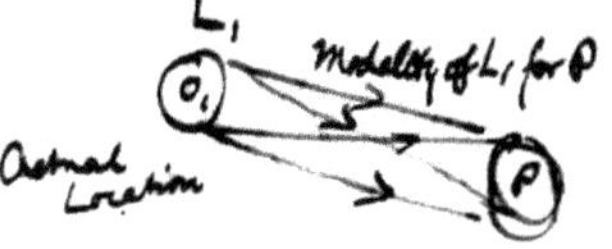

If it is, it may be in a slightly different position.

(O_1, O_2, ... O_n) = "Aspective Set of pure objects for standpoint P. in respect to B"

(The set of P objects) O_1 is modally present in P because associated with as being in L_1. You get another aspective set in P′ – (O_1', $O_{2'}'$...O_n')

(You never get an emergent Object B without a drastic simplification of all the L's etc. You've got to allow abstractively for a distinction of these positions. Question of how the pattern comes in is still to be treated.

We've got to realize that the physical universe is an extraordinarily complex affair. – Mustn't bring in too neat a little geometrical scheme and think we can do it justice. We need a very elastic scheme in order to come up with reality.)

Now very often though not essentially there are close relations between O_1 O_1'; O_2 $O_{2'}'$ etc. (closely related shades of green, or same shade). So you get an Aspective Set of Actual Locations for each Standpoint. The ingression of a pure object is not simply a case of simply being at a moment. What it is for any region is different from what it is for any other region etc.

Now how do we get patterns, structure. In Physical Science, under guise of Transmission, Continuity, etc. But Whitehead wants to be more general than that. All transmission is in a definite stream. Continuity represents principle of structural coherence and congruity. Structural pattern of basic object has to do with, firstly, a system of the L's – usually all one spot, or together make up a region, – a connected region.

(O_1, O_2, ... O_n) with their modalities is Analysis of the P-aspect of B.

What is it that's going on at P? These things not isolated facts; – O_1 with its modality at P, O_2 with its modality at P etc. etc.; but there's a Unity

1. This bracketed phrase appears, at an angle, in the left margin.

– They're welded into one entity – This fusion is what's going on there. If by an abstraction you look to the pure object you're dealing with in physical science, it is e. g. really electro-magnetic field at P as modified by the electron, which is happening at P.

In "Appearances" what's being realized is a unification and fusion of --- ?

What can we say of this Togetherness of essences in Becoming? – The three categories of last lecture.

Having a pattern means that there are simple, distinguishable relations between P-aspect of B and P′-aspect of B. These really associated and correlated together in a way which bears in itself a formulation in respect to [actual location ??] Simplicities of Correlation are related to Simplicities of Spatio-temporal relations. You have the realization of a plan – a relationship. You have the happening of an object in its subsequence.

But we want to ask what the object is in itself? What emerges from beyond itself – ? – "Wait a minute. First ---" About P-aspect of B. Is it merely that class of objects (O_1, O_2, ... O_n) with their modalities? No, it's a unit fact itself – not a mere class of entities. – In that case a slight alteration of any one entity wouldn't produce much effect. Wee alteration in one note and you spoil the music. If you were just apprehending a class of notes. Musician ought to say then – "Well, it's very nearly right." Unity hopelessly destroyed. Category of qualitative cumulation knocked out by Contrast. Violent difference obtained from smallest alteration of class of O's. As to Modalities – e.g. difference between a tune and a chord. – Spatio-temporal modalities quite different.

Another point before leaving P-aspect – The fusion doesn't really abolish their Separateness. Their modalities preserve them even while fused. When very closely alike the fusion may hide the separateness from us.

⟨Remainder of text written up left margin.⟩

Thus P-aspect of B is an entity which exhibits ⟨blank⟩ Unity in P itself eludes the analysis. – Otherwise we shouldn't know anything except as and when completely analyzed. Knowledge can start with the totality without having explicit analysis of the various objects. O_1 is modally present in P – means that it is fused into a Togetherness at P into an aspect of B. As separately at P it's a higher abstraction. The concrete fact is the P-aspect of B.

Hocking's notes

Saturday, Nov. 21, 1924[1]

1. Trend of Evolution.

Mass production of inferior types as condition of evolution of superior.

So these basic objects – the ultimate groundwork of everything.

1. It seems that Hocking should have dated this 22 November.

Last time was "hurrying through" outline

Whole nature of things based on duality of aspect:

1. The monistic totality of realization
2. The pluralistic aspect. has its basis in the world of potentiality

Neither is itself without the other.

No emergent entity is fundamentally independent. Independent only as being itself. as activity.

How does a pure object enter into the monistic realization in space-time.

This entrance is of a complex character.

No two term relation as subst-attribute is satisfaction.

The invention of "ingression" suggests the various types of relation

---

2. A basic object B, in process of realization, is a very complex fact.

For any region P, there is an aspect of B. called the P aspect of B. B is in a sense there. The "subsequence" of B, pervaded by B.

Another region, the actual location of B.

All sorts of thereness in the subsequence perception, memory, field, causality,

All these constitute the pattern structure now to be unfolded.

All difficulties are one difficulty arising from this false simplification supposing a simple idea is something just there.

11/21- 1924 - 2

Anticipation

3. Aristotle's notion of becoming. When a passes into b. a is not b *simpliciter*, but there is always the possibility of b ~~be~~ hanging about.

Whitehead only extends this – the electron is its field & nothing else than its field. Its location loses its sense if you take it apart from the field.

---

4. Same in science & poetry.

Poetry brings before you the concrete fact. Shelley is first strange. *Mont Blanc* – written undoubtedly after reads Kant.

Everything together, impressing itself. in woods in autumn. The idea of isolation is totally wrong.

Direct observation shows this.

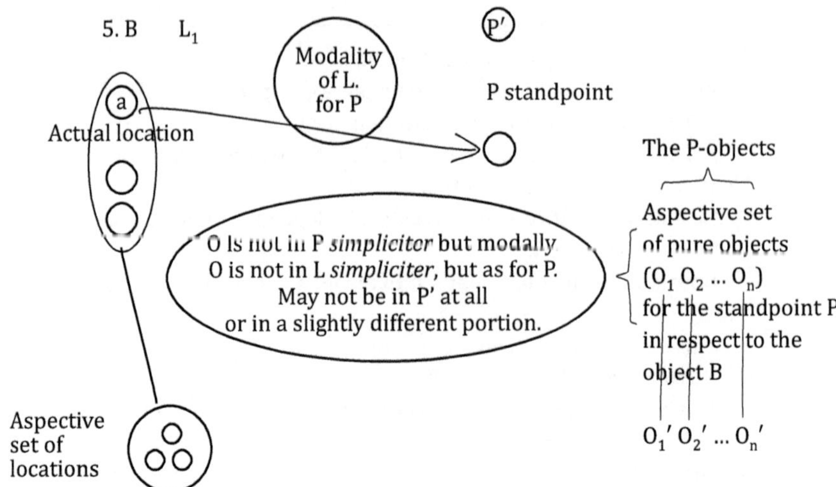

What is going on at P. – fusion – A unit fact – not a mere collection – or class. If you had a class, a slight alternation of some unit entity makes not much difference. If you have a unity – a slight alternation in a note makes a musician shiver. But the fusion does not abolish separateness

11/21 – 3

6. Relation between the P aspect of B & the P' aspect of B. – Simplicity of relation related to the simplicities elsewhere.

---

7. What is the object in itself?

Lecture 27

Monday, 24 November 1924[1]

Bell's notes

|51|
Whitehead – 24. Nov. 1924 –

Whitehead trying to stand Kant on his head; – without Kant's "Copernican Revolution". Formative not in sensibility of Subject, but in Object – Leaving Subject merely as knowing and as emerging from "it". In considering how you had pure objects entering into Realisation and how in this entering there emerges a "basic" object B, was first taking line that there is a Structural Pattern which is in process of being realised. The Reality is not one independent apart from its environment. – Apart from that it's simply a formula – a concept – an existent. It has its Concreteness and is a unity as in its environment.

So you have to work up detail of pattern by reference to what's beyond it. The embeddedness in the environment – Environment as having pattern embedded in it is the "Subsequence". This – the relational side of things – can be expressed in Spatio-temporal terms. Dealing with Spatio-temporal so far as we find the pattern of B throughout it is the "Subsequence" of B. Take it out of this and you're dealing with a mere Formula. "Formulation" is Environment as patterned – how the environment is patterned – Particular patterning of particular Environment – Reality as formed according to Formula. Formula is for someone else; formulation is what it is in itself. To describe Pattern you can only take any region within Pattern (a "Standpoint"). Total reality there (P) is far more than mere exhibiting an Aspect of the Pattern of B. – So we're dealing with an abstraction here already. "Aspective set of pure Objects" [Pure ?? – Aspective? – of Concrete Objects. I can see this better]. They are not at P. But are modally there. O_1 with modality of L_1; O_2 with modality of L_2 etc.; and the modality is the total Spatio-temporal relation of L's to P.

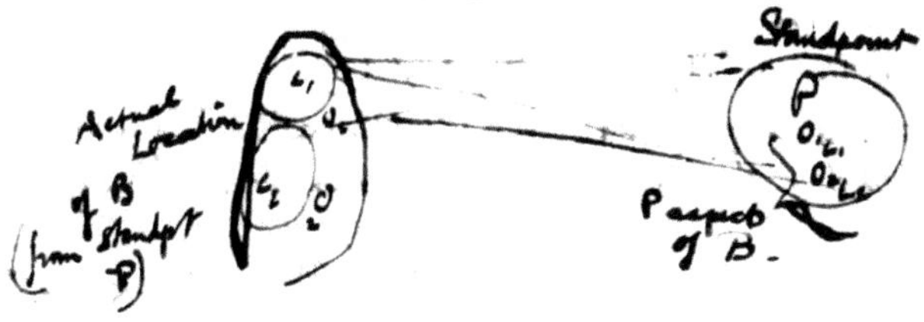

Rather more generalized view of Russell's [Lowell Lectures] "Perspectives"[2] O_1 is

1. Both Bell and Hocking concur that this lecture was delivered on 24 November, which was a Monday, although the normal schedule was Tuesday–Thursday–Saturday. The change might have been because of US Thanksgiving. The Radcliffe version of this lecture, from notes taken by L. R. Heath on 29 November, begins on |107| of her notes, page 455.
2. See note 2, page 102, lecture 25.

at L_1 in one sense and is at P in another. – Brownness of chair over there which I see. Think of theories of secondary qualities from Galilei down – a confession that you often are saying it's there or here and so bring in mind as *deux ex machina*. If you do that you (1.) make Hume inevitable. (Galileo gives a most beautiful statement, etc.) Then Kant becomes inevitable.

If you don't want to do it that way – if want to describe simply general categories that apply to what we know, – Whitehead follows Alexander or any thorough-going realist [Alexander is most interesting of recent ones]. You must find that description of "given" fact has been too simple, and must find a formulation which will get over the difficulty. Finding "Yellow" as description of something given there – quality of object *tout simple* is not good enough.

A mere class of things (the O's at P) – of isolated entities – is not what we mean by "realisation". There's in some sense a fusion of these aspects. Class-idea not quite right for the fusion. (Slight alteration of note out of a tune, etc., again). The fused unity which in some sense is absolutely altered. Whitehead considered three categories applicable to that type of fusion. The unity of these O's (modally) at P = "the P-aspect of B".

Then you have P′ aspects of B. Then ask what it is that makes the sort of pattern from which you get an important emergent object B – wherein the Emergence consists. Whitehead holds Realisation in a more or less. Where a decisive object then? Where a certain simplicity, unity, consistence, etc. – It's this self-coherence of the pattern that's always reproducing itself – the one thing. – This, Whitehead thinks, is foundation of Logic, too – and its applicability. Whitehead led to this very much by Sheffer's brilliant work.[1] Whitehead simply asking what we find (not meaning something out of head). The Actual location is the key to the pattern. These locations of the separate pure objects aren't ["absolute position"] any-where and any-how, but form a "Region". Total Superposition which is actual location of B, from standpoint P. This actual location of B will enlarge itself but will remain practically identical which articulates into a coherent whole the location of all the standpoints. There is a more or less of definiteness and reality about emergent objects. Hence this complexity here *re* variousness of location. The more concrete and definite the object the more determinate the Location.

Second point (or third)? Emergent object is the other side of the formulation. Formulation issues in the Object and vice versa. Formulation is Object considered with reference to ingredient details into which it can be broken up or analyzed. Object is Formulation considered as issuing in a unity.

1. This was Whitehead's colleague, Henry M. Sheffer, who earned his PhD in logic under J. Royce in 1908, and then returned to Harvard in 1916 to serve in the Philosophy Department. In 1913 he proved a new possibility among logical connectors in propositional logic, thereafter called the 'Sheffer stroke', which was incorporated by Russell and Whitehead into their second edition of *Principia Mathematica*. It may be to this that Whitehead was referring. Bell mentions Sheffer a number of times in his diary, along with R. Demos and others, whenever the more junior members of the Department socialised.

|52|
Now – how the Concreteness of the emergent object – its existence – Realisation. Two aspects of the Realisation – (1) How object can be considered real in itself (Environment reduced to systematic of simple background).

(1) then, question of Intrinsic reality of B and – To what extent is there any concrete individuality in object when you've reduced Background to mere systematic bones.

(2) is question of Extrinsic reality of B where background considered otherwise.

The intrinsic reality is the internal fusion into one concrete realisation of Value of "the Adventure of its Pattern".

[?] What Whitehead means by "Concreteness". There's no complete real object which is Totality of all reality etc. – There's no unconditional reality. The real is only so by reason of its position in environment. The "eternal ground or principle of Becoming" is not real on the level of the Concreteness of a finite emergent object. An entity which has concrete individuality in itself contains (a) in itself an individual achievement (a definite unity of Value for itself) and (b) there's an individual determination of achieving – Process is what it is because of it. What takes place has to take place with reference to that achievement. There's the Achievement in Transition – which is also an individual determination in achieving. And this embodying in itself (c) eternal principle, which underlies all achieving.

These three aspects always present. (b) it's forcing the process to be what it is, (a) is what's common to whole environment beyond itself. (a) Whitehead calls, also, "display" (b) "Determination" or "Formulation." in (c) you get what's perpetually general. Those three aspects are what Whitehead calls "Concreteness". It's the very nature of Realisation that it finds itself in the Embodiment of Emergent Entities. ([of the "Eternal" in the Emergent Entities?])

The very first principle of description of what we find in real entities is that the Concreteness of finite emergent entity is compatible with, and requires the conditioning by its environment. Apart from its Environment, Entity has lost individuality. Unity in something more than itself is exactly what you're always talking about.

|| This principle of the Necessity of the Environment. We lose individuality when we abstract from the particular Spatio-temporal root. – This is the anchorage in the Environment. Away from this you've got just something general– abstract, etc.

[But how this particular Anchorage without "Absolute" Space and Time]

Particular Multiplicity of Space–Time pulls together into a unity.

The formulation of Pattern --- ⟨blank⟩

There must be something identical to which every part of Pattern is Subservient. – Here the idea of Simplicity comes into a pattern. In proportion to completeness of Unity of formulation you obtain a proportional concreteness of finite entity. – An embodiment of the logical expression:– Fusion of Display, Activity, and Reference to general realm of Potentiality of which "This" represents the selective fusion.

Logic is nothing else than Science of Formulation considered by itself and without any taint of Empirical particularity. – The how in most general sense. From Logic you go to a particular formulation and then get to emergent entity. If you start from ⟨blank⟩
That's how Logic looks as against Realisation. – contrasted with "Existents" – Pure realm of Potentiality is it is the consideration of patterns which are possible or impossible. But of course Formulation is more than Logic. Has to have Coherence and Massivity:– a certain intrinsic richness of content, – and Endurance – a certain extrinsic extension. These come to how you get the actual Unity of Realisation. The actual fusion is in respect[1] to the key of the pattern. The Self-relevance of the whole pattern. Its Self-relevance is at $L_1 - L_n$. That Self-relevance of things there, to be taken up next; and then how it happens together for other things. (P's)

Hocking's notes

Nov. 24, 1924

Resumé

The emergence of basic object B.

"The environment conceived as having the pattern embedded in it, I call the subsequence of B."

–"Space-Time so far as it has the pattern of B throughout it."

Formulation is the environment as patterned. Formula is science – concept of formulation.

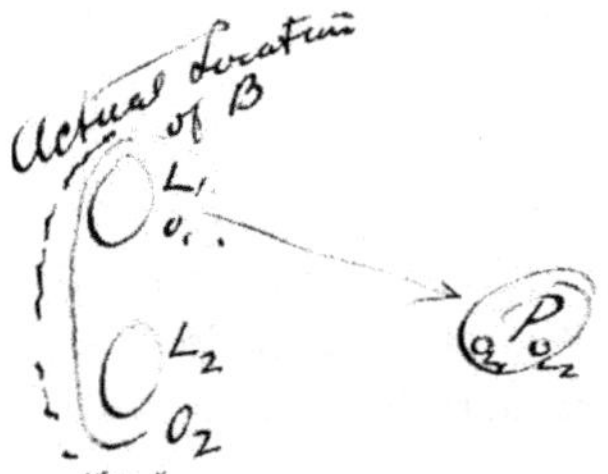

The graph is to describe the pattern standpoint being any region of the environment, you have an aspective set of pure objects O_1 O_2 ... O_n of B.

Generalizing the notion of perspectives in Russell – Lowell Lecture[2] to meet the difficulty of secondary qualities. Can't say they are there in the object, don't like to say they are in your head, & so say they are in the mind. If so you make Hume inevitable & Kant.

To avoid this, you must recognize that the description of the given fact has been too simple. (Following Alexander)

Logic has its origin in this unity of pattern. L_1 L_2 etc form a coherent region, the actual location of B from standpoint Ⓟ

1. The contraction here is 'resp.' Usually, Bell would provide a last letter – for example resp[t] or resp[c]. Without one we look to the 'to', following, which seems to imply 'response' or 'respect'.
2. Again, see note 2 on page 102, lecture 25.

The dotted line indicates the system of all the locations from all the standpoints P. A matter of more or less.

11/24 – 2

1 Intrinsic & Extrinsic Reality of B.

2. What do I mean by concreteness?

There is no unconditioned actuality

The "real" entity is real by position in its environment.

Concrete entity contains – as three sides

a. Individual achievement Something in itself

DISPLAY

b. Individual determination of achievement

FORMULATION

c. Embodying Eternal Principle which underlies all achieving.

GENERAL or UNIVERSAL

It is the very nature of realization that it finds itself in these three characters. embodying them.

Apart from its environment the object is an abstraction & loses its concreteness.

This unity with something more than itself is something we are always talking about.

We lose individuality when we lose anchorage in the specific spatio-temporal root.

Logic is nothing but the science of formulation considered in itself.

The emergent entity is how formulation looks in the realm of realization.

Of course formulation is more than Logic –

The actual fusion is to be found with reference to the region L. the key of the pattern – "the self-relevance of things there"

Lecture 28

Saturday, 29 November 1924[1]

Bell's notes

|53|
Whitehead – Sat. 29. November, 1924:–

How this Fusion which is Fact – this realisation of Unity (of Monism) – is to be analysed. The Emergent object. First point – In respect to the pure objects there is no such thing as position in Space-Time for a pure object. A meaningless idea. There's always a reference of a certain relation. A pure object turns up always as related to Space-Time. One region as a governing one in some sense; other as a Subsequence in some sense. Pure spatio-temporal relation between P & Q. An object there with the Modality in same derivative mode as derivative from P. P as "actual location of O" – from point of view of its "modal occurrence", in some way or other, at Q. There's always that duplicity. O_p always has this modality as derived from P. O is modally in Q as derivative from P.

We go back to question of "Object–pattern" later. Full realisation is an Object – an Emergent one. So far, a sort of Realisation at half-cock.

(2.) There are grades of realisation (concreteness) and realisation always lies in Individuals. The imprint of Individuality is the process in that mode. If you try to get Individual out of the process you get only a formula. Individual in a setting always. "The capture of the Eternal" in the finite mode. [Use of Rhetoric – "getting a bit sassy, when you don't want to enter into details."]

(3.) The more ideal Experience – mere being as such – is an essential plurality of isolated entities. – Here the imprint of Pluralism on Fact. When you begin to analyse reality you find these.

(4.) Realisation is the imprint of Monism. – A unity of all Being is achieving, in Realisation. This is not abolition of these Existences, but achieving under modal limitations, of which we're discussing meanings. An emergent unit in some sense relevant to all Things – by exclusion or other kind of union. – Some of the "all things" are there in a mere general envisagement. The unit entity embodies both particular relevance of general ground of Becoming to itself and --- ?

As selective by its relevance to whole aggregate of Ideals, it is Value – as that which has emerged for its own sake – The procuring of itself by itself through exclusion of Irrelevance.

1. The Radcliffe version of this lecture, from notes taken by L. R. Heath on what was likely 2 December, begins on |109| of her notes, page 456.

Now in slightly different aspect:– Aggregate of Modal presences in Q, viewed merely as an aggregate, is this:– When there is Q in realisation it doesn't occur as a blank region of Space-Time in which something happens. Not as if Space-Time occurs and then something puts something else into it. The Event is already prejudiced by what's already actual.

Relevance of Content of Q to the past. Q is in the Subsequence of P. There is an Aspect of Q – in the Content – what's happening in Q [one abstraction from total amount of happening in Q can only be expressed by its relevance to the Antecedents of Q]. This is not whole of what's happening in Q; but one abstract from that whole. This relevance to P (to Antecedence) has a principle of fusion in itself (the three Categories of Fusion again) (two of Culmination; one of Contrast). Appears to us palely as "Potentiality". There's no such thing as bare Space-Time. Q cannot be bare. So far as Q has its contents merely as relevant to Antecedents there's an "Actual Potentiality" of Q.

Discriminate actual from ideal potentiality! – Latter already dealt with as "Potentiality" tout simple. Modal presence of pure object --- ? ?
By ideal potential mean merely qualification of regions as being within relational essence of pure object. – By aptitude of the object either for modal location in Q or for actual location in it. – Pure object viewed merely in itself as Existent, pervades all Space-Time. That is, in a sense, the relevance of whatever's happening, to all ideal world – in "general Envisagement" – which is just the relation of every fact to all ideal entities re their general aptitudes. But for particular relation of particular ideal entities to particular Events we enter on graduated realm of ? ?

Ingression of various pure objects into emergent reality associated with occasion Q is prejudiced in some extent by what happens at P. – Is limited somehow by P. And this is a fact. What's really happening is the Emergence of Value from actual potentiality. This potentiality is part of the very nature of P. It's in Nature of P that there is Q in such and such Space-Time relation to it and with such and such prejudice from it. At P. – Q as with P as from Q. At Q:– P as from Q.

Coming down here heavily on "Internal Relations". Whitehead going back a little on his polemic against Predicate-Subject point of view. In knowing P., P itself has future already contained in it.

Viewing Space-Time conceptually is nothing but a plurality of bits with no ending; and yet all one. Yet it's in
|54|
a sense the field of Realisation. Every region, in some aspect, includes in itself, as being real, all of the rest, – only under some limited aspect. This is only another way of saying again that every Occasion bears in itself some direct relevance to what's beyond it. If merely atomic facts; then we know nothing of what's beyond it, and you're reduced to a bare skepticism.

That double reference (above) is essential fact born in about us whenever we begin to analyze anything that's happening. It's first business of any Metaphysical attempt to drag out that essential ∧Cross∧ reference.

There would point out that Q has an essential Cross reference beyond itself, too. (Same schema repeats itself with Q, R, etc.). The realisation which is inherent in Q is essentially ∧in itself∧ a meeting-point of the Past and the Future. Inherent in very nature of Q is reference to something not itself.

This is consideration of Q on its active side (towards R.) – on passive side (backward to P.) The Whole is in a way comprehended in Q in so far as can take Q as having any real individuality of its own. So whole principle of Realisation in a Q is their Integration of Ideal Diversity into Unity. The present occasion in its realisation is an occasion of the Integration into itself, in diverse ways, of its Antecedence and its Subsequence.

Integration of separate potentialities into a certain unity. But as matter of fact Whitehead holds strongly to position that essential Plurality inherent in Eternal Condition of things breaks out again in the Atomistic Character of Reality. ∧In∧ what is concretely realized you again find emergent Individuals – But these can't be separated from their Station in the Whole process.

This Emergence of Individuality is not a Hit or Miss Process. Realisation has grades of more or less. It is the subordinate realisation ∧(subordinate emergence of Individuality)∧ which takes place. There's an Integration even at the lowest – so you can talk about what's happening – emerging – some low grade of emergent unity – anywhere, in any event. That's the Unity, which makes it what it is. In general only faintly marked Unity. Only when some peculiarity ∧and definiteness∧ etc. of Pattern that you get the full and complete realisation....

Point of view > Whitehead says that what's realised is Value because it's the issue of a process of Selection – Exclusion – for itself and through itself. It's both the Issue and the Process – The Value and the Valuing. Extension[1] as the very essence of Reality. The Something being achieved for its own sake.

Three categories apply to this Emergence, then:–

(1) Idea of Unification – An Outcome

(2) Category of display (what it is for itself) – As a thing in itself and for itself

(3) Selection (conditioning future) – As determinative

1. In Bell's handwriting this word reads like 'extension' but, given the context, 'exclusion' may be intended.

Hocking's notes

11/29 – Whitehead 1.

A. Fusion. Realization of Unity.

How is it to be analyzed –

1. A pure object has no location in space-time always a relation of governing region to subsequent region. Q

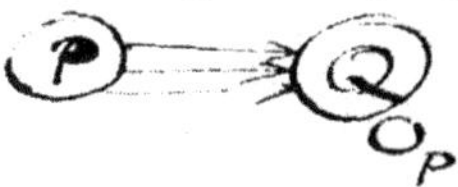

O is in Q as derivative from P, the actual location, – has the modality of being thus derivative.

2. There are grades of realizations

The full concreteness lies in individuals

3. Mere ideal existents – an essential plurality of isolated entities.

4. Realization is the imprint of monism – achieving the unity of all being – under modal limitation

To ask how anything is real is to ask for the unity in all things.

"Emerged by reason of itself, by exclusion of irrelevance" for itself – (hence value?) Emergence of value

---

Now put this in a slightly different aspect.

---

5. This aggregate of modal presences in Q occurs "with the prejudice" of what exists at P.–

The relevance of the content of Q to P. "derivative from" its antecedents in P. [Antecedence – Subsequence]

There is an "Actual Potentiality" of Q

6. To be discriminated from "Ideal Potentiality"dealt with from the beginning –

11/29 – 2

7. Internal relations

"Coming down very heavily upon this point of view of Internal Relations am I not" – "A little going back on my polemic against the predicate-subject point of view, am I not" –

P as from Q

Ⓠ

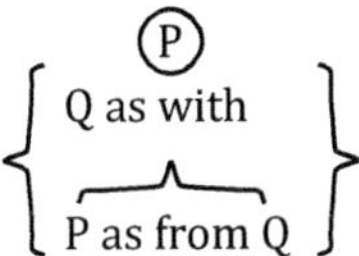

There is something in P which is Q (?)

"Can't get away from a plurality & a monism"
The more abstract your conceptions the more insistent[1] the separateness. as Space, Time, etc." As the field of realization space-time is a monism

There is nothing in realization without the complex reference to what is beyond – (as P is beyond Q etc)

8. 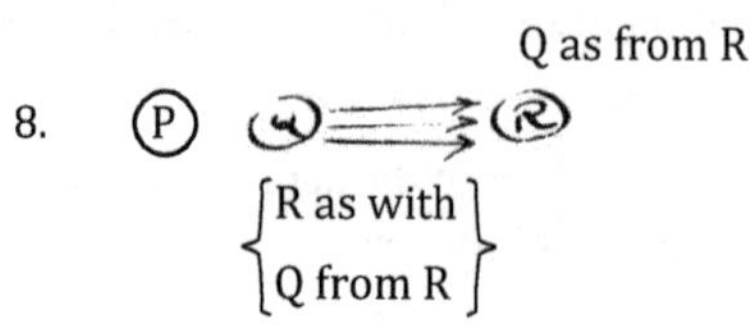

Q is a meeting point for past & future
"What is the past now? Is it a mere nothing?
You can't have any relations to nothing, you know."
So with the future."
"The whole is, in a way, comprehended in Q."

11/29 – 3
9. To the whole principle of realization is the integration of ideal diversity into unity

In diverse ways of its antecedence and its subsequence.

Emergent individuals cannot be separated from their station in the whole process.

"An event" – what is emerging there – implies some grade of emergent unity.

Very faintly marked in general

Only where there is some realization of structural pattern (relevant to past & future)

Do you get full & complete realization.

---

10. What is realized is value because it is the ~~out~~ issue of a process of selection.

The very essence of reality, the achieving of something by selection, for its own sake.

---

11. Three categories apply:
Ideal of Unification
display
selection

1. The first half of this term is not clear.

Lecture 29

Tuesday, 2 December 1924[1]

Bell's notes

|55|

Whitehead – Tuesday 2nd. December. 1924:–[2]

Abstractness of the Logical Diagram. (apology). Whitehead looks on Metaphysics as purely descriptive science, stating what there is that's absolutely general. Purely descriptive – because we've simply got to find these characteristics there. We're very apt to make mistakes – weave Fairytales.[3] – It takes us so out of our underlying line of thought. Therefore need of constructive criticism. As completed in head of some superman of future Metaphysics will be a simple and brief statement of this generality. With us it is a hesitating endeavor to get true general point of view out of Concreteness and has to be continually confronted with its consequences.

Every particular field of thought (such as a science) is in nature of an explanation because particular meanings have got to exhibit selves as particular examples – finding place and standing – in a system of general metaphysical ideas. To make a Scientific explanation is to see what's happening ~~as example~~ in setting as example of what's general in whole Universe. Every Science therefore help to Metaphysics and a criticism of it. You get to metaphysics by taking some body of originated thought and producing its implications backward.

Here's where Whitehead repudiates Epistemology as the one foundation of metaphysics. But it's the first source and Critic of the Metaphysics. System of Metaphysical thought has both got to satisfy nature of knowledge; and also – unless you can show how you know it, – must chuck it away (Hume sound here). Metaphysics as it would appear to a man who's spent all life meditating on Mathematics, Mathematical Physics, Logic etc. Endeavour to generalize ideas of science into Metaphysics– having regard to the necessary "fattening out" of this.

Essential interconnectedness of things in Physics already. As soon as you begin to deal with the finite as an Abstraction you're getting close to Absolute Idealism. Where Whitehead differs[4]

For them the Absolute is a Super-reality – Something more real than details of happening. Question whether Berkeley is a subjective idealist or an absolute idealist. Certainly has a super-reality in mind of God.

1. The Radcliffe version of this lecture, from notes taken by L. R. Heath on what may have been the same morning of 2 December, begins at the end of |109| of her notes, page 456.
2. In his diary for this date, Bell notes: 'afternoon tea with Whitehead'.
3. These two terms have been reversed as indicated by a curving line drawn by Bell.
4. Bell's arrow seems only to point vaguely towards the next paragraph.

When try to get ground of Reality more real than what's Becoming, you've got an Abstraction. It's not Super-reality but Subreality. Realty is always the Emergence into finite modal Unity. There are degrees of reality. But that's the ? ? ⟨blank⟩

When we analyze we get to Being which is really in a sense defective. Finds its reason in its function for this Realisation. The Realisation is always Unification in some sense. But the process is conditioned – what conditions it: is always a plurality – An absolute discrete plurality of Existents. – Isolated but can get together in sort of shadowy way via their "relational essence."

Function of becoming one is achieving unity "functionally" "modally" – i.e. limitatingly. The limited entity isn't merely itself; – is essentially referring to something more than itself. Has its concreteness only when seen in its whole circuit of reference.

Why Whitehead objects to treating Subject-Predicate as fundamental point of view.[1] Fundamental description of concepts which apply to this ⟨blank⟩ – must be reference to what's beyond itself. (Assuming system of Realisation as adequately described as Spatio-temporal). The foundation of the Content is the impress from Antecedents. Aspects of P from Q ["P from Q"] is to be conceived as in content of Q.

"Q with P from Q" – P has also as its content: That there's relevantly to be fitted on to it Q, and how P itself will "look" from Q. Itself as being relevant to Q is already there.

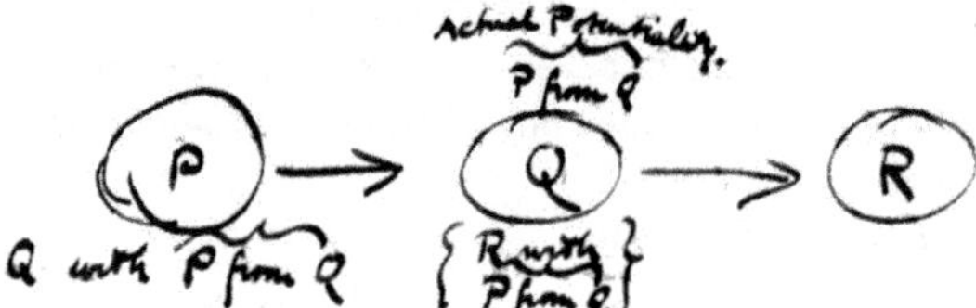

When Q is realised it's not a bare bit of Space-Time within which something happens. It's not an empty and disconnected occasion. There is a realisation in Q subject to an actual potentiality. Whatever happens is something which is interwoven in Essential being of P, and in Q's essential being, what it's going to mean for its future. Here we've merely got *disjecta membra*! Q is in some sense or other (going back on Aristotelian category of quality) something here and now for itself. It may not be very much. That you

?? can take Regions of Space-Time and talk of them for themselves, shows emergent unity from process of Unification. Realisation is through and through a process of Realisation of all that there is for realisation. – In some way or other touches everything that there is for realisation.

Why it essentially passes on to something else – Because in touching it has reference only to certain "modal aspects". It's not "pure objects" realised there but these as "from a beyond"

|56|

The emergent entity as being realised is in a sense all that there is of Reality for that [Q space]. The relevance of Q both to Antecedents and

1. Whitehead had attributed this point of view to Aristotle in lectures 14, 15 and 23.

Subsequents is transmuted there into something for itself. There's not three pieces to be conditioned. There's not mere Potentiality all by itself; active effectiveness (Emergence) all for itself. Therefore in a sense the emergent entity --- ? ? ⟨blank⟩

That is "capture of Eternal" – a particular "How" of having what is general. The process in itself has its own derivative independence. The unification comes in every possible way – not only with reference to "static" objects – but in a sense unifies Past, Future etc. etc. – namely as modally and in this particular way being essential to what it is in itself. C. D. Broad and "What we mean by the Future." Broad thinks it means absolutely nothing.[1] Whitehead doesn't agree. In very being of Q there is the Future (and the Past); – and that's why can talk of fitting on subsequent Experiences – You see the R as the realisation, for own sake, of what was modally and particularly already in Q for Q's sake.[2]

Now another line of thought:– Is there anything in our Knowledge which corresponds to this reference backwards and forward (in Schema). Talking about "this time yesterday". Question of the Medium here, is a very puzzling one. There's not merely P and its relevance to Q; but there's L in between. Why not say "P. modifies its next door neighbor; into[3] its next etc., etc."?

There is no next door neighbor [Ideas of Calculus.] To try and make it independent of some idea of direct reference is merely a system of fudge. To deal with Internalization by means of idea of passing through "Next door Neighbor" is nonsense. Even if you were to admit Point-Instants it wouldn't help you over much.

You've got to have general reference to "Beyond" in the event, to have any particular reference at all – any relevance of Past to Present and Present to Future.

Now what do we find in our Immediate Experience? The bigger the jump you make the more you get muddled up. You're not considering instances past or Future, but more generally what's "a little way back", etc. If consider what's 1/10 sec ago, you find you have difficulty to observe this – we only know Present as arising out of that. Sharply sundered Present is a Myth. There's a future, too, rising out of this present – even if we don't know all detail of its content. Content doesn't merely formally succeed. But very content is always relevant to the past.

Whitehead denies that there is any difficulty in finding the impress of actual potentiality. Most immediate deliverance of our knowledge. In some sense obvious that Q, in being its real sense for itself – ∧Actual Display –∧[4] E_Q (the unit – the instant – for itself). Starts with a real potentiality and aids with actual effectiveness. There is the "Actual Display" or "Achievement".

1. Broad defends this position in his *Scientific Thought* (1923, p. 66).
2. This whole paragraph is handwritten on the left hand side of the page, which Bell marks off with a line, and then continues next paragraph on right half.
3. This term is not clear.
4. These two words are inserted in the left-hand margin.

The event realizes for itself its own Enjoyment – has its own Experience as derivative from what's beyond itself. The Entity is within its own Experience and is the outcome of its own experience. – At same time the realisation of these periods depends upon realisation of ? ? ⟨blank⟩

The entity being its own good sense.

E_Q is an impress because it is P as an ? ? ⟨blank⟩

Similarly R with P from Q is an Impress implicating E_Q with R.

Emergence of E_Q is fusion of anticipation and --- ⟨blank⟩ to an actual experience. There's an addition not explainable purely in terms of past and/or future. It's involved in the realisation of Q. – this ground of realisation fulfilling itself into complete actuality under limitation of these modes.

What can be said generally of the Substantial Ground can be said of it under this standard etc.

What Whitehead's working up to:– Great question of what room for Freedom. Nothing here settles that one way or the other. The secret of what Freedom there is, if any, comes in fact that emergent Entity is more than the mere assemblage of past and transmission to future. –

In some sense or other is a creation, there, at Q.

In some sense or other the Entity chooses what it's going to be! There must be some selective basis – Now if[1] this is in the emergent entity --- ? ?

There is in some sense a Selection in fact of E_Q. How far this can be exhausted under potentialities of next Selection is whole problem of debate in question of Freedom.

Hocking's notes

Whitehead. Dec. 2. 1924.

1. Metaphysics a purely descriptive science –

Most general characters one finds there.

"A hesitating endeavor to get at the true general point of view in its concreteness"

"Has to be continually confronted by its consequences"

Explanation in special sciences, particular facts as special cases of the general –

All special sciences must find their standing in your metaphysics.

And then you get your metaphysical conceptions by starts with same special science & pushing its concepts back & back.

---

2. Repudiate Epistemology as the one foundation of metaphysics. It is the first critic & the first source of metaphysics. the first critic, because there is knowledge & you must find a place for it.

And then you can't talk about what you don't know.

1. It is not clear whether this is 'at' or 'if'.

---

3. Very near to absolute idealism when you take the finite as an abstraction. The slightest push would push me over. But where I differ – the absolute – is a super-reality. My point is when you try to get a ground of reality which is more real than the given, you get an abstraction; your super reality is an under reality. Reality is always emergence into a finite modal entity. Realization is always unification, whose conditions are always a discrete plurality – of existents.

<u>12/2 – 2</u>
4. The process of becoming[1] one is achieving something that is just that thing – Emergence of the limited entity. Not merely just itself but also referring to process beyond itself –

5. The fundamental description of emergent unity . not subj-pred. but references beyond itself. (P being antecedent of Q)

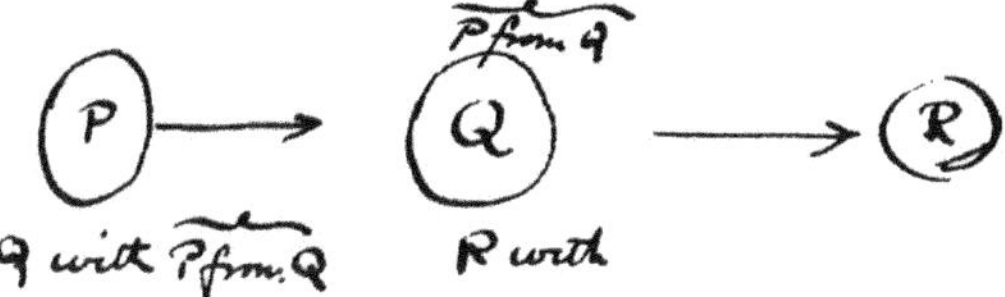

Here the causal process becomes <u>the</u> process

Q is not merely by itself but contains the impress of <u>P</u>

Whatever is real is the outcome of something, an actual potentiality, which is woven into Q – won't say it determines Q entirely.

So Q for R. "But then we have disjecta membra don't we, hang it all."

But in Q we have something for its own sake – going back to Aristotelian category of substance – quite aware of that.

Realization is a realization of everything there is for realization – Unifies some modal aspect of everything –
Transmutation into something for itself alone – the emergent entity.

That is why we call it a capture of the eternal – It is how anything has what is the general. The process has its own derivative independence. Q in a sense <u>is</u> the past and <u>is</u> the future. So you can talk of fitting on –

<u>12/2 – 3</u>
6. Is there anything in our knowledge that corresponds to this reference back & forward.

Talk about this time yesterday
There is a lot in between, like (L)

1. It is difficult to see whether this is 'being' or 'becoming', or the former changed into the latter.

Why don't I say that P modifies its next door neighbor? Because there is no next door neighbor. Cf. calculus.

If you are going to consider a system of transmission – can't do it by the method of passing through next neighbor.

Even if you allow that there are point instants there are none next door. If P is relevant to what is beyond, you must be prepared – unless you are going to fudge. So make the jump. And if you jump, you are fuddled up by the intermediate medium. You never consider what is merely next, but
next regions having some breadth.

The present is arising out of a past and passing into a future, and is relevant to both.

7. Go back to 

Q starts with an actual potentiality and ends with an actual effectiveness. And is for itself an actual display "realizes its own self-enjoyment"

P from Q is an impress,

Leading up to the question of what room there is for freedom – Doesn't think that anything predetermines this one way or the other. The secret of it lies in the fact that the emergent entity is more than the assemblage of the past or its leading into the future. Entity E_Q chooses what it is going to be.[1] There is in some sense a selection on the part of E_Q – & the question is how far the potentiality completely exhausts E_Q[2]

1. 'it is going to be' runs vertically up the right side of page, as Hocking had run out of space on this page.
2. These last two lines are written vertically up the left side of the page at the bottom.

Lecture 30

Thursday, 4 December 1924[1]

Bell's notes

|57|
Whitehead – Thurs. 4. December. 1924 –

Q (E_Q – Emergent Entity) – made up of its parts. But they of theirs etc. Vicious regress. Two and only two ways of doing it. You can say what you're dealing with are points.– Point-instants – Point-events they've got to be. They're not <u>really</u> events then, because <u>then</u> they'd have duration. You've got to have over and above these a mysterious external relation between them. In this way we proceed e.g. in all geometrical (mathematical) reasoning. Viewed from point of view of Metaphysics it's a Metaphysical fairy-tale. [But what of Leibniz's solution (the *petites perceptions*)]. What you really start from is the e.g. "Room" or the Event and then you begin to analyze and cut it up.

|| There is (secondly) the way:– The Room as a real fact (in Space and Time) is an emergent entity as one thing. – In being actual, its actuality is inherent in its Unity – you can't get the two apart. You may, in looking around, <u>miss</u> something of the wealth of Entities. But insofar as you have Reality you start with the Unities. You know the Unities as <u>contained</u> in the others. – No vicious regress here – "You can sit down on <u>that</u>; and we do". The room isn't a bare piece of Space-Time "on its own". – There is no such thing. The Concrete Reality here is that which is realized for the room as a whole. When you talk of the piece of Space-Time you're abstracting. The Entity is <u>emergent</u> from its ingredients. You abstract from these definite ingredients and think of the Roominess alone – the Space-Time part.

You must then allow for the <u>whole</u> fact of what is relevant to E_Q. And <u>then</u> you find (either from point of view of perception *genommen, oder von dem der Physik –o .von blosser Raumlichkeit*[2]) you can't describe the ingredients without reference to what lies beyond.

["Q" for an event used in a rather indeterminate way. E_Q to be unambiguous.]

Next pt:– (joining on to end of last lecture) – we have the actual factor Q – the E_Q which emerges is an addition not to expressed either in terms of Past or of Future. E_Q is the modal realisation of all that there is – <u>all</u> is in

1. The Radcliffe version of this lecture, from notes taken by L. R. Heath on what was likely 4 December, begins on |111| of her notes, page 457.
2. This is not the first time German is used (one term appears in lecture 5 and two terms in lecture 18) but it is the first time German is used for entire phrases. German: 'taken, either from the physics, or from mere spatiality'. As there is no indication of this in the Hocking notes, it seems Bell simply slipped into the language in which he had for several years transcribed Husserl's lectures.

some respect relevant to E_Q [Leibniz !!][1] and therefore in some sense is in E_Q as an Aspect. I.e. a mode of it – Modally in Q. E_Q is just itself, as outcome of the togetherness of all these. The various things are separately modally in Q and E_Q is the unity of them which emerges, – and emerges as an end in itself – It emerges not as a Unity *schlecthin sondern in dieser besonderen Hinsicht.*[2]

If you spilt it into " Matter and Form". Modal ingredients as thus fused are in the nature of the Forms, and it's just that mode of "Value" (Matter) – of realised achievement. This achievement is in itself and for itself – its own justification. What is fused makes a totality which is other than itself but which is its own unity. The unity of its own experience. This characterizes that reflexiveness of Experience on which Whitehead insists. Realisation of the unity is one of the facts in its own present environment. This is what Alexander means by "Enjoyment"[3] – Entity having itself for itself. Whitehead has put it otherwise (perhaps less helpfully) – [Leaves this out!][4]

In dealing with the Past there's "actual potentiality" – what E_Q accepts (Kant's "given") (You've got to have that element somewhere) – and that is outcome of the Past in the present. Then it also has its "Actual Effectiveness" – namely how what is begun.[5] Q is modally mirrored in Q as itself receiving some actual potentiality from Q. So Q as an entity fuses and bears in itself the whole of Space-Time – only not *simpliciter* but modally.

Then we come to several points more. – (a) Whitehead has dropped a point that he made great fuss about previously. – Retention. Illustration of the Tune. Tune not present until last note. But development of Event must be allowed for – in Psychology and elsewhere. – At end the earlier notes are present – not *Simpliciter* but modally as "there". But if you knock out idea of instantaneous point, you've got to have some Retention. – It's not just consummated and over with in the one instant. [" Silly little diagram". "Don't be afraid of being silly sometimes. You only end up by being pompous , you know".][6]

We are impelled by a kind of Hegelian dialectic to go beyond the mere structure of an Entity, to its Endurance. So far the *disjecta membra*:– "Structure" of Emergent Object e.g. more or less by itself treated.

What we want, to "produce" reality – Not by a conjuring trick, *gleichsam*,[7] but just to describe what we mean.

1. This bracketed comment appears, at an angle, in the left-hand margin.
2. Bell slips into German, again: '*per se*, but in this particular regard'.
3. According to Emily Thomas's article on Samuel Alexander in the *Stanford Encyclopedia* (http://plato.stanford.edu/entries/alexander) he calls 'enjoyment' our consciousness of the process of perceiving.
4. This seems to be Bell's comment.
5. The abbreviation in this case seems to be 'bey' or 'beg' but it is not clear.
6. This time the bracketed comment seems to be Whitehead's aside.
7. Another German term: 'as it were'.

|58|

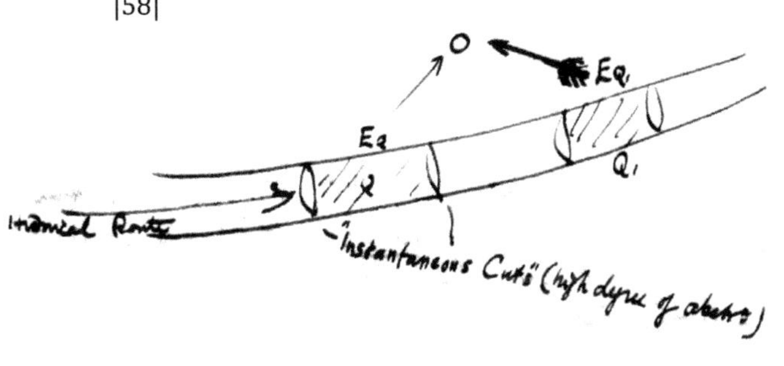

Each of these would be 3-dimensional spaces Something has been passed on – transference of something identical from to – equally there in any section you make. That's the "historical route" [Broad] and the Object "O" provides that route.

To feel that idea really you must in the first place note that the emergent Entity is a real addition and in its emergence is an emergence of all that goes with "its own subjectiveness ". In case of our <u>own</u> Experience all that's connected with adding to the "tone" of the fusing entities.

[Yes but what about "inanimate" objects ?][1] – Its "what the Unity is adding to the total fact there". Any Event must therefore represent the actual location which governs a simple coherent pattern – viz. the aspects of E_Q beyond itself with its added subjective tone, must issue in a pattern of actual effectiveness. This pattern and the actual effectiveness in which it emerges, must issue in a self-propagation. Certain identity of Pattern which runs through the Whole. As the pattern emerges from Q or Q_1 the dominant part of its actual potentiality is derived from the Q, Q_1. In Q_1 are then the aspects of E_Q. It reproduces the pattern of tone and effectiveness which is that of E_Q. So there's a reproduction of itself throughout. Then there are always the other entities contributing to actual potentiality there; and relation of other Entities must not be such as to destroy the "pattern". Root fact that we must accept as given:–

The process of Realisation is a process of self affirmation of an Identity which is realized as enriched with its past. In its self-realisation it realizes itself as outcome of "O" each time. – Always an essentially self-realisation as preserving the pattern "O" (a pattern of "tone" and of actual effectiveness). "O" may become more and more abstract. – What's common becomes so little as to be unimportant. (Problem of the darned Sock – the Human Body etc.)[2] But preservation of pattern. The difficult problem of an "Organism". That problem turns up over the living body.

Whitehead doesn't think problem of living body is essentially different from existence and permanence of an <u>Electron</u>. There always is in realisation the self-affirmation of a pattern for its own sake. – All that you can say of a molecule or anything else. When you get that full reality of Endurance – its that which you get.

1. It is not clear whose bracketed comment this is (perhaps Bell's).
2. This closing parenthesis is not provided by Bell, and while it works here it might also come at the end of the next sentence.

Hocking's notes

Whitehead
Th. Dec. 4. 1924 – 1

1. Emergent Entity
Run away from the unity –
Then the entity q.[1] is made up of its parts –
What are the parts made of ? Other parts
Infinite regress.
Got to stop it. Only two ways.
1. Come to points which are not events for then they would have parts.
Something different with mysterious external relations – basis of all definite geometrical reason
Metaphysical fairy-tale: has no basis –
What you really start with is the room – then you split it up – therefore there is
2. The room during a certain time is an emergent entity – there it is –
Its actuality is inherent in its unity –
It is given to you as one thing –
The room is not the bare piece of space-time –
Is the concrete reality –
Its space-time character is mere roominess – abstraction

---

2. (Joining on at end of last lecture)
E_Q is not to be expressed merely in terms of past or future.
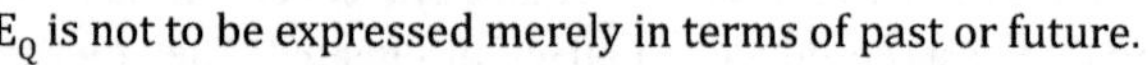
All the future is relevant to it and is in E_Q as an aspect, is "modally" in E_Q. not simpliciter.
If you ask what is E_Q? It has emerged as an end in itself & is therefore of the nature of value.

---

3. The "Actual Potentiality" – is what comes from the Past. Is "accepted" or in Kant's language, "given"–
The "Actual Effectiveness" is how what is beyond Q is modally in Q or mirrored there, seeing some actual potentiality from Q

12/4 – 2 ,
"There are two or three points we come to here but
there is this point" –

4. Retention. Illustrated by tune. "Made a great fuss about this in previous lectures" –

1. This lowercase q is as Hocking wrote it down, but elsewhere (in both Bell and Hocking) the symbol being used is an uppercase Q.

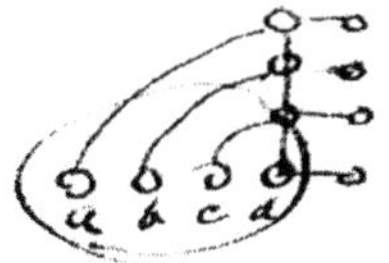

a must be in d^1, but modally not *simpliciter.*
Then there has got to be a going on (beyond d) a retention.
"Its no use being afraid of being silly –
You only end by being pompous – you know" –
"Impelled by a certain Hegelian dialectic to go beyond that" (point at d)

5. We've got to dig out the enduring object from the event. Tunnel – flux of time – instantaneous cut.

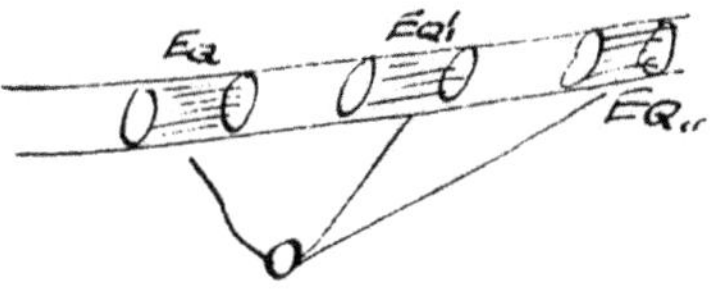

Each represents 3-dimensional spaces – Something has been passed on which is identical and equally present in any ~~section~~ segment. We will call this its historical route.

The object pervades its route.

"Its own subjectiveness"

Anything which is obviously for the unity itself is what the unity is adding to the total fact there

This pattern & the effectiveness in which it issues must issue in the same pattern running through the whole. Reproduction of the pattern of tone, i.e., of itself, throughout. Self-affirmation of an identity, enriched with its past, as preserving the pattern of tone. Problem of the sock's identity after being darned & darned, our bodies change every seven years, of the organism. Not essentially different from problem of endurance of electron – (no vitalism). In all realization we have the self-affirmation of a pattern for its own sake.

1. The diagram as sketched by Hocking (a copy once again of what Whitehead has done on the blackboard), and to which this clause refers, is not in Bell. This must be the first time Hocking includes a diagram that Bell does not.

Lecture 31

Thursday, 11 December 1924[1]

Bell's notes

|59|
Whitehead: 10. December. 1924. (Had been sick in interval).[2]

We've come to a stage in which we have to stop and look around. Point where the attempt to get beyond Cosmogony and found it in general metaphysical outlook, had better be stopped, and come back to Cosmogony in narrower sense (the order of Nature). Does what we have been doing lead to view of order of Nature such as we really find around us. If a philosopher says anything really new about it he must be wrong. All he has to do is to find the most general. – Of course "new" in sense of assumptions we haven't been aware of. The terrific permanences of: Matter in all its shapes and forms – there and indestructible. Does nothing but move about Idea of Evolution, Change, Progress in some form or other. Question whether we've arrived at a Cosmogony which gives us this sort of thing. Going to last day's diagram.[3]

A certain unity of pattern in the "Actual Potentiality" and a certain unity of pattern in the ~~Actual Emergent [?]~~ Actual Effectiveness and a certain unity of pattern in "the Ascent of Subjectivity" – The Achievement "for itself" "there".

This pattern "E" = The Emergent Entity. The Complete individual has individuality marked out in a two-fold way (both necessarily). "That pattern" is merely a concept. – Might have another just like. (two eggs etc.). You distinguish however by different historical routes. The emergent pattern marks out a route as having in itself a spring of Unity. – Giving it Substantiality and individual Independence. The one pattern is all you can do by way of the Concept. Mere route in itself has no survival unity (of pattern) in it. Quite obvious that Pattern is higher Abstraction than actual realisation within any portion of it. It's obviously inherent in this idea that the actual potentiality here (part of route) has two sources.

Essential to idea of Enduring Object: E., that pattern be such as to produce an Actuality which will lead to reemergence of pattern – That pattern secures its own survival. – The Value being realised is such as,

1. The Radcliffe version of a portion of this lecture, from notes taken by L. R. Heath on what was likely 11 December, begins on |116| of her notes, page 460.
2. Along with the parenthetical comment to the right of Bell's header for this set of notes, Bell also records in his diary for Saturday, 6 December: 'Whitehead sick'. So, this lecture is actually a full week since lecture 30, and likely on 11 December (as Hocking states in his notes) and not 10 December as Bell mistakenly records.
3. The 'last day', in the context of Whitehead having missed classes due to illness, must refer back to lecture 30.

when passes into actual effectiveness, to reproduce its own pattern along the continuous route.

The environment gives its contribution – at least as important as contribution of other elements. Big question (Whitehead put on to it by L. J. Henderson's *Fitness of the Environment*). (Later his *Order of Nature*)[1] Each organism to some extent creates its own environment – Previous life of organism has stamped itself somehow on whole present environment. How to account for enormous stability of Electrons etc. and for evolution of things in this order of Nature? Whitehead says by survival of fittest. You go back to where within order of Nature you had fitful stretches X X[2] of emergent enduring entities. We've finally got an order of Nature dominated by those entities which created an environment favorable not only to themselves but to production of other entities of same kind. An Environment i.e. created by a Society[3] of entities producing an Environment favourable to existence of them all. Environment has evolved as an Environment which secures selection of definite types. The plasticity of the Environment is the Key to the Evolutionary problem and that of Permanence.

Whole theory of Evolution considered largely from other point of view – as if Environment given and Organism had to adapt itself to this. This only true for isolated or feeble groups of Entities. This not a paradox but most matter of course thing. We're not here as examples of adaptation of early settlers to their environment in America, but vice versa. Why whole American race so imbued with idea of progress etc. – Race had to adapt environment to it. etc.

But lower – think of Brazilian Forest. – No individual tree could have grown without its environment. They produced an environment favorable to each other. That's really the Key! Those animals die out which passively fit themselves to environment. The great lizards e.g. [It's the restless ones that survive].

Physical Science gives us an environment not unlike outside view of great Societies. Electrons, Protons etc. – two or three kinds – large numbers – There builds up higher organisms from these.

Here you have cases of Emergent Entities which fit environment to each other. Electrons, Protons, etc. is just as much social beings as rest of us are. Then that other point:– The Past – its aspect living modally in the Present. So the very laws of Nature become adapted to Realm of possibility enriched by what has been.

---

1. Lawrence Joseph Henderson (1878–1942) received both his undergraduate and his medical education at Harvard, and following a few years of research at Strasbourg joined the Harvard faculty in biological chemistry, but with an interest in philosophical and sociological issues. His first books were *The Fitness of the Environment* (Macmillan, 1913) and *The Order of Nature* (Harvard University Press, 1917).
2. Presumably, the cross in the left margin marks a passage of interest to Bell, but once again we find him using these two 'X X' marked out in association with the phrase above. Was he marking a concept on which he wanted to invest more time, later?
3. This is the first time (of only a few times) Whitehead uses this notion of a 'society' of entities in these lectures, and which becomes so important in his *Process and Reality* only two years later.

Realized Values becoming ingredient in higher organisms. So that we have explanation of enormous permanence of Electrons, Protons, etc. – They've created a condition in which they are permanent. This may work itself out. – Equally possible that as Nature arose, it may decay. – Higher organisms may destroy lower or mere mass of them crush their organism. |60|

For Philosophy of Evolution. – You can't start it at some arbitrary starting point. Getting away from idea of starting from arbitrarily given stuff [It would have to be at definite point of time].

You must get behind even to an Evolution of the very order of Nature. Spatio-temporal system and the Sensibles – the Existents – which form the aboriginal "ground". – They although static ground of being in this order of Evolution – themselves are aspect for us of actual product of other orders of Evolution. Whitehead here saying something quite commonplace. Whitehead not talking of a time before there was time. But: process is not, even in order of Nature, continued in one linear process of time. – An Expansion rather than a Progression. The Cosmic Order presupposes an Expansion which stands behind (beyond) it. What's "set" for Nature is these "preválent modes of realisation". There are various stages. The Cosmological Stage presupposes other stages which provide very ground of Becoming, for Cosmological stage. What we have to take as given.

At each moment the *Elan vital* or the like realising itself etc. etc. – transplanting. – Very ground of Being rests on other expansive realisations which provide ---

which lacks that Value which it can gain in the cosmological order. "Its own access of cosmological realisation". Gone beyond, here, what we can actually observe with!! – but not beyond what has been found necessarily (East or West) – to give any rational account of Cosmological order. – Take e.g. the most straight-forward man H. Spencer[1] – starts with bits of matter – just there. In the East emergence of individuality in world much more like the one Whitehead has sketched out, giving it a Western tinge by making Spencerian Matter the result and by endowing the Emergent with Value for itself. Any endeavor to realize how process of things goes on either implies some blank assertion of something being there, or throughgoing Evolution. (Plato's doctrine of Recollection, Wordsworth's ode[2] etc. – however in West.)

What has grown in time is the importance of what's there. In one sense it's always all there. It grows and fades in importance in its own time. But. Growth of own importance within itself. As to its importance as adding its quota to realisation ~~of~~ in general.

1. Herbert Spencer (1820–1903) in his 10-volume *System of Synthetic Philosophy* (Williams & Norgate, 1862–98) establishes his 'principles of ethics' upon principles of biology, psychology and sociology.
2. Although Bell has recorded 'Woodsworth' surely 'Wordsworth' is intended, a poet Whitehead refers to repeatedly. In *Science and the Modern World* (p. 77) the ode Whitehead mentions is Wordsworth's 'Excursion', and perhaps that is the one he has in mind here.

Hocking's notes

Dec. 11. 1924

Stop effort to frame general metaphysical outlook & come back to the order of nature.

Have we been describing that: the "terrific permanences" – matter, etc. Evolution, change, progress.

1. What have we got at?
The idea of an enduring entity: an identity of pattern throughout an historic route.

Individuality marked in two ways neither sufficient. (a) Pattern. But two eggs may be quite like (b) Historical route distinct from any other route. Which alone has "no survival unity".

The pattern should be able to secure its own survival.

But the environment gives its contribution also, at least as important as the other.

The permanences a big question, Whitehead was put on some years ago by reading Henderson *Fitness of Environment*.[1]

2. Each organism helps to create its own environment
How account for the enormous stability of electrons or protons? Survival of fittest.

We have an order of nature dominated by entities which created an environment favorable to themselves and to the production of others like themselves.

The environment has evolved as one which preserves the existence of these types.

The plasticity of the environment the very key to the evolutionary problem & that of permanences. Environment not given as usually supposed. "The Pilgrim Fathers did not sit down and say, We must adapt ourselves to the environment." Trees cannot grow until there is a forest. They produce an environment favorable to each other. The animals which have passively fitted themselves to the environment such as marsh-animals died off at once, when the environment changed.

Electrons & protons just as much social beings as we are. They fit the environment for each other. And the very laws of nature become adapted to the entities. They have created the condition in which they (electrons & protons) have remained permanent.

12/11 – 2
3. Can't stop evolution at some 'final' point.
Nor start from an aboriginally given stuff. Nor from some point of time – very unlikely.

1. See footnote 1, page 135.

Not talking of a time before there was time – which is nonsense.

But not linear process – an expansion or[1] progression. an expansion which "stands behind it"

Stages of realization which presuppose others.

Cosmological stage – space-time.

Presupposes a deeper ground of becoming.

Which "lacks the value which can be gained in the cosmological order"

Have not gone behind what both East & West have found necessary to give a rational account of becoming. "most straightforward thinker" Herbert Spencer – pieces of matter.

East, less devoted to matter, more like

What I have presented here, but with a Western system & emergent value

What has grown is its importance: in a way everything was always there. And perhaps later fades in importance. Growth is of its own importance within itself.

1. Hocking has left a puzzling mark here. At this point of the lecture Bell has recorded: 'An Expansion rather than a Progression'. What Hocking uses looks like 'a' or an 'o' and then (almost) scratched out, but it is not clear what he intended.

Lecture 32

Saturday, 13 December 1924[1]

Bell's notes

|61|
Whitehead: 12. December. 1924[2]:-

Whitehead's "Philosophy of Evolution" – "i.e. of Change" [sic!][3] Review of some of ideas again – to show this as something working up to from Beginning. Every philosophy has an imaginative Background. Whitehead's is Cosmology and Order of Nature – shows in Title of Course.[4] Trying to embed order of Nature in a more general metaphysical background. Two ways of looking at Nature which give you two rather extreme lines (there are hybrids). Either your attention caught by the enormous permanences and make this the keynote of your thought – the mountains, atoms, electrons etc. In civilized era you note (Greeks) the geometrical, arithmetical, logical form. – Seem to get to something out of Time and Space – "Eternal". You then connect that imaginative background with the endurances in this Time and thus spatio-temporal flux comes to be a sort of condescension of Entities themselves of out Space and Time, and so at an instant you catch the true reality – apart from its enduringness, and thus your Metaphysics is one of Eternal things under Veil of temporal appearance. – The Metaphysics of the Geometer, the Logician, of the Arithmetician [the pure Mathematician]. Plato's inscription.[5] Spinoza. ∧and has haunted Science with scientific conclusion of Matter (not Aristotelian ὕλη) Having its being not only through Endurance but *schon*[6] at an Instant. ∧ Start with that emphasis and you get that type of Metaphysics.

The Logic of Classification is a classification of Stuff. Very important, but has warped Metaphysics. Important that metaphysics should have fresh impetuses. Trouble for this Metaphysics is time. Tendency always for

1. The Radcliffe version of this lecture, from notes taken by L. R. Heath on what was likely 13 December, begins on |117| of her notes, page 460. There are no extant notes for this date from Hocking, but they do pick up again on Tuesday, 16 December, although those seem to have been taken on his behalf by someone else.
2. In his diary entry for Saturday, 13 December, Bell notes: 'Was very sleepy in Whitehead's lecture'. Therefore, this is the date of lecture 32, and not 12 December as Bell mistakenly records in his heading. (If he noticed the mistaken date of the previous lecture, he might well have added the usual two days and made this second mistake.) There are no Hocking notes from this lecture with which we could compare. Both pick up again on Tuesday, 16 December.
3. This must be Bell's editorial insertion.
4. The title for the 1924–5 academic year (only) was 'The Philosophical Presuppositions of Science'.
5. In this oft-repeated story, above the entrance to Plato's Academy was the inscription ΑΓΕΩΜΕΤΡΗΤΟΣ ΜΗΔΕΙΕ ΕΙΣΙΤΩ, or, roughly, 'Let no one who cannot think geometrically enter here'.
6. German term for 'already' or 'really'.

that sort of view to ~~have~~ find Time and immediate things of experience as "Appearance".

Eternity rather runs away with you.

Whitehead makes a start (not the first) from very idea of Change. That ultimate Entity is in its own Essence is Transition, both internally and externally. (Externally:– always going to something beyond itself.)[1] So not independent of environment. All Concrete Things taken out of Environment are Abstraction – They're only concrete in environment. The other man can be as pluralistic as he likes. The man starting from Change and trying to be simply pluralistic must be in a muddle. The order of Nature is for us the unity of such transitional Ultimates. So we have exactly opposite difficulty – rather bothered with the Permanence.

(The real devil in Metaphysics is trying to slither away from your difficulties. Main thing is to acknowledge where you are). The enduring entity emerges and have to have theory of Modes – realizes all there is but under limitations. Whitehead thinks this can put in general form what is immediate experience of Man on the street. You also have to have Theory of "Actuality" – Aspect of the achieved in the Achieving (Past is in the Present) – "Theory of actual Potentiality". What is now Becoming in Becoming alters the Actual Potentiality and becomes very basis for passing on. This doesn't help question of endurance of course! But explains it thus:– What Becomes is something self-affirming for its own sake – "Value"; and self-propagating by impressing itself on a favorable Environment. So the order of Nature starts and has to take as its basis certain physical facts, obviously. (To-day, general Idea of General Electro-magnetic field. Three dimensions in Space and one in Time.) Electro-magnetic field is at first mere flashes in the pan, achieving no endurance. Then gradually self-propagating Entities emerge – Those with basis for Endurance. So our order of Nature gradually emerges – from actuality. First realising mere flashing Value to an order growing in its achievement of Value by its achievement of Endurance, and growing subtlety of Value achievable there through the Attained being the basis for --- ⟨blank⟩

But go back to Basis. – The Entities gradually evolve more favorable Environments – It's the Endurance itself which is evolved. – The slow steady evolution of "Environment" is what you have. It was a very advanced type of "Evolution" which first struck the world 60 years ago.[2] Environment thought of as fixed. – Now Environment is both given and plastic – in different aspects. In so far as Environment plastic it's the Evolution of Species rather than of individuals to be thought of. For the Species and Association of these produce those alterations of Environment favorable etc. to Individuals.

1. Closing parenthesis provided here.
2. Sixty years ago would have been 1864 and so this is, presumably, a reference to Darwin's *On the Origin of Species* (1859), which was first being discussed in the 1860s.

Not speaking here merely of organic things. Why this enormous number of Electrons? – This evolution of Similar Things! Must account for it by comparatively[1] few associations of few fundamental things.

Prout's Hypothesis re electrons and protons.[2]

An association of two types of things in enormous quantities. The real difficulty is, not to get Freedom, but to get Things that are alike. Wherever Environment is important Freedom sinks into background. Evolution for countless ages stood still until something happened to produce an environment. At basis of Physics (as seen today) two types of Entities together. Evolve into higher types you get your 92 Elements. – But not only 92

|62|

types of men! Wherever you get the Freedom emphasized you have stability of Environment and you get Freedom. Talking just as much of Molecules as of Men. – Wherever you get Freedom it is that you can presuppose Environment.

The Environmental type of Evolution is democratic. Where it lays emphasis on individual There is both. Evolution always has its competitive side – As between individuals and between species. But also has its side of mutual aid. Competitive side enormously overemphasized in 19th century. "Survival of Fittest"[3] (The Sparrow on London streets) – wrong ideas. Prince Kropotkin's book *Mutual Aid*[4] – Very general ideas – not can be brought down to a view of society. But can be to a view of Morals. (But sheer mutual ⟨blank⟩

Stable Evolution (Evolution towards Stability) is on the whole an Evolution in which the modification of the Environment is the important thing. An Evolution toward something new is on the whole a rapid thing and is devoted to Individuals. Your democratic Evolution is very slow, on the contrary. (America is example of slow evolution. Europe trying to do things quickly. America satisfied with its standard – wants its 100% Americans, – not trying for new types. Novel types of humans may be dangerous. The Enormous prescience of racial instinct – So a certain damping down of novel species of humanity – but enormous effort at change of Environment. Of course you have your "Experts" [Aristocrats!]. America will get there quicker etc.)

Plasticity of Environment obviously has its limits. Evolutionary process as far as we can observe it – takes place at an Edge between two Environments – between two fair stabilities and just enough Instability

1. The contraction: 'comp^{ly}' could also be 'completely'.
2. In 1815–16 William Prout (1785–50) published papers in which he proposed that hydrogen was the only type of fundamental element, and that all others were aggregates of this one type of fundamental entity (because all other known elements appear to be multiples of hydrogen's atomic weight).
3. Often associated with the Darwinian theory of evolution, this phrase actually originated with Herbert Spencer in his *Principles of Biology* (William & Norgate, 1864). Darwin himself then used the phrase in the fifth edition of *On the Origin of Species* (John Murray, 1869).
4. Prince Pyotr Alexeyevich Kropotkin (1842–1921) published *Mutual Aid: A Factor in Evolution* (W. Heinemann) in 1902 while he was in exile in England.

there. But the Sea and its strange attraction. – What hasn't the print of man upon it – Especially in an old land like England where everything else has stamp of man on it. You see something you might have seen 100,000 years ago.

Whitehead has assimilated in every way physical and organic evolution:– It's all an enduring and stability attaining organism which is being evolved. That is where this way of thought has this advantage over the other way (the one coming down on the permanences, the stuff of the scientific matter). Has one got rid of the given stuff? Obviously one hasn't. One's got, after all, this order of Nature has to get going, with its Laws of Nature – its three dimensional Space and its Time, and further:– underlying all is the Evolution of the realm of the "possible" "Existences". The Order of Nature presupposes a realisation of a realm of Existences – Eternal. This community of a spatio-temporal relationship of --- ⟨blank⟩

Aren't these the old matters of fact simply recurring in another shape. How can we get them in our Evolutionary Philosophy?

Lecture 33

Tuesday, 16 December 1924[1]

Bell's notes

|continuing on 62|
Whitehead: 16th December 1924[2]:-

"Cosmological orders." Whitehead contrasting two types of Philosophies throughout – (a) what starts with static Entities – (b) basing itself on change. (Change, Evolution and Emergence) Difficulty:– Looking out on facts of case. – Underlying change again static entities. – 3-dimensional space – and however as a Geometer you point out other "geometries" they depend on a definite Structure ∧of things∧. Even Eddington[3] etc. – start by putting down Equations. And equations must be in something.

From point of view of Philosophy doesn't make ~~what~~ difference what the Equations are – But something definite which might be otherwise but are what they are. This quite opposed to Evolutionary view – That you have transitional modes; and all matters of fact based on the one matter of fact. The only Eternal matter of fact is definiteness. The realization is realization of finites which are definite but in their very nature transitional. Wherever you get something definite which might be otherwise you're in a transitional --- ? ? ⟨blank⟩

Philosophy's too apt to be like the World-Elephant-Tortoise philosophy. Evolutionary Philosophy mustn't stop with Tortoise just because brain doesn't want to go further.

Point of evolutionary philosophy – Substance of Universe can't be expressed in static guise as mere "being" – but in energy pulling together the Eternal existents (themselves separated) – in modal forms.

The emergence is the fact of value as --- fusion ⟨blank⟩

The self-affirmation is nothing else than the energy of realisation within that Mode. Each mode is a fusion of all that is. But each part has ingression into it not *simpliciter* but modally – Particular Mode is selected; so generality of Existences are merely fused in that mode under aspect of envisagement barely, as realised in all experience.

Good Summary – but too fast.[4]

---

1. The Radcliffe version of this lecture, from notes taken by L. R. Heath on what was likely 16 December, begins on |121| of her notes, page 462.
2. The previous two lectures had mistaken dates, whereas the date given for this lecture in both Bell's and Hocking's sets of notes (and Bell's diary) are all in accord with the calendar.
3. Arthur Eddington (1882–1944) was awarded a scholarship in mathematics at Trinity College, Cambridge, and was taught by Whitehead. Like Whitehead, Eddington worked on the mathematical theory of relativity, and retained an interest in the interplay between the mathematical sciences and philosophy, but had also shifted into astronomy, the work for which he is best known.
4. This comment is written by Bell up the left margin, alongside the previous paragraph.

|63|

But selected Existences have major modes in fusion – issuing in formation of Value in that particular limitation. There's this particular limitation in this particularity of being realised. But no isolation of Modes. – Each Mode is itself transitional and goes over into bigger --- ⟨blank⟩

There's a community of Modes – a particular realised Community – As outcome of relational essences of objects as they're sharing in the immediacy of the matter of fact realisation. Transition of Modes is Transition of ? ? ⟨blank⟩

Wants to bring out that there's a definite real Space-Time – It's not merely a type of Community – but a definite individual Space-Time. – Not merely abstract field of potentiality but is the field of realisation. Weak point here – Idea needed pulling together. This brings back to a lecture of weeks ago re occasional and praeteroccasional abstraction.[1] – Idea of abstraction. Definite Beyondness of a definite Occasion is obviously the individual Space-Time that's inherent in the transitions of this realised occasion. Then abstraction not from this particular example of Relationship but from particular occasion within the Community. (of Modes).

There we come to idea of "Order of Nature" that's haunted all Western Thought – not merely as a type of order but as one particular one. We find that Order of Nature is in a sense a particular entity; and defines particular facts. Particular alternative ∧[perhaps]∧ spatio-temporal geometries.

Particularity and Selectiveness means that Order of Nature must be looked on as itself an emergent mode – having the processional, transitional character which relates it to wider community of modes, etc. The particularity of Order of Nature is very remarkable. Particularities of Space and Time already treated. Also in physical laws. We never get at law other than blank assumption.

Then: particular Species:– Electrons, Protons, Tendencies to ⟨blank⟩ But if general view Whitehead has put before us is correct: we ought be unable to isolate it as a self-sufficient entity [X?] – In same sense ought to issue from something wider than itself.

Much already said is Confused because unclear how far talking of physical field and how far about something wider. Point to be brought out:– Is not that impossibility of isolating Order of Nature is exactly the difficulty of all Science and Philosophy. Look up beginning of *Matière et Mémoire*.[2]

Order of Nature and then idea of Spirit as something going beyond this. Whitehead would call himself a "Realist" – but --- ! ⟨blank⟩ Realism run mad is trying to confine everything within the Order of Nature. – "Dear

1. Occasional and praeteroccasional abstraction was first brought up in lectures 12 and 13.
2. Bell's writing is a bit garbled, here, but seems to be this title, published by H. Bergson in 1896. In the first line of his Introduction for the English translation (Macmillan, 1911) Bergson says: 'This book affirms the reality of spirit and the reality of matter, and tries to determine the relation of the one to the other by the study of a definite example, that of memory'.

old Alexander!" over tries that perhaps. Tries terribly to get out of it – this imposes the difficult Space-Time chapters on him.

Whitehead says "The number 3 can't be quite so eternal as all that!" In complete occasion of our Knowledge there's something which refuses to be put under Causal law. Chapter on "Time" in[1] ⟨blank⟩

Time requires measure is rather superficial and attaches to what you observe in Nature (observed as transitional). Could imagine a person whose Specious present included from Norman Conquest to to-day. Religion has a being for whom whole time order of universe is one specious present.

There's a distinction between Transition of our Knowledge and the transition of the *Cognita*. Transition of our Knowledge is a transition within a cosmological order of wider extent and character of spatio-temporal cosmological order (of Nature etc.)

Those data of Rationality which insistently refuse to put selves under spatio-temporal guise (Arithmetic; Logic; Minds themselves) are really ⟨?⟩[2] has its root in a wider order; and has simply its aspect in this particular cosmological mode.

It isn't the case that all that is for knowledge can be tucked away with Order of Nature. Going beyond Order of Nature you're palpably not talking beyond your knowledge. Envisagement of Nature ⟨blank⟩

Some hope of seeing through veil of mine[3] somehow There are different grades of Realisation – The sheer abstract totality of Entities is one of the utmost abstraction.

|64|

(sleep)[4]

Potentialities for Order of Nature have already on them the shadows ⟨blank⟩

There are grades of particularity which have no significance for Order of Nature [Aristotle's τύχη][5]

Those who have realised relation to this part⟨?⟩[6] of Order of Nature. Thus the particular energy of fusion in the realisation of particular mode of Emergent Value, is a particular mode within Order of Nature itself a particular mode of the "Eternal" --- ? ? ? ⟨blank⟩

In a sense, Time [or "Process"] is deeper than Space-Time. Order of Nature is Spatialisation of process as retention and the temporalisation of passage in it.

1. While there are many possibilities (given what Bell left blank) but having just referred to Alexander, he may be referring to chapter II of *Space, Time and Diety*.
2. We have not been able to decipher this key term.
3. This term is uncertain, and Bell seems to be missing a lot amidst these sentences.
4. Bell actually leaves this parenthetical explanation in the upper left margin of this page.
5. A large question mark is centred above this bracketed injection. The Greek means 'luck'.
6. Clearly, this is an abbreviation based on 'part' but the superscripted ending cannot be made out. Neither 'particular' nor 'particularity' (both used nearby) quite works.

How that order is rooted in the [more Eternal ? ? ?] Envisagement of Spatio-temporal Experience from standpoint of wider Experiences which stretch beyond that order. Idea of "The Emergent" – which endures – as having in its origin a footing beyond that mere Order of Nature.

This of course brings up ⟨blank⟩

Endurance of individual even of highest and most important phase. You're never going to come to any consideration of it unless you have a philosophy. – Equally worried centuries ago as to where Soul came from; – not only where it goes to. Question of Emergent Origin of it is absolutely bound up with question of its Emergent fate. Whole of that question (can't be settled by Metaphysics) – depends upon having a Metaphysics, which shows you the Entity having an issue from something beyond Nature; below itself. Then question whether what emerges in Nature is somehow or other carries on to what is beyond Nature.

Whitehead deprecates ⟨blank⟩

[1]Metaphysics going about in a funk to-day. Burnt its fingers in past in two or three ways. But has its quota to contribute. B. Russell's disparagement of Metaphysics is not shared by Whitehead. Bearing that these general considerations have on questions of tremendous importance – but they can't settle anything – But can give elements that must be taken into account. Here e.g. Order of Nature set in a wider order and the Emergents have their foothold there etc. – Doesn't settle anything; but gives you a viewpoint that has to be taken into account when you are settling questions.

Hocking's notes

Dec. 16 – Prof. Whitehead's Lecture[2]

Summary of preceding lectures: the nature of passage, etc.

Cosmological Orders:

Every matter of fact is definite, and we must ask why it is just this and not that. Definiteness (the fact that every matter of fact is definite) is the only eternal matter of fact: realization is of finites, which are in their nature definite but still transitional.

Space-time is not a type of community, but it is a definite, particular community – (Why should space have three dimensions?) (It is too ~~m~~ much of a strain on one's credulity to believe in the necessity of the number three.) Explains his meaning by referring back to a lecture on "occasional

1. It is unclear why, but Bell keeps this last paragraph of notes for this lecture on the right half of the page.
2. These notes are included in Hocking's folder of lecture notes, but are in typescript – as are the notes for lectures 14 and 18 – and seem to be taken down by 'R. M. E.', who leaves the apologetic note at the end. Likely this is once again Mr Eaton, as clarified in note 1, page 68, lecture 18.

and praeter-occasional abstraction" – The order of nature is an abstraction, but an abstraction which gives a particular order – this is a particular space-time –

The order of nature is itself an emergent mode: it (as an order) has the processional, transitional character – it is related to a wider community of modes –

The order of nature ought essentially to exhibit itself as a (part) of a wider community – "A realism run mad is one that tries to confine everything within the order of nature." Whitehead suspects that this is what Alexander is trying to do

In the complete occasion of our knowledge, there is something that will not be put under the spatio-temporal – There is a distinction between the transition of our knowledge and the transition of the *cognita* – The transition of knowledge takes place within a wider cosmological order than the order of nature – The data of rationality, e.g. arithmetic, logic, etc. are emergent entities that have their roots in a wider cosmos, and simply their aspects in the order of nature – When you talk beyond the order of nature, you are not talking beyond our knowledge – "The potentialities for the order of nature have already on them the shadow of actuality of another cosmos."

Process is deeper than space-time – space-time is a particular mode of process: it is process under the form of spatialized retention and temporal passage – The cognitive process is a reinvisagment, from the stand point of a wider experience, of the spatial temporal process –

Professor Hocking: I regret the meagerness of these notes, but I find it impossible to listen and take a great many notes – The spirit of Whitehead is "too wide a cosmos" to be got into notes.

R.M.E.

Lecture 34

Thursday, 18 December 1924[1]

Bell's notes

|64|
Whitehead: 18th December 1924[2]:-

Certain points of view Whitehead has been repeating are not mere personal idiosyncrasies. Quotation from Bacon – from his Natural History.[3] Beginning of "Century IX". – "All bodies, though have no sense, have perception…" etc.

[Leibniz] – "a perception always precedes operation" – "else all bodies would be alike." This is Whitehead's "taking into account of the individual essences of things." "And sometimes in some bodies this perception much finer than sense" etc.

(The Weather glass) – It might be Bergson speaking.

In first place Bacon's ideas are purely qualitative. He talks always so and therefore he to a large extent missed the key to Modern Science which is based through and through on quantitative considerations. – Definite measurements.

Cf. Bacon with Newton re action at a distance, etc. Newton comes down on idea of Mass as common measurable element in different bodies. Lays down ideas re measurement of these, and law of gravitation (quantitatively formulated) – So you get a firm quantitative whole. But question of "at a distance", here. Whitehead's essential stretching out to a Beyondness. But also essential idea of a Pattern. An enduring object is essentially a pattern in space-time. Endurance during Transition is as essential as idea of spatial form. The fundamental fact is the functioning of an organism as purely spatial configuration of a structure. Our conceptions were founded by physicists for whom that idea of the Spatial Configuration was for their purposes sufficient. It's only recently that lines of thought come in which have disturbed that idea.

(a) Rise in importance of Biological Sciences. – Where you have to conceive parts in relation to the whole.

1. The Radcliffe version of this lecture, from notes taken by L. R. Heath on 18 December, begins near the end of |123| of her notes, page 463.
2. Bell records in his diary for this date: 'Whitehead excellent, but I was too sleepy to appreciate fully'.
3. The full title was *SYLVA SYLVARUM, or, A Natural History, IN TEN CENTURIES: Whereunto is newly added The History Natural and Experimental of LIFE and DEATH, or of the Prolongation of LIFE.* It was published by W. Rawley soon after Bacon died, in 1626. The quotation appears to be taken from the opening sentences of the chapter title 'Century IX'.

Read Jack Haldane's *Daedalus.* (Son of the famous Biologist, etc.– Father said son's book full of bad physiology.)[1]

B. Russell's *Icarus* following! We are entering on a physiological-biological-age. Scientific ideas founded on physics of 17–18th century breaking down.

But (b) breaking down within limits of Physics itself. – The electron itself is an organism – functioning through Time and you mustn't "spatialize" it. Bergson's service. – Contra knocking the
|65|
"functioning" – the transition out of things. Whitehead objects to Bergson's Anti-intellectualism – as fallacy of misplaced Concreteness. Error only introduced if you attribute to your abstractions a greater Concreteness than they really have. Bergson hasn't patience to trace out exact locus of error? It's not use of intellect at all which is erroneous. Karin Stephen on Bergson *The Misuse of Intellect.*[2]

Now want to consider how, when we deal with ∧these∧ useful high abstractions, the view of "Pattern" is useful. – Has been in Physics from very beginning. Whitehead's theory looks like action at a distance. In line with Newton and from him down reaction against action at a distance. Newton says somewhere that no one with grain of Scientific instinct can dream of action at a distance – There must be transmission. Hence Ether introduced in modern times – to account for transmission of light. – Then of stresses (Clerk Maxwell etc.). Recently transmission of influence of an electron.

Thought of science not phrased in Space-Time but in Space and in Time. You find that in Newton. That there is Space which can be conceived without thinking of Time; and there is Time as measurable succession which can be dealt with without caring much for Space. Then bring the two together and have Structures behaving themselves in Space at different points in Time. This has been a useful "high abstraction" for modern Relativity – although timeless Space represents some relations which are there (in abstraction from other things) – is not unique – There are alternative and correlative structures ⟨?⟩[3] name of "Space". And are alternative rendering of transmission etc. in "Time". And meaning of either must be adjusted to that of other – hence Space-Time systems.

Take such a system. – How is the idea of Electron as complex organism or pattern – unit pattern impressed throughout concrete Space-Time – rendered in this abstraction of the timeless space and the correlative

1. Jack Haldane – John Burdon Sanderson Haldane (1892–1964) – was a physiologist and evolutionary biologist, and the son of a physiologist, John Scott Haldane (1860–1936). Whitehead knew the family through Lord Haldane, a philosopher and politician, and John Scott's brother. For the 1923 debate between Jack Haldane and Russell, see 'Daedalus and Icarus revisited' by Charels T. Rubin in *The New Atlantis*, no. 8, spring 2005, pp. 73–91.
2. Karin Stephen (1890–1953) was a British psychologist. She published *The Misuse of Mind: A Study of Bergson's Attack on Intellectualism* in 1922.
3. We cannot decipher this term.

spaceless time. The pattern is the modification which the given Electron exerts on the total Electro-magnetic field. And Electron is nothing but this modification. How Mathematics deals with this. Takes the Scalar potential:– ϕ

and the vector potential:– (a_x a_y a_z).

There VforV any point p at a given time t it will have both potentials.

The contributions at an instant which a given electron makes to the field

Total is found by adding all the contributions of all electrons. – Principle of quantitative Cumulation. What is the pattern ~~of~~ that this one Electron imposes? – That the way the magnitudes are distributed through Space

[the ϕ' ϕ'', a'_x a'_y a'_z, a''_x a''_y a''_z] etc.

The contribution given us traveling outward with a definite velocity. – That of Light in vacuo (c) 3×10^{10} cm/sec. [Every Physicist has this on his heart like Queen Mary Calais.]

If you knock out the transition [Whole aim in teaching Mathematics is to eradicate Complex of inferiority re Mathematics].

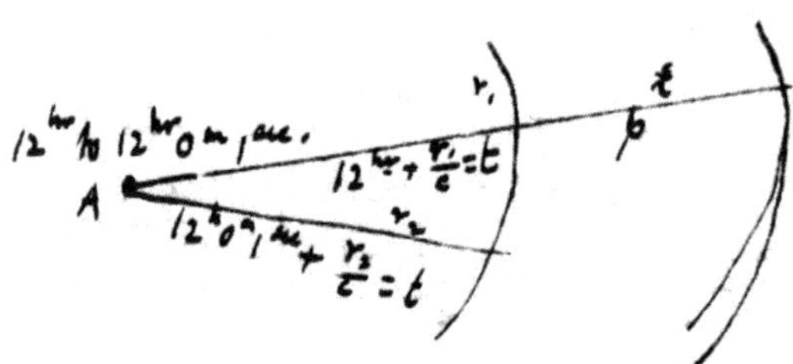

Rest of Space at that time is not receiving any contribution from that Electron during that period. Electrons not created simply for one second. If you think of A as statically always at that point. Then could knock out idea of Time because contribution never alters.

So idea of time drops out and you get Newton's Law of Gravitation.

Get complications if you get your Electron moving. The pattern comes in as a pattern of transmission – of the transmission of influence. There's a certain decay of intensity as you get further away. – Decay as inverse of distance.

There's the alternative point of view (Roger Cotes[1] – contemporary of Newton) – Says: drop question of how it happens but just say so it does, and measure it. If Cotes had lived now he'd be a Behaviorist. That's the way they Behave and that's the End of it. Foundation of thought for these is same as for Cotes. Newton said if Cotes had lived we should have known something.

There's a passionate repudiation among men of Science, of idea of Action at a distance. Continuity point of view is orthodox (Maxwell, Kelvin, Helmholtz etc.)

What is it that's philosophically really at the bottom of it. Went too far in

1. While Bell has 'Coates', this is surely Roger Cotes (1682–1716), the English mathematician who worked closely with Isaac Newton. The second edition of Newton's *Principia* of 1713 was edited by Cotes.

|66|
putting it into scientific Ether view. Consider pattern as a Whole – through Time and through Space. When you try to express pattern as a whole in terms of timeless Space and spaceless Time.

It's an obscure way of insisting upon it that Electron is an organism functioning. If you knock out idea of an Ether (Whitehead) you replace it by idea of an organism functioning. You've made an abstraction and an analysis and missed dominating fact of situation. Important fact is that you place problem of Electron on exactly same basis as that of living organism, on which rest of explaining the whole as an aggregate of parts you explain the part as an abstraction from the whole.

The concrete fact is the whole.

Another point is question of the Medium. An enduring entity is concrete only in its Environment and as soon as out of that you have nothing but an abstraction. Consider particular exemplification.

How abstract idea of Electron out of a medium becomes.

Hocking's notes

12/18 – 1924

1. Francis Bacon. *Natural History*. Cy 9.[1]

All bodies have perception. Perception precedeth operation

For else all bodies would be alike to each other. (i.e. there is "taking into account") (Individualities come into play)

"Sense is but a dull thing in comparison of it" "Sometimes at a distance"

Bacon vs. Newton. Quality v. quantity. Mass & measure.

2. The question of "at a distance"

An enduring object is a pattern in Space-Time.

Two ideas coming in.

The biological sciences: The organisms,

(Read Jack Haldane's little book *Daedalus* & B. Russell's *Icarus*.)

"We are entering on a physiological age" instead of physics –

Ideas are breaking down in physics – "Electron itself is an organism, functioning in time & you can't spatialize it."

Bergson's enormous service to modern thought – Bergson's error is in bringing an essential erroneousness into intellect. Its danger is that of misplaced in concreteness – Karin Stephen's book *Misuse of Intellect* very good.

Bergson may be silly ~~by~~ to tone down his anti intellect campaign – Whitehead wants to show that his theory in line with Newton's protest against action in distance though it appears to be running[2] that way.

1. See footnote 3, page 148.
2. While 'running' is plausible, in context, Hocking's handwriting is virtually unreadable.

3. How is this thought phrased? Not in Space-Time, but in Space & in Time. There are alternative abstractions equally right as timeless spaces.

But assume there is a timeless space & a time

Then how is this idea of an electron as an organism ~~of~~ VorV pattern rendered in this idea of a timeless space & a spaceless time. Say between 12 and 12h–0m–1".

The electron is nothing but the modification which it introduces into the electro-magnetic field.

Scalar quantity $\phi(a_x\ a_y\ a_z)$ & three vector quantities
potential potential

Its contribution is $\phi_1(a'_x\ a'_y\ a'_z)$

<u>12/18</u> – <u>2</u>

The electron continued –

Its effect on the field traveling outward with a velocity $c = 3 \times 10^{10}$ cm/sec.

The rest of space is not receiving any contributions from A as existing at that time.

If the electron never moves, you can ignore time, because the contribution never orders.[1] and you have Newton's law of gravitation in which the potential at any point is always the same.

If the electron moves you get complications.

A That is a detail –

But the pattern comes in as a pattern of <u>transmission</u>, with diminution of intensity with distance. (not inverse square)

(Roger Cotes a physical behaviorist, asserted action at distance.)

19th century passionately repudiated the view that you need to consider anything in between. can without[2]

Must take the <u>whole thing together</u> in space & time, an obscure way of insisting on the electron as an organism functioning. The electron is the functioning of the pattern.

Have to explain the part as an abstract from the whole, not vice versa

---

Another point, the nature of medium – The concrete entity cannot be abstracted from its environment. Consider the electron in its medium.

1. Hocking's handwriting clearly has 'orders', although 'alters' might make more sense.
2. It may have been intended that the terms 'can' and 'without' be added or fitted into what is above (although Hocking has no indication), thus giving 'need/can' and 'consider without anything in between'.

Lecture 35

Saturday, 20 December 1924[1]

Bell's notes

|continuing on 66|
Whitehead – 20. Dec. 1924[2]:–

Key idea to which Whitehead working up. – Idea that basic fundamental idea in Nature – Key to Nature is that of organism and not that of matter. Ultimate fact is an enduring organism (ultimate most concrete fact). That idea of Matter as of something *simpliciter* there – enduring etc. – is not the fundamental idea. Part of that ⟨blank?⟩[3] when you have idea of Matter as fundamental the functioning of it isn't so fundamental. Matter can find its realization at an instant, whereas Organism is essentially functioning. As At an instant the Organism is a very high abstraction – (Has simple relations perhaps, with other abstractions, etc: useful; but not very concrete.) Idea of Matter of course enormously valuable and valid.

This question of Organisms as enduring functioning is immensely important now from biological and psychological conceptions. Also: it is Key to interpretation of the new ideas which really have entered into Physics. True interpretation of modern ferment there is that idea of Matter, so brilliantly successful for past three centuries, now proving inadequate at the depth to which Physics has now dug.

No Philosophy is worth its salt which can't point out what is meant by Matter – just as no Philosophy is worth its salt which can't point out what is meant by an instant of Time.

But whole question is what is the ultimate idea. Enormous bearing on the perennial battle between Mechanism and Vitalism. Whole conglomeration – the body – to be looked on as simple aggregate of bits of matter? With Organism idea you've got a new situation entirely – simply another exhibition of organism you get in Molecule – and from there to Electron.

In so far as analyzable into parts these have to be interpreted in relation to whole. This Structure doesn't turn up first in human body, but: there's nothing else in nature.

Last time dealing with Electron:– in what sense an organism – It has Key of its pattern at a "there". Considering Electron simply as in

1. The Radcliffe version of this lecture, from notes taken by L. R. Heath on what was likely 6 January 1925, begins on |127| of her notes, page 465.
2. Bell records in his diary for this day: 'Excellent Lecture from Whitehead'. Bell's diary also confirms that this was the last lecture before Harvard's Christmas break.
3. Although the end of this line was left blank, it is not clear whether something is missing or whether he simply finishes this sentence on next line.

Space-Time. – Reducing Environment to systematic Environment of Space-Time. But the Environment is really one of other Electrons. Key positions which determine the Modality of the whole organism. (Conceive at first [diagram of points on board][1] as in a timeless Space.)[2] The Electron E we are studying doesn't remain what it would be in absence of this Environment. It will alter shape of Electron (instead of Sphere, it's a small Ellipsoid etc.) The way this then affects Electro-magnetic field etc. – its pattern of functioning is itself dependent on the particular Environment in which it finds itself. Thus if you tear it away from its Environment, it simply degenerates into a formula, which will tell you the type of pattern which will be the outcome of the Electron in any conceivable environment. It's the most abstract formula possible. – Simply the type of ~~outcome~~ ^pattern^ which would be the outcome of any Environment.

The concrete fact of the Electron is a certain individualization of the Energy of realization which issues in the realization, in varying environments, of a certain functioning pattern, which will adapt itself to the environment. But actual potentiality of E as here or there, enters into very nature of pattern produced by the electrons.

|67|

To get hold of the actual pattern you can't divorce it from its Environment.

Further, the Electron doesn't simply passively accept its Environment – it imposes itself upon it. – It affects the pattern of its Environment – its environmental electrons. Therefore (important as showing essentially organic character of Nature). You cannot determine Electron apart from consideration of the total to which it belongs – and vice versa. But this goes a long way to saying "Organism". Not quite got it, but are going that way.

How different is idea of what might call 18th century:– the billiard ball idea of an Atom. (of course not worrying much about atomic theory till Dalton – beginning of 19th century, but idea there only couldn't make use of it). Organism unintelligible here. The atoms affect one another's motion, but that's all. As soon as you want an Organism (a Totality to enter into very nature of Parts you can only lug in by scruff of neck some "Vital Principle" to put on top of your atoms). But this is most unsatisfying if simply brought in on top of purely mechanical, atomic view.

Whitehead simply illustrating by ideas of Electrons and lines of Electro-magnetic force etc. As matter of fact there is another point of view too! You may consider what is being realized intrinsically at any of the electronic points. Whitehead's answer:– Two things:– (a) "Actual potentiality" is Modal aspect of everything else which are neither simpliciter here or there; but have a mode here under aspect of being there. This is realized as a togetherness, having a definite Value and effectiveness. The two

1. Despite many diagrams sketched by both Bell and Hocking, this bracketed aside remains one of the few comments by Whitehead that he is working with diagrams on the blackboard. And while Bell seems to have been too sleepy to have recaptured any of Whitehead's diagrams on this day, Hocking does capture one or two 'diagrams of points on the board'.
2. Closing parenthesis provided.

ideas of Value and Effectiveness bring in a certain Bifurcation:– What is realized there is in the first place the Modal presence of objects "there" fused as together and as a Value. – What there is for realization there. When there's Cognition of it for its own sake is our immediate perception – apprehension – of what there is about us. I'm not realising simply Colour here or Colour there; but colour as being there, for me here. Modes set by Space-Time.

Also realization of the aspect of things as affecting Essentially transitional nature, though. – Determination, direction of transition, depends again on fusion of the aspects. But aspects as fused from point of view of Value.

P of v. Description from point of view of Effectiveness is what issues in physical description. [Whereas other are our ordinary apprehensions]. But this not hard and fast division – It depends on what particular effectiveness you are thinking of. Beyond mere physical field, however, most effective things you can get is the display. (E.g. [through arousing] detestation or attraction etc)[1]. When you come to ask for most ultimate simple elementary fact there is, Whitehead points to his "basic object", which is to be conceived as the emergence (enduring unity) of a structural plan of pattern which taxes its actual concrete realization.

More elaborate emergences then – Emergences of relatively isolated Organisms (Systems). But what do you mean by "Isolated"? Obviously simply important parts of it. But what do you mean by "Important"? – You can't get along without some quantitative method. On a qualif ⟨blank⟩

What is the quantitative attitude? It more the intensive ⟨unclear⟩

Therefore got to have ⟨blank⟩ Quantity of Intensive Civilization). Some quantitative basis [Slept].[2]

Apropos of Environment and Evolution. Progress of latter is essentially a problem of procuring a favorable Environment. At low stage of Organisation ? ? ⟨blank⟩

Another type of Environment:– in which each subordinates environment to purposes.[3]

Thus get emergent ~~2~~totality such that one entity is Emergent from Fusion of the Values – the aspects of its parts in itself.

Opportunity give for emergence of a higher type of Being (of Energy for Realisation). The organism in itself environmental concept

Idea of an organism in which whole-parts are parts for the whole.

1. Closing parenthesis provided.
2. This comment helps to explain why the previous passages are so broken up. This is the second time Bell has acknowledged his lack of attention.
3. These last words are not clear.

Hocking's notes

12/19 – 24[1] Whitehead

1. The key-idea which is now emerging the key to Nature is that of organism & not of matter.

Matter is at an instant: the organism is essentially functioning.

2. The key to the new ideas that have entered into physics. Matter proving inadequate to the depth of new situation. No philosophy is worth its salt that is not capable of saying what is meant by matter & an instant of time.

3. That the parts have to be interpreted in terms of the whole Does not turn up first in case of body – but there is nothing else in nature.

4. Electron continued. The environment usually not merely space-time, but other electrons. E does not remain what it was apart from that environment.

Changes its "shape" Say from sphere to ellipsoid. So its pattern, or the way it functions, is affected by its environments. Tear it away from its environment & it degenerates into a formula most abstract. You cannot divorce the actual pattern from the environment.

Again, the electron imposes itself on the environment. The environment is not merely given.

Ergo, you cannot consider the nature of an electron without considering the whole, nor the whole without considering the electron.

Contrast Lucretius' Atoms. Each perfectly happy and on its own. Influencing each others' mode of motion – can't get an organism out of that. Have to get something extra when you want a totality to enter into the nature of the parts. The organic principle must not be brought in on top of mechanistic principle

12 – 19 – 2

5. What is intrinsically being realized at a standpoint? Say in a brain?

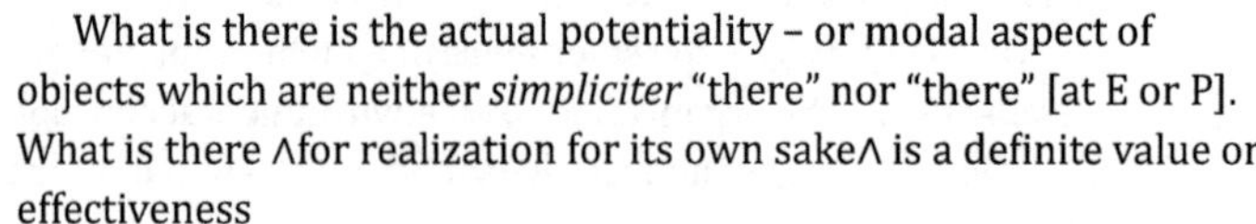

What is there is the actual potentiality – or modal aspect of objects which are neither *simpliciter* "there" nor "there" [at E or P]. What is there ∧for realization for its own sake∧ is a definite value or effectiveness

Cognition is our immediate perception of what there is about us. "Color for me here, as being there" – display of the togetherness of things.

1. Hocking records this lecture as if it were on 19 December, but clearly it is Saturday (not Friday) and both Bell's notes and his diary make it 20 December. Whitehead may have conducted a 'seminary' on the Friday evening, but Hocking's notes make it clear these were for Whitehead's lecture.

Value of things is permeated through & through with effectiveness. For us, one of the most effectiveness.

When you ask what is the simplest thing, I do not point to a little billiard ball atom, but to a basic object, an emergence, an enduring entity with a structural plan, essentially a pattern of things, – an organism spread through the whole of space & time, but with its structural key "there".

---

6. Relatively isolated systems or organisms – what do you mean by 'isolated'?

You have got to bring in quantity. A group is isolated when it is not being much affected by other things in respect to the properties that interest you

|12/19 – 3|

7. Favorable environment is one in which each subordinate organism has a function.

Unity emergent from the fusion of the values of the aspects of its own parts or ingredient cells.

Totality is an essential condition for the continuance of its ingredient parts.

This is an organism.

Lecture 36

Tuesday, 6 January 1925[1]

Hocking's notes

Whitehead, Jan. 6, 1925

Arrived at idea of organism –

1. The problem of vitalism is wrongly stated –

It assumes mechanism as prevailing throughout inorganic, & then asks whether vitalism supervenes, & with an organism of material –

If put in that way, the mechanists have it all the time.

2. What are the different attitudes of mechanism & vitalism?

Mechanism is arguing from parts to the whole.

The whole is an aggregate, the concrete fact is in the part.

Organism argues from whole to parts The part being what it is by the way it functions in the whole

3. The mechanist is materialist: starts with part as concrete & sufficient to itself

To understand say the human body, must divide it into its ultimate parts

Organism asserts you must bring the whole into your argument

4. Can you get a decision between them by observation, functioning of parts to preserve the character of the whole. Cells of body adapt themselves to preserve health of body, perform miracles to that end

But gyroscopic action also performs miracles

Gyroscopic compass will find its way back to polar axis when disturbed

So if you let mechanism in you have hard work to push it out.

5. Start with fact that we know we have an interest in our body as a whole

Mechanism leaves out my obvious interest in lunch[2]

"If you say your body is doing this & your Consciousness is looking on, that may be logical but doesn't strike me as sense"

---

1. From Bell's diary it is clear that he had spent the holiday with his family and friends in Nova Scotia. On 6 January he recorded his departure from Yarmouth on the steamer *Prince Arthur*, which reached Boston the next day. So, while there is no direct mention of Bell having missed Whitehead's lecture on Tuesday, 6 January, it is obvious why there are no Bell notes for this lecture. Neither does there seem to be a Radcliffe version of this lecture in notes taken by L. R. Heath. Fortunately, we have notes from Hocking, for what was Whitehead's first lecture after the holiday.
2. Whitehead gave these lectures from noon to 1 p.m.

Jan 6. – 2
6. On point of methodology, the vitalist has too easy a solution;– he has only to say "This is what the whole wants" – reintroduces final causes, whereas the whole of modern science has been built up on the Baconian basis –

Mechanists have here a strong point

If you are going to bring in organism you must do so, so as not to disturb the whole procedure of modern science.

7. Mechanism "presupposes a machinery of concepts" such as "matter" –

Matter is an entity which has simple location, you can say it is here in this place now, was there then, etc.

And this in independence of your view of space, whether absolute or relative.

It has qualities, by which we apprehend it.

There are relations between bits of it.

We are employing Aristotle categories –

Substance – quality – relation – place – date

Quantity – extensive – hence divisible

The ultimate simple becomes a density at an instant of time and a point of space

Can't stop at atom or electron, for these are wholes, and act as such.

Aristotle's categories do represent how we do think. Language conveys those thoughts.

If you are going to challenge those thoughts there is only one line to take. The ordinary man thinks in high abstraction, just such abstraction as is useful. Doesn't get down to what is directly before you in perception.

8. You must now make up your mind whether you are to proceed on an objectivist or subjectivist basis. The physicist would be inclined to take a subjectivist view, that what we have is a stress produced by a transcendent world beyond in your mental constitution.

Jan. 6. – 3
9. "Personally, I am going to proceed on a thoroughgoing objectivist basis" – not going to argue –

Hitherto objectivism has been half-hearted –

Knocking out secondary qualities, which are dealt with on a subjectivist basis.

Which gets you in a thorough muddle

All the things you perceive are to be included –

You've not got to start with matter, or stuff, which is simply there

You've got to have your whole world going

What is observed is obviously observed by the body. Put your finger in your eye – by playing the proper tricks on your body you can observe or obstruct the observation of almost anything. Awareness must be confined to awareness in relation to body. It is really the body as organism that you are aware of

---

If there is to be an organism, there must be some entities that are to be organized –

Can't organize in a vacuum. If we are going to endow that something with simple location, we get back to stuff.

Lecture 37

Thursday, 8 January 1925[1]

Bell's notes

|68|
Whitehead. – 8. Jan. 1925[2]

Mechanism–Vitalism last time. Suggestion that we discard idea of Material (Simple Location) at once. Clinging to idea of Organism – you mean:– What you're taking as your concrete unity of reality or experience is a whole which can be analyzed into Contents. – You may be losing reality in the Analysis, then, but won't be gaining it. Organism not essentially an organism of material, then. Latter presumes "simple location" as primary fact. Then "whole" is less concrete than its parts.

Question:– How are you ⟨to⟩ describe the Content which is discernible in the Whole. Whitehead takes line that in describing this content you at once have to come to notion of what Whitehead terms "Existences" – (a something in itself – Greenness e.g. – but not real). – They're "eternal" (not enduring.)[3] "Existences" not merely sense objects. Let's call them "eternal objects" then.[4]

Relation to general philosophical tradition in having recourse to such objects. Platonic ideas, of course, – William James's "Does Consciousness exist?"[5] – James plumps to a "logical realism" where he's dealing with Entities practically same as Whitehead dealing with here. Whitehead can't square his view quite with James', though. This world beyond what's Becoming in Nature is as necessary for James as for Whitehead.

The unit reality is what Whitehead calls an Event. That's the whole which is your ultimate concrete reality. The Eternal Object is not an ingredient in an event *simpliciter*. It is in Event always marking a connexion between two events. And the connexion between two Events is always that Eternal Object is there as ingredient in one Event clothed with the aspect

---

1. The Radcliffe version of this lecture, from notes taken by L. R. Heath on what was likely 10 January, begins on |129| of her notes, page 465.
2. This was Bell's first set of notes following the Christmas holiday, as he had missed the lecture on the preceding Tuesday (see footnote 1, page 158, lecture 36). Even so, on this day he had a dental appointment earlier in the morning, and admitted in his diary: 'Just made Whitehead's lecture in time'.
3. Closing parentheses were missing in both instances in this same line, and have been supplied at these locations.
4. Although Whitehead has been developing his notion of 'objects' (of various kinds) through previous lectures, 'eternal object' seems to be coined by Whitehead for the first time in this lecture, and then utilised through the next two lectures. It does not seem to be found in his earlier publications, but becomes widely used in his later work beginning with *Science and the Modern World*, published later in 1925.
5. James first published 'Does "Consciousness" Exist?' in the *Journal of Philosophy, Psychology, and Scientific Methods*, vol. 1, pp. 477–91, in 1904.

of another Event. Therefore any one event is a (doesn't like "Perception") (Apprehension, Experience all support cognition – which may or may not be there) "Prehension".[1] If cognition necessarily involved, Whitehead would be some sort of Absolute Idealist. An event is the Prehension of eternal objects with their aspects. Eternal Object by being prehended in one event carries with it Aspects derived from Events beyond. So each Event is everything there is, under a limitation.

James talks of the "Interactions" – But it also turns up in any account. Only question is How you phrase it and how you manage it. Take Leibniz's Monads (best example in modern thought of Organism as fundamental) – he tries to get away from intersections ("windowless") but they mirror.– etc. Hume's "Impressions" (must be something impressing). Kant puts emphasis on the Cognizing Organism. In trying to put this problem in what appears simplest light, one's not making the difficulty. It's there.

Whitehead tries to preserve a rather rigorous Realism. Puts emphasis of the function on the "Eternal Object" which is in one event with mode of being clothed with aspect of another Event. "Sensa" as examples. Only objection is that they've got tradition of being handled from point of view of simple location; and Whitehead doesn't want to get muddled through that idea. There are also, however, the apparent Shapes. What looks to me is the Shape as over there to me here. The Shape there isn't --- ?

(Broad gets in a muddle here). It's the aspect of Shape. Broad drops this. Berkeley does, too. Talks of thing *simpliciter*. But it's always the aspect of the thing you've got to deal with in Perception.

"Aspects" involve reciprocity of Events. System of Pattern of Aspects. Whitehead suggests that it's just this reciprocity of Aspects which is the (or: one) concrete fact on which our much more abstract idea of "Relation" is based.

Then there is the Pattern of Aspects which is prehended into Unity – which is the content of the Event.

Even this doesn't give you relation of Contrast. This Unit-reality is always essentially the emergence of Unit-Value for itself alone. The Contrast is an abstract characterization of what emerges beyond the bare content into the concrete Value. So that Relation lurks in the idea of Pattern, Value, etc. So Whitehead looks on the whole set of the Aristotelian categories as extraordinarily useful list of the Abstractions which are of importance for us. But highly abstract. Whitehead's list is nearer the Concrete.

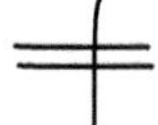

The Event essentially an Organism. Can analyze into three patterns –
The pattern of Aspects realized Value – "Display" (the intrinsic pattern)
Aspects of it for other Events

1. Whitehead is here introducing for the first time in these lectures a term, 'prehension', which will become important in his later work.

|69|
It is easier to explain Pattern in terms of Relation than vice versa. But that's just because of the habitual abstractions of our thinking.

We've left out here however just what is at basis of Materialistic idea – namely: "these enormous endurances".

[There is the "Beyondness" too – This is just another way of putting the "Aspects"]

No differentiation between Spatiality and Temporality so far in this morning's talk. But Space and Time must be two different aspects of universe. Same watch as a whole in every division of Time. – But not of Space. (only half the watch e.g.)

The only thing we've got, to endure, is the "Pattern" – also one other thing:– we've got in background, not perception ∧nor matter∧, but prehensive Activity. – A certain individualisation of the material finding itself in the Event. – The "Capture of the Eternal." The character of the Event is merely there for Appearance (Observation). But in Event itself it's an intrinsic character of the Activity. "Endurance" obviously means that we are considering some character which does discriminate between Temporality and Spatiality. It displays Spatiality and endures Temporality. Bergson's always talking about this:– "Spatialisation" (But Bergson's Anti-intellectualism based on fact that people have made mistakes, is rather premature).

If you're coming down on "Organism" as fundamental you must not start with parts, but with wholes. We're considering that character of an event which differentiates between Temporality and Spatiality. Doesn't want to assume there is only one way of fixing Temporality. But want something that will give you some definite fixing of Temporality.

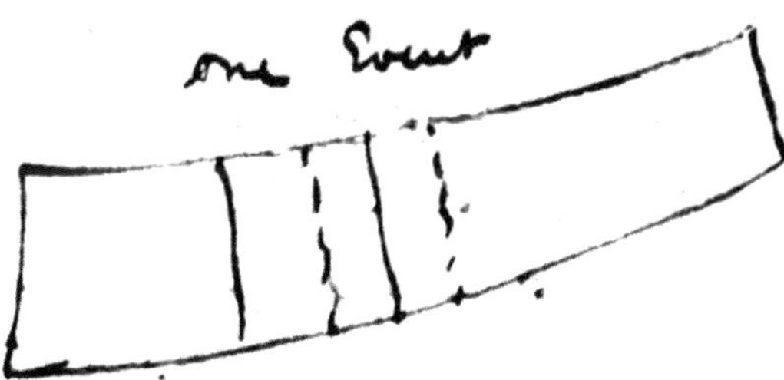

Overlapping part-events. – Analyzable in terms of this definite Temporal flux. Then the essence of Endurance is that this character must be nothing else than some partial identity – Pattern – Some partial content of Pattern of whole event found equally in

N.B. ? → every one of the part event. Endurance of the Pattern throughout one total Event, here – reproducing itself throughout. You have here a "life-history".

It is not Matter, but Pattern, which is concerned with the unity of the Event. – Some partial element of the Pattern – some characterization of it. e.g. which endures.

Emergent Objects which endure in themselves can vary in importance. "There are degrees of reality". There is no event in vacuo. Other Events in connexion with it. There may be nothing very intrinsic in the Pattern itself. – It may be constant reference to what is outside. – When the pattern is to be conceived as ∨a∨ Unity of the pattern of Environment.

When the Unity of Pattern is derived from inheritance ∧derived from∧ within the Event itself – derived from the Past, in the Present:– The self-reproduction of the Pattern.

Two other points:

(1.) Retention.

"Selective Envisagement". Value for its own sake is being retained for selective envisagement.

Prehensive activity in it ⟨blank⟩

(2) Essentially accumulated into ⟨blank⟩

Centre of Gravity now placed:– inheritance. But you can't get away from idea of Environment. Inheritance is a unity derived from own Past. But this is what it is --- by reason of the aspects due to the environment. Emergent object is the survival of the fittest.

The more concrete view.

An enduring thing is not this environment in there.

How the enduring unity deals with its environment.

Kemp Smith and Hoernlé in relation to Whitehead.[1]

Hocking's notes

Jan 8, 1925

1. In analyzing the whole, you may be losing reality, but not gaining it.

2. The whole question is, how are you to describe the content discernable in the Whole.

You have to come to (Universals) what I have termed existences, eternal not enduring.

Suppose we call them eternal objects.

W. James. in "Does Consciousness Exist"[2] plumps at once to a logical realism, what I mean by eternals.

This world-beyond is necessary for him as for me.

3. Connection between two events always marked by the eternal object, & vice versa. Eternal object never an event, *simpliciter*.

I take prehension as holding together without reference to cognition, which I am not sure is involved –

If so, I should be something of an absolute idealist

Always come across some kind of intersection –

Leibniz monads mirror the whole universe –

Impressions mean something – impressing –

The difficulty is there –

---

1. Clearly, Bell was having difficulty capturing the final thoughts of Whitehead's lecture in these last dozen lines.
2. See footnote 5 on page 161.

Keep a rather vigorous realism – edge off cognition –
And put the emphasis on the eternal object

4. Your organism is the event which is the prehension into unity of the aspects or pattern thereof.

Event a has aspects which it shares with event b. and vice versa. Reciprocity of aspects qua eternal.

Category of relation, etc. Aristotelian categories –

Useful abstraction, but rather high abstractions – get rather nearer to the concrete fact by the abstraction I have been setting before you

[Easier to explain pattern in terms of relations than vice versa, but –]

5. Event analyzed into three sets of patterns

Pattern of aspects
of realized values
of aspects of it for other events

1 – 8 – 25/2

6. <u>The endurance of objects</u>

Difference/ Space & Time has to be exhibited –

Pattern is the only thing we have to endure

Haven't got matter in the background

But we have prehensive activity

Displayed spatially & enduring temporally

You ought first to think of an event as a Whole

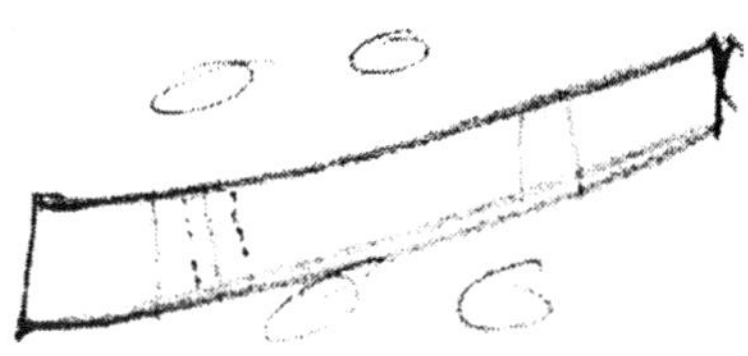

What aspect of it differentiates between Spatiality & Temporality

You can consider parts – segments –

There must be some ∧elements of pattern∧ aspect which is found equally in all segments

Endurance means persistence of this past[1]-pattern throughout the event

Things differ in the degree in which they dig down into this underlying prehensive activity

"The unity of pattern is derived from inheritance within the event"

You can't consider an endurance apart from its environment

Inheritance does not – require knowledge –

When I talk to Kemp Smith or Hoernlé never actually certain how much I differ

1. It is not clear whether this term is 'past' or 'part'.

Lecture 38

Saturday, 10 January 1925[1]

Bell's notes

|70|
Whitehead – 10. Jan. 1925.

Revision Lectures before Exams. Line of thought not disentangled from amplificatory details.

Start with notion of Flux, Expansion, *Durée*, Process, etc. etc. as fundamental starting point. Transition from Occasion to Occasion, –there's a certain Immediacy about it. And this immediacy of Occasion vanishes before you – passes on to otherness – a flux of immediacies. This at once has been the very fortress of the Sceptic.

What is it that binds together at all there – what binding Element that organizes in a way this mere flux of occasions. Two points of view:– (1) Knowledge is a comparison of Immediacies, and is after all an analysis of what is immediate in terms of Something in a sense outside itself – outside the transition. (2) The Self-determination of the Flux (the relation of Occasion to Occasion). You must find something that is adequate to the Binding. Hume presupposes an underlying mind. Although he provided for description and comparison of Occasions by the mind. – But not for any mutual determination by the Occasions. As a position which can't be refuted by mere logic Hume's is as good as any other not self-contradictory system.

But if you look on it as analyses of what is really presupposed in our praxis you can't be satisfied with Hume's Scepticism (as he wasn't himself). In each case of Knowledge, you've got to find something outside the transition – something identical. They've got to have a Being which is outside the "Realisation". Whitehead doesn't like William James' term "Logical Realism".[2]

As mere Being these would be nonentity. But as "part of the machinery" they are "Objects" (what stands firm in the flux) – sense-data are "Objects":– perceptual Objects. But these are rather complex entities. Rather now underlying "primary-Objects". Underlying things which bind together ⟨blank⟩ "Eternal Objects".

1. The Radcliffe version of this lecture, from notes taken by L. R. Heath on what was likely 13 January, begins on |132| of her notes, page 467.
2. Gerald Myers warns, in his *William James: His Life and Thought* (Yale University Press, 1986), p. 558 note 37: 'If James had not said explicitly, in *The Meaning of Truth* [1909], *A Pluralistic Universe* [1909], and *Some Principles of Philosophy* [1911], that he had adopted Platonism or Logical Realism, we might wonder if he actually had.'

– Realisation of Some form of Togetherness of pure Objects. The Flux is a transition of Togetherness. What happens is that the Togetherness of the VpureV Objects becomes Real – There's something for the unit Occasion. – What emerges is a Unit for itself. There is something for this emergent Unit which is in process of Realisation.

Unless you're careful this simply leads you down to Hume's difficulty – No relation between occasions thereby determined. But in very texture of our lives is interwoven relation between occasions otherwise than mere conceptual comparison of them.

It is there one is led to ask whether in describing Event as above we have really told the whole story. If so, then we're in Hume's dilemma. Then divorce between Practice and Principle. Better go back and find whether common Analyses are really what we do know of common occasions of which we have Cognizance.

Here initial insistence that in our Knowledge of immediate Occasions there's essentially a Knowledge of something beyond the immediate occasion.

Whitehead kept out allusion to one's "perceptual field". But Whitehead's view of Perception is that in connection with certain streams of Events there does emerge a self-cognisance. Awareness consists of – is Self-knowledge. So we must explain relation of our body to what we can perceive. What we can perceive is stream of our bodily events. What do you perceive? – immediate Knowledge there? – Perception of a Unity before us of things beyond ourself. But if it's the bodily event we are perceiving, what we perceive is in a sense here. That agrees with common-sense.

Neurologist will tell you tricks he plays in mapping nerves. You can see rats in two ways:– (a) produce a real rat. (b) take to drink. Knowledge in itself is Knowledge of the Body as a total Unit-entity. Whitehead believes he is with Spinoza here.

Then, though, that knowledge is a knowledge beyond the Body and therefore what the ingredients within are in a sense the occurrences without. There is no such thing as Simple Location (especially in connexion with "Eternal Objects") – This is fundamental for Whitehead's philosophy. To express them so is entirely to misapprehend their status. Looking at Whitehead's *Principles of Natural Knowledge* – he came down on this as essential point, though didn't bring it out well.

Fundamental fact not yet adequately stressed:– In all efforts to construct a Realistic Philosophy – That "Simple Location" is wrong concept to apply. – It
|71|
commits you to "Bifurcation of Nature" and this hands you over to Subjectivism. A chapter on "Concept of Nature". – We've got used to it in last 300 years. But take it up anew and find how paradoxical that position really is. Whitehead was in a great muddle about (in first book, and better in second or in *Principle of Relativity*) – to get quite clear as to what is to be substituted for "Simple Location". (You always find someone's been there

before!) – Whitehead said "Ingression". Everyone been there before gets some idea of Aspect or Perspective – Spinoza: "Mode". The green which I perceive isn't there simply – but is here under aspect of being there. Shows you that function of these eternal objects in the machinery is the connexion of events. – Any one[1] Into being ingredients in any of the others. Lends mode of being there to --- ⟨blank⟩

The Eternal Object conveys the other event into the one in question as an aspect. Sensa is one Kind only. More subtle Eternal Objects:– Shapes, e.g. (Broad missed a point here. Broad says shape of penny we see is an Ellipse.)[2] Whitehead's point is that what you perceive is an aspect of Shape and the essence of an Aspect is that it at once proclaims its complex character – proclaims itself as conveying to you – putting as an ingredient in event A, event B under an aspect – proclaims itself as analyzable into event B and relations of B to A. It is what sucks all Reality into a part. An event as a unification of Eternal Objects under aspects, and thereby of the events beyond, into the event here, is, Whitehead holds, just that connection of Events on which our whole life is based. And then the aspects are not before us as an aggregate, but as a pattern of aspects. And this pattern is nothing but the aspect in this event of the mutual determination of Events. Events aren't self-subsistent independent Entities. Duality inherent in an Aspect. – You analyze it into Element of something in itself (other event or eternal object) and other element as from your standpoint. That is why in dealing with an eternal object Whitehead insisted that you can subdivide their essence into two – the individual essence and the relational essence. Spatio-temporal aspect e.g. as relational essence – but thing is green e.g., in itself. An Event therefore is an Organism – which is an essential unity of a pattern of Aspects and these include "Aspect of Control" – mutual determination of events.

You can't get away from idea of Substance. But wants to get away from it as a passive Material. Wants it as underlying Activity, which is realising itself in this way. You cannot divorce the Substance from its feats of realisation. It is finding itself in its realisations of individual entities – Events. You may make your Abstractions of concrete Events as you like. But "mere activity" is a barren idea. Must think it as conditioned by the activity [X][3] – The "Existences" is the total realisation – not mere arbitrary collection. – Here is where Whitehead brings in "Envisagement" – In some way in touch with all these.

As in the Event it is that you get a Mode of Substance. No one mode can be taken by itself. – It's only real in the Community of Modes; and it is realising itself in --- ⟨blank⟩ It hasn't equal indifferent touch with all the Eternal Objects but a selective, preferential touch which arises from the

1. These two terms almost seem to have been added, after the note-taking, in the left-hand margin.
2. Closing parenthesis supplied, here.
3. It is not clear why Bell has left this marker.

actual individual status of that definite event. Here a lot of semi-poetical terms like: "Capture of the Eternal".

Then we come to further point that somehow we've left something out. – Namely the mere event has a unity, but precious little capture about it [What's been happening in this room in last ten seconds]. But if you make a proper choice you get a lot more of Unity and Endurance – some more emphatic reality, (and Whitehead always insists on degrees of Reality.) Then you've got to ask what it is that endures. – It must be in some sense a Pattern enduring. Again:– What's enduring is the unity – the pattern – of aspects (including Aspect of Control). That's what's being "prehended". Looking around for examples of pattern enduring, may be quite trivial, but it is something enduring for itself – because of what it is in itself. In order to get idea of Enduring you've dug down to something which differentiates between Space and Time. Spatialisation only falsification when mishandled!
|71|
"Aspect of display" which has to do with this.

It's not totality of Pattern which necessarily endures – But some elements of Pattern. This Pattern has double origin – From Environment; and its own inheritance; – from its parts. Endurance of a pattern derived from inheritance throughout its own historical route. Then you have a certain unity of pattern. Substantial Individuality – Individuality of underlying activity has a "Substantial" – i.e. a certain independence – realising ∧in∧ itself, under limitations, the functions which in their completeness have to be ascribed to underlying Substance *in toto.* The Envisagement and Selectiveness are within ? ? ⟨blank⟩

It depends on degree with which Endurance of Pattern can be ascribed within the event assisted by what lies without. So instead of being able to take only whole Organism, can validly take partial – relatively substantial – units out of it.

Hocking's notes

Jan. 10. 1925
Revision lectures

1. If you start from the notion of transition as a transition from occasion to occasion the occasion disappears in immediacy – and you have a movement from immediacy to immediacy – the result of the skeptic –

What binding element have you?

You have got to face that point.

2. On this two points of view.
a. Knowledge is comparison of immediacies, an analysis of the immediacy in terms of something outside the transition

b. The self-determination of the flux, – the relation of occasion to occasion

Hume presupposes something beyond that impress~~ions~~es

Have to get outside the transition –

To find something which is identical these elements that remain self-identical

For that reason I don't like William James term logical realism. There are beings, existences, objects which stand outside. As merely outside they might have no relation to the process, in which case we should know nothing about them.

But let us take them as parts of the machinery – as such we might call them existences, But prefer the term object – as what stands firm.

Sense data are objects – perceptual objects.

Then there are pure or primary objects – the underlying things that bind together these occasions.

3. What are these occasions?

The realization of some form of togetherness of pure objects.

But does this leave any relation between occasions other than comparison –

Do we land in Hume's dilemma –

Before admitting this, better go back and see whether our common ways of putting things are adequate

There is a knowledge of something beyond the immediate occasion.

What we perceive is the stream of our bodily events (?)

Jan. 10. 1925 – |2|

4. But if it is the bodily event that we are perceiving, viewed as a total organism,

What we perceive is in a sense here, you know.

Agrees with common sense. Neurologists mapping nerves. Two ways to see rats, produce a real rat, and take to drink.

[With Spinoza, he thinks]

But this knowledge is a knowledge beyond the body. The events within are, in a sense, the occurrences without. There is no such fundamental fact as simple location in the case of these eternal or fundamental objects.

Refer to *Principles of Natural Knowledge* – The essential point I pound down on – Reason I wrote the book. The fact I thought had not been adequately expressed in attempts to construct realistic philosophy – Simple location the wrong basis. Leads to bifurcation of nature, with all its difficulties.

5. What is to be substituted for "simple location"

Everybody gets some idea of 'perspective', 'aspect,' 'mode'. The green I perceive is not here simply, but is here under the aspect of being there.

The function of the eternal objects is the connection of events

6. Types of eternal objects – such as colors, shape, (here Broad misses a point)

Is the penny round or an ellipse?

What you perceive is an aspect of shape. Aspects are not before me[1] as an aggregate of aspects, but as a pattern of aspects.

The event in my head is nothing but a – connection, – not independent "as from" my standpoint.

But it "is green or shape in itself"

|Jan. 10. – 3|

7. An event therefore is an organism – unity of a pattern of aspects – including the aspect of mutual determination of events.

8. Control means that the procession – "after all you can't get away from the idea of substances, can you"? the underlying activity, which is realizing itself in this way –

"You cannot divorce the substance from its feats of realization, finding itself in its realization of individual entities, the events"

You can if you like think of

Mere activity – but a barren idea, is it not. You must think of it as conditioned by the machinery, in touch with the realm of existences.

Here one brings in envisagement – 'in some way in touch".

The underlying activity realizes itself in the events & therefore it is in the event that you get a mode of substance. It is only real in the unity of modes.

Not an equal touch with all the – eternal objects, but a preferential touch – Capturing the eternal – semi-poetic language – rather silly – but somehow get it across –

---

9. We have left something out –

The unity – the endurance –

Some more emphatic realization than others –

There is something enduring because of itself

(last lecture) You must have dug down to something that discriminates between space & time.

You have a certain unity of pattern

The individuality of the underlying activity has a certain independence – realizing under limitations functions which in their completeness have to be ascribed to the underlying substance *in toto*. There is a "substantial individualization" – a subordinate substantial organism.

1. This could be 'one'.

Lecture 39

Tuesday, 13 January 1925[1]

Bell's notes

|73|
Whitehead. – 13. Jan. 1925[2]

Back to question of viewing Nature through and through from Organic standpoint. (Contrast is Materialistic). Advantage of this point of view is that it admits of no jump whatever in principle as between living and inanimate. Same process through and through. – Objectively realistic, too. Also admits from very beginning (in principle dealing with physical entities) certain points of view which the Vitalists etc. clamor for in dealing with living matter. Namely this principle of Organism essentially involves that Whole at least *partially* conditions the parts, whereas Materialism looks on Wholes as aggregates of parts. You'd *first* have to look at *parts*. Point Whitehead wants to make is that this Antithesis collapses to some extent if you have thoroughgoing organic standpoint. In accordance with whole general principle the part (even molecules etc) is what it is by reason of being ingredient in *that* whole.

Long ago drew distinction between *basic* emergent objects and emergent objects of higher types. Infinite Regress here (emergent objects as composed of lower ones) *not logically* impossible perhaps; – but "instinctively" so. A basic object is essentially a pattern of aspects of eternal objects prehended into a unity in each event in which that object is manifested and enduring through a series of events. In discussing ingredients of a *basic* object you have simply the eternal objects. With objects of a higher type you have aspects of *basic* (or higher) objects. (If "higher", then at any rate lower than object in question). General principle remains the same.

Next point:– Consideration of the Environment as modifying the object.

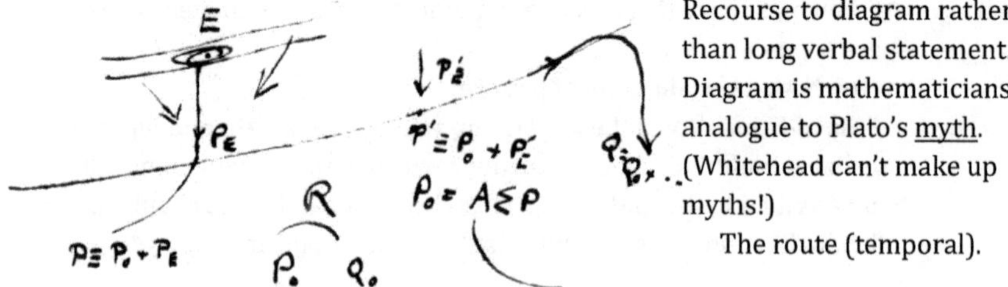

Recourse to diagram rather than long verbal statement. Diagram is mathematicians' analogue to Plato's *myth*. (Whitehead can't make up myths!)

The route (temporal).

1. What might be the Radcliffe version of this lecture, or portions of it, from notes taken by L. R. Heath on what was likely 15 January, begin on |135| of her notes, page 468.
2. In his diary for this date Bell records: 'after Whitehead's lecture ... W. asked me to go to lunch with him!' It is impossible to know for sure, but it was only a week till the end of first semester, when a number of administrative matters would arise.

Pattern has two roots (Inheritance of Aspects from life of object in past and Inheritance of Aspects from Environment).

Every way you can divide environment into Events has its aspect at "this point here".

Summation of all those patterns there.[1]

Suppose this physical object gets into a strikingly different Environment, we then find that as it is an effective object in Nature, earlier part of path is too far away to make difference, you'll have a pattern Q_0 coming along. Now under what conditions do you say "same" physical object? – When some common element of pattern of sufficient importance to be dominating. – This is saying: there is a certain conceptual identity. – Also necessary is a continuous historical route. Whether "same" or "other" depends rather on what you're talking about. – An orange vs the Irishman's sock – Very abstract identity here possible:– A pattern of a pattern of a pattern perhaps.

When you're dealing with types or species of objects you may not be interested in whether --- ⟨blank⟩ but whether they're both men, or electrons etc. – Then can drop continuity of path and simply demand underlying identity of nature – of pattern.

More definite example, now. Question of Electrons. At first people thought of these as little spheres – Lorentz e.g. at end of lectures at Johns Hopkins or Chicago. – working out of spherical electron.[2] Then Abrahams says Electron[3] will adapt itself to its Environment [he said: "Field of Force" etc] and will become prolate or oblate Spheroid (electron nothing but a pattern of Control [Whitehead spreads his arms out in three dimensions to show how he means it in the field][4] but how the points determined??) Under all such changes there must be some unity – "R". – Some method of qualifying both patterns P and Q.

The pattern that actually is propagated in a given Environment is the "effective pattern". Now another point:– The importance of the pattern of P of the event E as affecting its aspect in P depends on itself being part of the route of an enduring pattern. All the aspects
|74|
from this route at P will be in some respects the same. The really effective aspects are those which give life history of another enduring object. – The aspect of an object is of that object as enduring – functioning.

That's the general principle applying to all enduring objects in Nature. Whitehead gave his Explanation of the enormous endurance of electrons etc:–

1. As Bell drew the diagram appearing on the previous page, the curving line coming down from 'AΣP' touches on the word 'Summation' in this sentence.
2. The Dutch mathematical physicist H. A. Lorentz (1853–1928) published his *Theory of Electrons* in 1909, based on a series of lectures at Columbia University in 1906. Perhaps this is what Whitehead had in mind.
3. The German physicist Max Abraham (1875–1922) developed his theory of the electron in competition with Lorentz and others; he published it in a series of papers beginning in 1902.
4. This is a highly unusual case of Bell actually describing Whitehead's gestures, within Bell's own brackets (within Whitehead's parenthetical remark).

(1) they produce automatically a favorable Environment for objects of the same time; (2) they in some way favor the development – formation – of objects of the same Kind. – Principles of Birth and of taking Care of. – These account for endurance. Then you can get Societies and Species and when you get simple organism here will give you enormous and increasing Stability. Electrons very similar thus [what are they made out of? – I.e. question of matter. – Why not an infinte number already, with this "principle of birth"?]

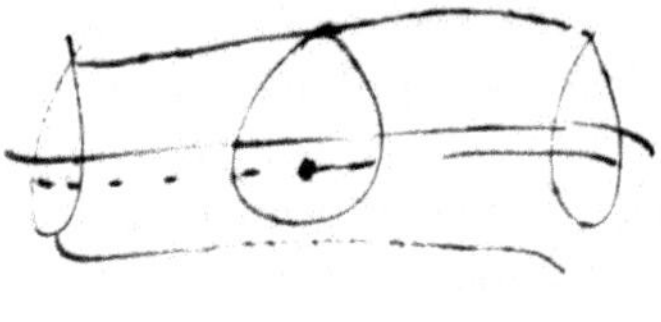

When come to higher objects you get an organism to be looked on as ingredient of other objects. Exemplification of Same principle, though.

Molecule in human body that would be P_0 outside, should be Q_0 in body. What the electron is – its pattern in there – is What it is by being within the whole body. As a matter of fact. – Physicist or Biological Chemist asks: Am I then to refer always to this totality? No! E.g. prolate spheroid may become jagged in form but still would be only another exemplification of sort of thing you had before.

Would be natural if you found that molecules had certain properties inside the body which they don't take outside. Take a piece of ∧soft∧ Iron – between Sun and Sirius presumably no magnetic properties at all. But put it between jaws of Electro-magnet! – Will exhibit properties it would never exhibit elsewhere. You cannot be certain that new properties won't open out as soon as you get new set of circumstances.

Take Eddington and others interested in Evolution of Stars.[1] Whereas mean density of Earth is about 5½ × that of water. It appears that inside respectable dwarf star you get about 50,000 × that. Then Eddington points out that whole situation re Molecules which keep one another at respectable distance under moderate pressure – Certain decent distances kept– even in chemical combination (because when Molecule broken up each atom goes off decently with its own electrons). In that pressure however all these pretty patterns broken up and you get an indiscriminate [note the mechanical thought here – not Electro-magnetic] mass of Electrons and Protons.

Next: the "obstructive aspect" of patterns – A point that Bergson goes into very much. – In the living body whole question is what response to Environment do you get? The living body gives quick response to certain aspects of Environment – Its pattern very responsive to Environment. This question of flexibility and adaptation coming in – and yet blocking out certain parts of Environmental effect, and letting only some through. Question then what you're going to respond to and what you're going to shut out.

1. Eddington had become Plumian Professor of Astronomy in Cambridge in 1913, and published his first book, *Stellar Movements and the Structure of the Universe* (Macmillan) in 1914. See footnote 3, page 143, lecture 33.

Hocking's notes

Tues. Jan. 13
(late missed beginning)

2. Environment as modifying the object

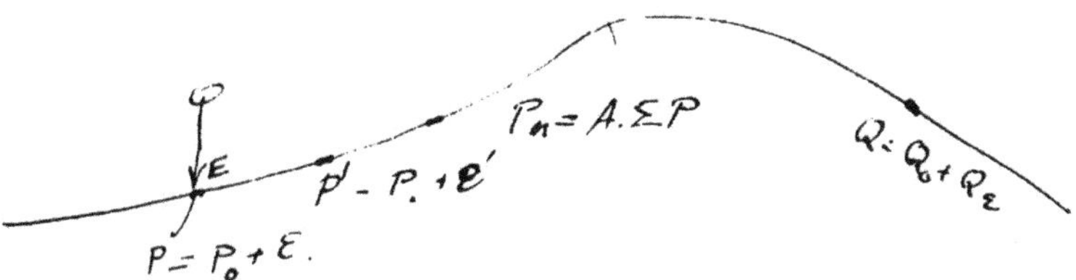

The object is the prehension into unity of the pattern P.

P has two roots – Inheritances of aspects from past & from environment.

Physical object propagates itself along the line –

But suppose it gets into a strikingly different environment. Then you will have pattern Q_0 instead of P_0. How then do you say the object is the same? Obviously when there is something common to P_0 & Q_0 sufficiently important. $\frac{R}{P_0, Q_0}$

Abrahams pointed out that the electron adapted itself to the field of force – because an oblate spheroid, not of a sphere.

To the man the same in a good temper & in a bad temper.

The common pattern will be called the "effective pattern"

|Jan 13 – 2|

3 Environment, continued

The importance of the effect.

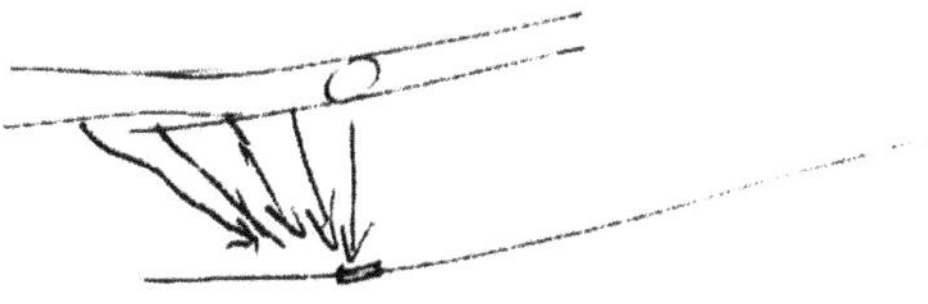

The enduring elements of the environments are the ones that count.

This applies to all enduring objects in nature

The enduring objects not only persist but tend to favor the continuance of others of the same kind. Will give you increasing numbers & enormous persistence.

Organisms as ingredients of other organisms.

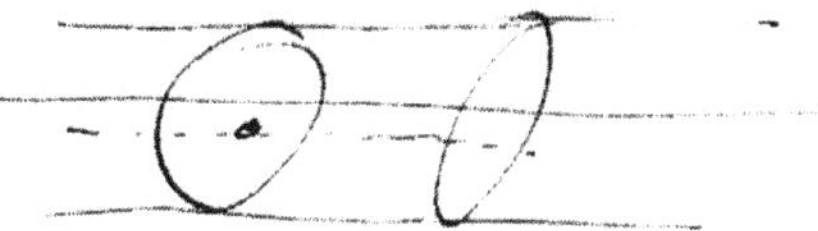

Human body & molecule in it. It is what it is, this molecule or electron, by virtue of being within the whole living body.

They follow the laws of nature, but it would not be surprising if they behaved in ways that had no exemplification outside the body. The piece of soft iron between the jaws of a magnet exhibits properties it would never exhibit anywhere else. Each element is what it is, but you can't be sure that new qualities won't crop out.

Or take densities in the North Star (?) 50,000 times that of water. Then the pretty pattern of the ⟨?⟩[1] is broken up, squashed together. New facts.

4. Environment as Obstructive. Bergson's Emphasis –

1. We cannot decipher this term.

Lecture 40

Thursday, 15 January 1925[1]

Bell's notes

|75|
Whitehead. 15. January 1925

On path of Vitalism. – This brought us up to question of VtheV Medium as obstructive. (from VthoroughlyV organic and thoroughly objectivist realistic point of view)[2] – radical here – so that cognition is simply the self-cognition of an event, and therefore not itself again formative. Psychological stream of Experience gives you the type of what an event is in itself:– a prehension of aspects – A prehension of what is beyond itself in itself. The aspects for prehension = potentiality – not aggregate [only] but patterned.

Intrinsic and Extrinsic sides to it Λ(total pattern)Λ then. Latter is its aspect in and for prehension of other events.

But extrinsic side is after all aspect of intrinsic side. The intrinsic side, then, goes to make other events what they are – Adds to potentiality of Events. Nature is what it is ΛpartlyΛ by reason of any one event being what it is – "Aspect of Control".

Certain futility here. Fundamental fact *re* an event is, as soon as you get it it's gone. That is why so necessary to come down on "Process". Tackle this through "Retention," "Endurance". Treated as time-succession of partial events in some sense or other a certain identity of pattern – (self-inherited aspect in subsequent phases) throughout.

This self-inheritance is Inheritance of totál pattern, whereas the inherited pattern is only certain sides of the pattern (pattern of the pattern etc) Therefore what is inherited depends on environment in a great extent.

But the Effective object in first environment is not [merely] R_0 it is P_0; and in other case Q_0.

P_0 e.g. = spherical Electron.

Q_0 e.g. = prolate Electron. The "effective" pattern may be looked upon as response of your enduring object to its Medium [Environment]

("Here just generalizing what is brought to us by the empirical observation").

1. What might be the Radcliffe version of this lecture, or portions of it, from notes taken by L. R. Heath on what was likely 17 January, begins on |139| or thereabouts of her notes, page 470.
2. Closing parenthesis provided.

There is a certain quantitative side to this pattern. – It leads to idea of total quantitative activity of Control – Science haunted by this idea – total quantity of effectiveness – call it "Intrinsic Energy"[1] Under "Relativity", idea of Energy being a bit modified. But there is a certain totality of and certain limitation to total Effectiveness. Certain intrinsic energy belonging to pattern P_0 and to Q_0.

Now, the more the Medium seems negligible the fainter its effectiveness seems to be, the less is the intrinsic Energy at P_0. The thing lapses into a state of minimum intrinsic energy.

(β) A "stressed" Medium – when approaching 0 call it "Unstressed Environment".

(α) P_E Q_E[2] at intrinsic energies then ~~if~~ $P_E < Q_E$ (where $\alpha < \beta$). Q has captured for itself some of the intrinsic energy of the organisms constituting the Environment. The response to the Environment, then, is "Capture of Intrinsic Energy". – Think of it in case of Electrons. – At P spherical and at rest. At Q either Kinetic Energy or deformation – Some response to environment will happen anyway.

Now: the Organism tends to organize itself so as to shield itself from Environment (to preserve its pattern). Hydrogen nucleus, e. – very responsive. So gets a nice little charge -e and then it's all right

Suppose you get this now in an unstressed environment. As soon as it is stressed the thing goes around somehow.
For that Reason Nature tends to have an Equipartition – to share out these things. Nature's build on Shock-absorber principle.
Electric condenser e.g.
State of strain or stress is roughly measured by V_1 the potential difference $\frac{e}{4\pi d} = V$

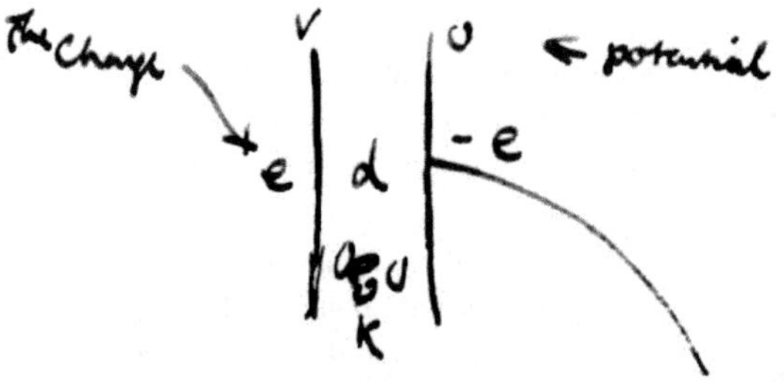

"Nature always seeking some form of minimum."

Medium rather void of any high type of organism. Now you go and introduce something like ∧piece∧ (molecules) of glass. Then you get "specific inductive capacity" coming in "K" – and –

|76|
and you get $\frac{e}{4\pi dK} = V'$

1. This phrase is added to this sentence (from below) but it is not clear whether it was meant to be added at the end, as appears here, or following 'by this idea'.
2. The blank left between P_E and Q_E seems to beg for a connector – perhaps 'and'.

Then V′ < V. One at once thinks of soft iron and its use in Electro-dynamics "to boost up the affair".

But now think how far-going the "obstructive" side at. Same sort of thing in any body in relation to its molecules. Body is portion of Environment of Molecule and Molecule is part of Environment of the Body. The Body is a very successful and elaborate organism. So that Molecule is highly responsive to changes of the pattern of the body. Molecule's changes act as a sort of Shock-absorber to preserve a sort of identity of pattern throughout. Elaborate organisms constructed on two principles:

(1.) that what is permanent $B_o = \underline{B}_p + B_a$. B_p = very important and permanent. B_a = sufficiently retentive to be an important element and yet does respond to environment.

What is common to the simple molecule throughout e.g. is the $\underline{B}_p$. B_a lets through a very few elements from environment – but alters with them.

When $\underline{B}_p$ is such that there is cognition the apparent world is B_a. $\underline{B}_p$ is the more deep-seated pattern. $\underline{B}_p$ is "enjoyed" in Alexander's phrase: and B_a is "perceived".

But only those parts of Environment appear at all which are relatively permanent (what you must know in order to have purposeful action – and keep B_o permanent).

The general principle on which the whole readjustment goes is exactly the same as the physical principle throughout. No new principle of "organism" brought in that isn't there already in case of Molecule.

Now for "soft iron" – goes other way around – not obstruction but boosting. Electronic coil with bar of soft iron inside. – You enormously increase "L" (coefficient of inductivity). – But the "L" comes into Electro-dynamics as a sort of principle of Inertia – you're checking something else! – the rise of the current. Whenever you're boosting up something you're shielding something else. But Organism is built up on two principles.
(1) (main) – Principle of Endurance, depending on "shielding". But where you want Cognition (where some very limited aspect is boosted up) you maintain a certain enormous permanence (B_p).

A qualitative measure of Self-independence in the underlying activity – You shield this in every way possible and very limited effects heightened. Physiologist tells you how responsive the Organism is. But the point is that the Stimulus must be the right one. – A certain small selection of stimuli it's very responsive too. Whenever you obstruct something you get [in other directions] greater intrinsic energy than otherwise would have. Organism is built up on those principles.

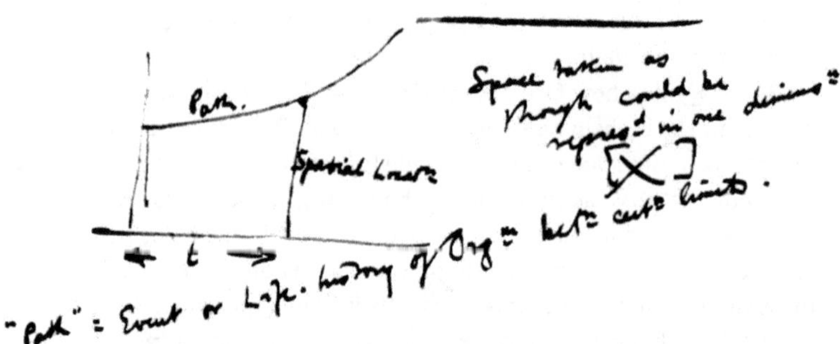

One other point:– Question of Locomotion. Now question as to route of Organisms' self-inheritance is left open. Question: What properties are you going to find in that. We're now at question of real foundational idea of Dynamics. – Question whether we can bring that too under same general head as so far. Important because of tendency for Mechanics to bring everything under head of Motion. – But now-a-days it's Radiation they are all interested in. This does not consider so much movement of Molecule as a whole and looks at aspectual --- ?

Quantum Theory and if try to bring it under head of Locomotion it gets so complicated that there must be something wrong. ("Message is same as message of Art").

Hocking's notes

January 15, 1925

A. Medium as obstructive.

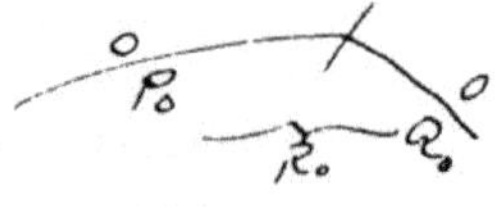

[Resumé. What is inherited depends on the environment. The same electron may be spherical at P_0 oblate spheroid at Q_0 but the effective pattern is the P_0 & the Q_0, not the R_0 alone. The object responds to the medium]

There is a quantitative aspect of effectiveness call it Energy of control, intrinsic energy

Stressed environment

Unstressed environment

Let P_E be the intrinsic energy at P and Q_E the intrinsic energy at Q

Then P_E is less than Q_E if Q is in an environment.

The response to the environment is the capture of intrinsic energy.

In Q your sphere will either move or be

Deformed – this is its response to the environment.

The organism tends to organize itself so as to preserve its pattern, to shield itself.

Charge •e is horribly responsive

What does it do? It gets a negative electron •e and shields itself. – e

Insulator

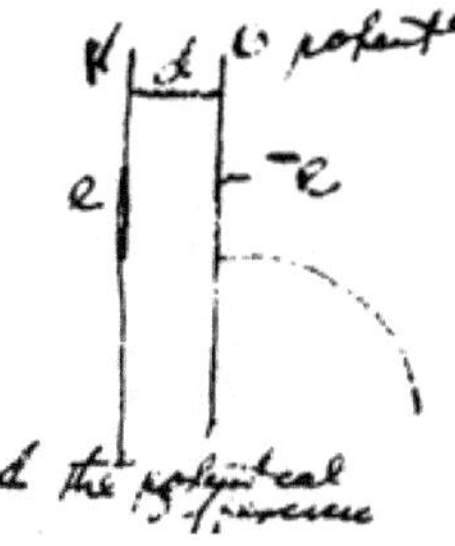

In an unstressed environment
goes around in a circle.
As soon as stressed it goes
around so fashion.

Nature is built on shock-absorber principle

Electric condenser $\frac{e}{4\pi d} = V$

Put in some glass molecules distorted.

Specific inductive capacity

|Jan 15 – 2|

$\frac{e}{4\pi dK} = V'$

$V' - V$[1]

Obstructs the stress

Same thing happens in the body

Body

Molecule

Two principles in construction of organism

$B_o = B_p + B_A$ B_p is very important

B_A is sufficiently retentive to be an important element

B_p is what is common to 

and 

B_A is apparent world, what is perceived.
only those parts of the environment appear
which have a certain permanence

L. When you are boosting something you are checking or shielding something else – soft iron.

Intrinsic permanence or 'enjoyment' Alexander[2]

1. While Hocking's notes contain only a hasty (slightly down-curved) dash, in Bell's set there is definitely a '<' (i.e. less than).
2. This line is written vertically up the left side of the page. All text after '1.' is written vertically up the right side of the page.

1. The principle of Endurance
   Which depends on shielding –
   Where cognition enters, something very limited is boosted up. B_A
   The physiologist says not that the organism is shielded but that there is a response to stimulus. But it must be the right stimulus
It is good playing a tune to a rose

1/15 – 2
2. <u>The question of locomotion</u>.

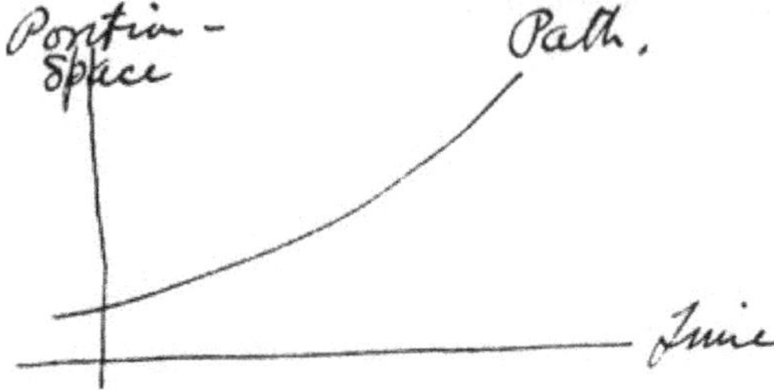

The foundational idea of dynamics,
Can we bring that under the same scheme of thought?
Bring all physical changes under the head of locomotion –
   (Nowadays, radiation takes the place of locomotion. Quantum theory prevents it from being brought under head of locomotion. But the general quality, I will consider next time).

Lecture 41

Saturday, 17 January 1925[1]

Bell's notes

|77|
Whitehead. - 17. Jan. 1925.

Started with frightening Physics. Getting back to Physics now. Working up to discussion of Motion – basis of Dynamics. That leads to question of what Motion is. How to be construed in Whitehead's Cosmology – and then of Space and Time.

Whitehead has been stressing assimilation of Space and Time. Beginning of it. Minkowski's famous address to Congress of innocent Physicians at Munich![2]

Paradox that waked people up. But overstatement. There is a difference between Space and Time. But modern physical research gives us an assimilation of Space and Time to Space-Time. Whitehead following this – keeping the Event as his unit, which is both spatial and temporal – Call it "Unit of Experience".

We've now got to consider (1) how Temporality and Spatiality distinguishable. The longer words on purpose here. Because, "Space" and "Time" are sharp-cut ideas got by abstraction from much more diffuse ones.

Where does difference become essential? (for Whitehead) – It is where he begins to talk about Endurance. Handled talk about "Process" so as to get on to --- ? ⟨blank⟩

What Whitehead said about an enduring pattern.

Event. Has its enduring pattern. Event is the life-history (within limits) of that pattern. You can divide to get a succession of (partial) Events if you can partition it according to one and the same rule of temporality.

An indefinite number of modes of partial divisions. But all follow same rule of temporality.

---

1. What might be the Radcliffe version of this lecture, or portions of it, from notes taken by L. R. Heath on what was likely 20 January, begins on |141| of her notes, page 471.
2. The German mathematician Hermann Minkowski (1864–1909) presented an oft-quoted lecture delivered before the Naturforscher-Versammlung (Congress of Natural Philosophers) on 21 September 1908, but that was in Cologne. This was published in 1909 as 'Raum und Zeit' in *Jahresberichte der Deutschen Mathematiker-Vereinigung*, pp. 75–88. It was this work that captured each event in space-time within the geometry of his four-dimensional space, each characterised by its own light cone. In 1907 Minkowski had also given a presentation in Gottingen published soon thereafter as 'Die Grundgleichungen für die elektromagnetischen Vorgänge in bewegten Körpern' (1908), *Nachrichten von der Gesellschaft der Wissenschaften zu Göttingen, Mathematisch-Physikalische Klasse*, pp. 53–111. (Minkowski was referred to in lecture 5 – see footnote 1, page 20.) As Minkowski had died in 1909, these are the only possibilities, and Whitehead must have had the Cologne address in mind.

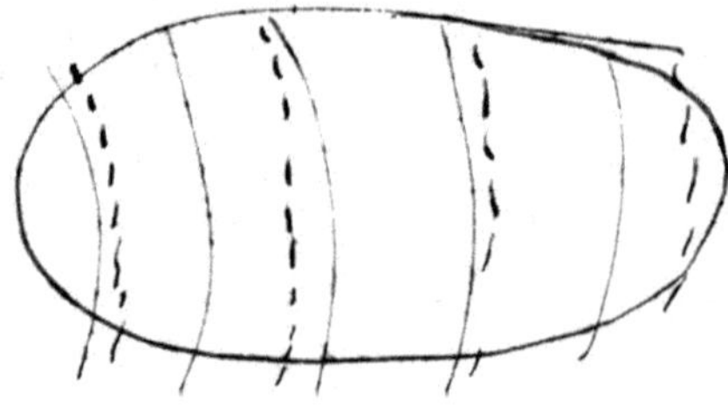

Point is:– there is a measuring to the rule of partition such that every subordinate Event which is thus marked out presents the same pattern as there is for the whole of it. "There is a pattern in the whole and in every part, provided --- ?" ⟨blank⟩

--- "Significant of a rule of temporality". A pattern therefore presupposes a temporal analysis. If you take the parts of the Event in any other way you knock the pattern to pieces.

"So endurance presupposes temporal succession."

Question as to the pattern and contrast between it and temporal succession.

The pattern is that Event as a unit – relates itself to Space and Time as a Unit. – (Then you've lost your successiveness). –So exhibits static ⟨blank⟩

Temporality gives you essential transition into difference. – Gives you the Process aspect – the divisive ~~and prehensive~~ aspect. It is the prehensive aspect that gives you Spatiality. Gives itself without the reference to otherness at all.

The essential transition is in the Time – the process. – One way of analyzing the idea of Process in terms of Cosmological ideas Whitehead has been presenting. The unit entity which has the most concrete unity of Realization is the Event. Analyzed:– partial Events. When you get your endurance you get the idea of the pattern which spatializes Space-Time and the transition (Endurance) which temporalizes it.

Therefore the Pattern is founded on the "Unity" of the Event; and the temporal Process founded on the Flux.

We come to "Patience" and "Significance" (words from Whitehead's books).[1] The "Patience" of the Space-Time continuum for enduring objects. What Enduring Object signifies in relation to Events among which it is located is its Significance.

Events must be patient of Emergent Objects, and no objects developed. (You can't turn Significance and Patience around.) But for significance there has got to be the background of Events.

You get a relation between Events and Objects here suggesting that "Internal-External Relations" have not been subtly enough defined.

The Events do not require the Objects but the Objects do require the Events.

1. Whitehead introduced the notion of 'significance' in his *Principles of Natural Knowledge* (p. 12), saying: '"Significance" is the relatedness of things. To say that significance is experience [here following Kant *vis-à-vis* Berkeley], is to affirm that perceptual knowledge is nothing else than an apprehension of the relatedness of things…'. He ended *Concept of Nature* (p. 197) with a clarifying note: 'The theory of significance has been expanded and made more definite in the present volume. It had already been introduced in the PNK…'. Then in *Principle of Relativity* he elaborated this notion further and added to it the notion of 'patience of fact' (pp. 14, 24–6).

Whitehead and Einstein's discussion at Queen Anne's Gate.[1] Einstein: "That concerns psychological Time. I have nothing to do with psychological Time." If you once go and lock yourself up in a private world (Whitehead) you never can get out of it.

|78|

Experience (taking account of) vs. Cognition of Experience (a further relation). In Einstein etc. – "the proper Time" of a body. Whitehead means that "proper Time" and psychological Time is same thing. Psychological Time is simply proper Time of a being which has cognition. [Aspects of it?] Those definite peculiarities of system of Reality relevant to very being of that object.

Whitehead was confused in *Principles of Natural Knowledge* about ? ? ⟨blank⟩

Now question: Is there only one Public Time? – only one temporal succession? Question of "fitting on" different experiences. Questions of Experience and Scientific Evidence etc. Wants to point out:– So far from question Relativity puts being an outrageous paradox, it's the first question to be asked and only recourse will be to observation and experience.

Owing to subjectivist points of view of modern philosophy, the "public world" – The fundamental question really didn't arise. "Public" world was there got by rather a fake.

Then we all know that general opinion of Mankind is that there is only one Time. Presupposition that there is only one is made almost unconsciously from our Ego-centric point of view. – Question of conjectural fitting on of one experience to another. Now turns out that these Times agree pretty well.

Whitehead talking not of measure of Time but of Order, which is the essential thing. Then you come to the careful, delicate experiments. – Due to advance of Technology. (Rot talked by "High-brows" that Technology has nothing to do with the human "Spirit").

One or two men of genius – Michelson and his inventions. There is now possible a series of Experiments comparing total Event A and total one B. – Of two objects. The difference expresses itself in motion with respect to each other. The meaning of Motion is a difference between Space-Time systems.

Long question is definite defining of Space. Definition of absolute Rest is rest in relation to its own system.

1. The home of Lord Richard Haldane (1856–1928) was at 28 Queen Anne's Gate, in London. He had invited Einstein to London in 1921 and arranged to have him lecture at King's College. Whitehead was Dean of the Faculty of Science for the University of London at the time, Chairman of the Academic Council, and had actively been teaching courses on relativity. He was also a close acquaintance of Lord Haldane. It was while Einstein was staying with the Haldane's at Queen Anne's Gate that Haldane arranged for Einstein and Whitehead to meet and talk. For more details on this encounter, see Ronny Desmet, 'Whitehead and the British reception of Einstein's relativity: an addendum to Victor Lowe's Whitehead biography', *Process Studies*, supplement 11 (2007), pp. 23ff.

If two objects have same spatio-temporality are at rest if no motion of one to other. But if two objects have different spatio-temporality A will be moving in B's Space and B in A's, etc.

Motion is the expression of a divergent spatio-temporality. Then we come to another point – really the foundation of whole Science of Dynamics.

Question of an object changing its Space-Time system. Unless this then --- ? ⟨blank⟩

Thus great fact to be considered is what are the laws and conditions which lead this Space-Time System. When A is not changing its System it is moving uniformly relative to B. When it is changing, it has Acceleration. Acceleration is danger signal of that change. That is why your laws of Dynamics always concerned, ultimately, with Acceleration.

Einstein says[1] You want some way of expressing your law in such a way that it doesn't matter which other system you use for reference.

Einstein wouldn't like all the talk of Space-Time systems.

Whitehead thinks orthodox Relativist is in a horrible muddle over Subjectivity and Objectivity. "Tensors" – simply the mathematical trick by which you get forms which are independent of the particular system you have taken.

Hocking's notes

Jan. 17, 1925

1. How spatiality and temporality are distinguished
   When we have to do with endurance.
   Go back to the enduring pattern.
   An event as the life history of an enduring pattern

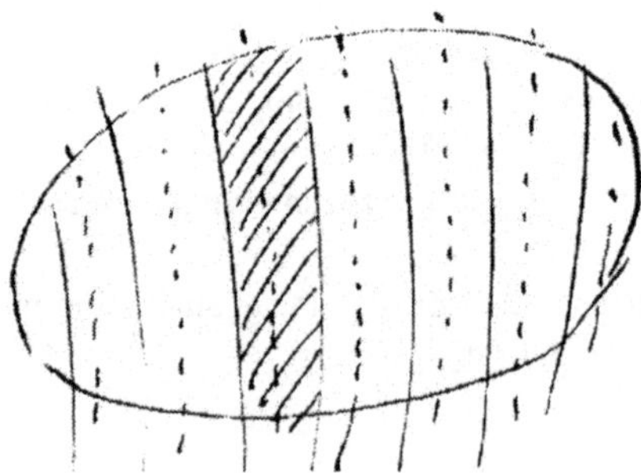

There is a pattern which is the same in every part taken in a certain way. [Or there is a way of dividing the event so that the pattern reappears as a whole – this is time]

1. Bell seems to have added this phrase, in the left-hand margin.

2. The contrast between the pattern & the temporal succession.

There is spatiality connected with each segment.

Pattern is founded on the unity of the event –

Endurance on the flux

3. The patience of the Space-Time continuum for enduring objects

The significance of something or other. Granted the significance there must be the patience, but not vice versa

"an internal relation if ever there was one" –

Events don't require objects, but objects require events

4. Conversation with Einstein in Haldane's study –

Einstein – "That concerns psychological time, I have nothing to do with psychological time" – But of course it is the essence of everything I have been saying that if you once lock yourself up in a private world you can never get out of it. The proper time & the psychological time are the same thing. The psychological time is the proper time of an entity that has cognition.

"In *Principles of Natural Knowledge*, I was confused"

1/17 – 2

5. Is there only one public time?

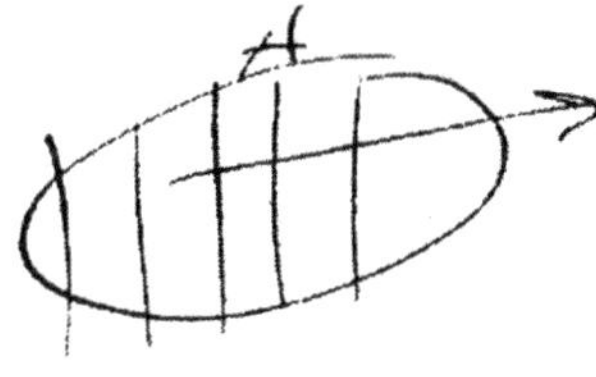

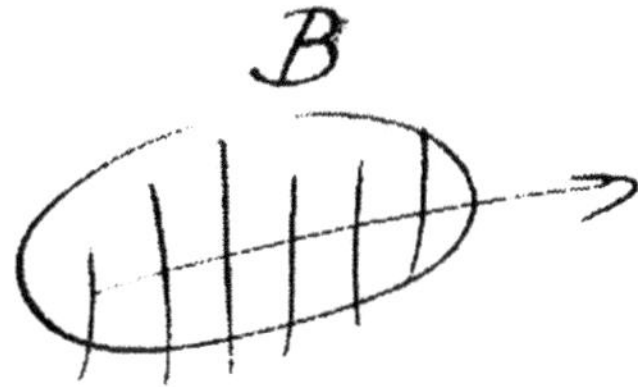

Is the temporality A the same as the temporality of B?

All questions of fitting-on are questions of scientific evidence – recurrence to observation & experience; This question didn't occur in dealing with psychological time, owing to the subjectivist bent of modern philosophy. Two private worlds no comparison.

Nor between public & private world.

General opinion is that there is only one time.

What is for me is for everybody. the subjectivist assumption

Question is not whether it appears long or short, but a question of order of succession

Come to delicate experiments & men of genius like Michelson

Meaning of motion is that an object is essentially at rest in the space proper to itself. Then if two objects which have the same Space-Time, they are without motion in ref to each. If they have different Space-Time.

B will be moving in A's space & A in B's Space-Time.

Then we come to another point – foundation of science of dynamics

When A & B are first at rest with reference to each other & then in motion, there has been a change in the

Space-Time system. Acceleration is the mark or danger-signal of this change in relative motion.

You want some way of expressing motion which is independent of which body you take as at rest

A, B, or C, or some other.

Talking about "tensors" – the mathematical trick by which you get forms wh. are independent form of the law independent of the system you have taken.

What is the general sort of thing our laws of motion express.

Lecture 42

Tuesday, 20 January 1925[1]

Bell's notes

|79|
Whitehead. – 20, January, 1925.

(Last Lecture)[2] – Laws of Motion. Root question got put in form in 17th century. With addition of Law of Gravitation was discovery which started modern science on its career. Discovery of proper ideas to apply. Idea of Motion of two Enduring bodies, when their Space-Time systems are different. Wherever you have that relative Velocity changing you've got as sign that one or both of bodies changing its Space-Time – relative Acceleration. Thus Acceleration comes to be looked on as Key to question of laws of Motion. That is Galileo's prime discovery, that it is a question of Acceleration. That traverses Aristotle's opinion that it's relative motion in itself that you have to look for, for change. There is a great deal to support Aristotle's opinion. But in trying to disentangle phenomena of Motion on that basis they got into hopeless muddles; etc.

Read Galileo's *Dialogues* and attitude of "the Aristotelian".[3] Aristotle would not have been with them. The Medieval people took a much more dramatic view of Nature – Everything with its rôle. There was a centre of the Universe which was proper end of Motion for these bodies; and an upper region which was proper end of Motion for Fire, etc. Whole thing had more sense of things moving towards a proper End. Modern theory – in a certain neutral attitude in which things are taking their chances. Corresponds with antagonism to "Final Causes" ^etc.^

The most important thing is always the imaginative background. Then details arrange themselves. Now jumping from Galileo to Newton ("Ahnmym" by Huygens etc.).[4] You find Newton has fixed on a properly neutral idea of Space and Time. But he has a unique Space and a unique Time. For Whitehead that would be taking any one of a number possible Spaces, with its correlative Time. Newton also pointed out that there is a definite physical quantity to be associated with each particle of Matter. Newton calls it "Mass".

---

1. There do not seem to be any notes taken by L. R. Heath for a Radcliffe version of this lecture.
2. Although this could refer to the previous lecture, more likely it clarifies that this was Whitehead's last lecture for the first semester.
3. Galileo's *Dialogo sopra i due massimi sistemi del mondo* (1632) was carried out by three 'voices' (featured on the title page), with Simplicio defending the Aristotelian position in opposition to the Copernican position being proposed.
4. It is clearly Christiaan Huygens (1629–95) to whom Whitehead refers, but it is not at all clear which of his works he has in mind.

For Modern Physics "Mass" reduces to manifestations of "intrinsic energy". Each particle of matter with its mass. Newton went so far as to call it the Quantity of Matter.

Then the way you can join Mass with Motion and Acceleration. Newton fixes attention on m v – which Newton calls the "Motion" (i.e. "Momentum"). Newton says this will remain unchanged unless some cause operating on it. Then fixes eye on Rate of Change (will be proportional to force acting on it – altering quantity or direction of it). Whole Value of that Formulation consists in fact that in looking for States of the Field which give you rates of change of that, question whether you get simple laws for your forces. Newton has merely taken simplest mathematical formulation he could think of. But Nature not necessarily working for convenience of Mathematician. Any other proportional thing. $m \times \alpha$ says Newton is the Force.[1]

But $m \times ra$ or $m \times qa$ etc. would do just as well. There is some physical quantity which scales up force by acceleration etc. It is no good trying "m^2" or the like because then m^2 would be the m we are talking about. Then on other side of equation there is some --- ⟨blank⟩ It is colossal luck that simplest things turned up trumps.

?] Newton had genius to tumble on to a fundamental thing apropos of whole question of Weight. Force between two particles

$$m \; \bullet\!\longleftrightarrow\!\bullet \; m' \quad = \frac{K\,mm'}{r^2}$$

r (proportional to product of masses and ∧inversely to∧ square of distance).

Found these two went beautifully together and explained all astronomical data. Newton outran his evidence a little – Saw Earth attracting moon and Sun the planets. At end of 17th century they had this neutral background and so instead of saying "the heavenly bodies" they said "Every particle of Matter" attracting one another.

Dialogues of Galileo on *The Two Systems of the World* – Simplicio's "centre of the Universe".[2] But Galileo himself rather wobbles on centre of one heavenly body. Newton swept aside all mention of centre of Earth, or centre of anything. Every little bit of the Earth is doing that attracting of the moon etc. – You've got the abovementioned "neutral stage". Newton's difficulty here, Heavenly bodies so far away that you can practically treat them as particles. But Newton wanted confirmation for particles on the earth. Question of mathematics of Sphere – Didn't know yet that for anything outside you can treat the Earth as a particle. It took Newton 15 years to do it. Dr. Glashier (Trinity – Cantab.) looked into this for Newton. Bi-centenary (in the 80's).[3]

---

1. Newton's well known equation is usually written with a roman 'a' to denote acceleration, but Bell seems to have used (following Whitehead?) a Greek alpha.
2. Missing close of quote provided.
3. According to the website http://www-history.mcs.standrews.ac.uk/Biographies/Glaisher.html, James Glaisher was elected to a Fellowship at Trinity College, Cambridge, in 1871, became a tutor and lecturer and taught at Cambridge all his life. That was, of course, Newton's college, and Whitehead attended Glaisher's lectures. By 1886 Glaisher was President of the Royal Astronomical Society and discussed Newton's theory in his Presidential Address. He scheduled a commemorative lecture at Trinity College

Thinks Newton had wrong conception of Mass of Earth (only about 10% wrong.)
|80|

One of interesting facts in History of Science that problems so difficult to find out, can later (after 200 years) can be put so simply that you could teach them to a large class of people without any mathematical ability. You still hadn't proved attraction between my watch and the piece of chalk. - This done in 18th century roughly by a man named Cavendish. Most extraordinarily difficult and delicate operation. Vernon Boyce ruined health over this (working at night).[1] Another interesting point concerned. All these attractions depend on mass of Earth. Newton didn't know what this was at all. Didn't know ratio of its mass to that of any other body. But $m\alpha = K(mm'/m')^2$ the m's cancel out! One of most colossal shots ever made and Newton right. - Said weight of earth probably about 5^3 times that of the water. We can't see what on earth guided Newton here. (Whether Newton took density of granite and thought of pressure it's under etc).

In Modern Times dealing with much more delicate measurements; and now a larger number of people of genius going into Physics. Therefore greater volume of thought increasing Experiments. Now, another thing being found out:- that α isn't quite right.

Found by very delicate experiments then instead of m v , you ought to take $\frac{mv}{\sqrt{1-\frac{v^2}{c^2}}}$

Now c turns out to be the speed of light (!) – of all things!

Poincaré's interesting remarks.[4] It's very good when experiment doesn't outrun theory.

If technique had been as advanced as to-day people wouldn't have accepted. luckily not enough known to show falseness of Keppler's laws.[5]

---

on the 1887 bicentenary of the publication of Newton's *Principia*, which (according to the editors of the symposium held commemorating its tricentenary) had to be postponed.

1. Whitehead is referring to the experimental work of Henry Cavendish (1731–1810), well known for his pioneering research in both chemistry and physics. What became known as the 'Cavendish experiment', performed in 1797–8, was the first to measure the force of gravity between masses in the laboratory. The accuracy of these measurements was not exceeded until the work of Sir Charles Vernon Boys (1855–1944), a century later. At the time, Boys was teaching at Imperial College, London, where Whitehead also taught, 20 years later.
2. This is what Bell has in his handwritten notes, but the equation should probably be $m\alpha = K(mm'/r^2)$, as previously on this page (and the alpha actually an 'a'). Compare with how Hocking records this in his notes.
3. It is difficult to read Bell's handwritten number (or he did not complete it?). It might be five, but Hocking records 5½.
4. See footnote 1, page 67, in lecture 18 for Henri Poincaré. In his *Science and Hypothesis* Poincaré warns that theoretical hypotheses and experimental evidence are not directly tied; to form a hypothesis is in some sense to outrun experience. But hypotheses can also be tested, indirectly, via their predictions, so (and perhaps this is what Whitehead had in mind), although theory may reach out beyond experimental results, it can get in the way for experience to reach too far beyond theory.
5. Also, it is not completely clear how the arrow connects with Kepler's (note Bell's misspelling) laws. Perhaps it is as an example of the looseness of fit between measurement and hypothetical laws. Kepler's first and second laws (that planetary orbitals are ellipses, and that planets sweep out equal areas in equal times)

Took about 200 years of dealing with m α to get all out of this. Now in most Velocities v^2/c^2 is negligible. So you get ⟨blank⟩

Now whole Physical Science – like the you want to say "Winder" you must say "window" – When you want to ⟨blank⟩ Obsessed with ? ? ? ⟨blank⟩

Another point:– According to Whitehead's way of looking at things, we're dealing with high abstractions:– We're talking of forces, particles, etc. at an instant of Time. The "at an instant of Time" is more fateful than the other. In 18th century Maupertuis[1] had an idea:– In actual path of a particle God had probably arranged things so that something brought off – some excellency achieved ⟨more⟩[2] so than in any other way. Last triumph in physics of "final causes". He found something there. – Method of Least Action.

Another thing: slightest knowledge of these people establishes correctness of idea that the instinctive assumption of an "Order of Nature" is certain rationality of adjustment in every part:– 15 centuries of Christianity[3] – including Mediaeval Rationalism. Scientific men always object to any explanation which explains everything straight off, without giving the "How" in Nature. Newton ends up with recognition of Wisdom of God, but keeps Him out of the reasoning. Modern Science keeps a Sunday mind and a work-day mind and keeps them separate.

Hocking's notes

1/20 – 1925

1. Laws of Motion

– Acceleration the key – Galileo – traversing Aristotle, who looked to velocity

Keeping on moving is keeping on doing something

The thing settled up by Newton.

– Had settled on a neutral & unique space & time

"For me that would mean taking any one of a group of ~~cor~~ spaces & its correlative time" –

– Mass. In modern theory means intrinsic energy.

To Newton the "quantity of matter"– fundamental

– M·V[4] Newton says this will be unchanged

---

seemed to fit observation quite well, although they each ran beyond those observations – as Poincaré would say they must – when posited. His third law ran well beyond observation when he theorised that the relative size of each orbital was determined by inscribing within the regular solids, each one constructed inside the other, in proper sequence.

1. Pierre-Louis Moreau de Maupertuis (1698–1759) developed what has since been called the 'principle of least action'. In part this figured into the early eighteenth-century debate between Newtonians and Leibnizians over the correct interpretation of the conservation of momentum, but for Maupertuis it also demonstrated the existence of an infinitely wise creator. Whitehead embraced the principle, but without the theology.
2. Bell does not use this word, but it does seem to be called for.
3. Bell's abbreviation in this instance is 'X^{ty}' and it seems plausible this is to be read as 'Christianity'.
4. Hocking records these as capital letters, but later on conforms to usual usage as lower case.

unless something changes. Fixed attention on
rate of change.

2. All Newton did was to take the simplest formula he could think of –
Any quantity involving acceleration (a)
Which becomes 0 , as a becomes zero, Would do as ma[3]
"It is colossal luck that the simplest thing you could think of turned up trumps" = ma.

To find that $f = K\frac{mm'}{r^2}$ is genius, as a formula explaining all the astronomical facts.

Newton outran his evidence somewhat when he said "every particle of matter"

Newton swept aside all mention of the center of the universe, the center of the earth, or anything – centers make no difference. Every particle attracts.

But Newton didn't prove that all the points in a sphere could be represented by a point at the center? Took 15 years?

And you still haven't shown that two bodies near surface of Earth attract each other

---

All these attractions depend on mass of earth

Newton didn't know but "by a colossal piece of luck". You can't find the mass by the mass of any particle for in $ma = K\frac{mm'}{r^2}$, m cancels out.
That is, a of $g = K\frac{m'}{r^2}$

How Newton guessed that the mass of the earth was 5½ times that of water, one of the most extraordinary shots anybody ever made.

1/20 – 2
3. We have found out that m a isn't quite right.
Instead of taking mv for your momentum
You should take $\frac{mv}{\sqrt{1-\frac{v^2}{c^2}}}$ c being velocity of light

$\frac{1}{\sqrt{1-\frac{v^2}{c^2}}}$ × rate of change of $\frac{mv}{\sqrt{1-\frac{v^2}{c^2}}}$

[Poincaré remarks on the good fortune that skill in observation doesn't outrun skill in theory. Kepler's law]

4. Physicists now likely to tell you mass is variable. mass = $\frac{mv}{\sqrt{1-\frac{v^2}{c^2}}}$
The plain fact is that the expression varies in value.

5. We are talking of very high abstractions – force at an instant of time.
Maupertuis had an idea in the 18th Century.
The last triumph of the idea of final causes.

The idea of least action, the most excellent path, established by God.

The instinctive assumption of an order of nature had got implanted in the European mind by 15 centuries of Christianity & the mediaeval rationalism, believing in the immediate interference of God.

---

6. Haunted by this point of view –
Modern science is not so haunted –

Second semester

Lecture 43

Tuesday, 10 February 1925[1]

Bell's notes

|81|

Whitehead: 10. February. 1925[2]:-

Starting with a different line of thought. Considering first now carefully how we get at exact ideas of Geometry ("Extensive Abstraction"). - Not generality of last term.

Summary of last term:-

What is Geometry? Whitehead starts:- Most concrete thing can get is an Event. Event has its own reality and is therefore the unit of Experience. Must start from the psychological field:- Our immediate apprehensions of what each of us is here and now. The bodily event is in itself not as inspecting from outside - not what the Surgeon sees when he cuts you open.

- But Totality in itself.

Another point of view - very different is to ask what it is that Physicist assumes re ultimate entities with which he starts - Electrons, Magnetic Field of Force, etc.

Details of what lecturing on worked out somewhat clumsily in 3rd Part of *Principles of Natural Knowledge* and throughout *Concept of Nature* and 1st Part of *Principle of Relativity*. Not so concerned now in following systematic details as in pointing out confusion etc. in those books as in pointing out what lies behind ideas there.

Also in Psychology of deductive reasoning. To that extent = lectures of Origin of Mathematics (most generally = Art of Deductive Reasoning).

Also:- Whitehead only knows three ways of getting up a subject: (1) to be "taught" it; (2) to write a book on it; (3) to lecture on it. Advantage of lecturing is that your pupils teach you as much as you them.

1. In the folder of handwritten notes taken by Hocking, the set labelled 'Feb–May 1925' do not seem to include anything from this date. The first lecture of the second semester for which there are Hocking notes seems to be that of 14 February. The Radcliffe version of this lecture, from notes taken by L. R. Heath on what was likely 10 February, begins on |1| of her second semester notes, page 473.
2. Bell notes in his diary that prior to Whitehead's lecture on this day he 'started [his own] lectures in Phil. 25b with a fair attendance. Later to Whitehead's lecture'. The listings for this semester show Dr Bell teaching a course 'primarily for Graduates' on 'Philosophy of Values'. So it went throughout this second semester, and in the early stages Bell often notes how tired he was, preparing his own course the evening before. It seems to show in this, Whitehead's first lecture for the second semester.

Physicist's Field of Force and the psychological field both give you same general apprehension of "field". Here and now I'm experiencing myself as one among other things, but as apprehending aspects of Everything else. I am a process of unifying totality of things not *simpliciter* but in patterns of modes of things; and mode of a thing takes form of a "spatial intervention" and it isn't the chairs or your bodies *simpliciter* that I'm unifying – but in modal aspects through an intervention of Volume – Held apart by intervention of Volume. That grasping of aspects are therefore the very nature of what my body comes to as a whole. Then the cognition also. – I appear to recollect what I scarcely seemed to notice at the time. – Very different from the Idealist. Instead of the Cognition being the fundamental fact, it seems to Whitehead the most variable thing about the Event. The Cognition is a grasping together of Aspects in which there may or may not be Cognition involved.

Physicist considers in his Theory of Electrons simply how the Electrons interfere with one another's life history in the Field of Force. So you have prehension of all the Electrons. – Prehended together into Field of Force.

It is the aspects of the Electrons as being in a region at a time. Fundamental fact or concrete unit is an Event looked on as a grasping together of a number of limited aspects of totality. It isn't a mere Multiplicity, however. Events can't be looked on as so nicely disjoined from one another. You get a pattern of aspects.

Not only aspects of a and of b but ∧also∧ of the relation between a and b. It is not the geometry that you got before you so. – It is by the "ingression" of particular sense objects etc. (Green as Boundary vs ∧covering∧ colour). It's the colour that bounds. So what you grasp is an aspect of something as specifically remote. Remoteness expressible as certain Voluminousness.

You then ask (after: Pattern of Aspects) whether there is a formula which can in an impartial way express the "Systematic Character" of the aspects of the Modes. Here Whitehead discriminates between the "systematic character" and the "physical" or "aesthetic" character. Former is the mathematicised subject. Whitehead differs from orthodox line of thought in insisting on the systematic character of Geometry.

The general Pattern from which at modal patterns can be derived. So: distinction between "real" and "geometrical" space. Former is assemblage of Events considered as in their systematic local relations. What Whitehead calls "simple location" doesn't occur there at all. But geometrical space consists of Events merely considered in the impartial formula, from which all motions can be derived.

Where the "simple location" comes in:– where you saw[1] ⟨blank⟩

The sense-objects (Green e.g.) may mark a boundary. What are you (in giving your impartial ??) What are you going to give ⟨blank⟩

1. This word is not clear.

|82|

– Voluminousness seems to Whitehead "Voluminousness" between you.

But the boundary? Question whether the boundary means a new assumption to another type of body. Or whether it is capable ⟨blank⟩ an exact impression in terms of ⟨blank⟩

An Event in itself is potentially bounded. Boundary not as a ??

An Event is an exact idea. Question is how to define it in relation between events. Defines for Whitehead idea of Exactment. – and to endeavor to determine what these its characteristics[1]

We must inquire of Experience to seem. Cognitive gelding[2]

What you need to arrive at Geometry.

What you do in ordinary everyday thoughts is to take all you "Know". So you are noting anything in a very full[3] set ⟨blank⟩ All sorts of ~~pre~~ infinite ⟨blank⟩

Idea is that of one "volume" [Event [4]] containing another. How much can be got out of that idea? – How much to be added to get anything out of it. – [Partially overlapping etc.]

["not separated from any of you by a set of points – [5]

But by a certain Voluminousness. This inclusion. – any point of B is some point of A. Euler Diagrams, etc. But that can't be it. Because has no sense unless you already know what you mean by a point. What do you mean by individual points being included in A. already! One reason why a point is so enormously useful. Enables you to new[6] idea of Inclusion under more familiar one of All and Some. Enormous simplification of Thought machinery. Is the point on same abstractive level as Volume? Whitehead doesn't think so. The "Containing" – J.S. Mill calls it "Connotation." – What makes colour of Poetry:– ∧Shy∧ Intervention of one thing for other is an assertion apparently incontrovertible [7]

If you make a new Symbol for every exact idea you'll cost yourself a lot for printing. So unless dealing with constantly recurring ? ? – symbols quite useless unless they have certain peculiarity

a contains b → aKb. "To" and "from" not to be used yet.

Way mathematics has arisen – Symbols like +, –, x, =, have exact significance. Can use these to make Laws from that grammar which express enormously succinctly ? ? ? ⟨blank⟩

1. The correct reading of these terms is uncertain.
2. The correct reading of these terms is uncertain, although one could argue that the abstraction needed to reach geometry constituted a kind of cognitive 'gelding'. It might be 'gilding', but this seems a much less colourful term.
3. This term could also be 'true', although that makes less sense in this context.
4. It is not easy to read the abbreviation, let alone provide the extended word. 'Event' seems likely given the context. Most of the key terms in the previous two sentences are difficult to decipher.
5. Both end bracket and ending quotation marks are missing, but since the remainder of the sentence seems to be missing as well it is hard to know where to place them.
6. This term could also be 'use'.
7. It is not clear how this phrase ends, nor how to transcribe the next few terms.

Your whole explanation becomes far more difficult than other works. – Conventional question:– in A. are you going to ? ? ⟨blank⟩ It doesn't matter what you do, but have to do something. If a in b – is ⟨blank⟩

The bane of reasoning is the special cases. Euclid has them. If you study carefully how mathematics evolved. Motive power for some of its best work is often to avoid special cases. – Ingenuity in refining original ideas to get rid of these. You'd say aKb to allow as special case where a and b coincide. b a "proper" part of a in *Principles of Natural Knowledge*. – Silly. In avalanche of new symbols in *Principia Mathematica*. Trying to get as few special cases as possible.

One or two assumptions in aKb. The relation is transitive. If you're thinking exactly you are bound to get to symbols. aKb . bKc . ⊃ . aKc.

Lecture 44

Thursday, 12 February 1925[1]

Bell's notes

|83|
Whitehead. – 12. February 1925.

Beginning discussion of Method of Extensive Abstraction – how the exact Geometry is deducible – what sort of Assumptions, Conceptions wanted for it. (Part III of *Principles of Natural Knowledge* and beginning chapter IV of *Concept of Nature*)[2] – may shock some properly trained in Symbolic Logic.

What we are after:– Distinction between ∧art of∧ strict logical reasoning (≡ mathematical reasoning) and criticism of Logical Reasoning. Logical Reasoning as Art is the Art of getting on (system from given Concepts and by strict adherence to explicitly stated hypotheses).[3] When you've stated your concepts as accurately as you can and your hypotheses and see what you can get from given concepts and given hypotheses, and whether anything you Know that is left out.

⟶ In this art if you (cf. Whitehead:– *Introduction to Mathematics.*) Art of reasoning:– (1) Avoid special cases. (2) Bring it to mind briefly. (3) Arrange symbols so that they bring out by their very nature a sort of picture of the logical relations on their formal side. – Order, and signs between them to form sort of picture of what is in your mind. So many philosophers write nicely but you can't tell whether they stick to their framework. Certain amount of Symbolism keeps your criticism automatic and possible here.

But for Criticism of your procedure in mathematical reasoning. – Criticise your Concepts in response to "independence". These "primitive groups". Then derived groups:– Any member definable in terms of the primitive group. Then you'll find [Beginning of *Principia Mathematica*] that the definitions of a body of reasoning are philosophically far the most important part. Primitiveness or definability are purely relative terms. You can always assume an indefinite variety of primitive groups, and define others in terms of them.

Then second part consists in Criticism of Hypotheses which you are using. There discrimination into purely logical Canons or rules of procedure. – "Primative propositions" of the *Principia Mathematica* – But they are not hypotheses of propositions, but declarations of procedure.

1. In the folder of handwritten notes taken by Hocking, the set labelled 'Feb–May 1925' do not seem to include anything from this date. What might be the Radcliffe version of this lecture, or portions of it, from notes taken by L. R. Heath on what was likely earlier the same day, begins on |4| of her notes, page 474.
2. Missing closing parenthesis provided.
3. Again, missing closing parenthesis provided.

Every ^case of^ procedure is reducible to special cases of these. E.g. If 'p and q is true', then 'p is true'. If you assert the one, you're prepared to assert other. When set down these Canons must be ludicrously simple. Affirmations of Reason about its own procedure

p implies q[1] General form of all reasoning.
p is true Question of how to construe this – construe it as meaning:
Then q is true Either p is true or q is not true.

Then paradox that a false proposition implies any proposition whatsoever. Art of deduction always consists in getting hold of a ^another^ proposition such that you are prepared to say this of it; that p implies q.

Some true from their very form – no matter what you're dealing with – Sheer logical deduction as distinct from "Material Implication". This construction is independent of particular propositions involved. Truth of traditional "Barbara" doesn't depend on old "Socrates" at all.[2] If X is a man. You may be interested in it for application to some one case. But that is not why it follows. This always happens with purely Logical Reasoning (as distinct from Induction etc).

"Bringing in a logical Variable" – Independence of particular Subject Matter comes in. $\phi(x, y, z) . \supset . \psi(x, y, z...)$ Whole essence of Logical reasoning is to dig out deductive methods from immediate subject matter.

Though $\psi(x, y, z...)$ really embodied in $\phi(x, y, z...)$ but we don't know it. – All arithmetic in two or three simple statements – But if I ask you whether $3^2 \cdot 954....+ 7^{12}$ = odd or even!!

(x1): f1(x1) + f2(x1) + f3(x1) $. \supset . \psi$(x1 ...) I.e. it doesn't matter what the x1's are.

G.H. Hardy just after *Principia Mathematica* appeared

(If 1 = 2: Russell is the pope of Rome. Russell and the pope are 2. 2 = 1\...)[3]

How are you going to criticize these. They may be true but trivial because hypothesis is false. You want to exclude these cases. So want to

1. There is no doubt that diagrams reproduced by Bell (and Hocking) must have been chalked up on the board by Whitehead, and with equations and as here with logical formulations it seems likely, as well. This classic syllogism is reproduced here as though Whitehead did write it out on the blackboard, while comments to the right may have been notes of what Whitehead mentioned orally.
2. Whitehead's use of these terms signals that he is referring back to Aristotelian logic in which 'Barbara' named one type of deductive argument – one constructed entirely of universal affirmations. In this early form of sentential logic, it was the logical connections between entire sentences that were being considered, like the sentence 'Socrates was a man.' Aristotle had used such sentences as clear examples of stating something true, and 'All men are mortal' as a universal claim. The careful constructing of such sentences into a deductive argument (such as Whitehead provided, above) could preserve the truth of the premises in the conclusion that 'Socrates was mortal'. When formalised with the symbolism developed in the nineteenth century, this came to be called propositional logic. The *Principia Mathematica* of Whitehead and Russell was based on the further departure taken by G. Frege (1848–1925) that introduced variables, which could be quantified, permitting a new way of extending propositional logic. This opened the way to their goal of providing a logical foundation for all of mathematics.
3. G. H. Hardy (1877–1942) is credited with bringing a French-style purity and rigour to British mathematics. He is reputed to have told his friend Bertrand Russell: 'If I could prove by logic that you would die in five minutes, I should be sorry you were going to die, but my sorrow would be very much mitigated by pleasure in the proof'. According to the story, Russell agreed.

find that this group is not <u>false</u>. – Criticize these with regard to logical consistency. – From all the rest you can't prove that the one omitted is false. If all but one will prove falsity of the one, then you've got a radical inconsistency.

<u>Then</u> you want to know how <u>little</u> you need to assume. Want minimum number of assumptions and that nothing ∧in there∧ can be deduced from rest in there. – That they are properly
|84|
independent. "Ordered independence" is a very secondary sort. You want it so that Order doesn't matter.

Then question whether Adequate Assumptions – That whole field you believe to be true can be derived from premises.

All this enormously important. (1) to be clear headed (2) in order to be able to get all the propositions you want.

You've made, say, 1,000 propositions you believe to be true from direct observation. Then first eight, lets say, would enable you to deduce all the rest. – or <u>any</u> eight. Important in Physical Science as well as Philosophy is to get the thing in this abstract or logical form. Importance <u>in Science</u> not often recognized.

You often find that, put into general forms, find that these forms hold for more than one field. This <u>Community of Form</u> suggests possibility of underlying physical law to account for it.

Informal suggestive use of Strict reasoning – Very important.

You find this sort of thing in applied mathematics. Same piece of mathematics turning up applied to wide variety.

Story (of Whitehead as undergrad of Cambridge).

Metacentre of ship

"You remove the ocean". All you think of is that there is a force upward in direction of arrow.

This Formal Criticism, though very important in general is <u>not</u> always important in a special application. If dealing with very familiar subject matter you know that the bundle of propositions you're dealing with are not inconsistent and the formal proof doesn't really matter. Again:– Question whether you've made too many assumptions doesn't matter in very familiar matters. What you <u>may</u> be wanting to consider is Whether your primitive concepts are adequate to express all you want to about the Science – whether you can define all rest in terms of them.

Many circumstances in which this would too much limit the subject matter etc. So in geometry. – what we mean by one Volume containing another is so familiar that we don't need this criticism. But <u>want</u> to see how far from <u>merely</u> one idea (of extending over). Whitehead <u>had</u> expected to be able to go further than he can. Defining a point. – By talking <u>at large</u> you can – but not exactly.

Whitehead paper for Paris Congress in 1914. Only one man say fallacy – Nic/Nee-Ko (Frenchman).[1]

1. Whitehead seems to be referring to the First Congress on Mathematical Philosophy, held in Paris in April 1914. His paper was thereafter published as 'La théorie relationniste de L'espace, *Revue de Metaphysique et de Morale*, vol. 23 (May 1916), pp. 423–54. And the 'Frenchman' was likely the philosopher and logician Jean George Pierre Nicod (c. 1893–1924).

Lecture 45

Saturday, 14 February 1925[1]

Bell's notes

|85|
Whitehead. 14. Feb. 1925.[2] –

Importance of Mathematical Logic for philosophy. Where formalism carried to extreme (not to excess). How to get at fundamental ideas.

To know how much exactness is worth putting in. Important that technique should not obscure aesthetics of subject:– Must be master of your technik. Method of using exact technique. Method of Extensive Abstraction etc to show how technik of exact reasoning is of importance to Philosophy. Middle Ages with too narrow a view of matter and used it with a little too much enthusiasm. The easier style of Locke, Berkeley, etc. has its advantages – but also disadvantages.... But question of difficulty of knowing whether you are not bringing in alien ideas under guise of obvious and unquestioned truths or propositions. Experience of greatest Mathematicians shows danger of pitfalls so. Almost all great Mathematicians of 17th and 18th centuries fell into pitfalls that now more careful methods of Gauss, Weierstrass etc preserve us all from.

Thought that a continuous function always had a differential coefficient – a definite rate of change. This took in Newton and everybody. It's not true that Continuity involves this. Those pitfalls reproduce themselves when you ask e.g.:– How does Geometry come from a few fundamental ideas.[3] Whitehead wants to know how far you can get ideas of Geometry from mere idea of Voluminousness – ("extending over"). Unless a philosopher has some idea and practice of sort of way to do it, he is apt to go astray when he gets to sort of thing as in Mathematics.

Last lecture was running over some of main ideas that Exact Logic can use. First thing is "Material Implication" – So simple, that word "Implication" means too much. It is so simple that it is hard to make people realize childish simplicity of it. People won't believe that Childish ideas

1. What might be the Radcliffe version of this lecture, or portions of it, from notes taken by L. R. Heath on what was likely earlier the same day, begins on |6| of her notes, page 475.
2. There is not as much commentary in Bell's personal diary on Whitehead's lectures during second semester as there was during the first. Entries for 12 and 14 February are typical, with 'Lectures as usual', but that for 14 February also acknowledges that Bell, with preparations for his own lectures, was 'Very tired'.
3. The plan for the fourth volume of *Principia Mathematica* was to have extended the logical foundations from arithmetic to geometry, and thus the basis for how physics handles space. Though Whitehead worked on this fourth volume over many years before coming to Harvard, it was never published. However, it is quite evident in these lectures that the challenges Whitehead confronted to reach this objective were not simply set aside, but had widened and deepened and were still being confronted. See Lowe, *Volume 2*, pp. 92–5.

are important. Material Implication is simply comparison of different propositions with reference to their truth values. When you put it down it doesn't tell you anything of great importance. – What are ways of getting two propositions together so as to have something more complex. Three ways:– Proposition and its contradictory. $p, \sim p$

→ Whole difference between modern Logic and Aristotelian logic is that Aristotle starts with analysis of Propositions (asks what are fundamental kinds and forms of propositions. – Important but Whitehead doesn't think it has to do with strict Logic and Reasoning, properly so called. Mathematical Logic really concerned with taking propositions and considering what propositions you can build out of them which throw light on comparison of their "Truth-Values" [Frege's term].)

Two other simple ideas:– "and" and "or". They are not all independent. Question of getting them down to fewest number is further and incidental one.

With two propositions p, q		Out of that you can build p and q.
p and not q.	not p & q.	not p & not q. When any one of these is
$p \cdot q$	p or q	true ∧(or false)∧ it throws light on the
$p \cdot \sim q$	$\sim$ p or q	relations of truth-values of p & q in relation.
$\sim p \cdot q$	$\sim$ p or $\sim$ q	For or use $\vee$. (Reason for Symbols is to be
$\sim p \cdot \sim q$	$\sim$ (p or q)	quite sure you mean just one thing.)[1]
$\sim (p \cdot q)$		Or in sense of either or both.

Solemn question is what significance ∧to∧ attach to writing two symbols together? – p q means p & q. But when you begin to reason p & q. each a whole collection ∧of symbols∧ and you want to keep them separate. So put a dot __.__

Then for superior or inferior dot relationship put two or more dots, :. __:__

Which of all above combinations which will throw greatest light on relation of

Truth-Values of p & q.

$\vdash . \sim p \vee q$ Now if you happen to know p is true $\vdash . p$.

Then you are justified in striking down $\vdash . q$.

Doesn't seem to get us very far. (Old objection against syllogism). But is basis of whole deductive system. Now suppose p is false.

Then either if p is false or q is true. $p \supset q$.

$\supset$ is "Implies". Not in ordinary dictionary meaning, but this attenuated image of it.

|86|

When you come to ordinary deduction:– p is a very complex proposition and made up by complexity of a lot of subsidiary propositions by a lot of "and's" and "not's". Relations of subsidiary propositions to one another.

All men are mortal. Socrates is a man. $\supset\bullet$. Socrates is mortal.

1. Missing closing parenthesis provided here.

[Extra dot means for any value of x.][1]

Real point of that is that it doesn't depend on Socrates and mortality etc.

"x" is a man; therefore "x" is mortal.

Such a structure that quite independently of material contents. Now suppose x isn't a man it holds. Then ⟨blank⟩

Very curious point that Reason in order to be powerful, has to be perfectly general. Whitehead doesn't think Greeks quite understood this.

$(x) . \varphi(x) . \supset . \psi(x)$ I.e. for any value of x either φx is true or ψx false.

"Material implication" (child's play) $p \supset q$

Sun in eclipse this a.m. "implies" equally that I am here and that I'm not here this morning. You can't nail the thing down to it. Proper triviality without Symbols. Idea of a "Propositional Function" too. Entertaining suggestion of assertion about some unspecified individual (x). [Not simple assertion about propositions]. Question of ambiguity [fundamental Logical idea].

Then there is idea of every.

Real importance of Deduction comes when you're using idea of propositional function x, and by inspection of its make-up you can Know that for every value of x will be good. Then other idea that for some value of x, ψx is true.

Whole question of Logic herein.

In dealing with premises in any field, there are various ways of criticizing them. Criticism of Premises as to Compatibility, Consistency, Independence – may not be of slightest importance for Philosophy. Of real importance to Philosophy is question what can be defined and how.

Want to see if starting with ideas: Events, Extending over etc. What can ΛoneΛ get out of that in way of Exact ideas of Space and Time.

You don't need idea of points to get idea of Extension – we mustn't bring in "Points" (not yet defined) initially. Extending over is not whole-part relation – use K for it. Suppose we say A hits B, or A loves B; ~~(A.K.B)~~ (aKb).

K is only going, really, to bear the significance that you put upon it. So any other meaning of K will do. I have thus automatically in thinking gone over to more formal or structural ⟨blank⟩

Then, next thing is to show:– What is Λthe set of assumptionsΛ we are going to make. What sort of assumptions can be made.

1. This explanatory note is added in left margin. It clarifies the use of the extra '•' in the line above. Note that Hocking has a slightly different way of symbolising this universal argument.

Hocking's notes[1]

Logic.

			$\vdash . \sim p \vee q$	
not	$\sim$		$\vdash . p$	
or	$\vee$	[either or both]	$\vdash . q$	
and	$\cdot$	[or juxtaposition]	$p \supset q$	[implies]

All men are mortal. Socrates is a man. $\supset$ S is mortal
x x

Reasoning to be powerful must be perfectly general.
No special cases. For every value of x
$(x). \varphi(x) \supset \chi(x)$
$(\exists x). \psi(x)$. There exists a value of x for which psi x is true.

Extending over. Mustn't use the idea of points to explain 'extending over'

K if you are going to use extend over to explain points.
aKb

"Laws of thought can be stated only because they might be otherwise, and are yet that. You can't talk of anything that might not be otherwise"–[2]

1. From the file of Hocking's handwritten notes, this seems to be the first lecture for which notes were taken in the second semester, even though these are undated and incomplete. Also, while the two-sentence quote at the bottom of the page seem to be in Hocking's handwriting, the early portions do not. On another occasion Hocking arranged for R. Eaton (like Bell, another instructor and tutor in the department) to take notes in his stead, but in those cases they were identified as such and had been typed up, as in lecture 33. No indication has been left in this case.
2. As mentioned above, this last interesting quotation seems to be in Hocking's own hand, even though the notes preceding it do not.

Lecture 46

Tuesday, 17 February 1925[1]

Bell's notes

|87|
Whitehead. 17. Feb. 1925:–

In order to guard against illegitimate intrusion of other ideas (than Voluminousness and Extending over) through general usages of language, strict method is necessary. – But needs to illustrate the lack of exactness that is permissible. Always a mistake to tie yourself up more than necessary.

What can we get out of those two ideas: K means "extending over."

aKb is to read "a extends over b". K is to be solely a property of Volumes.

The class of volumes is then the field of things to which "extending over"

is relevant. Field of things related by K. We want to indicate that

x is a member of F'K : – x ∈ F'K

Quite indifferent re extreme case, whether a can extend over itself. Here better not.

Initial assumption then

(i.) aKb . ⊃ . a ≠ b Now jot down simplest ideas we can get re "Extending over"

Unbounded divisibility of events and the "always something beyond" (including it.)

(ii.) a) a ∈ F'K . ⊃ . (∃x) . aKx

(ii.) b) a ∈ F'K . ⊃ . (∃x) . xKa

(iii) all the parts of b are parts of a. ∧Then a extends over b.∧ This doesn't come out of the others! – Tacitly assuming that a and b aren't the same.

a ≠ b : bKx . $\supset_x$. aKx : ⊃ . aKb .

[for every value of x] Dots used both for word "and" and for brackets and for superior brackets use extra number of dots.

Different nations have different ways of reading – right or left, up or down. Don't adopt any of these here – look for the ⊃ with the most dots and read from there.

Before that is the Protasis and after it the Apodosis.[2]

Then can analyze either of them.

1. The Radcliffe version of this lecture, or portions of it, from notes taken by L. R. Heath earlier on the same day, begins on |9| of her notes, page 477.
2. We would now call these the 'antecedent' and the 'consequent'.

(iv) aKb . bKc . ⊃ . aKc [<u>not</u> so for <u>all</u> relations. We have here a "transitive" relation]

(v) aKb . ⊃ . (∃x) . aKx . xKb

(vi) Want idea that a and b are in the same space. What we mean is that there is a Volume which contains both.

a, b ∈ F'K . ⊃ . (∃x) . xKa . xKb.

One advantage of the symbolism:– Not only rid of ambiguity of language, but if you want to know what your Existential assumptions are, simply look for your "∃'s". Then little worm in your heart:– whether you haven't put down some <u>inconsistent</u> things? – Even with the most <u>obvious</u> ideas. But here no need to worry about that much.

Then: assumptions get more complicated as go along. ii b and vi are obviously not independent. If a and b are the same then vi. reduces to ii.b. [ii.b. becomes special case of vi.]∧ In vi simply need to prefix x ≠ a. From present point of view "Independence" is of no importance whatever.

So here no need to add to the complexity.

Now let's see what we want:–

Ideas of Interaction, ∧Separation, Junction,∧ Dissection.

Genuine surprise to Whitehead to find that can define all <u>these</u> and yet <u>not</u> the "<u>point</u>" out of this. Whitehead started in 1910 on this. Nicod[1] saw error in attempted deduction of the point in *Rev. de Metaphysique*...

~~Separation~~ ∨Intersection∨ is quite easy.

(vii) – a intersects b if there is a[2] common part to a here. So get a symbol involving K. K_{in} ⟨blank?⟩

No reference to <u>K</u> It's the definitions which are really important.

$aK_{in}b$ =Df . (∃x) . aKx . bKx

and (iv) knock out possibility that bKa and aKb can hold. If there were <u>one</u> instance where aKb <u>and</u> bKa holds then a = b and this ⟨blank⟩ I.e. you <u>prove</u> unsymmetricality.

For <u>Separation</u> need no new symbol.

Also need to distinguish <u>Contact</u> and Separation

Can we define complete Dissection of ⟨blank⟩

You've got to get a class ∧of events∧ that each is contained in a. And to state that any two volumes belonging [volumes?] [δ] to δ are separated from one another. Want to

|88|

get rid of possibility of any part of <u>a</u> which ~~doesn't~~ ∨neither belongs to δ nor∨ intersects a member of δ (not sufficient to say the former)[3]

Quite enough to say it doesn't <u>intersect</u> any member of the class.

1. Although Bell has spelled this name 'Nicot', Whitehead is undoubtedly referring to Jean Nicod (1893–1924), who published in *Revue de metaphysique et de morale* in 1921, 1922 and 1924; he died from tuberculosis in 1924.
2. Bell clearly has the term 'no' here, but 'a' is surely intended.
3. Missing closing parenthesis provided here.

Idea of Function. – Approaching idea of Exactness – that a Volume at an Event is an exactly determined quantity and has definite boundaries. (for Geometry at least). – Whitehead wants to show that idea of surface as a geometrical entity and point is a more abstract idea than Volume, and can be defined in terms of the latter.

Where two volumes, considerable as a dissection of one volume 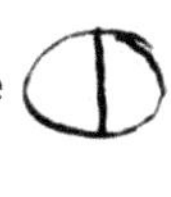 it's different from mere tangency where you have two distinct volumes.

We all assume that if aKb we can get additional set of volumes which, with b, completely exhaust a. If aKb then b is member of a class which dissects a. This ⟨blank⟩

⟨In this highly unusual case, Bell's transcription cuts off in mid-sentence.⟩

Hocking's notes

Feb. 17.

To express inclusion

F ≡ field. F . K ≡

∈ ≡ is a member of x ∈ . F'K

(i) aKb . ⊃ . a ≠ b, excludes case of including itself.
implies

There exists an event x

(ii) a ∈ F'K . ⊃ . (∃x) . aKx such that a includes x

⊃ . (∃x) . xKa also such that x includes a, something beyond

(iii) If all the parts of b are parts of a, then a extends over b.

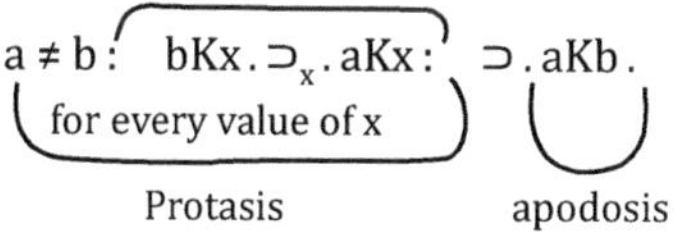

a ≠ b : bKx . $\supset_x$. aKx : ⊃ . aKb .

for every value of x

Protasis apodosis

(iv) aKb . bKc : ⊃ . aKc (transitive)

(v) aKb . ⊃ . (∃x) . aKx . xKb . (can always get something between)

(vi) a,b ∈ F'K . ⊃ . (∃x) . xKb . xKa . (Volumes in the one space)

[To make vi independent of ii 2, add a≠b]

Various Ideas to be expressed Dissection

Intersection

Separation

Touching

---

a intersects b if there is ~~no~~ a common part x

K_{in}

$_{a}K_{in}b \equiv (\exists x)\ aKx \,.\, bKx\,.$

Separation is the denial of intersection

$\sim aK_{in}b$

Now dissection (1) aKd. (2) no part not covered

i.e. none not intersecting some d

(3) If a contains b, then b is a member of a class of dissections that exhausts a.

Lecture 47

Thursday, 19 February 1925[1]

Bell's notes

|89|
Whitehead. – 19. Feb. 1925:–

Bearing of this method of procedure on Philosophy. Boutroux's studies in H. of P. (study of Socrates).[2] How would you prepare yourself for an attack by Socrates. Whitehead's method is that by which you ought to arm yourself for Socrates. Socrates asking: What is it you mean by "Volumes". Things that can't be defined but[3] only Known. Can only tell you their relations with one another. – Extending over, etc. Symbolism introduced simply to get rid of prolixity and pitfalls of ordinary language.

reasonableness[4]

Another point – harmonies there are these. – two ways of getting it:–

(1) to key down the intensities. – The second-rate way is to be afraid to commit yourself. (2) to deal with colors in themselves very vivid – but so to dispose them in relation to one another that you get harmony without loss of vividness (the Peacock's way).

Danger in Philosophical literary method is you get your harmony by sacrifice of preciseness. – Danger of "Wooliness". Literary method appeals to an intuition which is very necessary. But it keys down the logical intensity. Necessity of keeping exact logical critique going, to see that you've adequately selected your notions and accurately got their relations to one another.

Idea of Events lying to one another with an exactness of boundary. – Can we now deal in any way with junction of Events.

Dealing with exact ideas of Geometry to see what we can get.

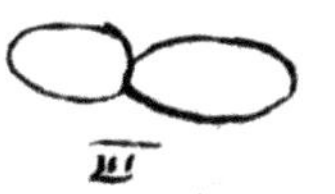

In I – a common part of x. y.

In II a common boundary (in events, 3-dimensional Boundary. E.g. this lecture room at 12:30.[5] Boundary has lost some of dimensionality [Duration etc.]).

1. The Radcliffe version of this lecture, from notes taken by L. R. Heath earlier the same day, begins on |10| of her notes, page 477.
2. Emile Boutroux's *Historical Studies in Philosophy* was translated by F. Rothwell and published by Macmillan in 1912. It incorporates five separate studies, the first of which focuses on Socrates.
3. This word seems to be 'by' in Bell's notes but the context calls for 'but'.
4. It is not clear why this word is left hanging, here.
5. Throughout the year, these lectures ran from noon to 1.00 p.m.

Then III – punctual, merely – Excessive loss of dimensionality. and finally IV.

When two Events taken together make up one Event, as one case. I, II, make up a third event: 2[1] IV means that absence of junction-enabling-them-to-get-into-one-event. Question of III? They're really two. How am I going to say that two events make one event. (Purely in terms of "Extension")? x and y together form one event when

$(\exists z) . zKx . zKy$ But also

$(\exists z) :: zKx . zKy :. zKw . \supset_w : xK_{in}w . \vee . yK_{in}w$

Another way is as in *Principles of Natural Knowledge*. But assuming III cut out, Whitehead knows no way of cutting it out by means of this definition except by presupposing existence of a point! As soon as you know what you mean by Junction of two events, you can get very near to idea of a boundary (as in II.) – Two non-intersecting Events which are joined.

Also can get this case II is adjoined. Two separate events with boundary in common.

Can find an event z so that x & z are adjoined and y & z are adjoined but z outside x and y inside it. Can we point to an entity which is the boundary?

– We've got it in our heads – and there on our diagram. Idea of walls bounding Room. But can we express it accurately, purely and solely with this idea of extension etc. If so we can make our geometrical ideas independent of other more casual ideas.

We want some unique symbol expressing transition from x or y to z. It is usual to bring in Surfaces or other deficiently dimensional things. – This brings that wretched postulational method. – Wretched Kantian idea that intellect creates ---[2]

This has for Whitehead as Logician same vice as Vitalism for the Biologist:– Too victoriously easy a method of producing an Explanation without knowing much of facts. – Explains nothing by explaining everything. You lose all critical grip.

Ex:– Let us suppose we know all about the fraction 0, 1/3, 1/2, 3/4, 1 ... 5/4, 3/2, 8/3 ... 2.

Here a "dense" series. Whatever two fractions you take you can produce a third one between them. But you can take some classes of fractions (e.g. all fractions up to ¾ vs. all from ¾ up to 2)

$0 < f < ¾ \quad | \quad ¾ < f < 2.$

Between those two classes – ¾ belongs to neither but bounds both. Is boundary of those two.

1. This seems to number the second item, but there is no '1' or '3'.
2. Bell seems to have lost the thread of Whitehead's point.

|90|

This is not the case in II. – You can't show a Volume which is the boundary of x . y, etc. But you can produce that sort of thing. Fractions whose square are < 2 etc. (Fractions):– $0 < f^2 < 2$ | $2 < f^2 < 4$

But here there is no definite fraction which bounds these. Eudoxus discovered Irrationals. Pythagoreans kept it a secret. – Thought it horrible – cherished secret. √2 isn't a fraction. Greeks hadn't such horror – because of their wretched arithmetical notion they had constant recourse to geometry to illustrate their arithmetical ideas. They looked on it as ratio of diagonal to side.

But that brings all the geometrical difficulties into arithmetic and you never get clear-cut problem to hand over to Philosophy and whole point is to get problem clear-cut for philosophy.

Now cannot we purely by reference to fractions find something which would have the boundary character. In list above there is an immediacy of transition but there is no "the boundary" there. If can reproduce advantage of having the ¾ in first case in the second case. Very simple. Took about 2000 years to do but is quite simple and knocks out all that talk of postulating a mathematically significant entity. It isn't the mere mark on paper; it must be the meaning of the mark [√]. You assume a unique meaning but can't quite see how.

The first class of entities defines its own boundary. But suppose we talk about classes of fractions. You can define what you mean by addition and by multiplication $\phi_{3/4} + \phi_{\sqrt{2}} = D\hat{f}\{(\exists f_1 f_2) . f_1 \in \phi_{3/4} . f_2 \in \phi_{\sqrt{2}} . f = f_1 + f_2\}$

[class of fractions of which f is any one of them]

(Assumes simply you know + for fractions themselves

So

Then have to show Def. × in same way – only alteration = f1 × f2 about.

Associative law, Commutative law, and Distributive law and then classes of "Unit" and "Zero".

Then you get (dealing with "classes" of fractions) --- you get "Real Numbers" which are really Classes of fractions. But have advantage of giving you none of these awkward gaps. – There is a definite ∧whole number – a∧ class corresponding to √2 and the other "gaps".

There is no such thing as postulating a mysterious entity which we "call" √2. But considering Logical Ideas which can be seen to apply to classes in respect to fractions – and picking out a class which supplies exactly what we want.

That knocks off his perch the man who wants to appeal to π, √2 etc. as evidences of "creative activity of Intellect". Intellect only finds what is there. He says simply: then:– Creation of a Class is an example of "creative" activity. But you've at least defined your central concept. – That of Classes – not √2 or π. Kantian's got to state it for Classes or give it up altogether. Enormous advantage of defining just where the philosophical discussion can begin. We can produce entities which exactly serve purposes of a Boundaries, etc.

Hocking's notes

Feb. 19. 1925

Junction of Events

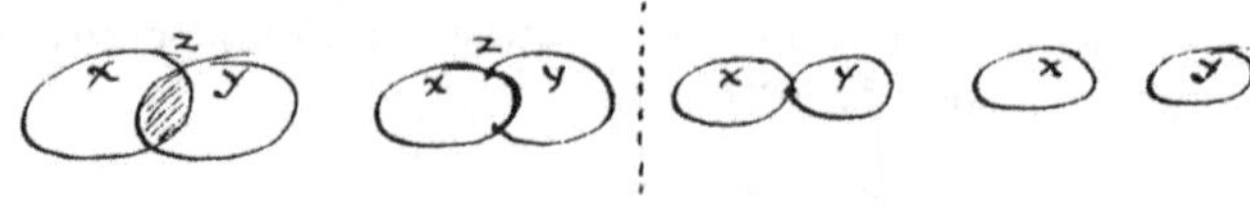

New Idea} x and y make one event   x and y do not make one event
$(\exists z) :: zKx . zKy :. zKw . \supset_w : xK_{in}w . \vee . yK_{in}w$

They make one event when there exists an event z, etc., above

Boundary

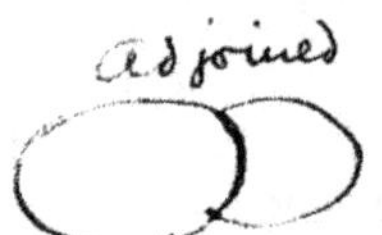

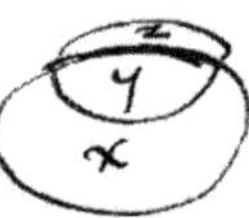

$(\exists z)$. x & z are adjoined
y & z are adjoined

(Evil that the intellect creates the objects it thinks about. Has the vice of ease in explanation. Same vice as vitalism. Explains nothing by creating everything. So can't create entity by postulating it.) | Wretched postulational habit

0, 1/3, 1/2, 3/4, 1, 5/4, 3/2, 8/5, 2. Between any two infinite other fractions. Dense series.

Compare all the fractions up to ¾. & All from ¾ to 2
$0 < f < ¾ \mid ¾ < f < 2 \mid\mid ¾$ doesn't belong to either
$0 < f^2 < 2 \mid 2 < f^2 < 4 \mid\mid {}^2\sqrt{2}$. There is no such fraction.

Get the diagonal of a square. There is your entity. But it is borrowed from an alien science. Bad policy. Don't quite know what you are dealing with.

Get a boundary without postulating some mysterious entity
Can't be a mark on paper. It must be the meaning of the mark.
Deal with classes of fractions – not fractions

$\phi \equiv$ class of . $\phi_{\sqrt{2}}$. $\phi_{3/4} + \phi_{\sqrt{2}} \equiv \hat{f}\{(\exists f_1 f_2) . f_1 \in \phi_{3/4} . f_2 \in \phi_{\sqrt{2}} . f = f_1 + f_2\}$
class of fractions such that

$\phi \equiv$ class of . $\phi_{\sqrt{2}}$. $\phi_{3/4} \times \phi_{\sqrt{2}} \equiv \hat{f}\{(\exists f_1 f_2) . f_1 \in \phi_{3/4} . f_2 \in \phi_{\sqrt{2}} . f = f_1 \times f_2\}$

Associative   Commutative   Distributive

That is what you mean by adding or multiplying classes of fractions the possibility of picking out members from each class and multiplying them

You knock off his perch the man who takes the √2 or π as instance of the creative activity of the intellect

Feb. 19.–2
That is, you almost knock him off his perch. If there is VanyV Creative activity Vit isV not in Videa ofV √2 but in the idea of classes. Kantian must either hold it in regard to classes or drop it altogether.

We can produce entities which exactly serve the purpose of a boundary, and show the place where the philosophical discussion begins.

Lecture 48

Saturday, 21 February 1925[1]

Bell's notes

|91|
Whitehead:– 21. Feb. 1925. –

Abstract of Whitehead's point of view in respect of procedure. – Whole doctrine consists in accepting doctrine of internal relations and taking it seriously. Point of the doctrine is that relationships of a real entity modify its essence. If relationship modifies the essence, you cannot then explain what the entity is without considering relationships. Doctrine of aspects. Entity is in itself a grasping together into a unity of aspects of other entities. Essence is constituted by, and modified by, aspects of other relationships. But this brings you up at once against what Whitehead considers nerve of any sound metaphysics – i.e. consideration of that limitation which is left in background too much – How an Entity can be its own limited self and not a vague all things etc. and yet have relations to other things.

How it grasps other things into itself under a limitation – and that is Relationship. Relatedness of real concrete Entity A to another B is the aspect of B for A. Calls that grasping together the "Prehension" (excluding any necessary reference to Cognition). A real Entity is a Prehension of all other Entities under limitation of Aspects. They are not an undifferentiated group; but you get mutual relatedness with a Pattern of Aspects. The aspects not an internal fact. A in being itself is also beyond itself. In being itself the thing grasps --- ?? ⟨blank⟩

Can these aspects and patterns of Aspects be exhibited in an impartial scheme which can express --- for all entities. The various types and orders of scheme here. There is a great inclusive scheme – the Spatio-temporal one which expresses how aspects are, which enters into all ∧the∧ patterns of aspects. therefore You get a partial expression of scheme of relation of events which has freed itself from any one Entity.

Viewed as a scheme is obviously of itself more abstract than the concrete fact; and relations of Entities as conceived under that scheme is therefore of higher abstraction than concrete facts. But remains a Key fact. In relation to that scheme Whitehead calls the Entities: "Events." Question: How to express this scheme in fundamental --- of ideas. Whitehead feeds his description of the scheme with ideas one after other and sees how far can go in expressing what this scheme is. (Much dropped out here, of

1. The Radcliffe version of this lecture, or portions of it, from notes taken by L. R. Heath earlier the same day, begins on |15| of her notes, page 480.

course – Idea of Process left in Background here.) As basis of scheme let us take idea of Extension:– one event including another – A peculiar and definite relation; Relata are to be only Events. To express all details of scheme in exact --- ⟨blank⟩

It has been found convenient to get beyond Events to idea of Point-Instants – This involves a higher Abstraction still. – Reference to Arithmetic:– Real numbers are higher abstraction than fractions and these than cardinal integers. We get great advantage in thought (for us if not for God) in these Abstractions. They simplify thought by enabling you to replace this Whole-Part relationship as referred to Extending over, by the more familiar one of all and some. And enables you to express ideas of Exactness much better (Exact transition from class of fractions $< \sqrt{2}$ to that $> \sqrt{2}$ and yet[1] no fraction which marks that transition. As matter of fact thought there gets itself on simpler basis (with whatever complexity of logical determination) by giving you higher abstractions where however you get a transition one).

Consider relation of two events to one another[2]

Going backward here – not following logical development – But when you are going to do the logical development: ~~but ha~~ what you have in back of your mind aKb.

We want to get to idea of points. You can replace "x extends over y" by "all points of x are one of the points of y" etc. But impartially facing the idea of "Aspects" in terms of volumes, "point" isn't clear in its nature.

Also "Intersection" "Junction" and "Complete Separation" discussed.

III is separated by adjoined. V is separated and not adjoined. Haven't discussed IV yet. In I, II, III, there is "complete" relation of togetherness. These three are peculiar in fact that there is a particular event z such that zKu. zKy and also that

$zKu \,.\, \supset_u : xK_{in}u \,.\, \vee \,.\, yK_{in}u$

$(\exists z) :: zKx \,.\, zKy :.\, zKu \,.\, \supset_u : xK_{in}u \,.\, \vee \,.\, yK_{in}u$[3]

Here really bringing in a new

|92|

idea or type of assumption about my ideas – this peculiar relation which some pairs of Events have and others don't. When you come to the All and the Some that peculiar property becomes rather blurred. Class χ whose extension is x and y (two members).

In IV Whitehead maintains essential duplicity between events going on in this room now and what went on in Thebes in 2000 B.C. – there

1. This word might be 'get'.
2. The diagram here is very similar to one used in the previous lecture.
3. Into this formalism we have introduced some corrections.

is, taking them together, only the class of these two. There is no ~~2~~ here.[1] Now take class of all points in x and y taken together $\tilde{\omega}$. [Then question whether numerically $\tilde{\omega} = \xi$. – Here I and II would differ from all the others! i.e. III would be with IV and V. [my own]]

Now we want to get hold of those Entities by which Exactness of Geometry comes. Think of it in points. Child and Euclid – paring down and down until you get down to being rid of all the parts and are left with a point. Now that's all very well; but it leaves you with (so far as "Point events" go) a class of entities which you can't quite see how to bring in.

That paring down is a sort of route of approximation to a certain ideal of exactness ∧and absolute simplicity∧. Suppose you want to get exact relation of what has happened in this room (for this hour) – or a certain verging to simplicity in concrete content if you reduce your dimensions.

for tenth of a second etc.

Now, however, are what we know about data consistent with this ideal. Our senses are always inexact. The fact that we know our senses are inexact shows we know there's an exactness which they don't hit. "You can't not be what isn't, you know."

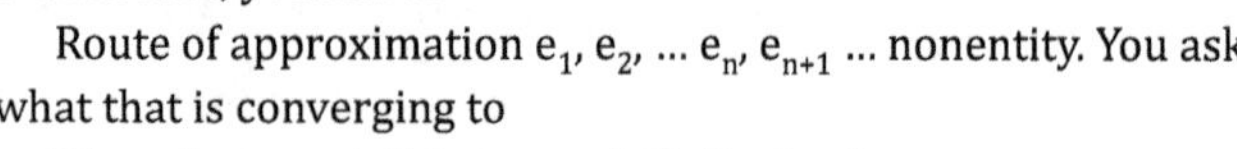

Route of approximation $e_1, e_2, \ldots e_n, e_{n+1} \ldots$ nonentity. You ask what that is converging to

There is no such thing as an infinitesimal.

Brought in unfortunately by Leibniz. Calculus discovered independently by Newton, not a philosopher, and Leibniz who was one. Yet Newton never said anything that today philosophically we have to reject. Saw how far to go and then stopped. But Leibniz being a philosopher kept on thinking – and talked of infinitesimal Entities. Right idea not for another two hundred years (see Weierstrass).[2]

By those mathematicians who do really understand what they are talking about, "Infinitesimal Entities" looked on only as a useful bluff. Any volume is of a certain size, and therefore no Volume leads to this idea of simple point. So this leads you up to nothing – not zero, but nonentity. Ought really to eliminate the blackboard and everything there.

What's the use of it? You have physical occurrences there simpler as you go along the line. Suppose there were a set of physical measurements q connected with e_1 – $q(e_1)$. (all quantities, e.g., in Smithsonian tables) Then you are getting down to greater and greater simplicity as you go along. These do converge towards something. A lot of them will ultimately converge to zero. All the lengths e.g. But some of these Entities are in form of Ratios between your measurements. Distance a train goes in Time, shorten time and take ratio of distance gone to time. Converges to idea of velocity at an instant. You get a law of convergence and see that it

1. With the strike-outs this sentence does not seem meaningful.
2. See footnote 2 on page 19, lecture 5 (4 October 1924), by Whitehead's first mention of the 'infinitesimal' of Leibniz, and 'that great German Mathematician', Weierstrass.

converges toward something perfectly definite – or idea of a Density at a point at an instant. The class of Limits which turn up along certain Routes of Approximation. A definite class of Limits.

Not only so:– but you can define Routes which in a sense have the same type of Convergence. E.g. The converging ○'s and □'s. In a sense they converge to the same nonentity. The convergence is the "same" meaning?
$q(e'_1)\ q(e'_2) \ldots q(e'_n) \ldots \rightarrow q(l) \leftarrow$ this is same class of limits as for $q(e_1) \ldots$ etc.
Hence the physical importance of the "point". You can free it from any definite type of convergence and get it to a "point". Can you, then, really define what you mean by $e_1, e_2 \ldots e_n \ldots \rightarrow$ Nonentity purely by idea of events extending over one another, and can you define the q1 classes purely with reference to extension? If so, you have got a geometry put on its feet a priori without reference to empirical physical facts.

Hocking's notes

Feb 21.
The doctrine of internal relations taken seriously. The relations modify the essence. You can't explain what the entity is without considering its relations.

How an entity grasps other things into itself under limitation. An entity is the prehension of all other entities under a limitation of aspects.

The business of half a session compressed into fifteen minutes.

The relata are only events Abstractions of point event.

With man it is not possible to deal with universe without abstraction. With God it may be possible.

For we take the idea of whole & part and are enabled to express exactitude.

Exactness of transition without an entity to mark it. Class of fractions whose squares are ⪤ 2 [1]

Feb 21 – 2

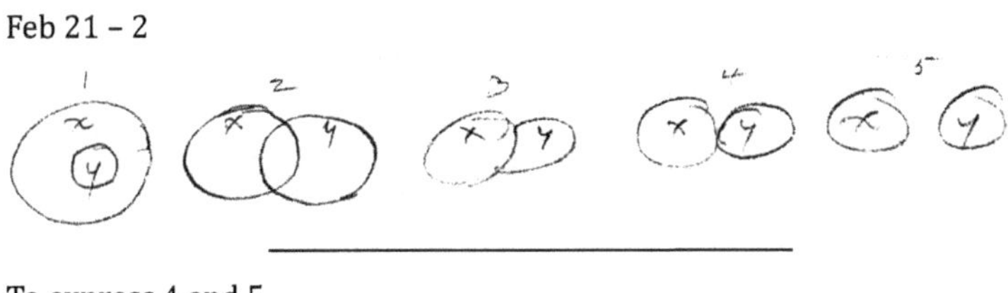

To express 4 and 5

1, 2, 3. There is always a z such that it contains x and y and there is no part of z which is not also a part of x + y.

1. This seems to be a symbol that means 'Class of fractions whose squares are less that 2 and class of fractions whose square are greater than 2'.

$(\exists z) :: \overbrace{zKx \,.\, zKy :.\, zKu \,.\, \supset_u : xK_{in}u \,.\, \vee \,.\, yK_{in}u}$

In 4, 5. There is no z which does not have an essential duplicity.

[you have to have first a general idea of what you want to express. Not reason along & then say, God bless my soul how interesting this all is!]

Idea of a point

Explained as a child by starting with volume and paring it down bit by bit.

Route of approximation to an ideal of exactness.

Senses are always inexact. We know this. That means we know an exactness they don't hit.

Unless we knew exactness it is nonsense to say senses are inexact.

Half the troubles of philosophy come from talking of nothing and thinking it is a kind of something.

$e_1\ e_2$ --- en $\quad e_{n+1} \rightarrow$

Leibniz brought in the idea of the infinitely small which doesn't exist. Newton who was not a philosopher [when he had got as far as he needed for his own purposes, he stopped thinking – a definition of a man who is not a philosopher] never said anything that needed to be corrected.

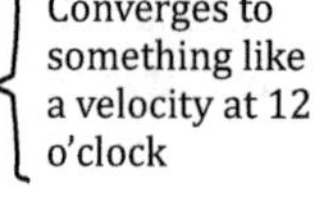

oughtn't even to write it Nonentity [not zero]

✓

Weierstrass took Leibniz' infinitesimals and they got into philos. & never got out A useful sort of nothing.

Any volume is of a <u>certain</u> size

The class of all such entities } $q(e_1), q(e_1)$ --- $q(e_n), q(e_{n+1}) \rightarrow q(e^0)$[1] { Converges to something like a velocity at 12 o'clock

Feb. 21 – 3

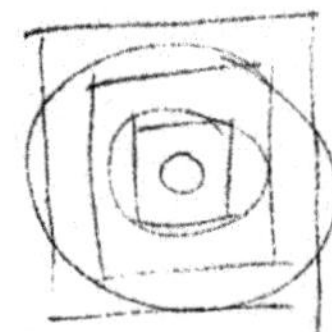

$e_1\ e_2$ --- $e_n\ e_{n+1} \rightarrow$
$e'_1\ e'_2$ --- $e'_n\ e'_{n+1} \rightarrow$ } Same convergence — nonentity

You can free it from any definite route of convergence.

Can you define what you mean by the relation of one to the other.

1. Hocking seems to leave a 'q(l)' with a zero directly above it, but we are neither certain if the 'l' should be another 'e' in this last element, nor whether the zero is meant to be a superscript.

Lecture 49

Tuesday, 24 February 1925[1]

Bell's notes

|93|
Whitehead – 24. Feb. 1925 –

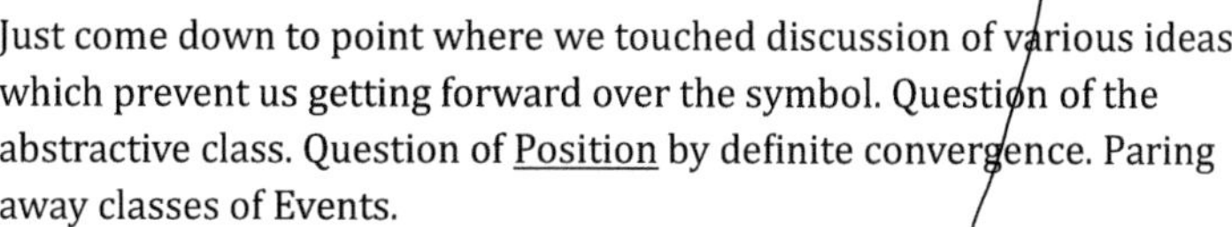

Just come down to point where we touched discussion of various ideas which prevent us getting forward over the symbol. Question of the abstractive class. Question of Position by definite convergence. Paring away classes of Events.

Volumes – must converge to a line, a surface, or a point. We've first got to put down quite exactly what you mean by an abstractive class. An abstractive class has ∧[here]∧ reference to idea of Extensiveness – what we've called K. Ab'K is Abstractive class belonging to K.

$\alpha \in \mathrm{Ab}'K$ [α is an abstractive class of K.] – This to be defined.

$\alpha \in \mathrm{Ab}'K = \mathrm{Df} :: x,y \in \alpha \,.\, x \neq y \,.\, \supset_{xy} : xKy \vee yKx :$

Then also $\sim(\exists v) : x \in \alpha \,.\, \supset_x \,.\, xKv$ If x and y are members of α.

(Then knocking out case where you've just got two names for ∨same thing∨) Knocking out case where core of Events to which the class approximating.

Now pitfalls:– As soon as pledged to be concise, see that you haven't included some silly trifling case that is nugatory but - - -. What about Class of no members at all. Then x,y $\in \alpha$ is always false and then the $\supset$ is always true. If α has no members then the second line proposition would be always true although $x \in \alpha$ always false – whole proposition true. But we've started with it $\sim$ (false).

[Can put in $(\exists\alpha) \,.\, x \in \alpha$]

Then question of equivalent class convergences.

Then more general idea. (Convergence to a line, or a plane, or point on a plane.)

Suppose α covers $\beta = \mathrm{Df}\ \alpha K c \beta = \mathrm{Df} :. \alpha, \beta \in \mathrm{Ab}'K : x \in \alpha \,.\, \supset_x :. (\exists y) : y \in \beta \,.\, xKy$.

However far down you get with your α there is a β which is inside it.

"However far down" That is where your interest comes in. But not for Logic:– There the important thing is any. That is really what Weierstrass found for limits. The "any" includes the "smallest possible". Latter appeals to dramatic sense – Yellow press stuff. Again always be on lookout to see that α, β same class, whether O.K. or not. Always look for the extreme cases and see whether you want them or not. Here the sameness not knocked out. A class "covers" itself then.

1. The Radcliffe version of this lecture, or portions of it, from notes taken by L. R. Heath earlier the same day, begins on |19| or |20| of her notes, pages 481–2.

α K-equals β =Df . $\alpha K_{eq}\beta$ =Df : $\alpha K_c\beta . \beta K_c\alpha$

We've got	K	K_{In}	K_c	K_{eq}
Holds between	Events	Events	Abstractive Classes	Abstractive Classes

We have also got ⊃ ∨ (x)[For every value of x] (∃x) , $\supset_x$ and uses of dots

There is the backbone of language – the propositions etc. Substantives – if you know about every fifth word you are lucky. – These for special topics. So ⊃ v ∃ etc. are general symbolism. K_{in} K_c etc. to special topic. You forget about it when beyond this particular investigation.

Minimum demands of relations between Entities which one thinks of as equivalent. Where always tempted to put down =. Certain minimum demands. Such a relation must be

(i) Reflexive aRa. (ii) symmetric $\alpha Rb . \supset . bR\alpha$

(iii) transitive aRb . bRc . ⊃ : aRc [Things equal same thing, equal one another]

K is iii, not i, not ii

K_{in} is i, ii, not iii

K_c is i, not ii, is iii (not necessarily ii but may be)

K_{eq} is i, ii, iii

Now ii and iii will give you i.
aRb . ⊃ . aRb . bRa
⊃ . aRa

Sort of use that Symbolic Logic is as auxiliary:– helping Philosophy. At this stage we ought to pull ourselves up and think what general ideas are which we want to have control us. Take elementary ideas of Euclid. Sir Thomas Heath's "Euclid in Greek".[1]

Euclid doesn't go behind the idea of Equality or Equivalence. But instead of this being one simple idea is variable and complex. You have got relations so that for some particular purpose any one of the bundle joined by those relations will serve equally well. Philosophy – 2000 years of it – has taken Equality just as naively as Euclid did.

Euclid says aRc . bRc . ⊃ . aRb (Things = to same thing etc. etc.).

If you add aRa you can get out the two things quite easily.

[94]

For suppose bRa . ⊃ . aRa . bRa . Euclid always presupposes Reflexiveness

⊃ . aRb . Euclid somehow thought Equality (ἴσος) meant same wherever you used it – instead of many different things, as it does.

aRb . bRc . ⊃ . aRc .

Then Euclid slips in another idea showing what he's got in back of mind: ("If equals be added to equals the wholes are equals." Also remainders equal etc etc.) Then he's got another idea – that of Addition – concerned somehow or other with idea of addition. Addition of two numbers is quite different from adding two Events together [My Whole-Part etc]. Addition, then, one of those ideas whose meaning depends on circumstances. Certain weakness in Greek Philosophy – They took Language as very good reflection of philosophical interests and then

1. Sir Thomas Little Heath (1861–1940) entered Trinity College, Cambridge, and became a Fellow at about the same time as Whitehead. Heath translated several works of ancient Greek mathematics, including the 13 books of Euclid's *Elements* (Cambridge University Press, 1908).

analyzed language. But they weren't sufficiently alive to tracks of different meanings in the one word. The Greeks knew no language but their own. Main advances of modern philosophy.

This idea of Addition has to be thought afresh for every field of thought. – Addition of Events where you've got real junction – so that you get a whole event out of them – a third entity of the same species, as resultant. Unless there is some basis for Addition, you can never get it out.

There is room obviously also for lots of subtlety in considering addition of Abstractive Classes.

Then funny little point – connected with last two Axioms of Euclid. Things which are congruent to same thing as congruent to each other – and whole greater than part. Point of historical interest here. – In Whitehead's school days not little Peptonised editions. – But whole Liddell and Scott. επιαρμονιζειν[1] One of key words in Euclid – one of greatest books in Greek. No reference to Euclid at all – To Xenophon etc.

Limitation of Greek Scholarship in 19th century.

Hocking's notes

Feb. 24.

Ab'K [The Abstractive class of K]

What do we mean by saying that any class α is an abstractive class?

$\alpha \in Ab'K = Df :: x,y \in \alpha . x \neq y . \supset_{xy} : xKy \vee yKx :$

To exclude a limit $\sim (\exists v) : x \in \alpha . \supset_x . xKv : (\exists x) . x \in \alpha$

this side of zero unnecessary

There is to be nothing to which these classes approximate

Definition of covering

α covers $\beta \equiv \alpha Kc\beta \equiv \alpha, \beta \in Ab'K : x \in \alpha . \supset_x .$ $(\exists y) . y \in \beta . xKy .$

However far down you get your α

You can find a β inside it.

["However far down" is transcribed by $\supset_x \equiv$ "for any value of x". The symbols do not reveal your interest.

Weierstrass discovered that "for every value" includes "however far down."]

"You can find a y which is also a β such that x ~~covers~~ extends over y"

1. Although this at first looks like a term constructed on 'harmonise' (which is in Euclid), it may actually be a reference to ἐφαρμόζω, which would be 'to fit' or 'coincide'. This seems to be confirmed on 19 March when Whitehead returns to these issues, saying: 'We've now got to enter upon idea of congruence. Euclid relies on idea of applying one body to another [unfortunate word --- = "fit on"].' See the end of |113| in lecture 59, page 274.

Feb/24 – 2

Get in the idea of equality K_{eq}[1]

α K-equals β ≡ $\alpha K_{eq}\beta$

$\alpha K_{eq}\beta \equiv aK_c\beta. \beta K_c\alpha$

Symbols

K, K_{in}, K_c, K_{eq}

⊃, ∨, (x), (∃x), $\supset_x$, ., :, :.

R. Relation of equality

Demands on the relation. It

(I) ought to be reflexive aRa	ii & iii
(II) symmetric . aRβ . ⊃ . βRa[2]	will give you
(III) transitive aRβ . βRc . ⊃ . aRc	i

K is iii and not i and not ii

K_{in} is i, and ii, not iii

K_c is i, and may be ii but needn't be, and is iii

K_q is i, ii, and iii.

Feb. 24. 3.

Come back now to Euclid's axioms –

See the importance of a very careful analysis –

1. Things equal to the same are equal to each other

Equality is not a simple but a variable conception.

Must at least be i, ii, iii.

If aRc and bRc. ⊃. aRb.

Add aRa, and you can get out ii and iii.

bRa. ⊃. aRa. bRa

⊃. aRb

aRb. bRc. ⊃. aRb. cRb. ⊃. aRc ≡ iii

---

2. Equals added to equals, the wholes are equal –

taken away from remainder.

Addition = Real Junction

Two entities engendering a third of same species.

---

1. While Hocking's handwriting suggests 'q', Bell makes it clear that this should be 'eq' for 'equals'.
2. Hocking seems to be using Greek 'β' within these two symbolic phrases, but perhaps they should be 'b', as elsewhere.

Last two axioms.

3. Things which are congruent to the same are congruent to each other

4. Whole greater than parts.

<u>Congruence</u>. –

<u>Boolean Symbolic Logic</u>

Russell –

Bosanquet –

Lecture 50

Thursday, 26 February 1925[1]

Bell's notes

|95|
Whitehead – 26. Feb. 1925

Came late.[2] – Whitehead giving essay topic.

"Auxiliary methods in Philosophy." Philosophy a science on its own – by also motive power and critic in every science. Every science deals in a field of Abstractions. It is for Philosophy to criticize those abstractions. Further it is for philosophy to point out how human life requires emphasis on other ideas not yet scientifically tested. We are only gradually covering ideas we use, with a network of Sciences. As complexity of life increases ⟨blank⟩

Philosophy is the discoverer of Sciences – and of ideas by which sciences can be recast, and made more useful etc. But to do this philosophy has to have its own ["Problematic"]. Has to consider the Sciences. – How far Ethics related to Physics, etc. But also has some things to find out on its own. These "auxiliary methods" – 1) this Strict Logical method we have been looking at. 2) Literary presentation. 3) History of thought – considering tonality of ideas from age to age. 4) To have an active interest in immediate general thoughts of your own generation.[3]

Idea of Philosopher retiring to neat little ∧set of Science Abstractions∧[4] of his own which is no use to anyone seems to Whitehead false.

Aesthetics, etc. – "All of them bear on Philosophy". Following one line doesn't mean devaluation of others. If following strict mathematical method. – It is quite easy, if you're expert enough at symbols, to use symbolic Logic as method of smuggling in new ideas without saying you're doing so. – Get into such maze of symbols that you fail to remember what you are doing.

Pitfalls of the literary method: – How you find yourself "landed" under charm of style. You have your choice of pitfalls. There is no method in any science which will supersede Common Sense and real Insight.

Considering Euclid's axioms. Had come to idea of two abstractive Classes which were equivalent to one another. Idea of "Equality". Euclid's first

1. The Radcliffe version of this lecture, or portions of it, from notes taken by L. R. Heath earlier the same day, begins on |22| or thereafter of her notes, page 483.
2. There is no mention in Bell's diary of his coming late, but on 24 February he acknowledges that he had 'missed a dinner invitation of the Hockings' and that he was 'up till about 4 every night over lecture et al'.
3. The abbreviation 'gener^{n}' could also be 'generalisation'.
4. It might also be that an alternative is intended: 'Scientific set of Abstractions'.

Axiom dealt with "Equality" from idea of Equivalence. This not one definite Relation but an Idea which would apply to whole class of relations – all which are symmetrical, transitive and reflexive.

"Equivalence" is only another way of taking notion of Species. Any two members of Species are "Equivalent" for any purpose for which you want a member of the Species as such. Euclid's first Axiom merely way of stating that he's got an idea which satisfies this type of relation. (Euclid forgets the reflexiveness).

Another idea is "Equals" as it comes from Addition. Now, have we anything like that? Yes! Idea of Junction. – two things which, put together, blossom out into a third one of the same kind.

Then idea of a test of Equality. Euclid's idea of Congruence. He is going to use his famous method of Superposition. This leads into difficulty, because presupposing rigidity of body. This doesn't follow from mere self-identity. You can make it a matter of primary knowledge when a body does or does not remain congruent with itself during its history. You've got to come down to that perhaps. But it's not unanalyzable. – You've got to have your Geometry first, and then see your Congruence in terms of its Structure.

I.e. you have got to have a non-metrical Geometry which gives you your Structure: – and then things with homologous relations to the Structure are congruous.

$\alpha K_{eq} \beta \,.\, \alpha,\beta$ both converging toward same type of limit (∧instantaneous surface∧, point or line etc). Whole species of classes with same type of Convergence.

There comes in a difficulty of dealing with this "same type of convergence". Suppose a class of Ellipses all just touching at extremity of minor axes.

Class α converging up to point p.

Could also have set β converging to same p but touching at ends of their minor axes. They have same convergence, but you have horrible exceptional case.

α covers β but no β may cover any α (you can get this case).

Now illustrating sort of trap you can fall into quite easily. Whitehead started
|96|
out by thinking could define the Point Instant by convergence of abstractive classes by a convergent class. Such that whatever class it covers covers it.

If $\alpha K_c \beta \,.\, \supset_\beta .\, \beta K_c \alpha$ This holds quite nicely except for special cases.

Exceptional case of tangency at the Surface. You can only knock out the exceptional cases by assuming you know what having a point at the Surface means, etc. by assuming you know all about the Geometry you're trying to define. Whitehead's paper – elaborate – showing how by being sufficiently elaborate you can hoodwink yourself. Whitehead found it

out, and only other one is Nicod[1] – (Perhaps no one else read it!) Very successful paper therefore! –

α covers γ but isn't covered by it.
But α and β cover one another.

Have to bring in additional notions (conditions, ideas). How treat these to elicit from them the geometrical elements. No use looking for them until you know what you are going to do with them when you get them.

Let us call σ name for some undetermined Species (looked at it in Extension: – That's the Species of these abstractive Classes you're dealing with.) $\alpha \in \sigma$ means α belongs to right type. I want sharpest, simplest possible convergence compatible with belonging to that Species. Call these "Primes" – Namely the Primes that arise from condition σ: – The σ-primes. Looking at it as a class – class of σ-primes: – Write prime(σ).

$\alpha \in \text{prime}(\sigma) = \text{Df} :. \alpha \in \text{Ab}'\text{K} . \alpha \in \sigma : \beta \in \text{Ab}'\text{K} . \beta \in \sigma . \alpha K_c \beta . \supset \beta . \beta K_c \alpha$

α is an abstractive class

Common Sense to have a short symbol for Class covered by both in this case: –

The joint class . $\zeta \cap \eta$.

Hence this is $\beta \in \sigma \cap \text{Ab}'\text{K}$

It is one of your difficulties to know how much symbolism is worth introducing. But you might want the fastest[2] type of Convergence you could get. "Antiprime"

$\alpha \in \text{Antiprime}(\sigma) . = \text{Df} :. \alpha \in \text{Ab}'\text{K} . \alpha \in \sigma : \beta \in \text{Ab}'\text{K} . \beta \in \sigma . \beta K_c \alpha . \supset \beta . \alpha K_c \beta .$

α then has most complex type of convergence possible under those conditions.

We want now the idea of all the Classes α which are both equal to one another and are σ-primes – Those converging to same point. – Call this class of "Geometric Elements."

Species of all classes are to one another and primes, is the point.

G_σ'K = bundle of Geometrical Elements obtainable out of σ. If G_σ'K then a series of points. Suppose P is a Geometrical Element got out of condition σ.

$P \in G_\sigma'\text{K} : = \text{Df} :. (\exists \alpha) : \alpha \in \text{Prime}(\sigma) . \alpha \in P : \beta \in P . \supset \beta . \beta \in \text{Prime}(\sigma) . \beta K_{eq} \alpha :$
$\beta \in \text{Prime}(\sigma) . \beta K_{eq} \alpha . \supset \beta . \beta \in P$

We want to state that all members of P are all sigma primes.

All members of P are primes[3] ---

All have same convergence – all satisfy σ etc. This Species which you there define, is your Geometrical Element.

Other considerations of Sigma – from consideration of Time. Differentiation between Time and Space. Because if you come down on idea of physical body – you come down on it at an instant.

---

1. As previously, 'Nichot' is undoubtedly a reference to Nicod; see footnote 1, page 208, lecture 46.
2. Bell's handwriting might be read as 'fattest'.
3. It is not clear which portions of the formalism each of these qualifiers refers to.

Hocking's notes

Feb. 26. 1925

Congruence presupposes the rigid body –

You can make congruence a primary idea –

But Whitehead thinks it can be analyzed up further –

Structure – Endurance of a pattern

Geometry first. Congruent things have homologous relations to geometrical structure.

When two abstractive classes are equal $\alpha K_{eq}\beta$. The point instant is to be the whole species of classes with the same convergence.

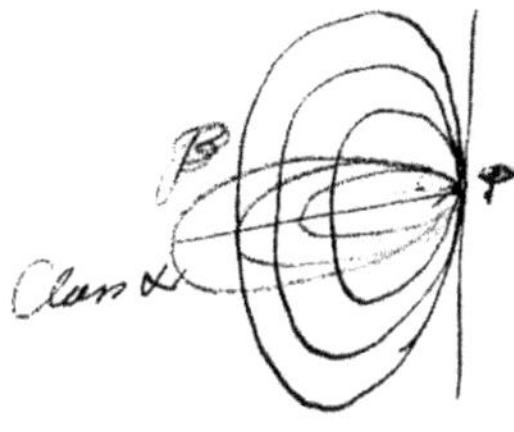

α covers β but β does not cover α

Thought at first that could define point as

$\alpha K_c\beta \,.\supset_\beta.\, \beta K_c\alpha$

But the two sets of ellipses give an exceptional case.

Have got to bring in some other condition.

How are you going to bring them in so as to elicit the geometric element?

σ. stands for species[1]

Let s be the name of an undetermined species of abstractive classes α, $\alpha \in \sigma \equiv$ alpha belongs to species

Now how to get the sharpest convergence compatible with belonging to that species σ.

The "primes" that arise from this condition will be called sigma-primes or Prime(σ), meaning

$\alpha \in \underline{\text{Prime}}(\sigma) \equiv \alpha \in \text{Ab'K} \,.\, \alpha \in \sigma : \text{(and also)}\ \underbrace{\beta \in \text{Ab'K} \,.\, \beta \in \sigma}_{\beta \in \sigma \cap \text{Ab'K}} \,.\, \alpha K_c\beta \,.\supset_\beta.\, \beta K\alpha$

Antiprime, $\alpha \in \underline{\text{Antiprime}}(\sigma) \equiv \alpha \in \text{Ab'K} \,.\, \alpha \in \sigma : \beta \in \text{Ab'K} \,.\, \beta \in \sigma \,.\, \beta K_c\alpha \,.\supset \beta \,.\, \alpha K_c\beta \,.$

$G_\sigma\text{'K} \equiv$ the class of all the geometric elements that can be got out of σ. Let P be such an element

$P \in G_\sigma\text{'K} \equiv (\exists\alpha) : \alpha \in \text{Prime}(\sigma) \,.\, \alpha \in P : \beta \in P \,.\supset_\beta.\, \beta \in \text{Prime}(\sigma) : \beta K_{eq}\alpha$
$.\, \beta \in \text{Prime}(\sigma) \,.\, \beta K_{eq}\alpha \,.\supset_\beta.\, \beta \in P$

necessary & sufficient conditions.

All have the same convergence

What are these conditions?

Have to get it from Time.

---

1. This is written vertically up the left side of the page.

Lecture 51

Saturday, 28 February 1925[1]

Bell's notes

|97|
Whitehead – 28. Feb. 1925[2]

Topic – How to produce Geometry by method of Extensive Abstraction.

State we've come to – Have defined what we meant by a "prime" class subject to σ.

The idea of some definite set of conditions *yclept*[3] σ (σπεζ·

Only abstractive classes can fulfill these conditions.

"Primes" are the set of abstractive classes which both belong to that Species and also closest ∨sharpest∨ approximation to least complex type of Geometrical Entities possible under σ.

$\alpha \in \text{prime}(\sigma) = \text{Df} :. \alpha \in \sigma . \alpha \in \text{Ab'K} : \alpha K_c \beta . \beta \in \sigma . \vee . g \in \text{Ab'K} . \supset \beta . \beta K_c \alpha$

<u>α</u> converges to a more complicated type of entity than β.

$\alpha \in \text{Antiprime}(\sigma) = \text{Df} :. \alpha \in \sigma . \alpha \in \text{Ab'K} : \beta K_c \alpha . \beta \in \sigma . \beta \text{ Ab'K} . \supset_\beta . \alpha K_c \beta$

What sort of Simplicity you must have if above sort of Idea is to be of any use.

<u>Regularity</u>.

$\alpha \in \text{Reg} . \text{Prime}(\sigma) = \text{Df} :. \alpha \in \text{Prime}(\sigma) . \alpha K_{eq} \beta . \supset_{\alpha\beta} . \beta \in \text{Prime}(\sigma)$

The first formula above could be completed.

$\supset \beta . \alpha K_{eq} \beta . \supset_\beta . \beta \in \text{Prime}(\sigma)$

[This anyhow]
Then <u>with</u> regularity

$\alpha \in \text{Reg} . \text{Antiprime}(\sigma) = \text{Df} :. \alpha \in \text{Antiprime}(\sigma) . \alpha K_{eq} \beta . \supset_{\alpha\beta} . \beta \in \text{Antiprime}(\sigma)$

Then can complete <u>second</u> formula:– $. \supset_\beta . \alpha K_{eq} \beta . \supset_\beta . \beta \in \text{Antiprime}(\sigma)$

Then if you are given regularity =

"Equivalent Routes of Approximation"

Think of approximation to a "point instant" to show sense of what we are now doing.

Condition σ so that any "Prime (σ)" has Route of Approximation → converges to <u>a</u> "point instant". I don't want to be bound by <u>one</u> route (The ⊙'s e.g.) I may want to do it by □'s

1. The Radcliffe version of this lecture, or portions of it, from notes taken by L. R. Heath earlier the same day, begins on |23| or after of her notes, page 483.
2. Bell laments in his diary that 'Hicks had promised to wake me (alarm clock on strike) but didn't. So missed my own lecture'. Yet he seems to have made Whitehead's lecture, two hours later.
3. This is not a foreign term exactly, but English from the time of Milton and Chaucer, meaning 'named' or 'called'. The sigma then seems to refer to the proposition Whitehead is constructing, but what is intended by the other Greek letters to the right is not clear.

Entity which represents all equivalent Routes.

If you ask what the point is – it is that which marks out the convergence as being ⟨blank⟩

It is all those Routes in the character of one Entity – marking out simply position to any degree of exactness you like.

Class of all these Routes of Approximation therefore. – If you know what you mean by class this entity gives it to you.

Think of all the Geometry --- G_σ‘K Suppose the class of ⊙'s is class of α's.

$$P \in G_\sigma{}^{\prime}K = Df :. (\exists\alpha) : \alpha \in Prime(\sigma) : \beta \in P . \equiv_\beta . \beta \in Prime(\sigma) . \alpha K_{eq} \beta .$$

For Equivalence see last day ≡ necessary and sufficient[1]

Now if I happen to know that σ is regular, I can knock out the (β ∈ Prime(σ)) Sometimes we want just the geometrical element from Antiprimes.

G_σ‘K $L \in G_\sigma{}^{\prime}K = Df :. (\exists\alpha) : \alpha \in Antiprime(\sigma) : \beta \in L . \equiv_\beta . \beta \in Antiprime(\sigma)\ \alpha K_{eq} \beta .$

Again if σ is regular in respect to formation of Antiprimes (β ∈ Antiprime(σ)) can be knocked out.

Sole merit of Symbolism is it helps you to make up your mind just what condition it is you want to find and what to do with them – where to look for them etc. – (E.g. for entities which have lost some of their dimensions [instantaneous lines etc. etc.])

Now how to find any conditions (like σ) which will be independent of the chances of physical appearance and belong to the "Eternal" side of the pattern. We must have something inherent in this complex of Events apart from their casual content. – It must be ∧something∧ pretty general – thoroughgoing – not any "dodge" – if all Geometry and *Zeitlehre*[2] out of it. It must be something fairly obvious.

So far: ideas of Extending over and Junction. Certain approach to idea of Structure. Whitehead started in 1910 with confident hope of getting it all out of extension. But didn't know relativity then. Then new hope when got Einstein's idea of muddling up Space and Time. It is obvious that differentiation of Space and Time can give you a key of ordering.

|98|

Idea of Process. Things which, in their very nature, imply transition to something beyond itself. Temporal transition vs Spatial transition. The differentiation between these two now to be considered. It at once strikes us: – Space is 3-dimensional. Time is much simpler. Suggests itself to us as 1-dimensional. When you consider Time-aspect of things you are thinking of it *sub species successionis*. What an enduring object is in itself. Anything with a life history. Then a rule of succession such that each of these parts as well as whole exhibited same[3] Self – a certain type

1. This comment appears in the left margin.
2. It is not clear whether Whitehead used this German term (literally, 'time teaching') or whether Bell, once again, slipped into German (for which we might read 'lesson').
3. This word is overwritten and unclear.

of unity of pattern of dealing with things. So, fundamental character of Endurance is not enduring beyond yourself but within yourself not only from moment to moment but within each moment. There is a rule of partitioning whole of life history according to temporal succession such that each part exhibits same pattern.

The universe as there for ∧immediate∧ display of pattern

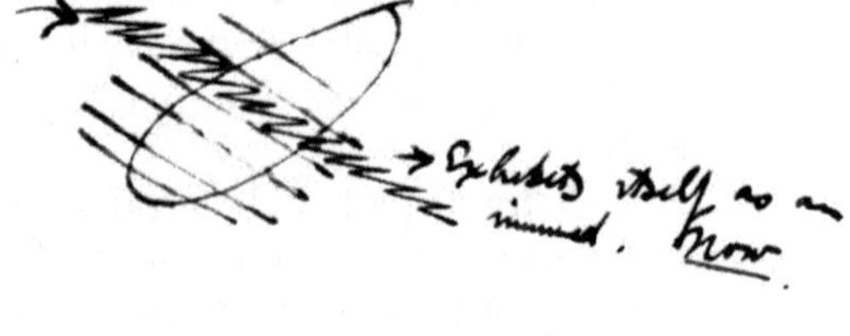

Very idea of Endurance divides universe into an immediate Now, and a Past and a Future.

Relativity and Leibniz. Leibniz not realizing Relativity of time?

Whitehead thinks he looked on Space as a relativity of External Relations and Whitehead's whole point is that they are essentially expressed as a locus of internal Relations. How entities in universe are related to one another via Aspects.

Then you'd find Descartes very much nearer Whitehead's view. Descartes says that Extension is of the essence of body and that wherever there's Extension there is something. There Descartes is very much more in touch with modern science than Leibniz. They really divided the truth between them. Leibniz in touch with Modern Science is that Modern Science looks on all Space and Time as a field of something going on. But in doing so, it drops idea of Leibniz's matter (the "body"). –

For Descartes' matter at an instant Whitehead substitutes therefore the event, which is not at an instant.

Whitehead asks What an event is and then goes from Descartes to Leibniz and says it is the grasping together of all the aspects of universe from one point of view.

But Leibniz hasn't shaken himself free of "stuff" idea yet.– Still looks on ultimate real thing as independent Substance with its own quality. This would knock out internal relations. Thus Leibniz solves his problem in only way open to him. Says Monads behave as though they were internal relations – Principle of organisation is the same. Whitehead applying, so far, to each immediate event the point of view of Leibniz. But beyond that. Where you have key to things you will have internal endurance in it. ∧Internal reality of event is E. as regarded as modified by these internal relations to aspects of all things.∧ Then asks: How can we express --- ? ? ? ⟨blank⟩

"Then get to idea of Event as having its own past. Relations to its own pasts and then to something outside of it."

We've discovered the essential law of succession which gives you the pattern. But has Descartes' view as to the e ? ? and Leibniz's as to what event is intrinsically in itself.

Descartes' differentiation between Duration and Time

In the universe itself Way mind conceives it – Got from a way of conceiving Motion.

He's thinking here of Measurableness of Duration

Descartes considers "general Expansiveness of Universe" as offered with measure of definite event. Plato already with some suggestion of this sort of thing in the *Timaeus*. Appeals to motions of heavenly bodies as something which helps you to measure Time. Bergson gets his ideas direct from Descartes?

Hocking's notes

Whitehead, Feb. 28, 1925 (Partial repetition, with introduction of notion of regularity)

(1) $\alpha \in \text{prime}(\sigma) \equiv \alpha \in \sigma . \alpha \in \text{Ab'K} : \alpha K_c \beta . \beta \in \sigma . \vee . \gamma \in \text{Ab'K} . \supset_\beta . \beta K_c \alpha$

Knocking out the contingency that β may approach a more extensive limit than α

(2) $\alpha \in \text{antiprime}(\sigma) \equiv \alpha \in \sigma . \sigma \in \text{Ab'K} : \beta K_c \alpha . \beta \in \sigma . \beta \in \text{Ab'K} . \supset_\beta . \alpha K_c \beta$

Regularity – The class of regular primes & regular antiprimes.

(3) $\alpha \in \text{Reg} . \text{Prime}(\sigma) \equiv \alpha \in \text{Prime}(\sigma) . \alpha K_{eq} \beta . \supset_{\alpha\beta} . \beta \in \text{Prime}(\sigma)$

By this you can infer from (1)

$\supset_\beta . \alpha K_{eq} \beta . \supset_\beta . \beta \in \text{Prime}(\sigma)$

(4) $\alpha \in \text{Reg} . \text{Antiprime}(\sigma) \equiv \alpha \in \text{Antiprime}(\sigma) . \alpha K_{eq} \beta . \supset_{\alpha\beta} . \beta \in \text{Antiprime}(\sigma)$

By this you can infer from (2)

$\supset_\beta . \alpha K_{eq} \beta . \supset_\beta . \beta \in \text{Antiprime}(\sigma)$

Different routes of approximation

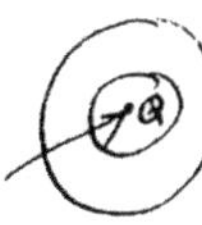

We want an entity that marks out all equivalent routes of approximation

---

All the Geometrical entities that have the same convergence – the class of then G_σ'K.

$P \in G_\sigma\text{'K} : = \text{Df} \, (\exists \alpha) : \alpha \in \text{Prime}(\sigma) : \beta \in P . \equiv_\beta . \beta \in \text{Prime}(\sigma) . \alpha K_{eq} \beta .$

Now for the antiprime condition $\equiv G_\sigma$'K.

$L \in G_\sigma\text{'K} : = \text{Df} \, (\exists \alpha) : \alpha \in \text{Antiprime}(\sigma) : \beta \in L . \equiv_\beta . \beta \in \text{Antiprime}(\sigma) . \alpha K_{eq} \beta .$

"No good looking for some niggling dodge" Must look for something so obvious you can't miss it.

Couldn't get the principle of ordering out of space alone. Tried it for years. Had to take time into account.

There are things which in their nature imply a transition to other things. This transition differentiates itself into spatial beyondness – and temporal.

Feb. 28, 1925 – 2

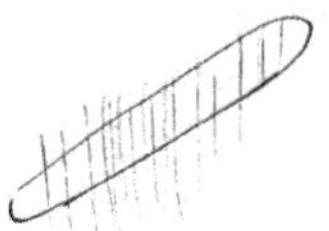

Fundamental character of endurance is that of enduring within your own history,– same pattern in each partition.

Rule of succession of spatialities

[Leibniz relativity ^of space^ is one of external relations –
Whitehead relativity of internal relations – The essence of a comprises the aspects of other things in a. [Descartes much nearer
Extension is of the essence of body – Leibniz' monadic idea also good.]

Internal relations knock out independent substances

Descartes says duration is in the universe itself while time is in the way the mind conceives it – (founded on motion – means its measurableness probably) Plato also in the *Timaeus*, though vaguer than Descartes, appealing to motions of the heavenly bodies as giving you something to measure Time. Bergson probably got his straight from Descartes.

Lecture 52

Tuesday, 3 March 1925[1]

Bell's notes

|99|

Whitehead – 3 March 1925[2]: –

Discussing how to deal with Time. Whole point is that we recognize Time is a sorting out of the aspects of the Universe, so that we perceive Events as patterned Simultaneously with us, or in Past or Future. Simultaneity is aspect of immediate realization for us – Events as involved in same immediacy as Event of which we are the self-cognition. This immediacy of pattern is disclosure in same immediacy a systematic Simultaneity – or "Spatiality", as being there (of other Events).[3] And unity of this systematic whole has a certain $\propto$[4] as being all that there is with that Spatiality. Events which come under that simultaneous Spatiality --- Grasp of Aspects into Unity. That Unity expresses itself as Universe which is realizing itself in same immediacy of Unity by --- ? Organism knowing itself as given in succession and grasping by aspect all that Simultaneity.

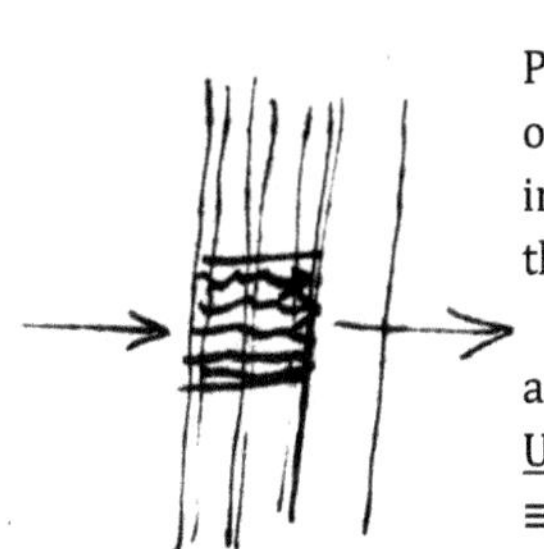

"Specious Present" – duration in itself. But with unity. Practical logic is Assumption of exhaustive partitioning of Universe by a Class of Durations, each supporting an immediate Spatiality. Succession of these Durations are the temporal flow.

Now what does that mean from point of view of direct assumption? We assume a type of Events with a certain Unlimitedness about them. Time ≡ T & Class of Durations $\equiv \Delta_T$. Then in order to express the idea we have to assume (just what man always has assumed).

(1) that only type of Event which could enclose a Duration would be another Duration: $x \in \Delta_T . yK_c x . \supset_{xy} . y \in \Delta_T$

(2) If two durations of same kind intersect, they intersect in a duration. [Trying to put down this shows Whitehead how he's never faced one thing with us. Has shown in Intersection of Events – Have Events in Common. But not that there is One Event of all that is Common to them. Can we state it merely and solely in terms of Extension? Can do it.

1. The Radcliffe version of this lecture, or portions of it, from notes taken by L. R. Heath earlier the same day, begins on |27| of her notes, page 485.
2. On this rather unremarkable day, Bell recorded in his diary: 'After Whitehead's lecture walked to Central Sq. & back with Federal Income Tax returns. Slept a bit in afternoon. In evening worked up diary, accts. Etc. Marked a lot of Phil. A papers (had had 2 weeks of them on my hands)'.
3. This closing parenthesis has been supplied.
4. It is difficult to determine whether Bell was using this 'proportional' symbol (which seems to make more sense), or did not quite close a '∞'.

How to do it. xK_cz yK_cz Then if any u it's in z. Then that only one z. "Complete Intersection of two intersecting Events".][1]

Now come back to "Complete Intersection" of two Durations. Ordinarily assumption it's another Duration of same kind.

$x . y \in \Delta_T . xK_{in}y . \supset.$ The complete ∧Intersect∧ of $x . y \in \Delta_T$.

In coming down thus on Simultaneity Whitehead is terribly old-fashioned. Modern Relativists ∨said to∨ have found it as having no basis in fact at all. But it is one of most fundamental notions in my mind.

[Idea of what is happening in London at this moment.]

The immediate spread of Events is my most immediate perception – I see events as spread out before me. See star. – I see an immediate twinkle indeterminably directed with reference to me and indeterminably distant. I see the sky spread out before me. Now task of science is to take these Indeterminates and --- ⟨blank⟩

To analyze system spatio-temporal patterns of immediate geometrical Spreads and their temporal succession. That is what Universe is for me as Self-cognizance of Event which I am. This aspect of the Universe in my Event isn't an Aspect of what the universe isn't, but of what it is.

An apprehension, on part of a portion of Nature, of what it is in itself. But what it is in itself expresses what the Universe is for it. But Universe has no other. – Nothing else but its parts for it to be for.

Then question of the Systematic Pattern which is thereby disclosed for us. If we once leave go systematic character of Space and Time, we leave go our great Sheet-anchor as to how Knowledge really is possible. Hume has difficulty because he has assumed detached cases of knowing.

But taking refuge in systematic (organic) nature of Universe you have difficulty that every individual event, is, in its very nature, modified by anything else.

Idea that you'd have to Know everything to Know anything. – Hume makes out that you Know too little. And opponents of Hume make out you know too much in order to be able to know anything. [See my Phil. 19 last year].[2]

– Now let's see what we actually do when we abstract. Fixing on partial aspects and saying: "Rest doesn't matter – That's true at least!" The two bits of

|100|

chalk there. What is necessary to be able to affirm that truth. The casual factor is the chalks, the room, the sunlight etc etc. These all unnecessary. But I'm supposing Spatio-Temporality of pieces of Chalk as "appearances"

1. This closing bracket has been supplied.
2. There was a course 'Phil. 19' the previous autumn, taught by Demos and Eaton, which likely covered Hume. And Whitehead may have been invited to provide one or more guest lectures, as happened from time to time.

[as sense-objects] – with a certain spatio-temporal relation with reference to each other. Can perceive the two pieces of color and their spatio-temporal system: Even if hallucinating!

How can you relate things to each other in a coherent whole, unless relationships which to-morrow's H[1] satisfy in relation with to-day unless the Relationships are already inherent in this room. Therefore if you drop a systematic character you can have no power of having any orderly relations between your various occurrences. Difficulty is to know how far to push claim for system.

Pleading for a certain casualousness. Ordinary Relativist want the measure-relations to be casual. Whitehead will point out that congruous, can only mean that they have homologous relations in the structure. And it's just that underlying geometrical relation which has got to be.

Not "the Self-identity" of the object is important – But its rigidity. Self-identity won't do. Pocket book doesn't remain congruent when is fold it⟨?⟩[2]

[Drowsy!][3]

[Congruence therefore is not self-identity.][4] Whitehead would like to see a theory of congruence worked out. We apprehend something which is only disclosed by the bodies in that condition. It really and ⟨?⟩[5]

Rests always somewhere in an immediate perception. For that reason Whitehead never able to understand Relativist theory. It sometimes assumes things which ultimately it really denies. No objection to simultaneity ? ? ?

"but it is rather difficult work."

Now to consider whole question of Abstractive Classes. The σ represented Species of Classes of Events. Corresponding with this will be sets of Classes.

$\alpha \in \text{cls}`\Delta_T \equiv \alpha$ is a class[6] of Durations.

We can get two types of abstractive Classes of Durations.

We must assume anything of 1st type must form an Antiprime (cls'Δ_T). For anything which covers it has got to be a class of Durations coming up to same time as Limit.[7] 2nd class would be a Prime. Because whatever class it covers must

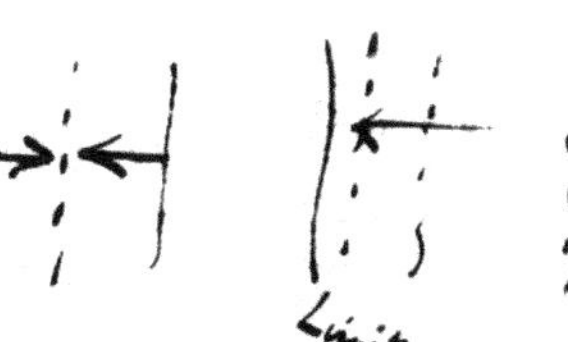

1. It is not clear whether this is an 'H' or the word 'fit' (as is suggested in Hocking's notes), or perhaps 'shut', though none of these make much sense.
2. It is hard to decipher this term.
3. This remarkably self-conscious remark is written sideways, up the right-hand margin.
4. This remark is written in the left-hand margin.
5. Once again, it is difficult to know how to decipher the term written here; possibly it is an abbreviation for 'definition' or 'dynamics'.
6. In the formalism at the outset of this sentence (and again four lines below) 'cls' seems to be used as a symbol or abbreviation for 'class'. While something similar is used in *Principia Mathematica* symbolism we remain unsure of this reading.
7. We have placed a full stop here, but it is not clear from the handwritten notes just where it should be.

come up to that limit in same way. (2)[1] is both prime and antiprime. But you could have another class in same relation to the line, coming up to it (on other side) – and quite independent of it.

We define "mode" as a class of K_{eq} antiprimes.

This is a ⟨?⟩[2]

You get idea of instantaneous moments – What Classes converge to. The limit to all instants and specious presents.[3]

A moment of time is nothing but an instantaneous 3-dimensional Space. In practice our apprehension of ⟨blank⟩

Entity is Idea of absolute perception[4] –

How do geometrical entities in an instant arise?

Then "Use of hypothesis" in Science and in Metaphysics.

Hocking notes

March 3, 1925

How to deal with time

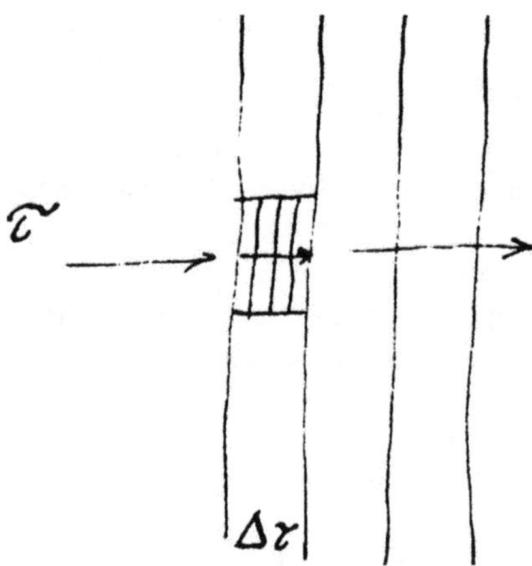

Temporal flow.

What are we going to assume about these durations? Δ_T. What mankind always has assumed. That only a duration can enclose a duration.

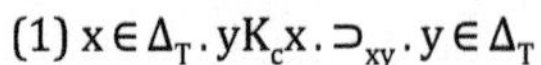

(1) $x \in \Delta_T . yK_c x . \supset_{xy} . y \in \Delta_T$

(2) If they intersect, they intersect in a duration of the same kind

We have defined intersection as having events in common.

But we have assumed that there is One event which includes all the events so held in common.

1. While Bell clearly writes '2', it is not clear if there is a '1' or whether it simply refers to the second type of abstractive class being discussed.
2. Once again, it is hard to decipher the term here.
3. The last three terms here are unclear.
4. Bell's drowsiness really has made several key terms difficult even to second guess.

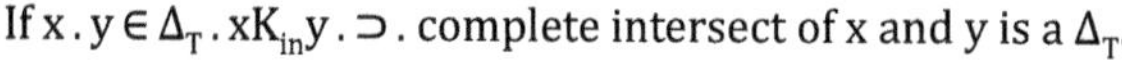

If $x . y \in \Delta_T . xK_{in}y . \supset .$ complete intersect of x and y is a Δ_T.

"Terribly old fashioned." The idea of what is going on in London at this instant, quite apart from ways of finding out about it, is one of my most fundamental notions. The spread of the heavens is immediately, now, for me

---

An immediate apprehension of what the bodily event is in itself – an apprehension of what the universe-event is for it.

If we leave go the systematic character of Space-Time, we leave go the sheet anchor of how knowledge is possible. With Hume no knowledge is possible because no occasion takes you beyond ~~you~~ itself. But take refuge in the organic character of the universe – everything is in its nature modified by everything else.

March 3 – 2 ✓
Does this mean that you must know everything in order to know anything

According to Hume you know too little –

According to opposite view, you know too much –

Can't fit on things unless the relationships which are to fit on tomorrow to today are already in today. These relationships must be already in this room. If you drop the systematic character, you can have no power of having any orderly relations between your experiences.

Geometry need not be Euclidean, but it must not be subject to casual changes.

Relativists want the measure-relations to be casual. Congruence can only mean that objects have homologous relations in a structure. That underlying structure must be systematic.

How measurement is possible.

Not a matter of self-identity? But rigidity?

Self-identity won't do – Open a book and shut it,

Same book but not congruent.

We apprehend congruence easily by means of bodies, but the bodies merely disclose what we mean. Always depend finally on some immediate perception. Two books can be perceived as very nearly congruent with each other.

Mar 3 – 3.
Let $\sigma = \text{cls}'\Delta_T$
a class of durations

There are two sorts of abstractive classes which are classes of durations.

This type forms an antiprimer and a prime.

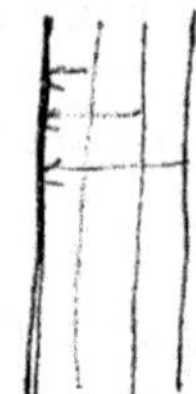

This type forms a prime. – Awkward to consider because a class from the left to the same limit.

Define a moment – a class of coequal antiprimes

If small enough – an instantaneous 3-dimensional space

In practice our apprehension of a moment is a duration within which transition is negligible.

Now how do the spatial discriminations arise?

Prefatory remark: The use of hypotheses – science & metaphysics.

Lecture 53

Thursday, 5 March 1925[1]

Bell's notes

|101|
Whitehead. 5. March 1925 –

Like particular crayfish the zoologist dissects. He doesn't care much for the crayfish (Whitehead does for his particular example here). But need to facilitate various notions that ought to be present to mind of Philosopher. Whitehead now using hypothesis – going beyond what is clearly and obviously fact, and making statements that we wish we saw quite as clearly as could wish. Doing what is necessary in all general theory. – Going beyond what one can see clearly and distinctly and forming precise hypothesis which goes beyond what can verify as immediate and direct description.

Question then as to use of hypothesis in pursuit of truth. (1) If your hypothesis is of a particular type – that admits of direct verification. (2)[2] If it is general – going beyond any definite occasions for verification; the particular verification cannot in general be a complete justification (That every 2 particles of matter will attract one another inversely … etc.) Whitehead looks on chief use of a hypothesis is to lead to cognitive apprehension of direct experiences which would otherwise be unnoticed. If a hypothesis has no bearing on some direct experiences (Whitehead with Pragmatists here), its use must be that it will modify, in some way or other, your experiences. Way it is will enable you to apprehend (have direct acquaintance with) something you'd otherwise miss.

Now, what constitutes a good and a bad hypothesis. This is question of the way our consciousness works. – We have vague apprehensions which we cannot formulate in precise detail. (E.g. last few lectures – re mathematics. Idea of $\sqrt{2}$ from time of Greeks on. – vague apprehension. But only labours of Dedekind, Cantor, and other mathematical philosophers of that school,[3] which have enabled us to explain with precision what that entity is. Combined with general Kantian theory of knowledge: – Creative activity of intellect etc. Intellect postulated or created an entity $\sqrt{2}$. Whitehead looks on this as Moonshine. How create. If I and you create, what sense in assuming that we've got same and common mathematics, etc. – You don't know what I'm talking about and vice versa).

1. The Radcliffe version of this lecture, from notes taken by L. R. Heath earlier the same day, begins on |31| of her notes, page 486.
2. This '(2)' has been added where it seems to be called for.
3. Geog F. L. P. Cantor (1845–1918) and Julius Wihelm Richard Dedekind (1831–1916) worked closely together through the 1870s on defining rational and irrational numbers.

There was a definite apprehension of an Entity without precise formulation of what that Entity was. We have now directly solved that difficulty. Basis of Hypothesis is having vague apprehension of a vague sort of relationship. There are items in our knowledge which we apprehend as Connected – a vague apprehension of connectedness. But formulation of precise connectedness doesn't enter into our direct apprehension. – There an hypothesis comes in – It directs attention to particular possibilities of connectedness. It prepares the mind for connections. (Whether this psychologically right etc. or not – Whitehead sure he's got hold of something)[1] – You do not perceive what you're not prepared to perceive. Hypothesis is founded on Vague apprehension – on apprehension that there are things to perceive. It then gives you hypothetical possibilities for perception: – Says: notice this and observe that. Gives relative importance to certain features. – Increases direct apprehension of things. Things perceived[2] thereby rise in importance. If false, but well-founded, prepares the way, by *gesteigerte*[3] apprehension, for the true hypothesis.

All that Whitehead's saying here is in Darwin's *Origin of Species* [beginning or end? – probably beginning]. Darwin discusses question whether he is doing harm by producing wild hypothesis. But says even if false hypothesis, if leads you to right type of observation, it is good. But real villain of piece is man who states a false fact. Because fact very often very difficult to verify again; and False fact prevents right science. By increasing your direct cognitive apprehension it increases direct knowledge of truth.

Chief use of hypothesis is as a discipline of cognitive apprehension. – Ex. in dealing with aesthetic questions – apprehension. Reading poetry at first – Platonic[4] in verbal statements but feeling that poet has gone beyond own direct experience. The immediate apprehension of wonder of Skylark came from the poem – from entertaining that hypothesis about Skylark.[5] But it led not merely to bare verification of the hypothesis, but ⟨blank⟩

The very felicitous phrasing is another and different thing. Substantial importance of the poem recedes – because has become part of your very being. In Art criticism generally Whitehead is helped by man who'll tell you what to look for. You don't go around then just verifying that detail is there etc. (pathetic people in European galleries). Fortunate people who have vividness and delicacy of appreciation far ahead of critics feel much of this

1. Closing parenthesis added.
2. The 'd' has been added.
3. It might have been Bell who slipped into German at this important point, *gesteigerte* being German for 'increased' (apprehension) or perhaps 'heightened'.
4. We are not confident of this expansion of 'P[lic]' but seems plausible, given Plato's warnings about poets being taken too literally.
5. Wordsworth and Shelley each have a poem 'To a Skylark', but from his comments (pp. 82–6) in *Science and the Modern World*, it seems more likely to be Shelley's.

nonsense. Much is – the man who writes isn't necessarily the man at the top in apprehension – but with certain amount of
|102|
that, and power to write. A lot of it is like telling you to notice the tail of the dog.

Of course you will notice that.

Another thing hypothesis does:– It increases your general harmony. If a wide hypothesis it may bind together into one unity vague apprehensions which are otherwise discrete, but which when bound together (by a hypothesis ∧[not?]∧ so general that it hardly comes up for any definite verification) – brings into open, by binding together a feeling that those things always did belong together – Appeal to something in you that is never risen even to vague apprehension.

Cognition is always of something which is there already – and if you didn't cognize it, it would be there as an uncognized prehension. That doesn't mean that Cognition is ineffective as a pure looking on. Cognition prepares you for your next transition – Is immensely important as to what will be your effectiveness in next second. Present creates future – and cognition as part of that. Hypothesis as heightening cognition is therefore directly effective agency. That brings you right up against dangers of Hypothesis. – It may lead to a Subjectiveness of Prehension:– of how the environment is affective/affected/effective??[1] Aspects of Environment fused with aspects of your own bodily past. You may have a bodily pattern which smothers the objective pattern of the Environment and clothes Environment entirely with ~~its own~~ something from itself. People observe what they wish to see. There it is. They are really seeing environment through their own bodily happenings – "Subjective Prehensions". Other, cruder, ways of getting illusions – alcohol etc. You've allowed your private pattern (inherited from past) (whose endurance really makes you) (this is very important then) you've allowed this to fuse with pattern of Environment.

Here is enormous importance of attending to claims of truth. – So manage claims of hypothesis that you don't interfere with objectivity of your perceptions. Necessity of candour – impersonality. Apprehensions may be motivated by fact that you entertain the hypothesis, but --- Logic is important. Imperiousness of Exact Knowledge reduces emotional element to a minimum. If you make a slip it is of the kind that someone else can point out for you. Use this discipline. ⟨blank⟩ assumption of to heighten without warping powers of perception. It is there that lazy hypothesis is bad. – actually impedes perception. Also egotistic assumption – association with emotional states – means association with aroused bodily states. You can't get rid of this lazy assumption which has often cloaked perception until ---, unless --- ⟨blank⟩

1. Bell includes all three terms; he seems to be unsure which of these three alternatives is what Whitehead actually said.

One of difficulty re perception of religious truths and difficulty of restating them, or really feeling them. Modern World has lost even feeling that they are vividly false. Lack of relevance. Some people quarreling about formulation of religious truth. But real question is that of the relevance. See acts of popularity that are brought in. How many people would go to stake for falseness of it. Whitehead in Greek-Russian Church in London as contrast. Modern service made up to interest and excite people. Ancient Church not so. – The relevance taken for granted. Cathedral at Freiburg-i.-B.[1] too. Vivid remembrance of Whitehead. Philip Wicksteed – Dante Scholar. (Unitarian Clergyman – New England type, "Medieval Religion as in Dante" etc. Conversation in University of London Club.)[2]

Dangers of lazy assumption – Kills apprehension. True reason for vivid Claim we make in any University for absolute freedom of hypothesis and discussion. It is practically the only preservative against lazy assumption and its slow poison. The real preservative of keen apprehension is candour in Criticism.

Hocking's notes

March 5, 1925

We are now using hypothesis – going beyond clear experiences – "quite obviously" –

The use of hypotheses (in metaphysics?)

Everything comes down to direct experience – ("here I am a pragmatist evidently")

Must have some being in direct experience –

It must modify your experience in some way or other – Enable you, perhaps, to apprehend something you would not otherwise apprehend.

What is a good hypothesis? If it leads to the right type of observation

Not a postulate. Creation of the intellect.

What I create you don't understand – No use.

A vague entity, like $\sqrt{2}$, is made definite by Dedekind, Cantor.

Essence of hypothesis – We have a vague apprehension of relatedness. We do not formulate it precisely.

Here an hypothesis comes it.[3] Directs attention to particular possibilities of connections. Prepares the mind for perception. "You do not perceive what you are not prepared to perceive."

It says, notice this & try that. It increases the direct apprehension of things, & if it is true it increases the importance of what is perceived.

1. Presumably this is Freiburg im Breisgau, an old German university town, and archiepiscopal seat.
2. Philip Henry Wicksteed (1844–1927) was known primarily as an economist. He was also an English Unitarian theologian, classicist, medievalist, and literary critic. For a time he provided extension lectures through the University of London, presumably while Whitehead was there.
3. At this point in the lecture, Bell has: 'There an hypothesis comes in'.

If it is false, it prepares the way for the perception of the true. Darwin *Origin of Species*.

1. Its chief use is to increase your direct apprehension of truth.

2. It increases your general harmony, binding together vague apprehensions which are otherwise discrete.

Making precise the vaguely felt unity.

May lead to subjectiveness of apprehension –

People who see what they wish to see – seeing the environment through their own bodily happenings.

It is there. Your private pattern

The imperiousness of logic reduces the emotional element. Hence the importance of logic – Esp. in re. religious truths. Difficulty of refashioning them. Of taking them as vividly false. Nobody can see that they have any vivid~~ly~~ relation or relevance

The ancient services appealed to a population already convinced of relevance

The modern service tries to excite a non-existent interest. (over)

|March 5 – 2|

– Wicksteed on Theology of Middle Ages

A man more like New England than Old England.[1]

"We shall never get back to talk about God & religion until it has all been forgotten and discovered again"

The danger of lazy assumption: it kills apprehension.

Claim for freedom of hypothesis and of discussion –

The only preservative against lazy assumption.

---

1. See footnote 2 on previous page.

Lecture 54

Saturday, 7 March 1925[1]

Bell's notes

|103|
Whitehead – 7. March 1925.[2]

In Mathematical discussion sliding from one to another attitude. – From direct metaphysical description of reality to sort of thing we have in Scientific Hypothesis. Primary use of Hypothesis is to introduce lucidity. It has got to have a basis, and this has its origin therein:– that we have definite and insistent Vague apprehension by which our whole proceeding governed. Language (a finite number of symbolism – rather vague concepts) hasn't succeeded in formulated vague apprehensions. – E.g. that we are living in a connected system. An hypothesis must be felt to satisfy and to arise out of that vague apprehension of what is the case in some field of our knowledge. A definite apprehension in which is embodied something vaguer than itself which we feel to be true. Goes further than we know – includes sort of relations which we know are applicable – but we don't know whether particular formulation is in that precise form true. – But it invites attention to these relations which are there to be known – and the question is a relevant one. We never use an hypothesis unless we have a sort of "instinct" – idea arises in mind etc. You wonder whether this isn't true. Then you get some instance where the hypothesis seems exactly to hit the case. The hypothesis has a vogue then. –

You get a place where you ⟨blank⟩

Take discovery of Argon. – It was assumed that by certain procedures you got something with all properties of Nitrogen. Then they found mixed with Nitrogen, another inert substance (Argon). That didn't say everything wrong about Nitrogen – only more exact.

So too[3] re Isotopes. When you go to a further degree of exactness and further niceness of experiment you ⟨blank⟩
When Ashton discovered Isotopes he didn't make all older chemistry wrong. – Simply showed limits of its accuracy.

---

1. The Radcliffe version of this lecture, from notes taken by L. R. Heath earlier the same day, begins on |33| of her notes, page 487.
2. Despite being 'till all hours on lecture preparat'n again', Bell records in his diary that he was working on 'Lectures etc. as usual', and when on Sunday afternoon he joined colleagues at a tea, 'the W. James', Mrs Whitehead & others there. Most pleasant … Mrs. W. asked me to supper.' 'Perfectly delightful little supper with them alone. Later the usual "Crush" arrived. I left fairly early.'
3. Bell has 'to' but context seems to call for 'too'.

"Lazy Assumption" over science on a colossal scale. Read in textbook (Impression of overwhelming evidence in favour of every accepted theory. All nearly done etc). All ^that of the 80's[1]^ toppled to pieces. Why?

Evidence so overwhelming only because only that in favour of orthodox assumption ever gets printed – or even investigated. (Danger otherwise of simple mass of aimless uninterpretable results.)[2] You put a man on experiment which ought to lead to a result in terms of recognized theory etc. what happens to other things? Probably x had cold in his head or instruments carelessly etc. etc. How does anything new turn up? Original men not so much subtle men, but men of strong character. Get hold of something that won't fit in; and worry at it. Not so much intellectual agility but mental lucidity. Finally manages to isolate something which you can't deny – of conditions of whose occurrence you can state – and which won't fit it. Sometimes what you want is the more delicate Experiment. – Today through advance in metallurgy and optical glass. With more delicate experiment you can isolate something which won't fit in.

1881 Michelson[3] and whole of Relativity. Interference bands wouldn't shift 1/1,000,000th of an inch. Then Michelson repeats it 10 years later with a better instrument and again later with another material of instrument. Then Scientific world thought it had to notice this. Then new and wider hypothesis to include this.

Another thing – sort of instinct: You feel this is a better hypothesis than that. – "Follow the nobler hypothesis". – Undoubted aesthetic sense coming in even in Science, and certainly beyond this. "Nobler" hypothesis – (1) Brings in the wider scope. Philosophy is after the nobler hypothesis (to see everything in rational relations to each other)[4]. Apt to get to the tragedy of the *deus ex machina*. Something brought in an explanation without detailed tracing of the relations. Sort of unexplained comfort in the general idea.

Idea of "Survival of the Fittest" – about 25 years ago – taken perfectly seriously as adequate etc. – It was getting into that position why elephant's trunk just that long or why that thing had no nose at all. etc. About 1900 uneasiness – desire that detail should be shown. – In Metaphysics calling in God, who somehow rises superior to and solves all your metaphysical difficulties. So Berkeley:– God called in to get him out of his Subjective Individualism. He's concerned for Objectivity of Nature.

---

1. Presumably, he means the 1880s. Francis Aston had distinguished isotopes using mass spectroscopy in 1919, but this work was based upon the discoveries by Ernest Rutherford in the previous decade and Joseph Thomson's discovery of the electron in the 1890s. In that sense, the Kuhnian 'paradigm' which held until the 1880s came toppling down.
2. Closing parenthesis added, to enclose a thought that, again like Kuhn, widely agreed assumptions also have a practical advantage.
3. In 1881 Albert Michelson began experiments, later in collaboration with Edward Morley, to measure subtle differences in the speed of light. Using increasingly sensitive interferometers they showed, against the prevailing assumptions, that the speed of light did not vary due to the earth moving through the 'ether', which led, eventually, as Whitehead says, to the revolutionary hypothesis of relativity.
4. Closing parenthesis provided.

If your consciousness of God grows out of your Metaphysics that is all right. But easy assumption of fundamental being to solve all difficulties "straight" is the bad thing. Your "nobler hypothesis"
|104|
mustn't fall into this error – Just the danger of the wider hypothesis. Mustn't lose its detailed grip – Here is where Logic is so important – Double importance. Logic more impervious to Emotions than most mental operations. Therefore independent of and guard against merely private prehensions.

In Science via mathematics it turns up. – In metaphysics:– working your ideas out by most precise logic possible.

[1]Now back to important point – Brought up by School of Philosophy at Cambridge. Passages from Broad's *Scientific Thought on Critical and Speculative Philosophy*. Gave high marks to the Critical Philosophy – Analysis of Concepts of a particular Hypothesis and to better then if necessary. Second procedure is more exciting, therefore more popular. But liable to very grave errors in leading to wide and general view. Broad's remarks are illustration of Pope on Addison – "willing to wound and afraid to strike"[2]. Whitehead entirely disagrees with Broad here. You can't get a fundamental connection ~~bet~~ of ideas of one Science without considering their connection with others. It is in their irrationality that you'll find best ground of revision. In trying to get Concepts of allied Sciences into harmony, you point the way to a reorganization within each branch of science. You will not get something you Know is true. But you will get "well-grounded" hypothesis; and you've taught yourself to be imaginative without any danger of running wild. The difficulty is to have firmness of mind to discard what won't hold water. Who is Broad to say where you're to stop in that way? Intense distrust of any wide metaphysical view – You've got to mistrust it; but it is one way of getting to truth. To refuse any of these is to do what Dante called the "great refusal". To think is most dangerous operation – First thing you do is to think wrong. But to think is our business. Refusal is cowardice. Constant Criticism of course is necessary. The way to correct is to speculate more. Constantly speculate and constantly verify.

North Carolina and teaching of Evolution – Ray Lankester[3] – Royal Society and the Norwich man shaking flints together. Broad's book reflects common habit just less baldly – refusing to think because of the dangers.

---

1. Hocking's notes suggest this is the location for '(2)', following the '(1)' that Bell does provide, above.
2. The *Epistle to Dr. Arbuthnot* published by Alexander Pope in the 1730s included 'Atticus' – from which this quotation is drawn – his character sketch of his one-time friend Joseph Addison.
3. Sir E. Ray Lankester (1847–1929), protégé of Thomas Huxley, specialised in the comparative anatomy of invertebrates. He is said by Ernst Mayr (*Growth of Biological Thought*, 1982) to have 'founded the school of selectionism at Oxford', after many years at University College London, and later in his career assembled his popular weekly newspaper columns into *Science from an Easy Chair* (1910). In this collection, and in J. Reid Moir's *The Antiquity of Man in East Anglia* (under Lankester's mentoring), there are speculations about flints found near Norwich which suggested the work of early man in the Pliocene – not exactly Whitehead's story, but see Hocking's version. A fierce proponent of natural selection when the theory

Those of us who (Whitehead self) have immense interest in Subtleties of a detailed investigation and in Symbolic Logic. Danger of detail-subtlety without having broad view too:– You get a lot of little investigations that seem to lead nowhere and have no relevance to anything.

You need broad point of view behind. – Something which if solved is going to be of importance and interest. No use working out details unless there is some hypothesis back of it. (Biologists – one specializing on one end of a worm, other on other). Team-work to-day. Fortune in finding bit of specialized work which works in as crucial illustration or test of some great hypothesis. – They may have no interest in this at all.

We use the interest of crossword puzzle type, to serve great general human ends. – (To beat the man who put it up.) One of the dangers of the modern world is that this teamwork leads many Scientists and public opinion of Science to neglect methods that really organize its general ideas.

Hocking's notes

March 7, 1925.

Hypothesis – inducing docility.[1]
Must satisfy a vague apprehension of the state of the case. Embody that in a precise form, going beyond what we know a "sort of instinct" precedes

Lazy assumption in science. Mid-eighties. Overwhelming evidence that everything had been cleared up, only the evidence in favor of an assumption ever turns up. You put young men on investigation which according to usual methods likely to lead to something. In a short life you can't be kicking over the traces. Original men not so much subtle men as men of strong character – stuck at points that can't be fitted in – get them into shape of definite experiment – finally manages to isolate something which won't fit in. Sometimes what won't fit it is found by the more delicate instruments and experiments, as in Michelson's experiment, repeated several times at intervals of ten years more or less, until the scientific world has to take notice

A sort of instinct, one hypothesis is better than another. There is an esthetic sense in science as well as in ethics, the "nobler hypothesis" –

Has two characteristics –

(1) It brings in a wider harmony – Philosophy is after the noblest & to see all things as issuing from some wider conception of the relation – rational –

Apt to lose the detailed relation – tracing out of how it happens – bringing in only a vague comfort. As "survival of the fittest" – which would

was still quite controversial, Lankester's weakness was speculations concerning early man, including the Piltdown hoax, according to Richard Milner in his article 'Huxley's bulldog: the battles of E. Ray Lankester (1846–1929)', *Anatomical Record*, vol. 257 (1999), pp. 90–5.

1. While this seems to be the word Hocking recorded, at this point in the lecture Bell has 'Primary use of hypothesis is to introduce lucidity'.

explain both why the elephant had a trunk & the other no nose at all. About 1900, we wanted the detail to be produced. So in metaphysics, God, as *deus ex machina*. Objection to Berkeley's use of God – to give a certain objectivity to Nature. God is another individual, & no more reason why you should see ideas in the mind of God than in the minds of each other – The easy assumption of the metaphysical entity who is to solve all your difficulties straight. Exactly the error the wide hypothesis is likely to fall into

(2) Given detailed grip, the wider the better, and in regards detailed grip logic is so important

March 7 – 2

Broad's reference to critical philosophy vs speculative philosophy trying to consider the universe as a whole

"Willing to wound and yet afraid to strike"

"Now I entirely disagree with Broad on this point" –

Can't get a ~~wider harmony~~ correction without trying for a wider harmony.

("Never get rid of irrationality –[1] because that would make an end to the joys of pursuing philosophy")

You can get an hypothesis which is well grounded – and which can then be tested in each occasion –

Be imaginative without running wild.

Constantly speculate & constantly verify –

[Question of principle, Whether – should have a grant of 5 and 20 pounds to test whether the chipping of flints could come about by natural causes. Ray Lankester wrote opposing the grant – ought to throw no doubt on the hypothesis]

Danger of detailed subtlety without a broad view. You get a lot of little niggling investigations that lead nowhere, no relevance to anything. You have to have something behind your symbols which when elucidated leads to something.

1. Hocking uses what looks to be an equal sign here, but a dash seems to fit the context better.

Lecture 55

Tuesday, 10 March 1925[1]

Bell's notes

|105|
Whitehead. 10 March 1925

Now back to question: How to deal with Space and Time by method of abstraction (Cleared up now is What Whitehead thought he was doing in all this.)

Question of time. By use of idea of Simultaneity Whitehead arrived at notion of a Duration. A duration is a certain type of event, with a certain Finiteness conditioned by "Simultaneity". Within that idea it is boundless – contains all there is subject to Simultaneity with certain Events.

Then: Relation of two Durations to one another. Not now talking of relations of all durations to one another. Now merely durations of same time-system. Two durations belonging to same Time-system can only have one of three alternative relations to one another.

Case I. - D' may lie completely within D (or vice vera)

(Complete Inclusion)

Case II. - D' may lie completely outside D

(Complete separation)

Case III. - D and D' intersect and the common intersection is a third Duration of same time system.

Excluded case where they intersect but not in a duration.

Whitehead thinks this possible for Durations but then they're not of same Time-system.

Abstractive Classes (in dealing with Primes and Antiprimes) which obeyed some definite condition Σ:– Condition that events composing class is all durations of same time system.

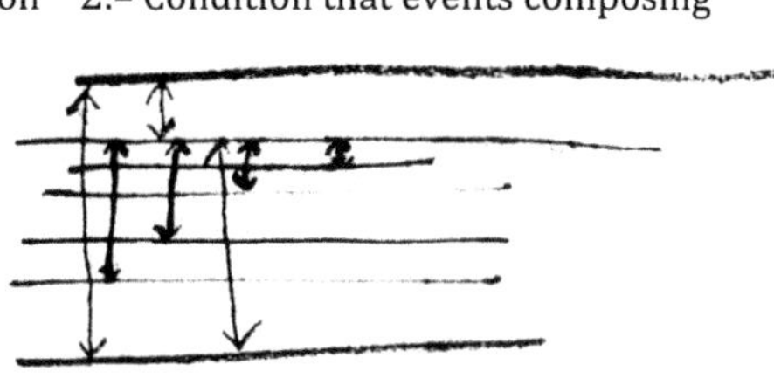

Now where class converging all up to one boundary. (Would give us trouble. Arises in dealing with primes). But with antiprimes

A class converging to some line not Boundary.

1. The Radcliffe version of this lecture, from notes taken by L. R. Heath earlier the same day, begins on |35| of her notes, page 488.

Now pencil abstractive Class covers ink one but not vice versa.[1] Pencil one is really what Whitehead calls an Antiprime.

So define a Moment as a Class of Antiprimes with the same Convergence (means just that they're all equal). The class of K-equal antiprimes is a moment. – Converges towards that ideal of exactness in which you'd have no process – but what Bergson calls mere spatialization. – Configuration of physical entities at a moment. We've got to assume with regard to moments, more than this? Properties all too obvious that you don't know or realize you're assuming them.

We assume an exact moment which marks transition from the Beyond M^1 and to the beyond M^2. Does this assumption involve anything beyond what we have already had?

M^1 (boundary). You can find Durations belonging to abstractive Class (pencil) intersects D and also ⟨blank⟩ You can find an abstractive class so that every one of its members is related to class D ∧?∧[2] in third way. Defining of boundary is quite possible – you assert such a class actually exists.

Now there is both M_1 and m_2 and only these two.[3] We assume as exact hypothetical formulation of what is vaguely present to us [We have in our minds an idea of Exactness for mathematical entities.]

Rendering of what is vaguely in our minds.

Any two moments of same Time-system can be taken as bounds of one duration. You've got to see that this, in any system you try, is either one of axioms or one of conclusions from this. Apt to take *Das Umgekehrte*[4] as equally obvious [?].

Plato vs Aristotle on question of what is meant by a "Point" – comes to exactly same question as Moments. T. L. Heath's impact.[5] (From the Euclid in Greek – Notes on early definitions).

New idea of position in Time brought in and added to duration etc. Question whether to deal with this by introducing new type of Entities – Points, Instants, etc.

or --- ? – ?

Can you mark what you mean by position in some other way.

Heath refers this back to Aristotle. Aristotle brought up idea of place.

---

1. Assuming Whitehead normally presented his diagrams to the class on a blackboard, it is not apparent how Bell is distinguishing between pencilled lines and ink ones.
2. It is not clear if this is Bell's '?' being inserted from below, or Whitehead's.
3. The use of upper/lower case m and of super/subscripts is not clear in these lines.
4. This is another momentary switch to German on the part of either Whitehead or Bell, where *das Umgekehrte* is used for 'the reverse'.
5. It is not clear that Bell does intend 'impact' here, but it is clear that Sir Thomas Little Heath (1861–1940) earned a first-class degree in both mathematics and classics at Trinity College, Cambridge, while Whitehead was still there. Before Heath entered the British civil service he completed important translations of Archimedes' work (1897) and *The Thirteen Books of Euclid's Elements* (1908).

|106|
Only by motion in space do you then get a line.

Aristotle: – Even an indivisible line has Extremities etc. Aristotle makes definite logical mistake there.

Fractions: – Can find always two fractions which differ from one another by less than $x/\sqrt{2}$ but no one is exactly

$2/3, \quad 1, \quad 5/4, \quad x/\sqrt{2}, \quad \sqrt{2}/1$

Whitehead here agrees with Plato – Put points on same line as durations. – and regards them as geometrical fiction. Plato had some knowledge of mathematical construction.

All talk of "fictions" is misleading. ∨Aristotle's??∨ Plato's meaning in it.

But Aristotle was in a strong position with his Common Sense. A sort of "Habeas Corpus" demand. – If you mean anything by "Beginning", produce what you mean. All the important ideas here produced by early Greeks. Later things is just muddle over "fictional" "points" or the like.

Naturally when you are dealing with one Time-system, Moments thought of as forming a series. How about this in terms of what we've already got? – Or, must it be brought in blankly as something new as Aristotle with his "points"?

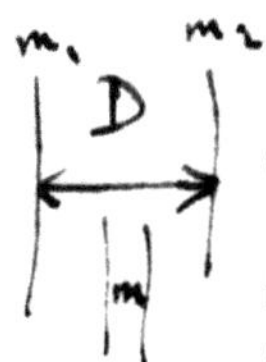

m_1 m_2 define D.

Now I can define what I mean by a moment lying between m_1, m_2. Can say a duration with Case I relation to D.

Almost always the small end of the abstractive class that is very interesting. As you are getting to limit of accuracy of your own observation.

We come here to one of very few cases where consideration of the big end is of importance. Want to define an m lying on side of m_1, remote from m_2 and outside D.

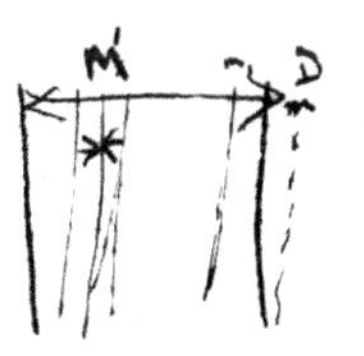

You can find an abstractive class converging to m_1, which includes a duration such that M' lies between its 2 bounds and not m_2. (Pencil one)

When dealing with big end of an abstractive class needn't worry about abstractive classes at all; really talking about existent entities.

Serial Relation. Whitehead would be surprised if could deal with it alone on assumptions Whitehead has already made. In serial relation we have left our very essential difference between Past and Future – that we're going in one way and not in other. – We'd have a line.

We're then obviously dealing with a very high abstractions with regard to idea of Time. Notice: although Whitehead dealing with "Vague Apprehensions"; Whitehead not dealing with paradoxes. But look at bundle of Whitehead's "Assumptions" and each of them wakes suspicion. – Seems less obvious than the vague apprehensions. Definite Notions are less obvious than whole thing together. – This is merit and character of the precise theory.

Shows you the Alternatives that lie hidden in a precise theory.
Basis of what Keynes says re Probabilities.[1]

Hocking's notes

March 10. 1

Case i Case ii Case iii

D' of the same time system

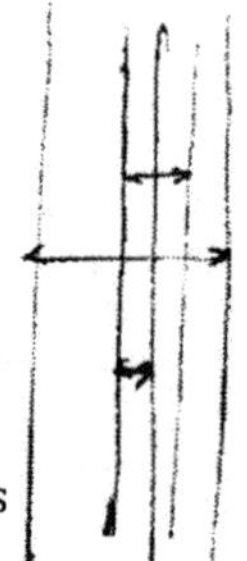

Get rid of special cases by dealing in antiprimes –

Define a moment as a class of antiprimes with the same convergence.

A duration may be included between two moments as boundaries.

Now come to a little point of disagreement between Plato & Aristotle, on what is meant by a Point. Comes to same as what is meant by a moment.

Euclid. First note under first definition.

A point is that which has no part, indivisible

σημεῖον[2] means mark. But the mark is not the point

We want to get position without size.

O.K. but position is a new idea.

Aristotle had said we can make no distinction between a point and the place where it is –

No accumulation of points he says can make a continuity, a line, which can only be made by a point of. But how can a position move.

Plato objected to recognizing points as a class of being at all, and regards them as a geometrical fiction. (But a fiction stands for something. If it is a mere make up it can have no relation to geometry)

Aristotle said produce something as what you mean. Habeas Corpus

---

1. John Maynard Keynes (1883–1946) entered King's College, Cambridge, to study mathematics (while Whitehead was still teaching mathematics at Trinity) and after work with the British civil service (at the Treasury, while Heath was still there) published his *A Treatise on Probability* in 1921.
2. It is difficult to read the ending Hocking provides for this Greek term, but the one provided does mean 'mark' and is found in Euclid's *Elements*.

Now we come to another point. Do the moments form a series?

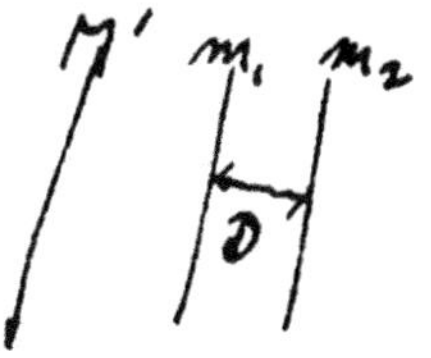

A moment <u>m</u> between m_1 and m_2 can be defined.
A moment M' outside D and on the side remote from m_2
You can find a class containing m_1 and not m_2

We have not included the essential direction from past to future.

Lecture 56

Thursday, 12 March 1925[1]

Bell's notes

|107|
Whitehead – 12. March. 1925 (Thursday)[2]

Concept of Time elaborated by means of "moments" and other conclusions. This, like every exact theory, goes beyond our immediate vague apprehensions. But Whitehead thinks it the most direct doctrine to be founded on those general and vague apprehensions. On other hand we have general apprehension of geometrical loci. But this founded on extraneous ? ? All defined to our immediate apprehension by color-play etc. etc – sense data. How to establish theory of geometrical loci in terms of what we consider to be our fundamental entities (Events)?

Thirdly, we have on hand certain very special observations (difficult to make – requiring instruments like Michelson's Interferometer. – Astronomers' instruments). These explicable by doctrine of alternative Time-systems. Then you get (introducing necessary mathematics) theory which satisfies all demands. But must ask: Has such a hypothesis any support or contradiction? (1) either from our general metaphysical analysis; (2) or with general ideas we have which have arisen naturally without reference to other one.

As to (1):– Relationship to general metaphysical ideas is also its relation to common sense. Then --- ? Question of harmony of idea to --- is very important one. What is our metaphysical position with reference to Time? Time is bound up with notion of Endurance. Time is idea absolutely necessary to specify a meaning for Endurance. That means a rule of Succession and this is Time. That Rule of Succession does not depend on this pen-case – it is discovered or utilized by the case. – May be shared by other things. With reference to it the endurance of the case has its significance. Simultaneity also required for this pattern [case] during its life.

1. The Radcliffe version of this lecture, from notes taken by L. R. Heath earlier the same day, begins on |37| of her notes, page 489.
2. In addition to noting in his diary the fine weather and 'Usual morning programme', Bell also records for this date that he hosted his own dinner party: 'Then in town getting Place-card materials, Confect & etc. for my dinner. The party came off very successfully in Evening. 6:30 – "Old Grey House," Anderson St – Prof. & Mrs. Whitehead, Cousin Fanny Cole, Mr. & Mrs. Douglas, Dr. & Mrs. Pratt, Mr. and Mrs. Dustan. Sherry in Sitting room first. Then dinner in "Candle Room". Thence to Jordan Hall. (Arrived a bit late because of blockade near Mechanics Hall). Heard last 2 movements of Smetana's Quartette ("From my Life") E minor. Leginska's Poems after Tagore.... Then Franck's glorious Quintet in F minor with Leginska at Piano. [N.Y. String Quartette – all Czechs].... Later got ready for Phil. A.'

But this doctrine quite neutral on question whether any two endurances using the same rules of succession – or whether they are consistent? This makes it a perfectly sensible question – Not perfect nonsense, as it at first strikes you. On other hand, consider evidence against it. Mankind thinking and observing for 3,000 years and more on other scheme; and only in last 20 years have they found out anything about it. Any theory produced must produce a reason why it hasn't been observed.

Secondly there is a very insistent experience that our Experiences do run in a unique fashion – it is not merely a conventional matter with us how our Thoughts run. – But that is allowed for in the theory. – What we are aware of is in that time-system whose Enduringness is our own Endurance. One point not there allowed for:– (changing velocity changes Time-system. But such change below threshold of our experience of time-quantities).

But immediate Experience knocks on head one way of putting it introduced by Poincaré[1] (great genius and therefore great vogue of his saying – one of those trenchant ideas Frenchmen love). – Theory that it is purely a matter of convention which system we use. – Wrong! We don't choose by sheer convention that system according to which these two pages are of same size. We have there a unique apprehension of relation.

Many persons who ought to know better think that Poincaré means simply re inaccuracy and inadequacy of our perceptions. Poincaré would allow for wild differences. We have definite and not conventional experiences; and it is those we have really to deal with, explain and rationalize. Result of all this is There is no rejection of the hypotheses arising from our immediate apprehension or general metaphysical position.

As to (2). By doctrine of diverse Time-systems (enabling definite explanations of otherwise inexplicable experiments) – do we harmonize any diverse fields. Harmonizing doctrine of Time with our apprehension of geometrical loci. – This doctrine gets us out of difficulty at once – It enables us to show the loci – brings in doctrine of separative and ordered position within the space of the moment.

On the whole the doctrine comes out of the battle very well. Let us then elaborate it and use it as an instrument with which to confront facts.

What are the prime loci with which we want to deal? – Punctuality, Straightness, Evenness (Euclid:– Straight Line lies evenly between its two points), Planeness.

1. Henri Poincaré (1854–1912) from his early work on differential equations was sceptical whether the axioms of geometry should be considered either analytical (and thereby deduced from logic) or *a priori*. In his *Science and Hypothesis* (1902) he warned more widely that many of the beliefs underlying science are convenient conventions. (See also footnote 4, page 191, lecture 42.) Whitehead had already made his reactions to this clear in *Concept of Nature*, pp. 121–4.

All these (Whitehead) as arising from intersecting of Events in any Time-system. Parallelism[1] will be derivative of relations to each other of moments of the Same Time-system. Here bringing in Euclidean space, of course.

Not necessary doctrine, but the simplest. Other systems give you another uniform Space-Time system – but with an extra Constant, and more complicated – and this constant

|108|

→ "Metaphysics is endeavor to make Common Sense precise over whole field" Special sciences is ditto for special fields.[2]

might enter anywhere into our physical determinations. Mathematics has been worked out for a sort of Elliptic Time-system. But as mathematical notion that is O.K. But as to its application to Time XX

There Rest in a time-system will be simply that "This" is making ∧unbroken∧ use of one consistent Time-system with reference to endurance – Relations there are unequivocal.

Rest in one time-system is uniform motion in straight line in another Time-system. Another point to be noticed (one speaker last night didn't notice it).[3] The 3-dimensional spaces so far spoken of are those obtained as a limit as you approach a moment. When dealing with instantaneous spaces you are Spatializing the universe. – There is no such thing as movement in an instantaneous space.

Therefore another idea has got to arise:– idea of a "timeless" Space:– i.e. a space which is neutral as regards all particular moments of a given Space-Time system and is a general timeless Space arising in connection with each Space-Time system. Have we any apprehensions which give us anything re a timeless space? People have gone on so happily with this that they don't know when they are hopping from the instantaneous to the timeless 3-dimensional Space [Troland[4] last night]. How are you going to correlate.

How to correlate position of P in M space with that of P' in M' etc. etc. How are you going to take your cargo of positions from M to M'. (Putting it all on blackboard is misleading). – See this thing – turn around – see "it" "again".

1. Expanded from 'paral^{m}'.
2. This is inserted, at a slant, in the upper left corner.
3. Whitehead's comment seems to refer to a philosophy symposium on the previous evening, Wednesday, 11 March. And, although Bell's diary for Tuesday and Thursday in this week simply records 'usual morning programme' ... for Wednesday he notes: 'Did some Phil A. papers. In evening Philos. Club Symposium. In Emerson D on Determination of Present by Future (Hocking, Lewis, Troland, Whitehead). Lasted 2½ hours (seats very hard before end of that time).'
4. In Bell's note about those who participated in the previous evening's symposium (see footnote 3 above) this is likely Leonard T. Troland (1889–1932), who earned his PhD in psychology at Harvard (while still one department with philosophy) and taught advanced courses at Harvard during this period. In addition to books on 'mind' and 'motivation' he would eventually publish, he had co-authored with Daniel Comstock *The Nature of Matter and Electricity* (1917). Troland had earned an earlier degree from the Massachusetts Institute of Technology, where Comstock taught physics, and both came to work, curiously, for the company that produced Technicolor motion picture film.

How Know same position in same space?? – If not, all our ideas go to pieces. Whitehead agrees with Troland that ∧in this point…; – that∧ Motion-Rest is fundamental.

This represents a body at rest.

This amounts to this:– For your apprehensions.

That fundamental property of being at Rest. If body utilizing this Space-Time system it is not at rest in it. Every body is at rest in its own Space-Time and that gives you a correlation of various points in ? ? ?

P, P', P'' are instantaneous points. the line then represents a locus of instantaneous points. Therefore your timeless point. Your timeless Space therefore looked upon (Diagram lets you down) as a locus through Time.

This locus is the timeless point of physical Science. This timeless point of physical Science is not so simple as Euclid's – it is a focus of numbers of things. Now how to represent Motion. Come to an alternative Time-system.

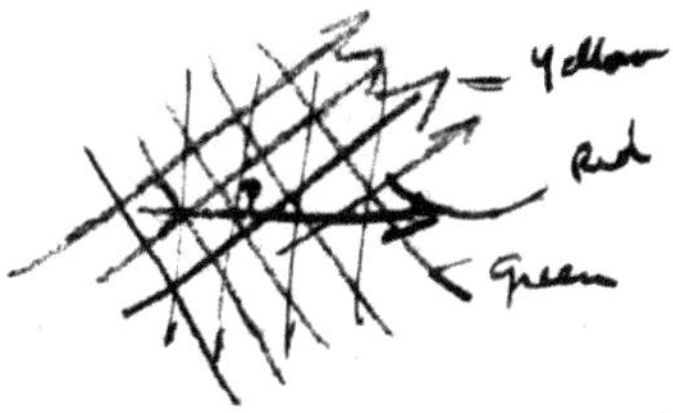

– Its timeless points

Point P at rest in the ink system but moving in pencil one. So P moving in the timeless space of the Green system. The motion through the yellow points are what we apprehend as Uniform motion. Every uniform motion indicates rest in some Time-system but not the one I'm using.

Thus complete reason why rest and uniform motion in Time is at same level. Although Aristotle made a mistake in assuming that force was wanted, Galileo had good reason but never gave reason – just blankly assumed it – that uniform motion is same phenomenon as rest – needs no extra force to explain it.

Clark Maxwell's *Matter and Motion*[1] alone tried to give a reason.

This gives coherent meaning to what you are asking about in regard to your Laws of Motion etc.

Next point is to go back and see exactly how we are to deal with points etc. And how to deal with moments of different systems intersecting.

Hocking's notes

March 12, 1925

Alternative time-systems.

Time is bound up with endurance.

Patterns may be found in any moment through

Are two different endurances using the same rule of succession? A sensible question

1. Maxwell's *Matter and Motion* had originally been published in 1877. It had been reprinted in 1920 with notes by Sir Joseph Larmor (to whom Whitehead has also referred).

Considerable evidence against it.

A new theory must produce a reason why it has not been observed before.

What we are aware of is the time system of our own endurance.

Knocks on head one way of putting it – introduced by Poincaré – that it is purely a matter of convention what system we adopt.

We have definite, and not conventional experiences

But there is no rejection of the hypothesis of alternative systems a priori by our metaphysics

Question whether we can harmonize different

[Missed]

Gives us the widest harmony.

Rest in one time system is uniform motion in a straight line in another.

A timeless space, – neutral in regards all moments of a given time-system.

Otherwise how correlate a point in one moment with a point in another?

You have a body which you recognize as at rest [route AB].

This fundamental property of being at rest gives you the correlation of the positions

---

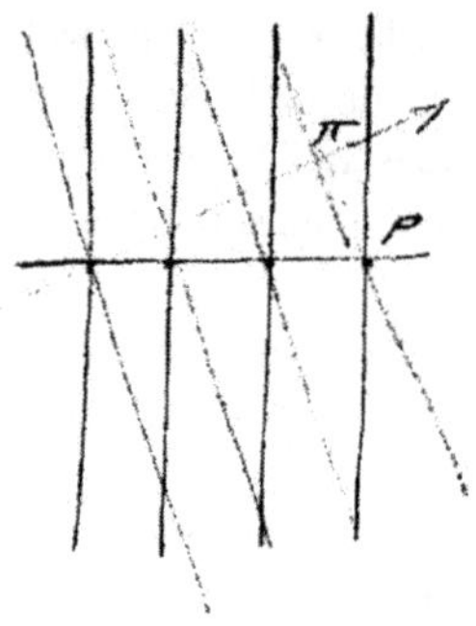

Introduce another time-system, the body moves [along πP]. The same physical fact seen from another system.

Lecture 57

Saturday, 14 March 1925[1]

Bell's notes

|109|
Whitehead. 14. March 1925[2]

Whitehead's discovering that "Event" as most concrete etc.... fitted on with idea of Duration etc. One temporal succession will not enable us to really explain the structural ideas that arise within geometry. But a group of alternative successions will enable us to do that.

We have a vague but insistent apprehension of temporal succession and a series of these. Within ∨that∨ idea of exactness which also haunts us. But in preciseness Whitehead's gone beyond what we immediately apprehend, but thinks it really just exactifies that – what is felt in it.

Re alternative ~~apprehensions~~ Temporal structures no such immediate apprehension – Quite the contrary. Our geometry, too, goes beyond our immediate apprehension but just exactifies that. But by our ideas of Motion, Uniformity of Motion – Ideas of Progression together with ideas of geometrical configuration --- [??] ⟨blank⟩

Suppose five Time-systems. Moments of different Time-systems L, M, N, P, Q. To have any geometrical structure we must have intersection of the moments. – In same Time-system they don't. Diagram here apt to be deluding. Moment is fullest least you can have.

[Abstractive classes in moment M.]

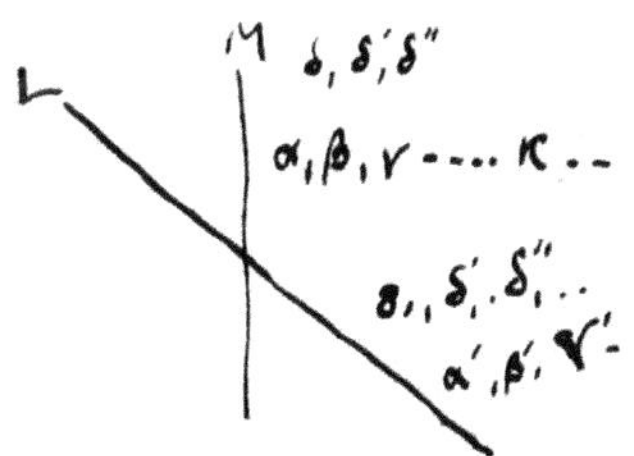

Abstractive Classes converging to one sharper point. But if covered by any one of them they are covered by all.

Idea of intersection of two moments is then the set of abstractive classes (k, l, m, n) which are in common to L and M.

Intersect as locus of all abstractive classes common to the two.

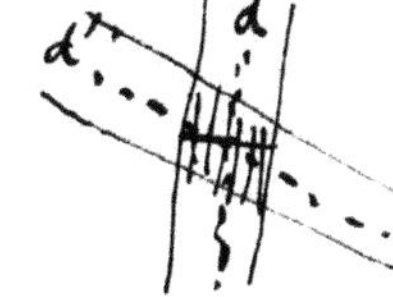

Why in dealing with "intersects" don't I go back to original idea of Durations? Two durations belonging to the two Time-systems?

Whitehead hasn't vaguest apprehension of different Time-systems, while standing here. – Idea

1. The Radcliffe version of this lecture, from notes taken by L. R. Heath earlier the same day, begins on |39| or thereabouts of her notes, page 489.
2. Bell's records in his diary for this day: 'My own lecture: Whitehead's ; then 2 hours going over Scholarship applic'ns. At 4.30 … dep't meeting (Philos'rs only) on Scholp applic'ns. – Till 6.30. At 7 in to Harvard Club.'

of simultaneous whole:– All going on in world during any Second of his lecture, e.g., seems quite clear.

So no talking of intersection as if one unit event like this. – It would be making least strong part of theory fundamental – Though logically ought to. – But can't apprehend any such one event of intersection.

Intersect of L and M LM. (Common set of Abstractive Classes lying in each and both of those).

Here come to great jump. – This locus viewed in relation to M is the plane in M. – The reflexion of the property of L in M.

Now with LMN. LMN = LM. MN

Two cases there will be – one where the planes intersect; other where they're parallel.

(a) when LMN is not a null class, there is straight line (not a plane. LMN ≠ LM ≠ MN.) For this Whitehead is following his geometrical intuitions and stating them as properties of alternative Time-systems. Otherwise. What Whitehead is knowing in his geometrical intuitions? – Really, the mutual properties of alternative Time-systems.

[Here example of how you have to have a lot that is due only to your Symbolism. – Here the order, which isn't significant. LMN = LNM = MNL etc. – not stating anything re original field but re Symbolism].

(b) LMN is null. Then how they do not intersect – Only case where LMN is null. There Whitehead has nailed up his flag to Euclidean geometry of Time. (A hyperbolic element of time is quite possible – Elliptic one is quite impossible). – Euclidean assumption then rather shaky. – Here you have got a straight line.

Then LMNP = LMN. LMP (Two straight lines.) Straight lines may be coincident, may intersect but not coincide (both special cases) – general one: – (3)

They may not intersect (general case).

(1) Straight lines coincident Group of positive-points which is that set of abstractive classes. – Here get a point.

(2) Straight lines intersect but do not coincide

(2) Straight lines[1]

Point is the idea of Complete Simplicity in Geometry. – It is mere position.

|110|

One feels that everything has some position. How interpret it? Most fundamental way comes from apprehension that any one time moment in reference to all the other moments of the Time-moment.

General position of a plane in M.

You get almost idea of absolute Position here. For straight line three Time-systems is necessary.[2]

1. From 'Then LMNP' to this point, the layout of Bell's presentation is confusing. The '(3)' comes at an odd juncture, and this latter '(2) Straight lines' may be superfluous.
2. 'Necessary' is uncertain – Bell's abbreviation is unreadable.

Taking instantaneous m for granted

Why then say that point has simplicity. – What mean by saying point absolutely simple. k. – Given from 4 – guardian and relevants in φ[1]

There is a particular ⟨blank⟩

Either y doesn't intersect k at all ⟨blank⟩

No moment of any other Time-system can divide up those classes.

"Position-Points" – "Puncts" – sounds difficult and scares people.

Idea of simplicity as --- ? ⟨blank⟩[2]

How Magnitude also to go in. Other point of view ∨re point∨ of diminishing extension. – You think of a simple Position-Point as by paring away and away etc.

"Without magnitude" means "without spatial extension*"

Let any abstractive class X fulfill all conditions ζ

"Point-Event" is the set of abstractive classes which compose a. z – premises

Anything that satisfies condition ψ covers α will cover --- ?? ⟨blank⟩

A set of K[3] – equal classes.

What it represents in the b

This represents idea of "without parts"

One further simplicity not usually mentioned with regard to points but taken for granted: – "Physical Simplicity" and a certain manner.

An abstractive class was like a chinese box-nest e_1, e_2, e_3 ---
→ indefinitely

q(e), q(e_1), q(e_2 ---) comes up to a set of "limits".

A certain measure of simplicity – this:- ⟨blank⟩

K -equal abstractive classes converge up to same set of physical limits.

The surface e.g. ⟨blank⟩

Over position-point we presume further "physical simplicity" – That any set of --- are all identical and all identical with physical limits of corresponding point-event which converges up to it.

Hocking's notes

March 14, 1925

1. Space is not a sufficient structural principle.

2. Time is the only other thing we have.
   Hence event is the unit.
   Explains the meaning of endurance.

---

1. These Greeks letters are unclear but seem to refer to a diagram Bell has not included, although it almost seems he has left space for one. (And there is an additional diagram in Hocking's notes.)
2. As will already be clear to the reader, Bell seems to have been struggling. From this point on sentences are not completed and the referents of Greek and other letters are not clear, and so on. We have simply transcribed as best we can, with this warning.
3. It is not clear whether this is meant to be kappa.

3. One temporal succession will not enable us to explain the structural ideas that arise in geometry. But an indefinite number will allow us to do that

4. Alternative time systems present themselves to immediate apprehension only via geometrical structure. In interpreting geometrical structure that way we are going beyond. We apprehend one.

5. Suppose L.M.N.P.Q moments of different time systems. They intersect.

Let us get exactly what we mean. No difficulty.

M will contain a number of abstractive classes

$\gamma, \gamma', \gamma''$ etc

$\alpha, \beta, \gamma. \delta, - \kappa$ ---

representing convergences to something sharper than the whole moment.

L. also contains a number

$\delta_1, \delta_2, \delta_3$ ---

covering α', β', γ', ---

Intersection of two moments must be <u>the set of abstractive classes in common</u> to L&M., $\kappa, \lambda, \mu, \nu$ ---

Why don't I go back to the notion of duration

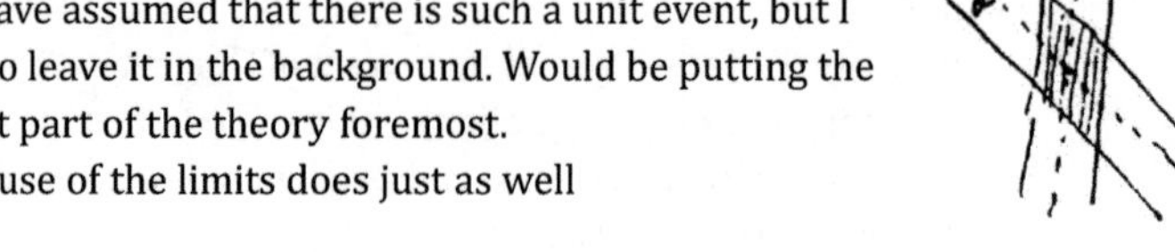

Haven't the slightest idea of an alternative time-system. I may have assumed that there is such a unit event, but I prefer to leave it in the background. Would be putting the weakest part of the theory foremost.

The use of the limits does just as well

Mar. 14 – 2

6. Intersect L&M, write as LM, etc. LMP.

Here we come to the great jump I make

I have been making you all along.

(1) $L\hat{M}$ corresponds to being a plane.

This the origin of geometrical notions.

(2) LMN = LM. MN

Two cases: planes intersect or are parallel

(a) LM. N is not null, is a straight line

LMN ≠ LM ≠ MN

Cut out the notion that MN might be a plane within LM. Using my geometrical intuitions.

(b) LMN is null: Then there is a plane L' such that MN = ML'

Intersected by two different moments of the same time-system.

In doing this have nailed up my flag to Euclidean geometry. Perhaps a hyperbolic geometry is possible. If there is a hyperbolic geometry of time.

(3) LMNP = LMP. LMN. Two straight lines.
Obviously several cases of which the important are
(a) straight lines are coincident
(b) straight lines are intersect & not coincident
(c) straight lines do not intersect (general case)
In case (b) you get a point, a Position-Point.
Consists of the innumerable abstractive classes common to L,M,N,&P.

Mar. 14 – 3

7. The ideas associated with a point
Simplicity.

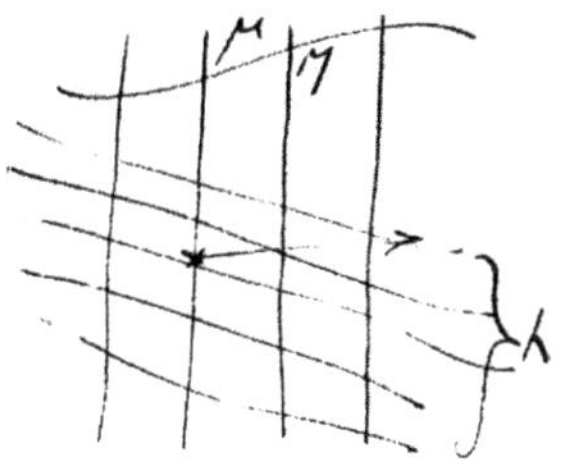

Position in M comes from position in λ, i.e., possible parallel planes up & down.
A sort of idea of absolute position
Why do you stop at four intersecting systems and regard point as simple?
If you take any fifth time system Kappa κ contains a moment Q. It either
(a) doesn't intersect κ at all or else
(b) coincides with it.

Euclid calls it without part

---

8 – Let x be a position point.
α any abstractive class satisfying the condition ξ[1]
~~Regular~~ Point event is the set of abstractive classes which compose a ξ – prime.
Anything that satisfies the condition ξ and covers α is covered by α
Represents the utmost getting rid of extension i.e., "without parts"

---

9. A further simplicity not usually mentioned – "physical simplicity"

1. It is not clear whether this is meant to be 'zeta' or 'xi'.

Lecture 58

Tuesday, 17 March 1925 [1]

Bell's notes

|111|
Whitehead: 17 March, 1925:-

Idea of ultimate simplicity – not only geometrical and temporal, but also physical. Idea of simplicity of position (Euclid), and of magnitude. (These two come down from Greeks, vaguely) Physical idea comes in with every book of physical constants. (Idea of something at a point at a time). This hypothesis performing one of those services an hypothesis ought to perform. – Taking ideas found in use and showing exactly how they lie with reference to each other.

The point (instantaneous point-event – or with Greeks the point in space) – Discovery of idea of point-simplicity as terms in which all exactness in geometrical relations can be expressed. This idea of taking point as fundamental for exact expression of geometrical relations may be guessed as one of very first discovery of Greeks.– Egyptians hadn't come down on this so. Really birth of geometry in getting this straight. – Even if not quite satisfactorily worked out. General vagueness which hung about it didn't detract from its importance; and lack of logical analysis didn't so much matter. – Of course latter matters a very great deal for ontology in wider philosophical connection. Greeks spotted irrational numbers but waited for Dedekind and Cantor[2] to give exact logical analysis. This was an event much more for philosophy than mathematics. Analysis tones down that air of excited mystery. Typical example of exacted mystery is $\sqrt{-1}$. First introduced by mathematical physicists quite pragmatically, because it worked. But if read comment on it in first 2/3 of last century you'll find that excited mystery.

Now we Know. Advantage at end is that you know exactly what you're handing over for philosophers to talk about.

When we know what we mean by instantaneous points, we still have idea that our moments give us instantaneous 3-dimensional space. This is nothing but concept of configuration. But idea of change in an instantaneous space is ridiculous.

Instantaneous space M_1 with a point P_1 in it, M_2 with P_2 etc. You may see same general pattern in M_1, M_2, M_3 and you may observe something has some function of place in pattern P_2 as P_1 in M_1 etc.

1. The Radcliffe version of this lecture, from notes taken by L. R. Heath on what was likely earlier the same day, begins on |43| of her notes, page 491.
2. See footnote 3 on page 241, lecture 53.

But that is all that there is to it. But question of movement, and of relation of space M_1 to M_2 to M_3 not yet given.

[Whitehead's point is that these M's are *überhaupt*[1] very high abstractions, behind which we must go to something more fundamental. – If you don't you've brought in no principle of Comparison for the M's whatsoever.][2]

We don't observe in instantaneous spaces, but through a duration. If we say P is at rest what we observe is pattern through the duration and if we observe the pattern as at rest fundamental notion of being at rest gives you a whole locus of instantaneous points. – Same element of the pattern as enduring. Notion of "At rest" is apprehension of endurance through duration.

Our apprehension is of "at rest". When we ask what we mean by Pattern? "Enduring" means not beyond but within the history in question (of which you have apprehension). This has reference to a mode of slicing up the history into temporal parts in each of which the pattern is same as in the whole. (Desk all the while there while Whitehead lecturing).

Idea of having to have rule of Simultaneity. (Mustn't take head now and feet three seconds later– as in Futurist picture of dancer etc. [good!][3]) A pattern is at rest in that Time-system which it utilizes, – which is required for or presupposed in the expression of its endurance.

The locus P is to be taken as meaning timeless point that is used in physical science. (Greeks:– Most highly endowed in logic and in immediate apprehension of Nature. – Show what is most immediate and natural way of rendering things, unsophisticated by theory). But Greeks muddled timeless enduring point (complex locus) with that of instantaneous point which is ideal of simplicity. Physical idea of simplicity needs latter. Geometrical point is a complex locus.

1. Position point = intersection of two Events (This is a locus too –)
2. Point-Event = regular point --- = simplicity of Magnitude-Extension
3. Geometrical point = a locus of Event particles

In 2. you get your minimum of extension.
In 3. you get your concentration of Rest.
∧Apprehension of∧ Rest shows that we, observers, are utilizing another Time-system. In *Principles of Natural Knowledge* Whitehead let himself be fooled on point concerned with this.

1. It is not clear whether Whitehead introduces this German term, or (as in other cases, more likely) it was Bell; the term means 'after all' or 'in general'.
2. Closing bracket provided.
3. This seems to be an editorial comment by Bell, but what either Bell or Whitehead might have thought of the Futurist influence on dance during the previous decade is hard to say, except to note that Futurism, from Italy and France, had also made its way onto British stages.

|112|

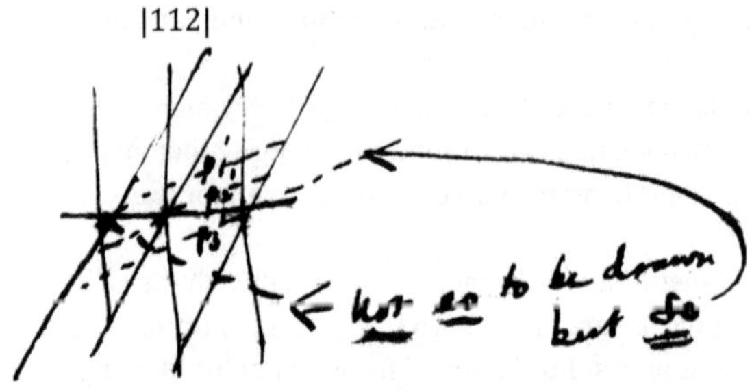

That is to be identified with uniform motion in a straight line. Uniform motion and rest give us same or <u>different</u> according as you're referring it to <u>another</u> or to its <u>own</u> system.

A point-event is the way in which we express in precise form the ideal of <u>exactness</u> which we have – the patience of this universe of events for <u>exact</u> discretion of the relations.

The geometrical point represents the way in which universe of events is patient of Endurance. We had another definition of straight lines (by intersection of planes).

What <u>is</u> it that you observe from What you see in quick observance [except for speed of light etc.] <u>Then</u> you start off to canvass such <u>Former</u> event-particle never analyzed – all past.

<u>Physical</u> constants at --- (?)

We think of <u>our</u> present as projected into space of other person.

That is what you hear so much of conceptual tone,

"Mathematical Constructs". (Every body has own jargon ~~here~~)

X S.[1] rather objects of ~~use~~ of term. – Like use of Vitalism in Biology. But anything that means exactly.[2]

"Oh! It's conceptual" – as though that were end of the matter! How is your concept related to immediate experience!

Quite obvious that only dealing with those space time systems.

A way of explaining Michelson method by FitzGerald.[3]

Where you've got another Time-system. Change of Space-Time system changes Geometrical ∧Pattern∧ in which we get our Geometry.

– Class of abstractive classes

Now another point – Back to more fundamental instantaneous "point events". How do they aid in formulation.

Point-event P inhering in E.

E. = a set of equal standard classes.[4]

So α = every ⟨blank⟩

Point P inheres into event P. if any of its abstractive Classes E_n such that it's completely contained[5] in. ⟨blank⟩

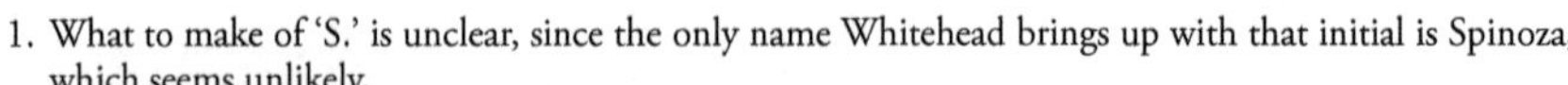

1. What to make of 'S.' is unclear, since the only name Whitehead brings up with that initial is Spinoza, which seems unlikely.
2. It is not clear how this term should be deciphered.
3. In his 1889 paper 'The ether and the earth's atmosphere', George Francis FitzGerald (1851–1901) – as did Hendrik Lorentz (1853–1928) two years later – conjectured the 'contraction' of an object at high speed, which would account for the null results from Michelson and Morley's series of experiments.
4. These comments make more sense while looking at Hocking's more complete version of this diagram.
5. It is not clear how this term should be deciphered.

In any abstractive class you can find one in any left in any sum – not summed up.[1]

Then have at once idea so simple that --- ⟨blank⟩

Event "E" can be completely and uniquely defined by total aggregate of

Points-instants which lie in it. So obviously an idea, that should have[2] time to go. Points that lie in E' because of K, lie in E too. So FK-e'. Can get rid of K.[3] Way of expressing geometry wobbles uneasily between the two ideas. Whose essence of a point is no extension. A point is quite "unspacey". Then (starting with point as essentially) we've got to as that is ⟨blank⟩

That relation of the points ? ⟨blank⟩

Points, and relations between them. – Neither spring out of the other. Whitehead has very strictly stuck to Descartes expect that Descartes didn't associate Space with Time. Essence of Space (for Descartes) is really difference of Essence of people around him. Then subterfuge to get over this.

Whitehead basing in modern physical science – that joins from symmetry) Exactly same kind of modern rendering of Descartes vortices extending over in time.

Electro magnetic field? Leibniz says space is relations between bodies in space. Then the point is a logical cock-up or construct from two or more.[4]

Hocking's notes

March 17, 1925

1. Different varieties of simplicity.

Position. Magnitude. Physical-Atom.

The concept in terms of which all exactness is to be expressed.

Imagines the Egyptian have not come down self-consciously on idea of a point.

Philosophy as giving things the right names.

2. Excited Mystery

The typical example √−1

Had the quality of working –

"There's no reason of getting excited about it. Perfectly humorous matter"

So with analysis of point.

You know exactly what you are handing over to the philosophers to talk about.

1. It is not clear how this term should be deciphered.
2. The last four words are unclear.
3. Much of the preceding several lines is unclear.
4. The last two of these especially confusing lines are written up the right-hand margin.

3. Instantaneous 3-dimensional space
Is merely concept of structure
very different from change.
Change in such a world is nonsense.

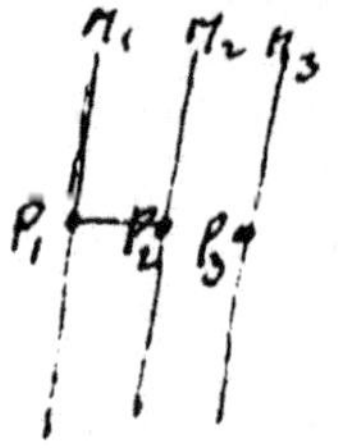

We do not observe instantaneous spaces.
We observe the continuity –
We do not observe that P has the same
function in an identical pattern in
all three moments –
We observe the pattern as enduring, i.e., at rest
The fundamental notion of being at rest
the apprehension of endurance.

4. The timeless enduring point. The Greeks muddled it up with the instantaneous point event.

The Geometrical Point ≡ Locus of Event Particles
Point event ≡ regular Point = Simple Extension
Position Point

Mar. 17–2

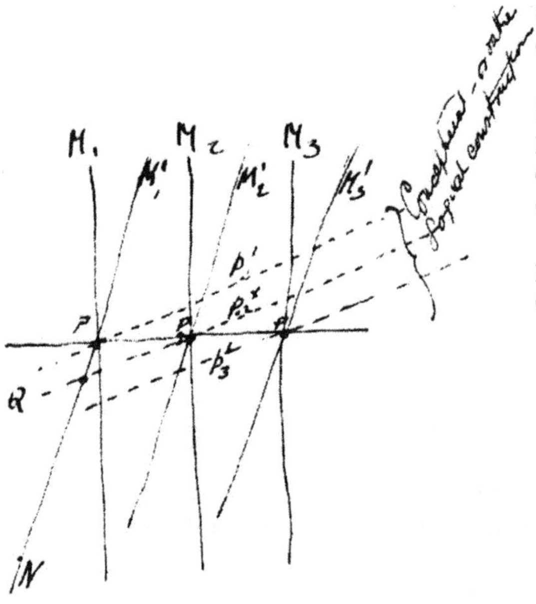

⟨Conceptual – or rather Logical construction⟩

[Calls attention to mistake in previous figure in *Principles of Natural Knowledge* in direction of dotted lines]

"Point event expresses the patience of the universe for exact observation"

Looking down a straight road P.N. you start off for the point Q, what you reach is P_2.

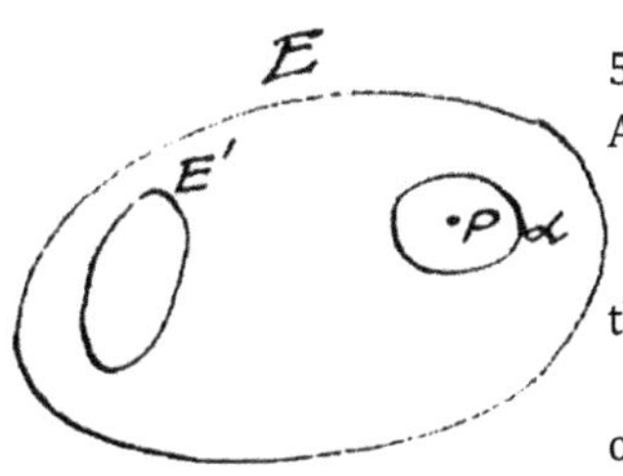

5.

Abstractive class α conveying to P.

It lies in the event E.

If you can find an event E' which contains one of the converging class~~es~~ near the tail end.

The event E can be uniquely defined by the totality of points that lie in it, or inhere in it.

---

6. Suppose E' is a part E, such that EKE' then all the points in E' are also in E.

Then you can replace the notion K by the relation of part & whole.

If you start with the point as fundamental – the point as Descartes thought of it as if no extension. The most "unspacey" thing.

You can't find what the spaciness of space comes from. Need auxiliary apparatus.

Points & the relations between them.

We have stuck rather closely to Descartes except that we associate space with time – Space he has is essentially occupied – so he had his vortices which pushed around the planets – occurrences –

Lecture 59

Thursday, 19 March 1925[1]

Bell's notes

|113|
Whitehead. 19. March. 1925[2]:-

Title for paper: – Derivation of the Exact Geometrical Elements from our general Apprehensions of Spatio-Temporality.

Literature: – Whitehead's own work: Lectures. *Principles of Natural Knowledge* (done in detail but some of conceptions put much more clumsily)

(With less mathematics) in *Idea of Nature.*[3] Again in <u>earlier</u> chapters of *Principles of Relativity.*

Then Broad's *Scientific Thought.* Eddington has <u>another</u> way of looking at matter. (*Space, Time and Gravitation*)

[Whitehead can't make it exactly clear to himself. Eddington jumps some of difficult points. Assumes congruence and measurement without pointing out any structure with regard to what measurement is made: – Eddington is spokesman for ? ? school.]

Another consideration (almost perfectly mathematical) is Robb. (A?)[4] – *Space and Time.* <u>Weyl's</u> preliminary chapters too. – Also to all philosophers *passim.*

Whitehead's criticism on general handling of problems of Space and Time in philosophy is that by not having sufficiently detailed and exact analysis their arguments are apt to have flashes of enormous insight; but as we progress to a more exact analysis one can see that many of difficulties and obscurities are that they are under one word talking sometimes one thing, sometimes another. – Advantage of living later in time: – Inheriting distinctions that hadn't been made in Aristotle's day, e.g., and which wouldn't have been made if these men <u>hadn't</u> written.

Maintaining reverence for past and at same time sturdiness of criticism. We have an advantage that <u>Aristotle</u> hadn't got. You also get flashes of genius in Aristotle that even best commentators have got force of. Early commentators of <u>Leibniz</u> e.g. had view all on <u>one</u> dimension of his thought

1. The Radcliffe version of this lecture, from notes taken by L. R. Heath on what was likely earlier the same day, begins on |48| of her notes, page 493.
2. For the preceding Wednesday evening, Bell's diary records: 'In evening Prof. Hocking at Canadian Club (In Conant Hall) (Talked about Phil. & Sci. in Rennaisance & Relat'n bet'n Religious & Scientific interests etc.).... Thereafter up till very late preparing lecture etc.'
3. This is surely *Concept of Nature.*
4. Also listed in Hocking's notes as A. Robb, and probably Alfred Arthur Robb (1873–1936), the British physicist who published *A Theory of Time and Space* in 1914.

– therefore selected and published all ideas of Leibniz bearing on what was in their minds.

When Boole and Peirce brought forth their ideas, then whole mass of Leibniz's manuscripts[1] thrown aside by the previous ones, become of enormous importance. Then Couturat comes to the manuscripts and writes his *La Logique de Leibniz*.[2] Distinction between duration and time now always attributed to Bergson, is in Descartes. But Original commentators didn't feel importance as Bergson does.[3]

When you talk of Geometry vaguely; we now distinguish three different (if closely allied) meanings. You can be talking vaguely of general 4-dimensional Space-Time continuum – Using "points" by which we make exactness in dealing with it. ʌ"point-events"ʌ[4]} Another meaning of Space is instantaneous 3-dimensional space of any particular moment. The points of such an instantaneous 3-dimensional space again are the (instantaneous) point-events – Representing maximum of simplicity.

Then we get Space of Physical Science. – Timeless-space – as sort of receptacle holding the events. Points of such timeless space require for their definition an appeal to our apprehension of rest and motion.

Also pointed out that we get an harmonious conception if we make hypothetical identification of Space within which a pattern is at rest with the Time-system which the pattern utilizes to express its Endurance. This not out of immediate apprehension, – but a harmonizing by hypothesis of various distinct Experience.

Then the point turns into a locus of ʌpoint-eventsʌ – Has now lost that idea of utmost Simplicity. Muddling of a point in timeless Space with the Point instant, which has got that Simplicity. These former called Spatial point. If you have alternative time-systems then the spatial point ---

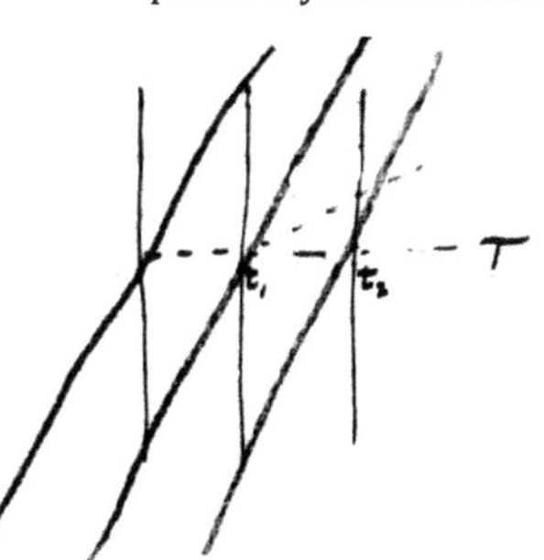

Time as measurable from t_1 – t_2.... But in other Time-system (changing from one assumption of rest to another.) If what you mean by Time and Distance is same in the two cases then your formulae in Relativity are nonsense.

If we've got a theory of congruence which enables us to configure measurements of Time in different Time-systems then --- ⟨blank⟩

1. This (and two lines below) assumes 'mss' is the abbreviation for manuscripts.
2. Louis Couturat (1868–1914) published his *La Logique de Leibniz* in 1901; it attracted Bertrand Russell's attention. In 1905 his *L'Algebre de la logique* provided an introduction to the algebraic logic of George Boole and Charles Sanders Peirce. Couturat was perhaps the strongest French advocate of Whitehead and Russell's *Principia Mathematica*.
3. While Henri Bergson had only recently published *Duration and Simultaniety* (1922), this concept was key to his widely read *L'Evoution créatrice* (1907) and had already been introduced in his doctoral dissertation *Essai sur les données immédiates de la conscience* (1889). Descartes had considered 'duration' an attribute of substance (*Principia*, LV–LVII) from which time is a generalised abstraction.
4. This is where Bell inserts 'point-events' but with no punctuation (e.g. a dash) to indicate where exactly it was intended to be added on.

Also want theory of congruence to configure measurements of Space in the two timeless Spaces.

If no standard of Configuration at all then formulae from one assumption of what is at rest to another is nonsense. That length of this room greater than of table and one red intenser than another – Can't configure them – no standard of relevant Configuration.

We've now got to enter upon idea of congruence. Euclid relies on idea of applying one body to another [unfortunate word - - - - is "fit on"].[1] This |114| presupposes that you know what you mean by moving a length to another. The idea of the rigid body has some point there. Whole question of rigidity again presupposes that Spatial properties of the two are always brought down to direct configuration of a body as here and there.

Got to come down to question of How you can measure – what that means. Only method of Configuration is by seeing a geometrical structure and noting that elements configured have a certain analogy of relationships in that structure. – No configuration then, unless you have a defined structure to start with.

Methods of differentiation: – by rest and motion as giving you point loci.

By --- ? Still a further question of Structure not yet dealt with.

Anything which is at rest in the ink is always recuperating[2] series of points in the pencil and that locus of points is a straight line. A few elementary ideas of 4-dimensional geometry.

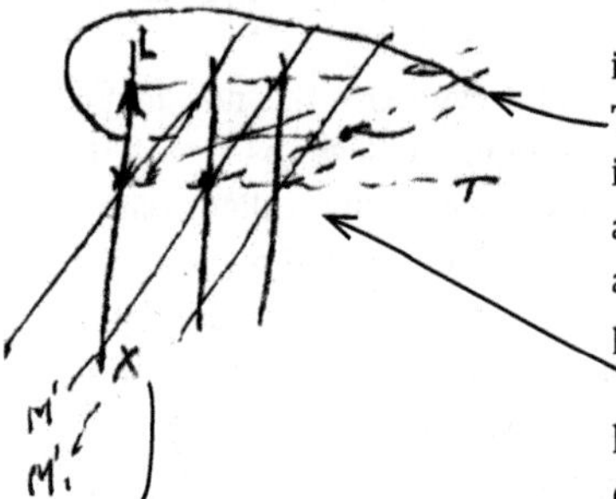

L T M' M X

This locus of points has a corresponding locus in any instantaneous Space of any Time-system. Thereby made an assumption verified (vaguely) in geometrical experience. Whitehead had another way of defining a straight line (It is not altogether matter of definition) (Intersection of planes).

I'm not defining a straight line then. (might have used this as definition) here but claiming a definite physical property which claims at least connection with experience.

We need to define right angle.

Here Whitehead agrees with Euclid – All right angles are equal. You can't state this as a postulate unless you know what a right angle is.

All symmetry as flowing out of uniform relations of one Time-system to others.

x is a point but looked on as intersection of M and M' is a plane.

And the line L stands for two things here: – ? ⟨blank⟩

1. In lecture 49, almost a month earlier, when considering some of these same issues, Bell records Whitehead providing a Greek term which was a bit of a puzzle. A colleague and Greek professor, Bruce Robertson, suggested an alternative Greek term, which seems to be confirmed here: ἐφαρμόζω can be translated as 'to fit' or 'apply'.
2. This has been expanded from 'recup^g' but seems to be an unusual term (if correct).

Think of it as a line however: – $L_{TT'}$ it is the projection of point in T' on to moment in T. Then any number of parallel planes $M'M'_1$ etc. – a whole family of parallel planes. They both arise from relationship of the two Time-systems T, T' as represented in simple moment. $L_{TT'}$ & $\widehat{M'M'}^x$ – very peculiar relation there. Can we spot anything in our geometry?

We find we get exactly what we want with "normality". (Rectangularity) – Relation of falling body to plane which it strikes. Moments of falling chalk intersect 3-dimensional space of floor etc. at right angles – Series of planes parallel to floor.

Very peculiar relationship which ought to find some expression – and must be symmetrical. Symmetry all the way around. Now question of measurement. (Read Descartes on Duration and Time Part I (Veitch) of *Principles of Philosophy* §57)

But:[1] When Descartes says "a mode of thinking" – he thinks he's settled something. But this only raises question: – Why on earth is this mode of thinking relevant to the world of things and facts? Why this mode of thinking should be of importance?

Whitehead trying to show it rooted in differentiations of Structures we find in immediate apprehension.

The geometry of everything here is --- ?

Hocking's notes

Mar. 19. 1925 – 1

[Eddington assumes measurement without studying congruence or assuring structure. A. Robb. *Space & Time.*[2] The philosophers passion.]

Descartes made the distinction between duration & time attributed to Bergson – These great men have many more flashes of genius than subsequent philosophy happens to have been able to work into a coherent system.

Veitch[3] trans. Part I. *Prin. Philos.* §57

1. 4 meanings of space
   a. 4-dimensional
   b. 3-dimensional instantaneous – point instants –
   c. Physical – timeless –

2. We need a theory of congruence which will enable us to compare measured time in one time-system with the same in another – If you have no standard of comparison, the formulae which enable you to go from one

1. The following comment seems to have been added separately, in pencil, at the bottom of the page.
2. See footnote 4 on page 272.
3. Hocking's notes have 'Beach' but it was (as in Bell's notes) the Scottish philosopher John Veitch who did translations of Descartes' *Discourse*, *Meditations* and at least selections of the *Principles*.

assumption as to rest to another assumption are intrinsically nonsense

We have now got to enter upon the idea of congruence – Euclid relies on "applying one body to another"

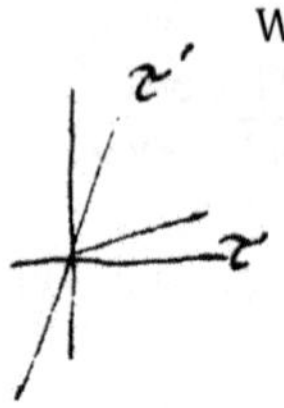

I maintain that the only method of congruence is by seeing a geometrical structure and observing that the elements compared have an analogous position in that structure.

3. A Prior point. What we mean by a right angle.

Lecture 60

Saturday, 21 March 1925[1]

Bell's notes

|115|
Whitehead: 21. March 1925

Now really to come to grips with question of Congruence. Elements which have analogous relations to same general structure. You can't therefore start your geometry with Congruence. But with that Abstraction by which you reach your structure. General idea of measurement, quantity, magnitude. Predicates belonging to lines – Such and such station in structure. Dropping Geometry and considering abstract idea of Measurement:– first determination of ideas we have to deal with is that of a class of incompatible predicates. γ[2] = a class and its members c_1 c_2 … c_s etc are predicates which may or may not apply to entities and if one does the others can not. – Idea of a class of incompatible predicates – an infinite number of members of high type of ∞. You can't define class by simple enumeration. Must be some ∧general∧ determining character or concept [Predicate of the predicate] application of which shows that c_3 belongs to that class γ. Here to point of W. E. Johnson[3] – "determinate" and "determinable" [?] (Whitehead thinks Johnson in a muddle).

Set of entities A, B, C each of which has one and only one of these predicates applying to it. C_s is a predicate of B. Fact that C_s belongs to γ. C_s has γ-ness. So the determinable is not directly a predicate of B but predicate of a predicate of B. Point of this class is incompatibility of the members of the class – Not logically incompatible, but merely [material] contraries. Raises a curious cosmological problem; Johnson has done a service in calling attention to it. But Johnson muddled in that γ applies to C and C to B. B can only have one C but one C can apply to many of the A, B, C, D, E, F…'s.

When finally get to ideas of measurement, A, B, C … will be the concrete quantities and c, c_2 … c_s will be the magnitudes.

Now idea of Matching is best abou⟨?⟩[4] one to keep in one's mind.

1. The Radcliffe version of this lecture, from notes taken by L. R. Heath earlier the same day, begins on |50| of her notes, page 495.
2. It is not always clear in Bell's handwriting whether this is gamma or a 'v', but in Hocking's notes it is definitely shown as a 'γ'.
3. William Ernest Johnson (1858–1931) was elected a Fellow of King's College, Cambridge, in 1902, and taught there for 30 years. The distinction to which Whitehead refers arises in chapter XI, 'The determinate and the determinable', of Johnson's *Logic, Part I* (1921).
4. It is not clear what ending has been given to this abbreviation (if it is an abbreviation) and thus not clear how to expand it.

$A = B \to \gamma$ means: same predicate cs applies to A and B.

(in respect to) Red is a colour. (to be a colour is to be a γ)

A is red. Can say A is coloured. (But this is high degree of abstraction.)

Johnson by looking on determinable as applying distinctly to determined obscures question. [But ... --- ⟨blank⟩]

$A = B \to \gamma . \supset . B = A \to \gamma$ (no question of order there). The order comes in in our way of thinking about, or symbolizing it. Knocking out Order is merely knocking out irrelevant matter which our way of thinking about it brings in.

$A = B \to \gamma . B = D \to \gamma . \supset . A = D \to \gamma$

This transitiveness of Matching depends on incompatibility of the characters. Otherwise B = D in another of the c's. Look at Whitehead's "Relativity" – He carelessly fell into trap and introduced Incompatibility after this.

$A = A \to \gamma$ (Reflexiveness). But not any entity so. Because no predicate of the γ. type may apply at all.

These general ideas of equality are not nearly sufficient to give you the numerical relation (Idea of Magnitude) To do this you have got to have some notion of what you mean by Addition (Addition of Magnitudes is derivative of this) Also "greater than" (Euclid introduces this with greatest cheerfulness). Also must introduce something corresponding to ∞ divisibility. Finally the "Axiom of Archimedes" (to be explained when we come to it).

Addition: – In special case before us, the things to be matched are stretches of straight lines. In what sense can we talk of addition. – We have to have a junction. Two stretches in same straight line – one starting where other ends. Idea of addition will have to be made up of C as compounded of A and B.

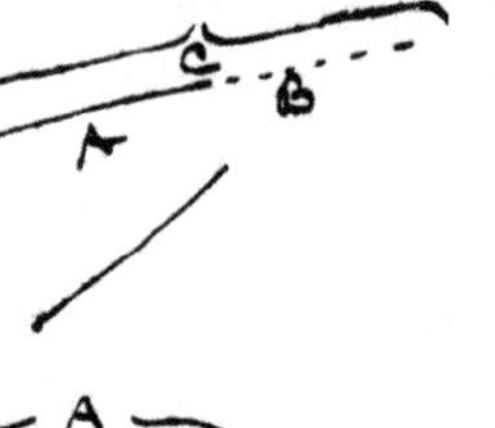

Adjunction here.

Next question: – "Greater than" – Must obviously be connected with idea of Whole and Part. But not quite so simple.

Euclid's idea not founded entirely on this although he phrases it so.

Fundamental idea

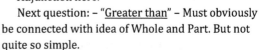

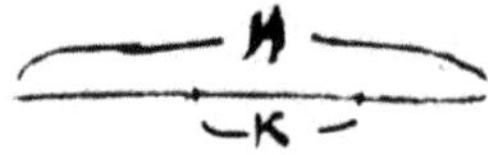

$A > K \to \gamma$ when $A = H \to \gamma$ + K is part of H.
matches

If philosophers don't criticize selves by careful and tedious reference in Logic, they're apt to

|116|

think all ideas of > following out of such a nice little definition without further trouble. But two more assumptions there: –

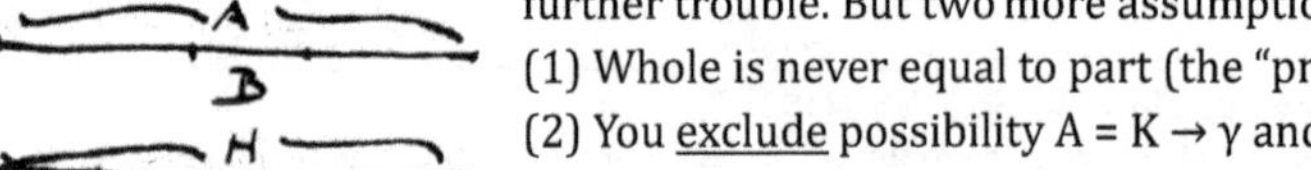

(1) Whole is never equal to part (the "proper" part)

(2) You exclude possibility $A = K \to \gamma$ and $B = H \to \gamma$.

Even with all that you can't work numerical measurement without ideas of divisibility and Archimedes' idea.

Now "Divisibility": – Let us say a sequence of stretches when stretches are successively adjoined and are equal.

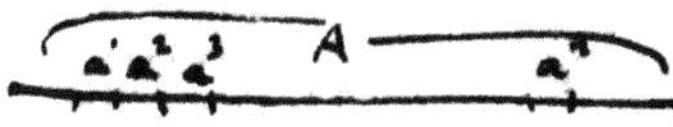

$a_1 = a_2 = a_3 \dots = a_n \rightarrow \gamma$.

Whole stretch call "resultant" stretch of the sequence.

Supposed c' applies to any one of these and A is c. Then relation of c to c'.

[c] = n[c′] or, equally:– [c′] = 1/n[c]

That requires this hypothesis or condition. If A is any stretch and n any integral number, then it is possible to find a sequence of n terms such that its initial stretch A is unjoined at one end and An unjoined at other. – Can divide A into n equal parts. (Whatever number you take).

Axiom of Archimedes. So obvious that very creditable of Archimedes[1] to find it out.

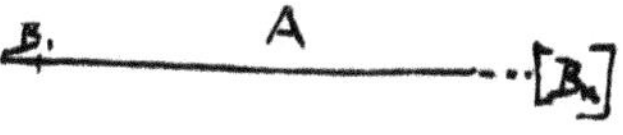

If you have any stretch A and any other B part of it; then it is possible to continue a sequence of which B is first term and find some number n large enough for last term of sequence to fall outside of A. There any number beyond n will do it too, of course. This is fundamental. But for this you might imagine Stars such that their light would never reach us.

Now all necessary for regarding γ as class of entities which are quantities with their magnitudes; then numerical measurement is possible. In various philosophical questions fact that philosophers haven't always conditions for numerical measurement clearly in mind that they go astray. You may have many of the conditions satisfied but[2] not get near measurement. E.g. in hedonistic calculus. Obviously something in idea of addition of pleasures but just what? Philosophical treatment influenced by idea that if we knew enough it was numerically possible. Whitehead would think not.

Now question: how we are going to --- ?? γ

Mathematics found out, whether only one such class γ (only one class of magnitudes) – There may be alternative and inconsistent ways of doing it. There is a whole set of γ's. Any one of them will satisfy these conditions as well as the other. From point of view of Math (Poincaré) it is quite conventional (from mathematical point of view) which you adopt. – Some are simpler than others.

From old metrical geometrical view point there are three groups. One will give you metrical geometry of Euclidian type. Another will give you metrical geometry of Lobatchewsky's hyperbolic space; another that of elliptic space.[3]

1. Bell has used 'A.', which we take to mean the name, rather than referring to the diagram.
2. Bell has 'by' but the context suggests 'but'.
3. Nikolai Ivanovich Lobachevsky (1792–1856) is primarily remembered for his 1830 work on hyperbolic geometry in which he demonstrated one possible non-Euclidean geometry by replacing Euclid's postulate

Therefore distinction between different types of geometry is not so much geometrical; but on which system you choose to express your measurements. Halfway to Einstein. This twisted by overenthusiastic to say that in measuring you're dealing with concept which the measurer has forced on his material – material subjected to it only by Volition of Observer. No! Those relationships are there in the material and are discovered by observer who says these are good enough for my metrical purposes. But if not there then they wouldn't help him to find out anything. A muddle arising out of Kantianism. Columbus instead of sailing West to India and discovering America might have gone South and discovered Australia. Quite open to him to discover either continent and be satisfied with. The pure subjectivism which makes measurement arbitrary doesn't follow from alternativity of possible measurement system.

(2) Poincaré says: it is a matter of convention which system we adopt. Whitehead ventures to doubt this. As a matter of fact we do apprehend one definitive dominate system of magnitude which applies to Space-Time. A great many Mathematicians reading Poincaré miss the point. The alternative γ's aren't just a little inconsistent. (Some are but many wildly inconsistent). It is not a matter of Convention whether Standard "Yard" in bureau of standards is > or < my distance from State House. But you could find Mathematical System which would make the two equal.

Another question which gets confused with it. Our senses – and even with genius of Michelson – always inexact measurement. You can't always tell exactly which one you've got hold of. But this only leaves open a very small bundle of alternative γ's. Becoming "more exact" means identifying measure relations he has he can more exactly measure the lengths. There is always the perfecting exactness. Within your limits of observation you always measure perfectly determinedly. –

When you look at definite Structure of things there is for us a certain observer's γ which is the one that exhibits certain dominant relations in the Spatio-temporal Structure.[1]

Hocking's notes

21 – 1
Mar. ~~19 – 2~~

Congruence.

Thinks W.E. Johnson in a muddle in relation to Determinable and Determined. "Of course one thinks oneself not in a muddle – that is a fundamental law of thought"

---

regarding parallel lines with one assuming they would diverge. Bernhard Reiman demonstrated in 1854 that other non-Euclidean geometries are possible assuming two such lines when extended will converge.

1. This last sentence is written up the left-hand margin.

Quantities	Magnitudes	C, etc., being
A, B, C ---	applying to Quantities	incompatible
	$\gamma = (c_1 \text{ --- } c_s \text{ ---})$	predicates
		[2 cannot be 3
		or 2.1, etc]

MATCHING

$A = B \rightarrow \gamma \quad \supset \quad B = A \rightarrow \gamma$ ("A with respect to γ")

means that the same predicate cs applies to a & b.

Red is a color –

A is red –

$A = B \rightarrow \gamma.\quad B = D \rightarrow \gamma. \supset. A = D \rightarrow \gamma$ because this simply means having the same predicate c [In book on Relativity, fell into the trap & introduced the idea of incompatible predicates one sentence too late]

$A = A \rightarrow \gamma.$

ADDITIONS. GREATER THAN. INFINITE DIVISIBILITY. ARCHIMEDES

Mar. 21– 2

1) Addition

Matching of straight lines

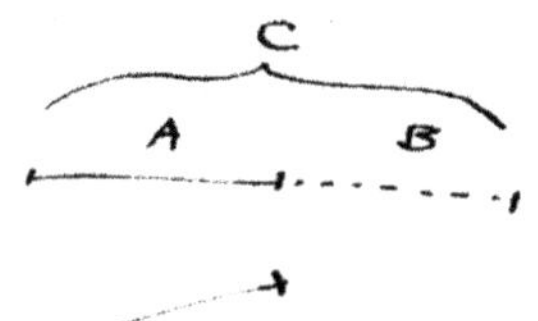

Addition here means adjunction.

---

2) "Greater Than" connected with "Whole & Part"

A > B

when

$A > K \rightarrow \gamma$ means $A = H \rightarrow$ and K is part of H.

Two assumptions:

The whole is never equal to the part

A

Impossible that

$A = K \rightarrow \gamma \,;\, B = H \rightarrow \gamma$

3. Infinite Divisibility A (predicate c)

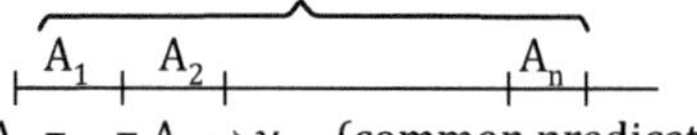

$A_1 = A_2 = \quad = A_n \rightarrow \gamma.$ (common predicate c')

Call the whole stretch, A, the resultant or sum

$[c] = n[c']$ or $[c'] = 1/n[c]$

If A is any stretch & N any integral number it is possible to find a sequence of n terms whose initial stretch A, & final term A_n just fill it up

4. Archimedes Axiom

A
B

If you have any stretch A and a stretch B which is part of A it is possible to continue a sequence of which B is the first term, which will yield a term Bn which lies with its last boundary outside of A. Otherwise there might be a star so far away that light will never get to us.

This gives you your class γ, as a class of magnitudes. Try this on the addition of pleasure for sake of Hedonistic Calculus

Mar. 21 – 3

An "interesting fact" discovered by math.

Whether there is only one class γ.

There may be alternative and inconsistent methods of measurements.

$\gamma_1\ \gamma_2\ \gamma_3$ may all satisfy these conditions.

Poincaré – it is quite conventional from the point of view of math. which you adopt.

---

3 groups of predicates.

a. Euclidean geometry – a group of γ's any one of which will give you a Euclid geometry

b. Lobatchevskis[1] hyperbolic space –

c. Elliptic space –

This has been twisted to the remark that in measuring the observer has forced his concept on the material –

This forgets that the relations are in the material & discovered by the observer.

The alternative possibilities are also there –

Poincaré says it is a matter of convention – I venture to doubt it.

We do apprehend one definite dominant gamma which applies to Space-Time.

These alternative gammas not merely a little inconsistent , some are wildly inconsistent.

It is obviously not a matter of convention whether the distance here to Washington is greater or less than the standard yard.

The inaccuracy of the senses has no bearing on it – Michelson can keep becoming more exact

---

1. See footnote 3 on page 279.

Lecture 61

Tuesday, 24 March 1925[1]

Bell's notes

|117|
Whitehead. 24. March 1925[2]

To come back to Magnitude again. Addition (P + Q to get R). You do perceive some situation which can be expressed as a 3-term relation in which P + Q have a special status – and indicating a third unique element R. P + Q must be in some sense or other of same kind – and R of same kind too. There must also be (as last time) – if addition which results in application of number. If only some of these conditions satisfied you'll get a looser meaning (e.g. in theory of value). ∧Question of∧ Intensive and Extensive quantities. Whitehead's view is that all extensive Quantities presupposes addition in strict sense and is derived from Extensivity of Events.

"All and Some" gives you.

All measurement of intensive quantity comes from perceiving some (perhaps quite arbitrary) correlation to the All and Some or Whole and Part. Mere intensive quantity originally puts your intensive quantities on a Scale and it is getting measurement of the scale which is the difficulty.

Take measurement of light – Double from two candles? All you can see is that light is up on the scale. Then you correlate this with the two candles. You won't have correlated it with bright patches (concentrating with lenses etc.)[3]. Always by reference to Congruence, "Continued in", All and Some, and Whole and Part and Numbers.

[Correlation with something which can be dealt with by].

Congruence to arise from Spotting an analogy of relationship of two things of same type – spotting their analogous functions in the geometrical arrangement of straight lines, planes etc. Congruence arises from uniformity of relationships between Space-Time relationships. "Uniformity" is that difference in position is irrelevant to expression of these relationships. Position always refers to particular moments of Space-Time system involved. This and That, e.g. representing four Space-Time systems – four moments each from four Space-Time systems. All difference of position means the configuration of the Space-Time moments involved – Always comes down to lapse of Time ultimately.

1. The Radcliffe version of this lecture, from notes taken by L. R. Heath earlier the same day, begins on |53| of her notes, page 497.
2. For this day it was 'Work as usual' for Bell, but he noted in his diary for the preceding Sunday: 'In evening around to Whitehead's again – taking G.L. & Paul Herzog. Quite a crowd there ultimately.'
3. Closing parenthesis supplied.

Differentiation of moments into planes is via ? ? ⟨blank⟩ therefore all uniformity means that if you have expected relations between two Space-Time systems in one way by a set of their moments; by taking an analogous set of other moments ⟨blank⟩ Common relationships can be expressed independently of position. In dealing with congruence in space.

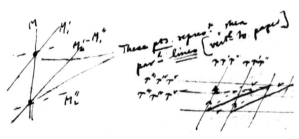

If I take a system of parallel lines in a plane each bearing in itself relation to three Time-systems. Whitehead's point is that congruence expresses that here you have merely difference in position; and therefore ⟨blank⟩

The v's are parallel. This has an analogous function to that. Mere difference of position, else.

Prof. E. B. Wilson's paper[1] years ago gave Whitehead this. But this has difficulty: – no way of getting congruent bits on lines that aren't parallel. No good saying: "Oh! We discern congruence." We ought to be able to point out something in what I'm talking about. The mere fact that mind wanders off to "Congruence" when you see this and that - - - - doesn't give you anything.

– As if you could forecast acts of Republican party by going to Africa and studying elephants. – Like whole business of secondary qualities.

Question of congruence from one moment of an instantaneous system to another of the same system. The congruence has got to get itself reflected into the timeless space.

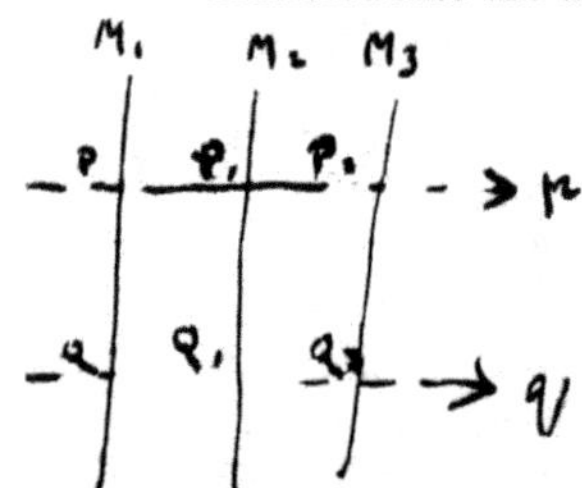

PQ = P'Q' = ---

That suggests as to whether we can't found our ideas of congruence on something a little more obvious to us. Two particles mutually at rest remain at same distance from each other. This assumes that being at rest is not primarily defined as being at same distance.

Two particles mutually at rest in any one Space-Time remain at same distance in any other Space-Time.

No necessary conclusion (as usual) that ? ?

The famous Lorenz-

|118|

Fitzgerald thesis[2]:– The moving molecule gets flattened in direction of its motion. This general principle now enables us to get all we want in

1. Edwin Bidwell Wilson (1879–1964), an American mathematician, wrote a textbook entitled *Vector Analysis* (1902) and with Gilbert N. Lewis wrote the paper 'The space-time manifold of relativity: the non-Euclidean geometry of mechanics and electromagnetics', *Proceedings of the American Academy of Arts and Sciences*, vol. 48 (1912), pp. 389–507.
2. See footnote 3 on page 268, lecture 58.

first diagram. Instead of how use of a higher dimension gives you help in dealing with a "lower" one. We've there come back to old Euclid's idea of things transferring themselves.

[Newton's end point.]

There of course that gives me idea of definite distance between two points in absolute Space. Definite Time-system is between ? ?

Now to lines not parallel. Have I any elements concerned with lines not parallel which indicate any simple things for Time-systems. The plane and the family of lines perpendicular to it.

Then project these on left hand line representing M: – Then they'll be lines.

To introduce idea of complete symmetry: Inside the Space M. Then upright line has to have perfectly definite relation with the plane. The 4 base angles are market leaders. Now can we get and define what we mean by congruence in lines not parallel. Now take line BC and bisect it in D and erect perpendicular. AB & AC congruent because of symmetry there. Founded on symmetry arising from perpendicularity. This often thought of before. – Often urged that true way of proving *pons asinorum*[1] is by appeal to Symmetry.

Whitehead regards this as good from Metaphysical point of view.

Difference between your metaphysical principle and your logical premises.

Question of getting smallest number of principles and premises. Minimum of Assumptions in Logic. You may find your reason for believing "it" true in some one case is ⟨blank⟩

So here: – ∧General∧ Appeal to Symmetry (from which get all your congruence) is the general principle that we want.

Symmetry arises when you are dealing with those relations as between Space-Time systems expressible in space of one moment of one of them; and simply express relevance of two systems independent ? ? ⟨blank⟩

Another little difficulty crops up at once:–

We can define obtuse and acute angle. (No matter of measurement but of lesser and greater – within and outside.)[2]

1. Proposition 5 in Book 1 of Euclid's *Elements* came to be known as *pons asinorum*: the angles opposite the equal sides of an isosceles triangle are themselves equal (and its converse).
2. End parenthesis provided.

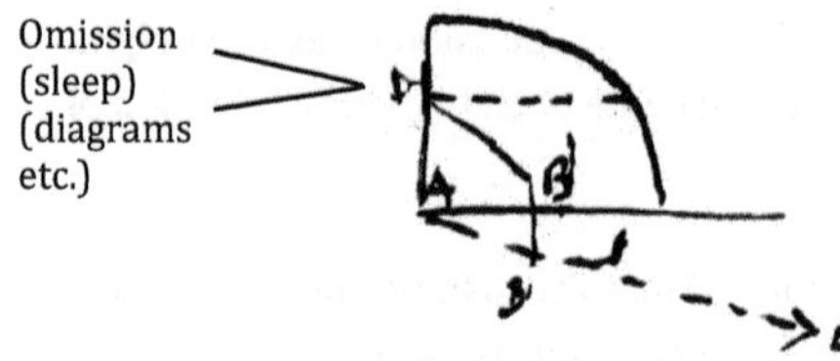

As soon as you may assume a square you'll angle <u>right</u>[1]. Now by particular case of *Pons Asinorum* <u>Then</u> in <u>clover</u> with 3-dimensional space.   AB = AD from 1st space and AD = AB' in other cases.

Difficult to bring out properties in lower grade of dimensions by appealing to a higher one – Congruence in 2-dimensional by appeal of 3. Projective geometry as e.g. – Getting out of your plane and can get back to it again. Interesting fact <u>re</u> independence of hypotheses (van de Walle[2]) – where greater <u>number</u> of things you are dealing with makes all the difference.

Hocking's notes

Mar. 24. – 1

<u>Addition</u>. p + q = r.   p & q must be in some sense of the same kind & r.

Extensive & Intensive quantity. "My suspicion is that all extensive <u>quantity</u> presupposes addition"

<u>Return to Congruence</u>. Congruence between Space-Time systems –

<u>Mar 24 – 1½. |</u>

"But it has a limitation doesn't give us a ghost of an idea how to get congruence on lines not parallel. No good saying, We discern congruence. We must point out something in what we discern that is the foundation of congruence"

Parallel lines

Congruence ✓
suggested by Wilson[3]

[Secondary qualities have the same relevance to experience as elephants in Africa to Republican Party]

---

1. Here, and in the text to follow we see the effect of Bell nodding off. As previously, he is conscientious enough to note that he did (in left margin). Whereas Bell jotted down some text across the diagram, we present them separately. Compare with Hocking notes.
2. We are not sure who Van de Walle might be, although it could be a misspelling of Johannes Diderik van der Waals, who received the Nobel Prize in 1910 for his work on the thermodynamic equation that bears his name. To derive this equation one assumes the particles involved move independently, and the more of them there are, the better.
3. See footnote 1, page 284, of this lecture.

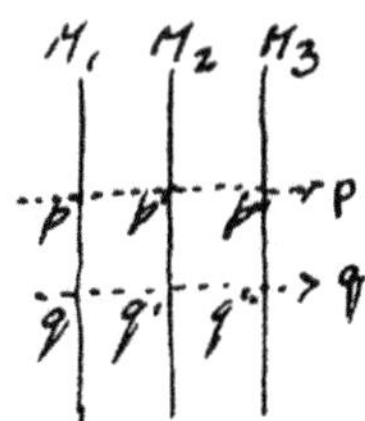

What is congruence between one true-moment and another. Must be found in the timeless space.

PQ = P′Q' when p is the path of a point at rest and also q.

Can't we found our notion of parallel on something a little more obvious? The above way by intersection is best as most general, but in special cases, being at rest of two particles will do. Not having defined rest as being at the same distance.

Assume two particles which are mutually at rest in any space time [Something here falls by the bound– giving what you want for the Lorenz–FitzGerald transformation].[1]

Now how to get to lines not parallel

Bisect BC in D. AB is congruent with AC. outside of parallels[2]

Founded on the symmetry that arises from perpendicularity. *Pons Asinorum*. Euclid I, V.

But you don't want to multiply appeals to symmetry unnecessarily.

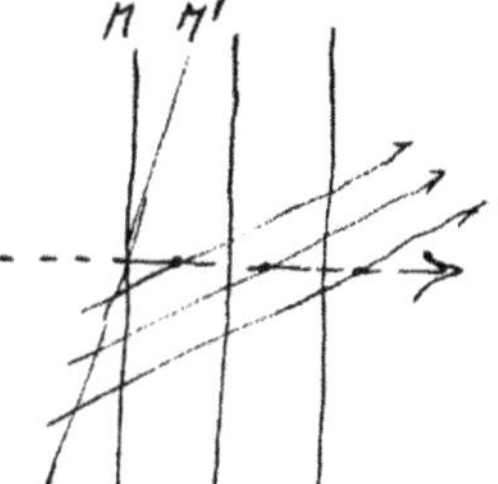

Have we now got all the premises we want?

But start with ↓ and find congruent segments.

Can't bisect the angle, because that assumes congruence of angles.

We <u>can</u> define obtuse & acute with only C lying within.

and the diagonals

Consider a parallelogram

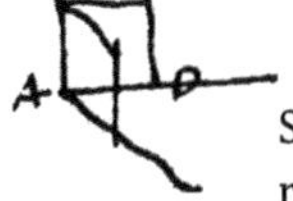

Square as a parallelogram whose diagonals are at right angles.

A lot of congruences at once

Construct squares in planes on another dimension.

Calling on higher Dimensions Than you want – the general rule of method.

1. See footnote 3, page 268, lecture 58.
2. Ending of this term is unclear, and it might be 'parallelogram', as further down.

Lecture 62

Thursday, 26 March 1925[1]

Bell's notes

|119|
Whitehead. 26. March. 1925:–

How we can discuss congruence of Time-stretches. Your anger or boredom during an hour – increasing in intensity. Comparison demands more than mere serial arrangement and more than mere difference in quality etc. There is a method of Configuration. We ought to have a theory which gives a basis in nature of things for what we directly apprehend. If we can't do that it not only throws us back on a theory of secondary qualities, but divorces these from real world. Unless shows us something in Universe which we're directly apprehending, tells us nothing about it.

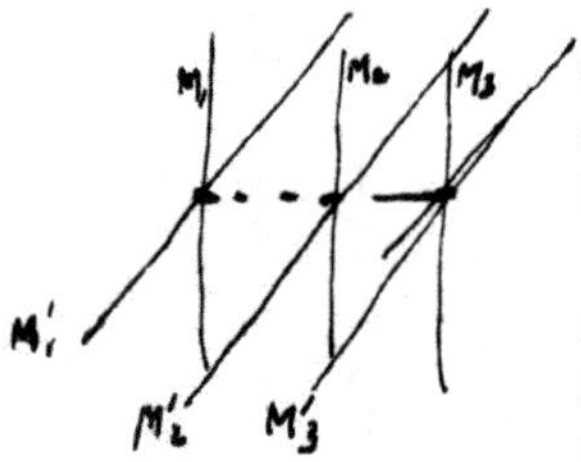

Is there anything in our System which enables us to configure the two lapses of Time as equal to or different, and give us an analogous place in the Structural System?

We see at once something which could give us a method of Configuration. What do you perceive in respect to time? [Whitehead here in direct agreement with most philosophers in getting Time out of motion] You have a direct apprehension of uniform motion and of its uniformity [sic!] and tracing of equal spaces by this is your apprehension of Time. At rest in System T but moving in system T'.

Now come to a difficult point – taken always as paradoxical. But no new paradox beyond the different Time-systems. Question of relative velocity. Question helps us to another point. Configuration of Time-lapses within one Time-system is okay. But Configuration of Time-lapses in different Time-systems not yet accounted for. Of course you can configure them; and hard to see difficulty when haven't laid down principle upon which yet.

We at once know that all velocities are configurable. You are traveling in a moving carriage and roll something along floor. We have immediate apprehension of configurability of Velocities. Also know what we mean by Configurability of all Lengths, in different meanings of Space. Since

1. The Radcliffe version of this lecture, from notes taken by L. R. Heath earlier the same day, begins about midway on |55| of her notes, page 498.

"Waves" are definitely configurable, therefore Time is configurable. Then can ask ourselves: what assumptions re magnitudes of Velocities will enable us etc., etc. Velocity of a fraction a second will turn out in very different number, in miles per hour. – Numerical result depends on Time-units, even if keep Space-systems same.

Can always choose two Time-lengths and say I'm going to consider them equal. But how get a coherent system here? Question of relative Velocities; and this shows a lot of distinctions we don't ordinarily make.

two cars moving with velocities v, u.[1]
Q gaining on P with velocity u – v.

P'Q' = PQ + QQ' – PP'
P'Q' – PQ = QQ' – PP' = ut – vt = u – v
(P'Q' – PQ)/t = u – v   Anything which disturbs that must be nonsense. Mathematicians have habit of worrying other people. – Apt to put things in a worrying way.

– Don't always see quite what they're doing. Pleasure of making mathematics mysterious is one of the aboriginal pleasures – Nothing in relativity which knocks out that   and if there were it would be nonsense. Important to be able to recognize nonsense when you see it.

All here referred to Space in which road is at rest. T_r (Observer's space). Any particle is at rest ---   Space-Time system of Car P and that of car Q. Velocity of separation in the T_r system of Q from P = u – v ("Differential Velocity) ʌin T_rʌ. But in T_p system there is the Velocity of Q. [P is at rest]. Velocity of separation of Q from P in T_p system. [All your geometrical elements have different meanings here. – Points etc.] Again a third one:– Velocity of separation of P from Q in the T_Q system. Here a new problem. – Suppose we have congruent measures of these Velocities. Are these three the same physical quantities? Not obvious, because we are dealing with different meaning of our words.

Whitehead takes line: (1) to be as near immediate apprehension as possible. (2) where diverge from common sense, do so, so as to conciliate definite scientific observations with immediate apprehensions.

|120|

When the T units are also congruent to each other: the last two are equal (Definition, but also common sense and usual assumption)[2]. (If e.g. the watches of the two observers regulated to show congruent T-units). But neither of these two velocities will agree with the first one. Explained by saying that their meanings of distance are different.

Of course all these distinctions fade away at anything less than Velocity of Earth in its orbit – and genuinely with velocities like that of light.

That "equality" there looks like a pure and absolute definition. But there is really a substantial assumption.

1. It is not clear if these are 'μ' and 'v' (although this is simply velocity and not frequency) or simply 'u' and 'v', which we have used.
2. Closing parenthesis provided.

Suppose I start with T_o then T_p, T_Q, T_R etc:
By this ⟶ second

I fix up what I mean by a second in each of the others.
But then comes the substantial assumption (really is matter of relative Velocities).

That when I've got unit of T in T_p and in T_Q, then assume that Velocity in T_p will be = to Velocity in T_Q.

V_{PQ} = velocity in P due to rest in Q

V_{QP} = velocity in Q due to rest in P

Nobody has ever doubted this. But it is a very real physical assumption. It doesn't come out of any metaphysical point of view. Elementary Science here perfectly consistent with our metaphysical outlook.

You will find a great deal still indeterminate as to relations between various Time-systems. Decided by bringing in further considerations re symmetry.

[Leaving this for fear of confusing issue]

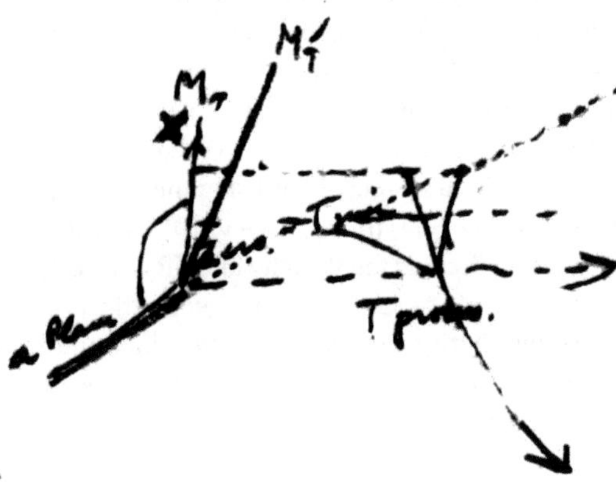

a line – road to a building you see in distance e.g. X.

The plane is e.g. a wall blocking the road.

Then take another Space-Time system. – Somebody going with uniform Velocity along the road. Its moment of time will be $M_{T''}$. It will intersect M_T in a plane and rest in T' will mean movement in M_T...

Car at rest in T' $\quad V_{TT'} = V_{T'T}$

Nothing magical about it. – It may be all wrong. But if want to show that, it is just here that the assumptions are made.

Suppose now some other moving relatively to both these. (Aeroplane). When you know motion of Aeroplane in T space and know congruence of T T' to express it in T' system. Suppose aeroplane flying over road too.

Naturally would say: $V_{T'A} - V_{T'T}$ This is what is denied.

Really: –

$$V_{T'A} = \frac{V_{TA} - V_{TT'}}{1 - \frac{V_{TT'} \cdot V_{TA}}{c^2}}$$

This is what is denied. Where c is a Velocity very nearly that of Light in *vacuo*.

This is practically negligible except in enormous Velocities. Gives explanation of why we have never noticed difference.

Slight difference in all our dynamical formulas. – Verified all along the line in finest measurements in all our ⟨blank⟩[1]

1. Bell's notes for this lecture seem to end here, mid-sentence. And Hocking's also seem to be incomplete.

Hocking's notes

March 26. 1925 – 1

[Metaphysics is finding good reasons for things nobody can believe – H]

"The work is all done before you begin to reason" – i.e., finding what you are talking about.

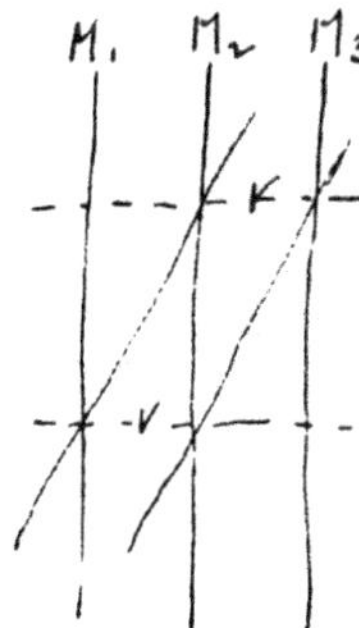

But this isn't what we directly perceive. Most philosophers agree that we get time out of motion.

Descartes, for instance.

Uniform motion means equal spaces are traced in equal times.

When you are at rest in system T' you are moving in system T.

And covering equal spaces in equal times. Descartes then is a special case of the general rule.

Have as yet no means of comparing times in different time-systems.

Now we know that velocities are comparable

Roll a ball over the floor of a moving car –

We can compare the velocity of the ball with that of the train over the track.

You can always choose your units so as to make any two velocities the same?

Two cars P & Q with velocity u & v.

Q is gaining on P. After the time t.

u – v, the rate of separation is $\frac{P'Q' - PQ}{t}$

"Any thing that disturbs that is nonsense."

"Mathematicians like to worry non-mathematicians –

"The pleasure of making mathematics mysterious is one of the aboriginal pleasures of the human mind"

"It is very important to know nonsense when you see it."

Space-time system of the car P and of Q. T_p. T_q.

Velocity of separation in T_r is U – P.

T_r the time of the road

Differential velocity.

In T_p the velocities of Q [P being at rest] = v^s of Q from P in T_q system

When the time units are congruent, the rates of separation as measured from P and Q are equal.

But neither of them will be the same as u–v. Why not?

Because their meanings of distance are different

<u>3/26 – 2.</u>

Suppose I start with time-system T.

I settle what I mean by a second in each of the others by the method above.

Whose essence is:

V_{pq} = velocity in P due to rest in Q
V_{qp} = velocity in Q due to rest in P
$V_{pq} = V_{qp}$.

Then we assume that all the other adjustments will come all right. Nobody ever doubted it.

We will now pass off from the question of congruence leaving out further questions of symmetry.

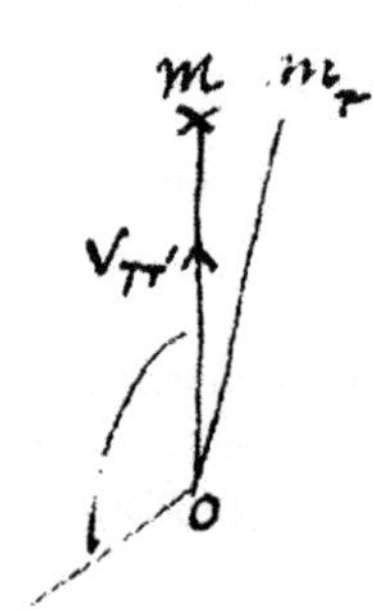

Say man at rest in T' is moving along OX in T.

Lecture 63

Saturday, 28 March 1925[1]

Bell's notes

|121|
Whitehead. 28. March. 1925[2]:–

$$\left.\begin{array}{l} x = vt \\ X = (v - u)t \end{array}\right\}$$ This wrong – as shown by most exact observations

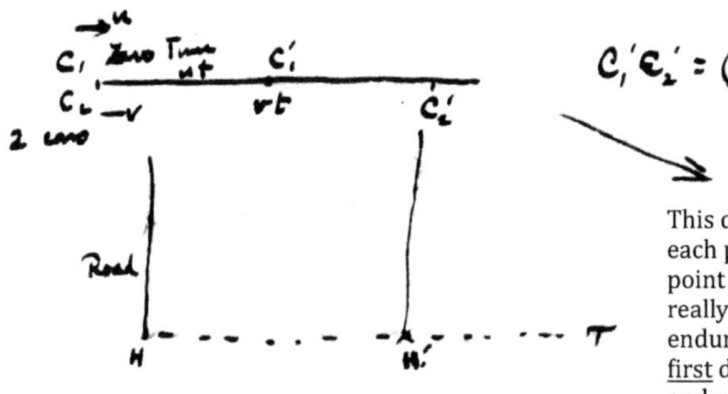

$C_1'C_2' = (v - u)t$

This drawn in impartial Space. But each point drawn in instantaneous point of instantaneous time. But really the thing of any one instant endures down through time T. The first diagram is second looked at endwise on (from T)

We needn't bother about the two cars.
Let us measure the Point-event P from zero-point. x – ut from moving car
If man in house says P is at distance xt from me, man in car says it is x – ut away. Now in second diagram.

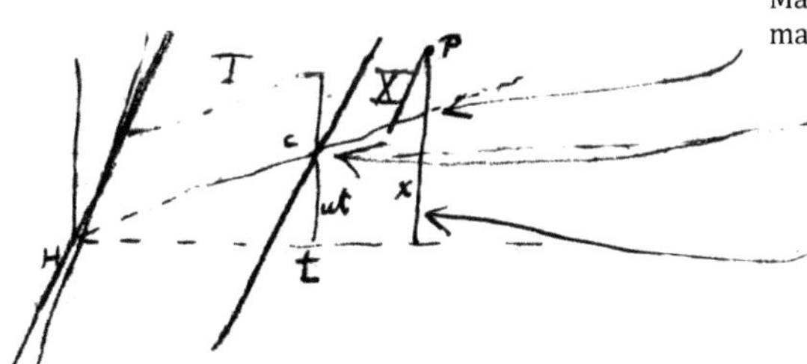

Man in house says here's man in the car.

But for man in car, here's the instantaneous road.

Event-particle P to man in house says event-particle is at that distance x.

The man in car says I'm in this point c' and X is distance away.

1. What might be the Radcliffe version of this lecture, or portions of it, from notes taken by L. R. Heath on what was likely earlier the same day, begins on |58| or |59| of her notes, pages 500–1.
2. In his diary, to his oft-repeated 'Work as usual', for this day Bell adds: 'Lectures going well these Days' (presumably, his own).

T and t different too. They're talking of different things. Man in car taking himself as at rest. No more reason to take house as at rest than car. House anchored to a bigger thing, that's all.

Car gone up ut on road according to reckoning of house. Question now what's relation between x, t, and X, T.

There is, now, a maximum velocity c. When anything is at rest in a Time-system, it must move in any other Time-system with a velocity less than c.

Relation between the two

$$X = \frac{x - ut}{\sqrt{1 - \frac{u^2}{c^2}}} \qquad T = \frac{t - \frac{ux}{c^2}}{\sqrt{1 - \frac{u^2}{c^2}}}$$

Most absurd formula, no one would have guessed it; but it has trick of doing what physicists want. It comes out in two ways, (1) as Whitehead has done it: – by finding a more general structure of type of thing Space-Time is, or (2) by taking Maxwell's equations and finding sort of formulae you want to make those equations sense.

So long as your u is moderate these formulae make little difference. With Velocity of earth in orbit, a wee effect – so wee that it takes Michelson's technical genius to discover it.

Wherever effect does show, formula has turned up trumps. If true, then "Distance" hasn't got a unique meaning – It all depends. Now take Law of Gravitation – Attraction as square of distance – Which distance? Rest → system for both of them is O.K. but won't work for astronomy [simply can't measure them in this ? ?]

Take motion of Moon – Earth and Sun both pull. You have got to have it formalized impartially as between every Space-Time system. Then you have got to think out a new Law of Gravitation which uses something more general than idea of Distance (but holds for that). There came one of Einstein's strokes of genius. "Tensors" (Einstein didn't invent them – Green[1] and some Italians did that – but Einstein saw their use here). (There are two or three ways of thinking it out).

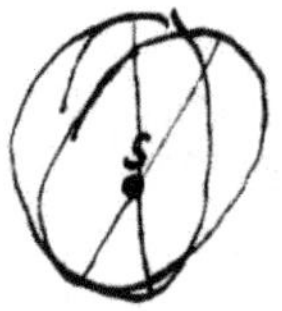

Suppose Space-Time system, of sun, with Mercury going about it in elliptic orbit itself revolving. Major axis slowly turning. The small terms which turn up in above formula directly account for this revolution.

Again: – If anything moving with critical Velocity c in any one Time-system – must move with same in any other. Light moves with that in vacuo, according to Maxwell's expert:– and with approximately that in Air. Michelson's interferometer. Light moving in direction of Earth's motion should move faster than at right angles to it. Doesn't happen.

Again: – 100-years' puzzle. Water very quickly through tube.

1. Bell is referring to the self-taught British mathematical physicist George Green (1793–1841), and likely to Gregorio Ricci-Curbastro (1853–1925) and his pupil Tullio Levi-Civita, who published together their 'Méthodes de calcul différentiel absolu et leurs applications' in *Mathematische Annalen*, vol. 54(1) (1900).

|122|

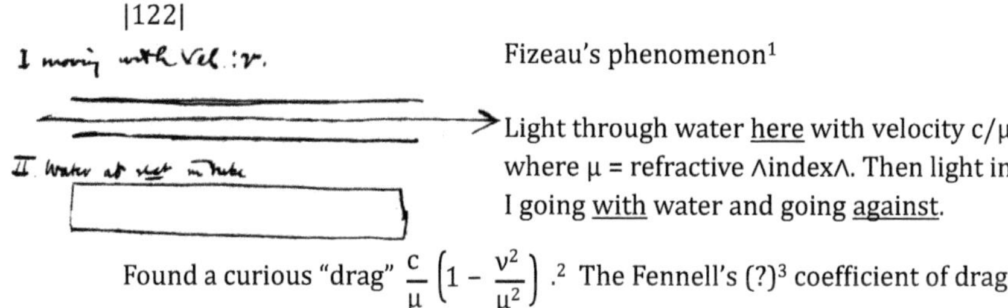

Fizeau's phenomenon[1]

Light through water <u>here</u> with velocity c/μ where μ = refractive ∧index∧. Then light in I going <u>with</u> water and going <u>against</u>.

Found a curious "drag" $\frac{c}{\mu}\left(1 - \frac{v^2}{\mu^2}\right)$.[2] The Fennell's (?)[3] coefficient of drag.

Fresnel gave an explanation of it that helped somewhat. Again the little terms explain Fresnel's coefficient of drag. Peculiar – because light is not passing through the <u>water</u>, but the <u>Ether</u>, and is only refracted etc. by molecules of water.

<u>Again</u>:– Motion of electro-magnet in neighborhood of a gravitational body. Ray of light slightly bent in going around Sun. How effect shows on Astronomers photographic plate. <u>Here</u>, only on <u>average</u> the differences you get are explained by tiny quantities in the formula.

<u>Then</u>:– Take Maxwell's Equations. Expressed in terms of Distance and Time. There are also Newton's Laws of Motion:– mass × acceleration = force. Newton knew clearly that you might as well take car as at rest, as house.

Newton's Law perfectly happy from point of view of old formulae; but need correction from <u>new</u>. Idea of acceleration doesn't alter its <u>meaning</u> from old formulae. But if you take Maxwell's formulae, then you'll find somewhat different forms of your Equations. That makes law of Electro-dynamics to take different forms depending on what you take as at rest. If you could find an Ether at rest, as Michelson and everyone else expected, you could say <u>that</u> is Space to which, if you refer everything, you will get your formulae in that simple form.

That is why Michelson's experiments were such a blow to the ether. But suppose you ask relations between T, t, X, x then exactly same expression of Maxwell's laws allowing for those equations with the $\sqrt{1 - u^2c^2}$

<u>Larmor</u>[4] found out that these equations were those which kept Maxwell's formulae "invariant," but like a true Cambridge man said simply here's a good dodge for solving problems. It took <u>Einstein</u> to see here basis for entirely new view of Space-Time relations.

---

1. A generation before Michelson and Morley, the French physicist Armand Hippolyte Fizeau (1819–96) conducted a range of experiments to measure the speed of light, including what has come to be singled out as the 'Fizeau experiment', in 1851, to measure the relative speeds of light in moving water. The dragging effect he measured was far lower than expected.
2. Bell placed placed a '?' in this equation, in place of the denominator inside the parentheses, which has been corrected following Hocking, who places μ^2 there.
3. Bell is correct to question whether he heard this name accurately. Whitehead is referring to Augustin-Jean Fresnel (1788–1827), who, in 1818, proposed 'aether drag' on the speed of light, which could be calculated by a mathematical coefficient, seemingly confirmed by Fizeau's experiment. We substitute the correct name hereafter.
4. Sir Joseph Larmor (1857–1942), a British physicist and mathematician, building on work by Lorentz, showed by 1897 and in his most influential work, *Aether and Matter* (1900), that Maxwell's equations were invariant under a two-step 'transformation' he proposed.

Einstein didn't produce mathematical theory of Tensors
[nor] these formulae } But did point out significance of whole lot. The "team work" – popular views quite wrong. Different people in isolation. Fizeau, Michelson, Mach ("Dynamics" – or "Mechanics" – created great sensation), Larmor, people with Mercury.[1]

It is not Einstein *in vacuo*, but Einstein coming up just at right point in history of thought. $m\ddot{x}$ = force becomes ridiculous in new forms.

Einstein saw what he had to substitute for the $m\ddot{x}$. Each type of physical force to be taken on its own!

Other, more subtle points which turn up. Duplication of dark and light bands in Spectroscope (e.g. from Atoms). Some of this slight duplication has now been accounted for. Sommerfeld (German mathematical physicist) did that.[2]

Whitehead has been endeavoring to show them consistent with a view of Space-Time which while not compelling is just a little --- ⟨blank⟩

It was always interpreted in terms of older view because nothing suggested opposite, which seemed to be paradox anyway.

Take an Event-particle E and see how all others are patterned out with reference to it according to this theory.

– Really 4 dimensions here.

"Moment of Simultaneity" – two uprights in two Space-Time systems.

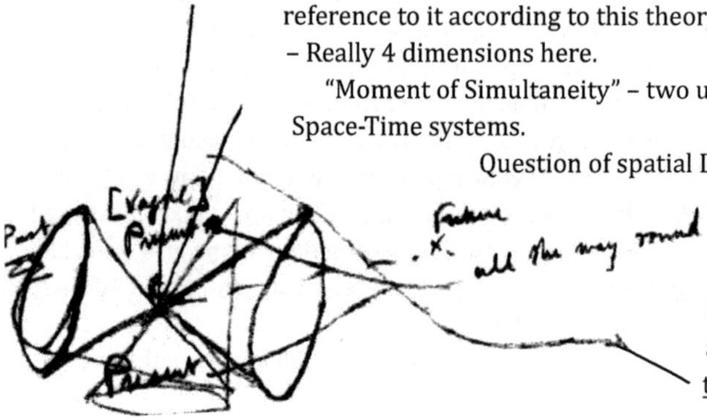

Question of spatial Distance or Temporal Distance. (Whitehead looks on them as quite distinct – They are regarded from one another in limit by this locus)[3]

1. Previous footnotes have identified most of these 'Different people, in isolation'; but as yet have not identified Ernst Walfried Josef Wenzel Mach (1838–1916), who published the *Science of Mechanics* in 1883. Karl Ernst Ludwig Planck (1858–1947), best known for his postulating that electromagnetic energy could be emitted only in minimal 'quanta', was quick to support Einstein's special relativity, proposed in 1905, and a few years later to bring Einstein to the University of Berlin. Others working on special relativity at that time would include Larmor, and also FitzGerald and Heaviside in Britain, along with (if not alongside) Einstein and Lorentz. The 'people with Mercury' refers to those who tried to account for Mercury's orbit precessing more slowly than Newton's laws predict (known for some time) and would have included Poincaré, Minkowski, de Sitter, Cunningham and Silberstein. Only Einstein's general theory of relativity (incorporating gravity into the theory of relativity) was able to do so by 1915, and Whitehead's own alternative theory of gravity. It was Einstein's work, however, that was propelled into the public's eye by the astronomical observations of Dyson and Eddington in 1919. In recent decades historians of science have echoed Whitehead's point that while Einstein did not work in a 'team', neither did he work in a vacuum, and they have debated extensively the relative contributions of several of these named to the theory of relativity.
2. Arnold Johannes Wilhelm Sommerfeld (1868–1951) introduced the fine structure constant in 1916 to account for the relativistic effects on atomic spectroscopic lines.
3. Closing parenthesis provided.

Hocking's notes

3/28 – 1 Saturday –

1. What all this comes to –

First put down two pieces of elementary mathematics dogmatically.

Compare the ordinary formula with the new one.

---

milestone moving cars c_1 & c_2

c_1 velocity u

c_2 velocity ν

at end of time t

$c'_2 - c'_1 = (v - u)t \equiv X. \equiv x - ut$

This is the ordinary formula, & This does not satisfy the most refined observations.

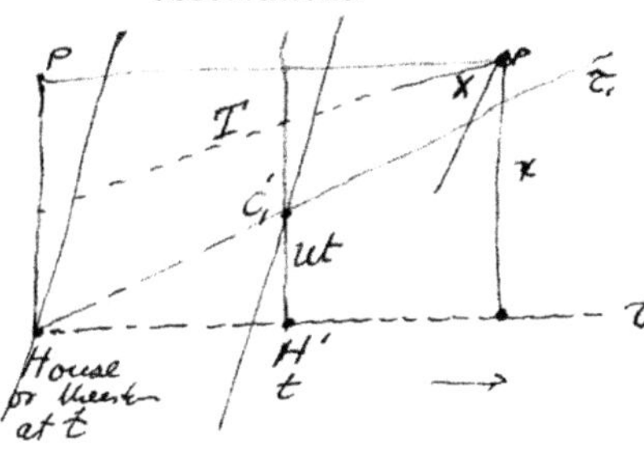

Each point is represented as instantaneous in instantaneous space

[x = distance of a point p in road]

Path of man in car

His time system is different & doesn't agree in measurement with

What is the relation between

x, t

X, T

$$X = \frac{x - ut}{\sqrt{1 - \frac{u^2}{c^2}}} \qquad T = \frac{t - \frac{ux}{c^2}}{\sqrt{1 - \frac{u^2}{c^2}}}$$

You can get this in two ways.

Trying to conceive a more general apprehension of Space-Time.

Whitehead's way

By taking Maxwell's equations and finding formulae of transformation from one assumption of rest to another.

Law of gravitation as formulated by Newton valid for the case when 2 bodies at rest in reference to each other. but not for case of bodies moving each other.

Have to get an impartial way to refer to every space-time system.

3. 28 – 2

The empirical evidences of the formula

1. Mercury ⟨?⟩[1]
2. Michelson–Morley
3. Water in a tube, (1) moving ,

(2) at rest send light through water –

---

1. There is a mark, here, which looks like something scratched out, but it might just attempt to capture the precession of Mercury's orbit.

velocity is c/μ in water
μ = refractive index of water
Send it through two ways to see whether light is carried by water.
Going against the water, velocity is

$$\frac{c}{\mu}\left(1 - \frac{v^2}{\mu^2}\right)$$

"Coefficient of drag"
~~Fresnel's~~ Fizeau
Exactly explained.

4. Light in neighborhood of gravitating body.

5. Maxwell's equations expressed in terms of distances & times. [x over "distances", t over "times"]

Newton's: f = ma. Acceleration means same whatever you take at rest.

Maxwell's don't take the same form when you substitute for x, x – ut, the law reckoned from a moving body is not the same as when reckoned from a body at rest. This making it no longer arbitrary which to take at rest. Is it a resting ether that accounts for this, giving the stable Space-Time system. Michelson's experiments gave a blow to this. The modification of Einstein makes them transformable.

[Cambridge man had found these out. Einstein didn't even produce theory of tensors – He did point out the importance of the whole thing]

<u>3–28 – 3</u>

Cooperation
One man making pure math –
One making one experiment etc –
"It isn't Einstein in a vacuum it is Einstein coming in at the right time"
The new equations make f = ma wrong $m\ddot{x}$ = force.

"What I have been endeavoring to do is to show that these formulae are consistent with a view – Though it does as a fact bring in a certain amount of paradox."

E = event particle
Four dimensional cone

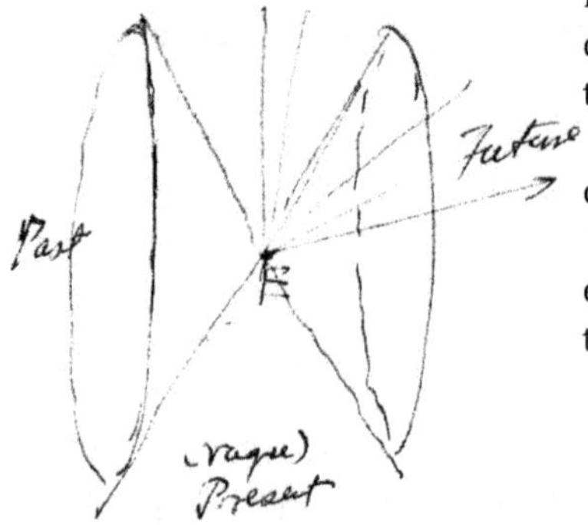

Past
Future
E
(vague) Present

Different loci of rest having each a corresponding moment of simultaneity – in the vague present

Present & Future are separated by the limit of the cone.

As you get close to this limit, to be enduring & to being spatially distant come to the same thing.

Lecture 64

Tuesday, 31 March 1925[1]

Bell's notes

|123|
Whitehead: 31. March. 1925[2]

Struggling (unsuccessfully) with Mathematics of Relativity. All 4-dimensional investigations are a strain on the imagination. Between:– from general metaphysical intuitions and from physical (definite scientific) observations, we arrive at an hypothesis which satisfies them and goes beyond them to a logical unity of theory.

This whole hypothesis presents us with idea of alternative modes of progression in time – alternative transitions of temporality – A difficult idea: – not obvious, even to himself. Can show it not ∧definitely∧ inconsistent with metaphysical principles, though – with our ordinary ways of looking at things.

Leads us to distinguish that general passage or development of things from particular Time-systems which simply manifest one aspect of this passage. "Passage" (whether fortunate word?)[3] Whether elements of transition suggests essentially linearity or not. We want idea of getting ahead – of Becomingness – and yet get ahead of linear idea of progression. This distinction, in some form, haunts past philosophers. Plato talks (*Timaeus*?) about how Time gets constituted. Descartes: – distinction between Duration and Time. Duration is in the external nature but Time is how we think about it. Bergson of course.

That distinction in itself, which is regarded by the new doctrine of Relativity is at least not foreign or repugnant to philosophy.

In what does the distinction consist? Bergson here on whole right; but phrases it so that you never can be quite sure what he means. This is why Russell dislikes him so. Bergson has a merit greater than clearness even. – Philosophical Originality – Putting things in ∧which he feels and sees∧; whether he can make them clear or not.

Bergson says *Durée* is indivisible. Brings in Divisibility by Time. Right to this extent: – A Time-system is a definite scheme of Divisibility. Passage of Nature or Universe can exhibit itself in various divisional schemes.

1. What might be the Radcliffe version of this lecture, or portions of it, from notes taken by L. R. Heath on what was likely earlier the same day, begins on |62| of her notes, page 502.
2. In his diary for the preceding Sunday Bell adds that after dinner: 'Around to Whitehead's. Just a half dozen people there. To bed very late.' And on this Tuesday Bell records: 'Lectures etc. as usual.... Lunch with Whiteheads – very pleasant.... Didn't get away until after 3:30.'
3. Closing parenthesis provided.

Bergson says Passage as whole is indivisible. Really: not bound to one scheme of divisibility. Each scheme exhibits some aspect of the general passage of Nature, but doesn't exhaust latter; but shows some aspect of the passage which is really there. Divisibility is holding up Nature as whole and parts of Events. Differentiation of Nature into Events has its foundation in, or exhibits as parts of its essence, these alternative Time-systems.

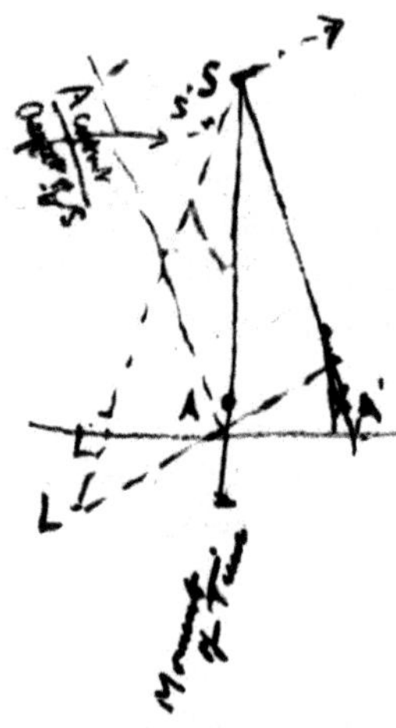

Two points here: – (1) There are paradoxes and difficulties in this idea of alternative time systems. A = a man on Mt. Wilson observing a star. – Using Time-system in which the Earth is at rest. Light travels, and therefore he just sees the star there. But asks what is happening "at this moment" there. Suppose one of the "runaway" stars (α Groombridge)[1] Now suppose observer there looking at our solar system and wondering what is going on "at this moment". But A', where he sees it, is in the Future for A. What S-man sees comes ⟨blank⟩

Light could go from L to S and he sees at A.

The aspect of L's phase of A's existence, for S. (S-aspect of A) In the future the Sun will actually get to A'. As matter of fact, man at S sees it as though there ⟨blank⟩

A content/an Aspect of S

He sees it in wrong position.

Real point is that you have inconsistent view of simultaneity A', really future to A, is where S sees it.

You have to deal with stars long off, – otherwise not enough to notice and deal with. But principle at work in this room too. But too infinitesimal to notice. The future A' has already some ∧real∧ significance even in A's scheme of things, while A is only at A.! The future is not.

Some procedure of expressing future is in relation to ? ? ⟨blank⟩

The difficulties out of modern doctrine of Relativity not first raised here. – Were already there in another form. Same old difficulty.

Whole difficulties in idea of utmost of consequence.[2]

In very nature of flux of Time itself (within the one Time-system) we have just

|124|

same problem and difficulty. Kant in Max Müller[3] 143–9 in 2nd Vol. of Kant.

Under heading: – Principles of Pure Understanding: "Axioms of Intuition". etc. – Then reads – re "Extensive quantities" – Necessary procedure by the parts.

1. Stephen Groombridge (1755–1832) began compiling a star catalogue in 1806. In 1842 one of the stars in his catalogue, 'Groombridge 1830', was discovered to have a very high proper motion. Perhaps this is what Whitehead refers to as a runaway star, and thereby 'a Groombridge'.
2. The last words are unclear.
3. Friedrich Max Müller (1823–1900) became Oxford's first Professor of Comparative Philology. In 1881 he published a translation of the first edition of Kant's *Critique of Pure Reason*, in two volumes.

Two senses of "Preceded" – Preceded in Time? Of course. Also of reality. – That a is substantially independent of b, but b dependent on a.

Then Kant's "intensive quantities" in 2nd Anticipations. Then talks about intensive quantities – but with a side kick to Extensive. "Every sensation … has a degree".[1] Between reality and nonentity an infinite series of degrees. General remarks re Quantity in general – No part of number indivisible:– called "Continuity". Time and space are quanta of Continua because no part of them not enclosed within Limits– and is again a Space and a Time. Space consists of spaces only Time of Times. Points of Moments are only limits; – determining places only.

There Kant has said what Whitehead has been trying to say. Points and moments are mere places of limitation etc. This is exactly what Whitehead's been saying. What Whitehead calls "durations" D ⟨blank⟩ To find some object that stands for mere Time-place is "Position".

Two sides to Kant: – Kant's general thesis and his working out in detail. Pritchard:[2] – In detail any amount of inconsistencies in Kant. But you've got to have some point of view like Kant's about Time and Space – Parts and Wholes.

Real difficulty is when Zeno comes on the scene. Zeno made unfortunate choice in dealing with Motion in space – His difficulty comes much earlier. He's getting Whole and Part by mixing up Space and Time. Zeno has high merit of being refuted once a century for 2000 years. [Law in Egypt. Juries and native judges terrorized there. Acquitted three times of Brigandage – expulsion from country].

Zeno comes in exactly here. If Time (the "moment is simply a logical construct from relation of various Times to each other). Whatever turns up then, on a passage into future. Any entity:– of no entity can it be said: This is now being realized. Everything presupposes something else and so on *ad infinitum* to nothing. And therefore no intelligible account can be given of the flux of Time unless you are prepared in some way or other those two statements of Kant.

Question is: How is process possible? If you conceive it as a temporal transition into non-existent, you can't get it going. Nothing you can point to into which a transition – Nothing you can point to as a creation. That has really never been taken very seriously. We do live, etc.

Concepts that have hold of a lot that is right; but that hasn't got over other spot of difficulties.[3]

---

1. Under his Transcendental Analytic, Kant's *Critique of Pure Reason*, his Analytic of Principles includes 'Axioms of Intuition', where he declares phenomena 'extensive', and then, secondly, 'Anticipations of Perception', where he characterises all perceptions as 'intensive quantities'; and where Whitehead finds the statement he quotes: 'Every sensation … has a degree'.
2. Harold Prichard (1871–1947) is primarily known for his work at Oxford in moral philosophy. But earlier on, in 1909, he published his only book: *Kant's Theory of Knowledge*.
3. The last words are unclear.

In some sense or other you've got to have an existence, in some respect simultaneous with past and with future. Unless you can, then Zeno has you.

Hocking's notes

Mar. 31–1

1. "The relativity hypothesis presents us with the notion of alternative progression in time – not obvious to myself, only by a miracle that I have made it obvious to you –"

Get rid of the linear idea of becomingness.

Plato in *Timaeus* –

2. In what does the distinction consist between different views of time –

"Bergson has the merit of philosophic originality even greater than clear men – seeing the real philosophic problem, even if he can't get it clear"–

Bergson says *Durée* is indivisible – difficult

A time-system is a definite scheme of divisibility.

3. Paradoxes in the idea of alternative time systems.

A. A man on Mt Wilson observing a star, S, whose light has taken some hundreds of years to come.

A constant of aspects of S

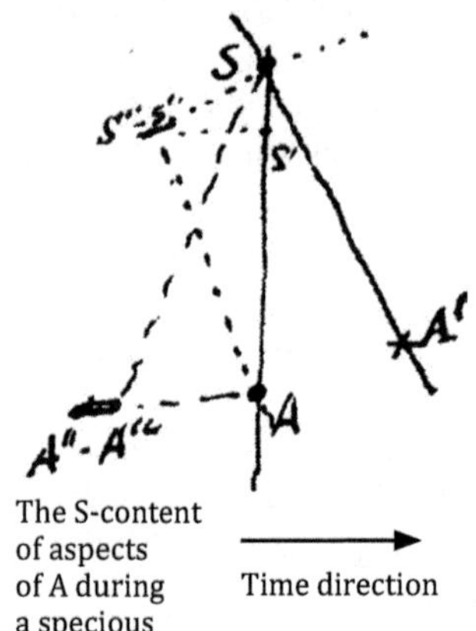

The S-content of aspects of A during a specious present.

Time direction

But what at this moment is going on in the star. The question has some sense.

Observer at S is also wondering what is going on at our sun. Which he locates at A' which is in the future (?)

You have inconsistent views of simultaneity – A' is future, but you can't draw lines into non-entity.

Can't put off as mere thought, as I can think Jack Giant Killer.

4. Difficulties were not wantonly sacred[1] in modern doctrine of relativity. Were there before in another form.

Take the old idea of the flux of time.

Kant's analytics, Müller, 143, 149

Axioms of intuition –

All phenomena are extensive quantities

1. These terms cannot be clearly made out.

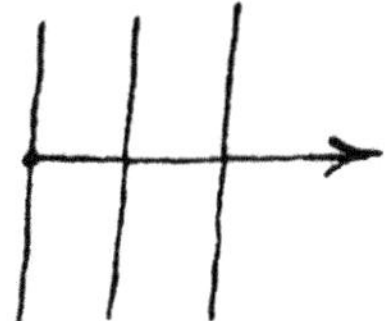

"Drawing on intuition"
So time – a successive synthesis of part with part
And these parts are antecedent to the whole!
"Points & moments were places of limitation" –
Can't compose time or space of <u>them</u> but only of times & spaces.

Whitehead is in perfect agreement with Kant, who is not here expressing his own particular ideas.

You've got to have some theory of the parts & the wholes

Zeno made an unfortunate choice in dealing with motion & space – muddling up time & space together

Mar. 31 – 2

Egyptian law. If you have been acquitted three times for brigandage you must be expelled from the country – (lest you kill the judge & the jury).

So Zeno refuted often enough – suspect something permanently true there.

Zeno: How are you going to move forward into the future? How is process possible?

If you conceive it under the guise of a temporal transition into the non-existent, you can't get it going – there is nothing you can point to into which there is a transition, or is then and there created.

Lecture 65

Thursday, 2 April 1925[1]

Bell's notes

|125|
Whitehead. 2. April. 1925[2]:–

Problem of "little c" still to be taken up. For present Metaphysics.

Last Time landed (principles re extensive quantity admirably formulated by Kant).– Main question: how is extensive quantity possible? Inherent contradiction seems to turn up in dealing with Extensive Quantity. – Reason why Bradley[3] et al. apt to tell you Space and Time unreal (Because when you try to make your view of them coherent you land in an inconsistency. Hence must be illusion).

We've tracked down one of the roots – All extensive quantities comes back to Space-Time. Kant's argument[4]:– Parts antecedent to wholes and parts and wholes alike equally extensive quanta. Qua extensive are divisible, therefore every part itself a whole, etc. – and hence vicious infinite regress. etc. Whitehead believes whole thing can be put on its legs properly. Certain aspects have been wrongly handled. – Fallacies of misplaced concreteness are responsible.

The vicious infinite regress – What does the "antecedence" mean. – Means that Part-Whole Relation here enters into very Essence of the Whole. – It is an internal relation then, so far as the Whole concerned. Spatio-Temporal relations taken as a whole (for Whitehead) "internal", whereas general view is that these are the very examples of External relations. Whitehead suggests this as root of whole difficulty. Kant's doctrine has sense only in taking Part-Whole relation as "internal" for the whole.

The infinite regress is Zeno's difficulty. Whitehead believes we have got to put the Part and the Whole on an equality. The Whole is of the essence of the Part as well as vice versa. This has appeared under doctrine of aspects in Whitehead's lectures of last few months.

This view of generation. View was so luminous to Kant because Kant was last great philosopher who was a Mathematician. Calamity that Kant's followers hadn't Kant's intellectual equipment. When Kant wrote his paragraph, he was bursting with Newton's "Fluxions". [Question where

1. What might be the Radcliffe version of this lecture, or portions of it, from notes taken by L. R. Heath earlier the same day, begins on |64| of her notes, page 503.
2. In his diary for this day Bell notes: 'Walked home from Whitehead's lecture with him, questioning him'.
3. See footnote 3 on page 39, lecture 11. The inconsistency of space and of time as dimensions is presented in F. H. Bradley's *Appearance and Reality* (1893).
4. Kant offers his 'proof' under what he calls 'Axioms of Intuition' within the 'Transcendental Analytic'. And 'his paragraph', to which Whitehead refers three paragraphs below, may well be the second paragraph in this section of the *Critique of Pure Reason*.

Newton got that "Fluxion" idea – "flowing" quantities.). Whitehead's own early delight in Kant: that he was putting in more general terms what Whitehead was getting in his science (classical mechanics). – Was disappointed in Caird[1] because he missed that – talked as though Kant had been educated in Oxford "Greats".

Shelley is chiefly [Hogg's Shelley etc.] through and through a man of Science[2] – Wanted to be a chemist – wasn't enough chemistry to keep him interested, so wrote Poetry. Commentators don't know this. But is a Scientist.

You have got to make what lies beyond any Event (Part) then part of the essence of any particular Occasion. – Must get a theory of the Beyond so. We are here obviously at one way of tackling one great metaphysical question which turns up in various forms. Whitehead ought to have added Pythagoras to Spinoza on this question:– How does the Unlimited become Limited? The totality only exemplified in limitations. Spinoza rather jumps at the difficulty of how the Modes with which we have to deal get into the Substance. Spinoza says a lot about both but the Pythagorean difficulty:–

It's only another aspect of same difficulty with Hume on Causation: we have present occasion and how does it tell you about anything else. The "Togetherness" as "habit" (inconsistently). Then there is our old friend Zeno who asks, in effect: How is generation possible?

Tendency to say: All VtheV trouble comes by abandoning Start from Points and Moments with [Plato and Aristotle's little tiff in the Algebraical treatise] external relations between them.

You have got to throw Kant over somewhere or other. Whitehead wants to do it at point of generation. – Kant was wavering between an external and internal view.

Can't you, by starting with points and moments, come back to all the traditional things? If the relation is external it means it is not of the essence of the entities. Therefore there is meaning in the idea of a point which has lost its Space. A moment then which has lost its Time would be conceivable as moment too. A moment has essential relation to other moments – to its Time – to Time in general.

Again:– There is something very queer in a point being occupied. If you say it is a point with its Content that you are dealing with. But the pointiness of the point makes an absolute theory of Space-Time necessary. Now, no respectable person ventures to assert this today. But that Absolute Theory comes from point-moment

1. For Caird, see footnote 3 on page 38, lecture 10. In that lecture, 16 October (and in the two following lectures) Whitehead focused on a quotation from Caird concerning Kant. After more than 25 years at the University of Glasgow, Caird returned to Oxford as Master of Balliol College.
2. Percy Bysshe Shelley (1792–1822) showed keen interest in science already while at Eton, but (unlike another romantic poet, Coleridge) did not pursue it later. He became friends with Thomas Jefferson Hogg (1792–1862) while the two were at Oxford, and though Hogg never completed his four-volume *The Life of Percy Bysshe Shelley*, two volumes were published in 1858.

|126|

etc. With a relative theory a position must be a way of talking of a relation. (That is what the relative theory means) You then mustn't stop with a point or a moment. Must ask:– What are relations you are talking about.

Whitehead thinks there has been a muddle in retaining ideas belonging to one frame of mind and taking them over into something belonging to another frame of mind.

Now, on theory of extensive abstraction to show how a point does refer to a definite relationship (Dedekind's idea of a "cut" comes in here[1]).

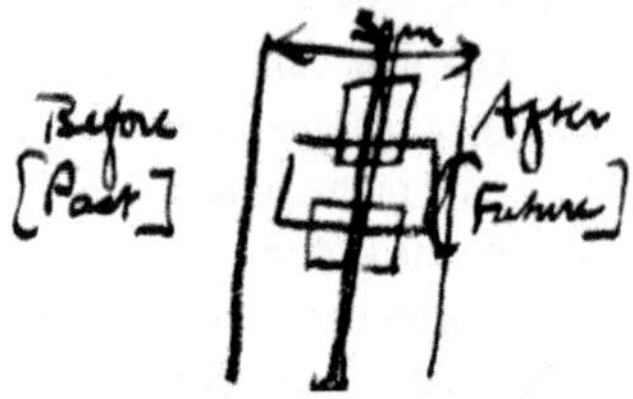

What is that moment? It is the relationship not any one of Past to Future but particular concrete relatedness of that past to that future is this moment. But "the past" has too much of Pythagoras's "unbounded" in it. What we deal with are past events, past occasions. Past Events on one side of line in relation to future Events on other side. There is an indefinite number of ∧relatednesses∧ of past events to future events which yet come under idea of that unbounded past to that unbounded future.

By idea of Simultaneity we got relatedness of [pencil] relation to [ink] relation. – I brought in a certain unboundedness to the "Moment".

We only get hold of the Unbounded by discerning ⟵ "discerning" = good word
System. In dealing with internal relations. – Logical puzzle then that you can't know anything without knowing everything.

Whitehead's answer is that we discern system. Duration has a certain unboundedness via System. How, now, to express that relatedness of Contiguity between unbounded past and unbounded future. [This is just that Ad-junction already defined] Two with no past in common, but together making up one event. There is nothing there but Relationship between two durations. That particular relation between those particular durations.

Why then all the bother about "abstractive classes" and moments. – Point is that I don't care a hang about of the particular durations – Want to get rid of the particular little cases (Don't know which ones to take – Want whole unbounded Past and Future). So say whole span "Span" can be divided in innumerable particular ways – and thus get a whole set of durations. Then take idea of a whole class of those durations. Only entity finally common to the whole class, is not any one duration but every member has inherent in its essence that particular instance of relatedness between two of its members. But I have still got a particular mode of approximation. We recognize that always a closer approximation possible.

1. The first reference to Dedekind was in lecture 53 (see footnote 3, page 241). He published his notion of a 'cut' as a partition of the rational numbers in 'Continuity and irrational numbers' (1872) as a method of constructing the real numbers. The paper was originally published as a separate booklet, *Stetigkeit und irrationale Zahlen, Vieweg: Braunschweig*; an English translation included in *Essays on the Theory of Numbers*, W. W. Beman, ed. and trans. (Open Court Publishing Company, 1901).

We want to approximate any one mode of Approximation with any other one. There we brought in idea of equal classes, covering one another.

Now we have got rid of particularity. And what that bundle of procedures does, is to appeal to the systematic character of Time and thus to show you how to define transition between this indefinite past and that indefinite future, without reference to any particular contiguous concrete events. The systematic character is the entity you are dealing with. We've not got a detailed hold of the past. We're thinking of a certain up-to-know with an indefinite beyond. We appeal to System via the details which we do definitely apprehend.

Wants to bring out that: as soon as you've done this; the spatial relations between the moments, appear external. But particular character of external relations is really retaining the particularity and keeping the System in the Background (tacitly presupposed). The System presupposed the particular Occasion ---? ⟨blank⟩

Ignoring the System which is presupposed in any example of Relationship; but keeping the particularity of the Boundedness.

Why shouldn't we drop whole way of looking at Space-Time as relations between points and moments and look on latter as relations between Quanta.

Hocking's notes

April. 2 – 1

Doesn't Extensive quantity contain in itself an inherent contradiction

As basis of the alleged unreality of time & space in Bradley – something wrong about it. All extensive quantity tracks back to space & time –

Take Kant's statement that the parts are antecedent to the whole – therefore every part itself is a whole with antecedent parts. Vicious regress.

"Previous" means the relation of part to whole enters into essence of the whole. An internal relation so far as whole is concerned. I believe I am nearly alone in holding that the relation in space-time are internal: they are usually taken as the very example of external relations.

Both must be put on same footing – The whole is of the essence of the part, Just as the part is of the essence of the whole. (Cf. doctrine of aspects) Kant's idea of antecedence "was bursting with Newton's fluxions when he wrote that".

Whitehead's delight in Kant. One of the reasons he turned to philosophy. Caird did not seem to know this – treated Kant as a product of Oxford Greats School

April 2 – 2.

You have got to make what lies beyond any part or event as of the essence of the part. The beyond is necessary to organize your ideas of space & time.

How does the unlimited become limited?

Pythagoras – Spinoza. The infinite is only exemplified in limitations. How do the modes get into the substance is the Pythagorean difficulty.

Zeno asks how is generation possible?

Aristotle's idea. We ought to start with points & moments & avoid all these difficulties – Points & moments with external relations –

But if your relation is external it is not of the nature of the term. Then the point has lost its space, or a moment its time.

But a moment has an essential reference to other moments & you are wrong in regarding the relation as external.

Then there is something wrong in the notion of such a point as occupied. Makes a perfectly absolute theory of Space-Time necessary. If you don't hold this theory, a point must be a way of talking of a relation between things.

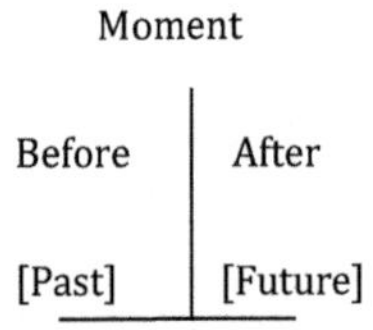

Moment

Before	After
[Past]	[Future]

What is that moment? I say it is that particular concrete relatedness of that past to that future. We do not deal with the past passim – but with past events. In their relation to future events –

We only get hold of the unbounded via system. We discern system. Apart from system you can't know anything until you know everything. My answer to that puzzle is, you discern system. As in the example of simultaneity.

How are we going to express this relation of the unbounded past to unbounded future?

Lecture 66

Saturday, 4 April 1925[1]

Bell's notes

|127|
Whitehead – 4. April, 1925[2]

Combining points of view of Extensive Quantities as all these (not under aspect of Generation) with point of view that Succession/Extension is, after all, a generative process. Difficulty of Zeno is particular motive point here. The Space-Time continuum on one side as qua eternal, on other hand as scene of generation. Qua eternal – at beginning of course, as field of potentiality – as this unbounded and divisible ["potentially divisible" is redundant – Idea of divisibility is itself an appeal to potentiality.][3]

Whitehead going on line that atomic side of things is what finds its home in generation. Generation, process, must be atomic (Zeno). On other hand we have got to preserve divisibility as in nature of things. The difficulties inherent in Metaphysical theory never in Scientific hypothesis:– Supposed continuity, etc. but also a sort of atomism. Under influence of quantum theory, atomic view growing. Jeans[4]:– you've got to take your Time in chunks. In dealing with these problems – conciliating these two things, one is not ~~out or~~ spinning something out of touch with scientific thought of one's age.

In this extensive continuum (Space-Time) qua Eternal, you lose all distinction between Time and Space because it is all there. Our particular view of Time derives therefore from generation. Even particular direction of Time. But idea of Extensiveness and extensive relations as being ultimately internal:– grasping aspects of each in a whole. In Space-Time qua eternal you get the potentiality of generation. General --- in envisagement of underlying activity, of those relations which, although eternal, must find their realisation in --- ⟨blank⟩

What is lost in "Potentiality" is the particularity. Space-Time as home of all potentiality. It is matter of fact which makes measurement possible [?][5]

---

1. The Radcliffe version of this lecture, from notes taken by L. R. Heath earlier the same day, begins on |66| of her notes, page 504.
2. On this day Bell simply notes in diary 'Lectures etc. as usual'.
3. Closing bracket provided.
4. Sir James Hopwood Jeans (1877–1946) became a physicist and astronomer, but began in mathematics (along with G. H. Hardy) at Cambridge, and was elected Fellow of Trinity College while Whitehead was still there. In the July preceding Whitehead's inaugural lecture at Harvard, he chaired a symposium (a joint session of the Aristotelian and Mind Societies) entitled 'The Quantum Theory: How far does it modify the mathematical, the physical and the psychological concepts of continuity?', in which Jeans' 1914 *Report on Radiation and the Quantum Theory* was discussed.
5. This bracketed '?' is more likely Bell's than Whitehead's.

– You measure not what might be but what is. As having all possibilities in it, it loses all Value. All Value in limitation. Plato mistaken in putting all his Value in the Eternal and considering realization as loss of value. Whitehead thinks there is a sort of half-way house – which is the business of theology to explore. Idea of "Potentiality" is pure abstraction.

Distinction between that which lies behind particular facts of generation and generation itself can't be got away from. Some people dislike the one or the other. But..! In Science, with its tilt towards the actual. But Science can't get on without its ideal conditions which aren't always realised. Metaphor of "obedience" to "Laws" overworked. Science always searching for something underlying the particular occasions – the Law, etc.

Something not itself in flux, but exemplified in every flux. Every Mathematician has emphasized this and most biologists and some philosophers have kept it in background as far as possible. Mathematician drilled in thinking about what is left when you abstract from all particularity.

The "Bounded" as the flux of particularities of matter of fact. – A particularizing of the unbounded potentiality. Pythagoras discarded the problem of the "How" here. – Of the generation of limitation. Plato with refined form of the Pythagorean idea. Plato with his adventuring of ideas in the flux of generation. Philosophy as the search after the Eternal.

Then you get the Biologist – Aristotle – son of a doctor. His primary interests are in the flux. Plato rather put the flux into the ideal – Aristotle vice-versa (of course, many-sided. – Did Logic too). Aristotle did quite right – Plato's thinking rather breaks away wild. But Aristotle can't explain generation without appeal to potentiality. ΛButΛ Potentiality is a positive fact. Not absolute nonentity. If potentiality is nothing, it can't help you. To call it Potentiality is to point at it by contrast – namely with reference to generation [like the word[1] "Protestant"]. There is no such thing as an absolute negation.

Why Whitehead brought Spinoza to attention is because Spinoza exemplifies so well the positive side of potentiality. With eternal attributes and generation as modes of these. B. Spinoza leaves the Λstatus ofΛ modes in the Substance in very considerable obscurity.

Descartes is father of all duality – bifurcation. Had mind as receptacle for ideas and matter as basis of generation. – An extraordinarily easy schema to handle – and a good guide to immediate problems of Science of his time. – Another case of erroneous statement – imperfect view of truth – where one of them better guide to science of his time – Galileo vs Inquisition. Both wrong [??][2]

Hume reduces that side of Berkeley to a minimum. Has a vague reference to

1. Bell definitely writes 'work' here, but 'word' makes more sense in this context.
2. Once again, we take this bracketed reaction to be Bell's.

|128|
impressions as though coming from without. Ends up with an impasse. But modern philosophy starts by considering Hume – How to overcome him.

Kant:– Has his things in themselves – not in Time and Space – But he also has stable forms for mutual experience in addition to things in themselves. Also the postulates of his Practical Reason lead to something, back of.

Kant's Excessive Subjectivism lands him in hopeless difficulties. What is wanted in Philosophy is to attempt to do same sort of thing Kant does, getting rid of Subjectivist basis of it.

This is what Hegel tried to do? And Hegel always maintained that he was endeavoring to sum up what is inherent in all Philosophy. But Hegel got in too much the idea and too little the generation. But Hegel and ⟨blank⟩

The hypothetical elements should come out of a critical consideration of the fundamental questions of existing state of knowledge. Bradley represents revolt against Generation and Bergson one in the opposite direction. Any attempt at conciliation represents some sort of a fairy tale.

You can do because of Physical ⟨blank⟩

We're used to terminology of Latin.

Essence of ∧Generative Process∧. How very naturally it gives you an atomic version of itself. (1) The instantaneous moment is a high abstraction.

In considering the realization of a pattern – As though realized in a moment. This is illogical. Togetherness of Ingredients (Entities with "Duration")[1]

To express what it is as real, we have got to take a duration.

(The joke about N or M in English church Catechism)

M is the aspect (grasping together from one aspect of a totality of pattern which is immediate)

What is there for me here. – Immediate, now.

That grasping of the total Universe (in sense perception) as now is only a reproduction in the bounded, of a character of the unbounded. The extensive continuum, qua eternal, is, in a sense, all there. Our specious present is again with the character of the given. Every philosophy has to start with the "given".

Bergson dislikes the "spatialization" – is essentially a man of the flux. In so far as it's muddled itself up with taking the isolated moment as the basis of all reality.

Whole of duration there as given. But as given, retains in itself transition. The generation of the given is atomic. The transition is within the given.

1. In this case the opening parenthesis has been provided; the three terms inside are all quite heavily overwritten by Bell.

Differentiated into ∧temporal∧ relation of transition – and a spatial one of potential transition. The given is not effective. Basis of all Subjectivism. There is an aspect of the truth that does readily lend itself to Subjectivism.

The whole of the given is for ∧M∧ under aspects (of colour etc.) The relation being internal can be looked on as forming what the object is in its aspect in M. "Effective potentiality" for M due to A.

Whole transmission theory of physical Science – M can't be itself without adjusting itself to what it's received from A.

Hocking's notes

April 4, 1925

1. Realization is a generative process.
   Qua generation – Selective actuality of realization
   Qua external – space-time as the scene of potentiality

Divisibility is in the actuality –
   We have got to preserve that –

These are the elements of thought we have got to play with. Also in science – continuity & atomicity – always haunting.

And under the influence of the quantum theory the atomic aspect has become more urgent than before.

Trying to conciliate these points of view – the business of philosophy.

---

2. Qua eternal you lose distinction between time & space in the time-space continuum, because it is all there. Time has to do only with generation Space-Time as the locus of all possibility But as the home of all possibility – no value.

Here Plato went astray – regarded the eternal as the home of all value –

"But there is a half way house which it is the business of theology to explore, –not my business –

---

3. Ideal conditions exemplified in every flux, but not in flux. Mathematician emphasize it. Biologists try to keep it in the background. The mathematician is trained to deal with what is left when you have abstracted from every particular thing –

So Pythagoras worried about the infinite, how does the unbounded become bounded – the discoverer of mathematics discovers this problem.

Generation is limitation.

Plato also had the venturing of the idea in generation.

Aristotle, biologist, son of a doctor, wants to put the ideal into the flux, whereas Plato wanted to put the flux into the ideal

Fails to explain generation without an appeal to potentiality. But potentiality is a positive fact. If it is nothing it is something, it doesn't help you.

April 4 – 2

4. Potentiality – is so called in contrast with actuality. But a contrast implies a positive element.

Spinoza does justice to this positive side of potentiality – Space and time as attributes of the eternal were possible.

Descartes "the father of all bifurcation." Mind the receptacle of ideas. Matter the basis of generation – Very good as a guide to the science of his time. As Galileo and the Inquisition both wrong, but Galileo's idea the guide to the science of his time.

Berkeley & Hume – Mind of God. Hume – vague reference of impression ends in skepticism. Lands in a impasse –

Kant but extraordinarily useful. But he has stable conditions of organization of experience – Excessive subjectivism lands him in perfectly hopeless difficulties. We must keep an eye on Kant. Try to do the same sort of thing he does & get rid of subjectivist basis.

Hegel tried to do this. The idea in generation – The unbounded becomes bounded – trying to sum up what was inherent in all philosophy. He got in very much too much necessity – The idea too much & the generation too little – But Hegel & Pythagoras shake hands.

Where I think Hegel missed it. The contribution of philosophy to thought, an adjustment of Eternal to actuality. Has to bring into accord the hypothetical adjustments of each particular region of experience, as in logic, science, etc. Philosophy must have its hypothetical elements, taking account of the hypothetical stages of each department of thought.

Bradley represents a revolt against generation

Bergson a revolt against the eternal side. Speculative element – must have some element of fairy tale in it. If any philosophy seems not to have it, it is because we are used to its terminology.

April 4 –3.

5. The naturalism of the atomic view.

I have been insisting that the moment is a high abstraction.

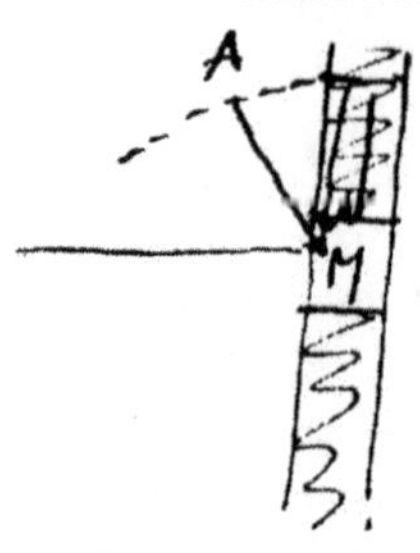

M. ≡ Man. { What is your name N or M(NN)

The aspect of a pattern grasping together of a totality as all "now"

"A reproduction in the bounded of a character of the unbounded"

The given, spatio-temporal

The whole of the given is 'for that' – a subjective aspect. It is how all that is related to that – internally.

Transmission from A to M. M cannot be itself without adjusting itself to what it has received from it.

Lecture 67

Tuesday, 7 April 1925[1]

Bell's notes

|129|
Whitehead – 7 April 1925[2]:–

Quantum Theory and demand for what is an atomic theory of Time. Whitehead not sure whether Science wants an atomic theory of Time or not. At any rate its asking for it. Question whether metaphysically we must tell it: You can't have it. Danger of Metaphysics clothing itself in viewpoint of Science of the past. But Metaphysics ought to *verwerten*[3] scientific results of period – and use its own viewpoint to get a bit ahead in vision.

Then of complex of Events extending over each other – Then you get divisibility. But you'd only get Atomism by making Temporality something more than mere Extension and Divisibility. Here exact logical method might be a help. In geometry had to bring in "Simultaneity" in addition to ⟨blank⟩

Difference between Temporality and Extensiveness. As it was brought in in our logical investigation it was brought in as rather a second rate thing to help us out of a difficulty. (2) Idea of Extension doesn't tell you which way Time's going. That has to be dragged in by scruff of neck. There is a track for Metaphysicians in that. Accepted as one of those little and obvious things. It is just the task Metaphysics to realize in itself – and "put over" – that it's those obvious things that bite down into very nature of things. Always there. Therefore obvious of course.

All these specific features of Time don't come under title of Extension.

Idea of Extension is way in which you get whole complex of Events qua System of internal relationships. How each events modifies character of systematic character of --- ⟨blank⟩
Not individual characters but systematic character of Events that Extension takes up. How the real is involved in a complex system of mutual interrelatedness. It is more than the actual – it is concerned with the field of reality: What is there for realization – Stands between completely ideal world and matter of fact.

Points out definite relationships with the actual facts; but also a System of relatedness which goes beyond the actual facts. Events as thus related in the future are the field of complete abstract potentiality. How, also,

1. The Radcliffe version of this lecture, from notes taken by L. R. Heath earlier the same day, begins on |71| of her notes, page 506.
2. Again, Bell simply records in his diary 'Lectures etc. as usual'.
3. These uses of German are more likely Bell than Whitehead, and in this case might be translated as 'to exploit' or 'to utilise'.

mere ideal objects can be related to what is actual. In addition to complete abstract potentiality you get ∧actual∧ potentiality. ("Eternal objects pervade all Space-Time"). This makes relation of Extension – an "occasional abstraction": – expresses definite relations to the actual occasions. But in this extensive complex you don't have any distinction between Space and Time and also the forward-moving feature of Time finds no recognition.

Hegelian Dialectic in some form (that you must move beyond any mere abstractions to more concrete something) – applied.

Temporalization of the Community of Events is just another way of talking of its Realization – That is what's imported into the community by Realization. Future qua merely potential isn't temporalized except very vaguely by its actual potentialities.

Realization is Individualization of each event in Togetherness via peculiar pattern. Qua relevant to Reality --- ⟨blank⟩

An event in present realization is real for itself; and this becoming real for itself is Temporalization.

Here we bump up against (1.) Atomic question (2.) Whole subjectivist view–point. Sound rule: – That any type of idea which persistently tends to force itself on your modes of expression (even if over-emphasized by Descartes and Co.) has something in it. Whitehead always convinced that objective point of view is safer. But you've got to express whole of subjectivist view-point within the objective one. My psychological field is there. You've got to express the Subject of Experience as one element in your objective Universe – and one that can't be apart from that universe.   Here a myth in diagram again.

The Subject as a parallelogram in Space has transition and divisibility within itself. In being real this Subject is essentially the realization of a duration as for itself, – as entering into its own being. Whitehead has already slipped in that transition, is in that Subject. (Away from idea of Subject etc. at a moment. It is not the moment but the duration).   The Subject is the pulling together of a duration from its own view point – viz: as entering into its own essence. Therefore you have E (event) ~~within~~ entering into Subject not as being merely Subject but as being itself, and |130|

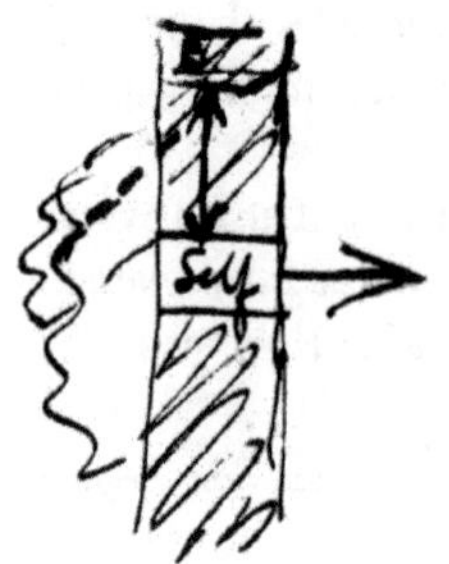

Event is how eternal objects get their manifestation for Subject. They enter into Subject only as in and thru these events. Events are aspects of Eternal Objects and vice versa. Essence of Subject? – is what. Sounds formal; but only put in language what we always do. Something standing out before me – Houses, men etc. And I am just the apprehension of that. My apprehension of realization of a whole Simultaneity.

I'm sure of the table as well as of myself.

But that is too simple. – There is a very indefinite antecedent Past out of which the present has emerged. It is the aspects of that past having aspects in present duration through limitations of potentialities owing to that past.

Potentiality you can think of equally well as giving you something or as excluding something [as Bergson knows].

Whatever can be abstracted, can be known. We think of as Sense-perception. Descartes calls this "*inspectio*". Whitehead prefers to keep inspection, doubting whether "intuition" is a good translation. You apprehend the realized spread in your duration as arising out of the effectuality (actuality as effective) of the past.

Just what physics does. – Sees molecule moving up. Then sees retarded potential and then what is actual in Electro-magnetic field has to become real subject to effectiveness of the molecule. The "transmission" idea of physics is nothing but ⟨blank⟩[1]

Molecule so tenuous because Physics thinks of antecedent Past simply in terms of those conditions of mobility of things which it lays down. The pattern which makes the ? ? of the subject requires duration – but no reason why it should not have minimum of Duration – also no reason why it should so.

Metaphysician's not duty to say what might be.

⟶ Becoming real is not the production of the real Element of Time via its parts. – An [contradicting here Kantian Extensive Quantity] extended Subject through a Duration does not become real via the parts of Duration. Subject which becomes real has in its essence a transition of Parts. – Doesn't become real because of the transition – but what is real has in its essence transition. Kant on intensive quantities. Whitehead agrees – but quite inconsistent with the first Kantian dictum. If you throw over that you then get your atomic temporality of ⟨blank⟩
But what becomes real is not divisible but divisible in being real. But the real must be there to be divided. Clear-headed trenchant thing to say is that Future is nonentity. But transition can't be a relation to absolute nothing. You see your transition within the given.

Atomic view of Transition which looks on realization, as realization of a Subject qua Succession of Durations. What succeeds has in itself Transition – But the Temporality is the Succession. And this brings in the direction of successions. One reason why Whitehead feels Einstein's point of view (lack of uniformity of Space and Time). Eddington says mere Geometry at one place is different from that at another. Whitehead dislikes that. You divorce the psychological field – what is actually before you – from "Nature." Takes world of psychological projection out of field of physics. But then you get into difficulties – Hard to know what you are talking of. The ∧Physical World∧ becomes a mere scheme of thought. Hardy (mathematician[2]) talking of physical universe as a "Scheme of Thought". But it is obviously a "scheme" about your own psychological field. Then leads you straight into solipsism.

---

1. The remainder of this line is blank, but it is not clear in this case whether this is intentional. Bell leaves no dashes nor a question mark.
2. See footnote 3 on page 200, lecture 44, for G. H. Hardy.

Einstein's way of dividing things up is, however, consistent with what Whitehead has been giving us.

Hocking's notes

April 7. – 1

A.[1] Professes a state of muddle, due to the state of the subject.

1. Science at present is asking for an ~~quantum~~ atomic theory of time. Shall metaphysics say to science it can't have it? This is not the function of metaphysics. Can we find any ground for it.

2. Consider events as extending over one another. You can only get it by making temporality something beyond inclusion, extending – over. This distinction between temporality and extensiveness has already been suggested. The question of the direction of time is fundamental – Can't slip it in in a nicely constructed sentence. The idea of extension doesn't include time-direction.

3. Mere extension soon demands that you should go beyond it. Hegel's dialectic has some application. A set of abstractions demands, in order that you shall land in no-meaning that you go beyond them.

4. The temporalization of extension, via realization of the potential. The individualization of each event into a peculiar togetherness. The future, qua relevant to reality, is merely for something which is real. An event as present is real for itself. It is this becoming real which is temporalization.

5. But here we bump up against the atomic-ness[2] of things, also the subjective view. The subjective view has got to be expressed within the objective view. It is there – the psychological field. You have got to express the subject as one element in the universe – nothing apart from that universe

Indefinite Antecedent past out of which the present has emerged

The subject has to realize itself qua the influence of the past on it. Limitation

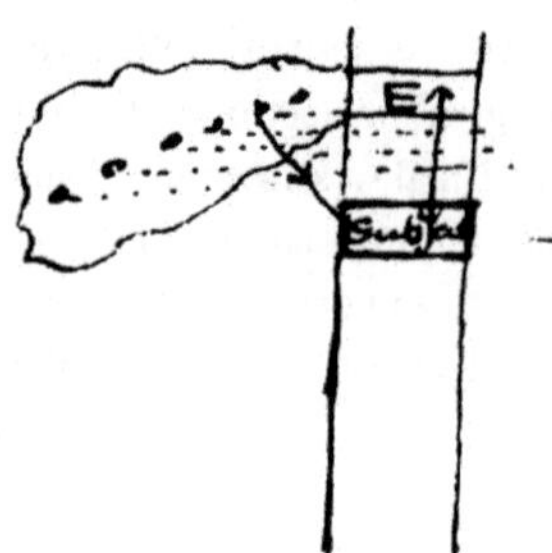

A subject is a parallelogram in our myth. – having divisibility & transitions within itself

→ Its reality is the realization of something as entering into its own being. "The pulling together of a duration from its own viewpoint, i.e. as entering into its own essence."

1. Possibly, what looks like 'A.' is 'W.', for 'Whitehead'.
2. This ending is almost impossible to read.

The event, E, is within, enters into the essence of the subject, not as being merely the subject but as being itself. The subject is what that grasping-together is. "I am the apprehension of that" – you people, the President's house, all these things, of a whole simultaneity. The whole duration as realized for the subject

April 7 – 2

Atomicity

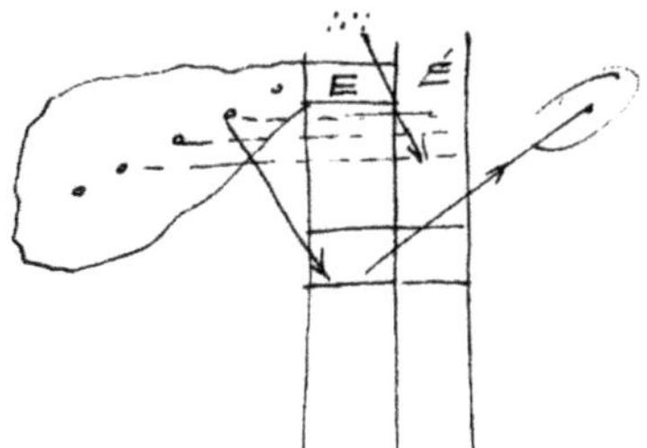

Realization may require a minimum duration –

There is nothing in a moment.

The becoming real is not the production via the parts of the duration – contradicting Kant.

"The transition is in the nature of what has become real, but it hasn't become real because of the transition."

The second extract from Kant was inconsistent with the first (on extensive quantity): if you throw over the first, you get your idea of atomic quantity.

The time transition must be a transition within what is already there. The simple thing to say is that the future is simple non-entity. There is no relation between something & nothing. You see the transition within the given, the trolley car going before you.

The atomic view of succession, E, E',

Eddington's view takes the psychological projection out of physical nature. Einstein said he was not talking about the psychological field.[1] The physical universe simply becomes the scheme of thought.

But it is a scheme of thought about your psychological field – nothing else to be thought about – land you straight into solipsism.

1. Whitehead is referring back to that conversation he had with Einstein at the home of Lord Haldane; see footnote 1 on page 185, lecture 41.

Lecture 68

Thursday, 9 April 1925[1]

Bell's notes

|131|
Whitehead: 9 April, 1925:–

Got to one of those awkward points that bring up all sorts of imaginable difficulties so you can't shirk them. When you try to run to the continuity idea you get into various difficulties. Continuity idea natural to objectivism. What you have is a part out of something else. But you get the Zeno difficulties. And then in Natural Science you meet the atomic.

Running atomic view gets you into Monadism. In Natural Science the two don't clash – You get the Atom in the field of force. Tendency in Philosophy to run the two as contradictory. And then you have to bring in a (non-religious) "God" – *Deus ex machina* – whose sole purpose is to rise superior to difficulties of metaphysics. (So Berkeley in spite of his religious fervor. Especially Leibniz)[2]. The real point is how to deal with this balance of Atomism and Continuity. (Gets itself reflected in Philosophy in Subjectivism vs Objectivism.)[3]

In dealing with relationships of Events (Started with these quite happily as most concrete things). Then get "extending over" and a Community of Events and a Totality (with past as well as future in it). This gives you something not sufficiently differentiated. Then had to discuss what Realization meant. What is merely potential becomes really together and emerges into this real entity. This brought in the Time idea. – But not as a mere generation (as if Time is merely generating extensive aspects of events. Then Zeno gets you at once). The moment is only a relation between events. What is real is togetherness of the Content of the Events.

Taking view that Cognition is Self-cognition. Then right to Descartes' question: – what is an Event for itself – Intrinsic content of a Subject for itself. When you differ from a philosopher it is not after he's begun but before. Descartes presupposes what language asks you to presuppose:– viz. that he's to take himself as an independent Substance and ask what its qualities are. Right away he runs up against Cogitation – and this involves "*Inspectio*" of what is other than itself. (Thus Descartes raises question of Epistemology in its most acute form at once). But Substance – as that requiring nothing else --- and then having its subjective passions etc.

1. The Radcliffe version of this lecture, from notes taken by L. R. Heath earlier the same day, begins on |75| of her notes, page 508.
2. Closing parenthesis provided.
3. Closing parenthesis provided.

– and finding at once that in make-up and very character of mind itself (*Cogitatio*) he has *Cogitata* – things that are other than Mind.

First thing he finds, really, is that idea of Substance and Quality doesn't apply. So Descartes made it apply. [Descartes is "wrong" because he's such a big genius – Knew quite clearly what he meant. All the world engaged in controverting Zeno, Descartes, Bertrand Russell – not because they're the worst – but because they are the best.]

Cogitata cannot be represented as qualities of Mind. Descartes took "secondary qualities" (Galileo already). – Said those were private passions of mind – and then brought an *inspectio* which got you to things other than the mind.

But (Whitehead) *inspectio* is simply Self-Knowledge – and true way of putting it is way which really starts from idea of internal relations – That is true way of expressing balance between Subjectivity and Objectivity.
– viz: – The relationships of the real emergent Entity with other Entities modify its own essence. – In asking what it is in itself you can't avoid taking into account what other real things are. Relata are part of essence of Relationship. So if you are giving account of any Subject you've got to give some [Question – what kind of "Some"?] account of things than itself. Any real Subject is one among other things. And all these are in a community of internal relationship. This is the first general fact about Reality. (Newman: – "To talk sense is to talk with Aristotle"[1]. But it is because he is so nearly right that it's worth opposing him. E.g. in bringing in Subject-Predicate idea as fundamental way of looking at things. It is this that brings Descartes into his difficulties).

You can't take Realization as one definite thing without any grades. Mind is not something standing behind events but realizing itself in events. Mentality as one aspect of what is being realized.

As being real the Subject has to be taken atomically. It is in itself Spatio-temporal.

Back to fundamental point of view – Realization is becoming of Limitation. How does the particular Subject realize the embodiment of the Whole in that particular limitation?
|132|

⟨The following pages are undoubtedly the messiest set of notes, diagram and arrows Bell produced. The text has been pushed down onto one single page to preserve the 'connectors' with as much clarity as possible.⟩

(1.) Embodies the Whole as modifying its own essence. The Whole is as Now (Duration Now) There is nothing in the Whole which is not referred to the now.

The most concrete now is the various Spatio-temporal events.

The now as in the essence of Subject is [take any definite object-event]

1. John Henry Newman (1801–90), whether in *The Idea of a University* (1852) or almost 20 years later his *An Essay in Aid of a Grammar of Assent* (1870), was always disagreeing with but always in conversation with Aristotle.

the object-event as lending its aspect to "Eternal Objects"

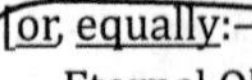

[or, equally:-

Eternal Objects as lending their aspects to Object event].

Whitehead is "Trying to analyze out the sheer idea of Appearance." It is not merely the object Red but red as there and it is the object under the aspect of red that is being there.

[Cf "specific intentional correlates"]

But what is in the Essence at Subject is the Object Event as being there – and Eternal Red as being in the Object Event.

That is how you have Appearance.

But the Universe is not merely a now. Not merely a Now which has internal relations to Subject. So bring in the Past [drawn quite Vaguely, on purpose]

The x's represent what is for the Subject qua realized. The Object Event is for the Subject in a more superficial sense. Past is for Subject not only as qua entering into Subject's realization but qua its realization entering in sense. Past is for S. not only as qua entering into S's realization but qua its realization entering in so.

The Past as having that relationship to future, as issuing into the future.

The past as entering into the Future has its aspect here It is the past qua Condition ["Actual Potentiality"] which is its aspect in the Subject. How Essence at Subject is conditioned by Past of Object Event is correlated with [Subject is always, from its own point of view, is always in the present] – Includes aspect of Present as issuing from the Past. "Subject has an *inspectio* of its Present as issuing from Past." This is determination of Subject's own essence by its internal relationship with the Past. Particular correlation of particular element of present with particular element of past.

This all put much more concretely in second or third lecture of this session (Schoolboy catching cricket-ball.)[1] – Whole question of "perceptual object". – Sense-data as issuing from the "Strain of Control".

How Present is for itself includes modification by Past.

That's not so far from ∧what∧ Norman Kempt Smith's been saying in his recent book.[2] Might have done better if he'd drawn a diagram. Whitehead has sympathy with Smith when reviewers all said him unsatisfactory.

1. Whitehead first introduced the example of a schoolboy catching a cricket ball in lectures 7 and 8.
2. Although Kemp Smith is mentioned in passing at the very end of lecture 37, no footnote was provided. Norman Kemp Smith (1872–1958) is primarily remembered for his commentary and translation of Kant's *Critique of Pure Reason*, but Whitehead's reference to his 'recent work' must be to *Prolegomena to an Idealist Theory of Knowledge* (1924). It is this work to which he referred in the Preface of his 1927 *Symbolism*.

You also have Continuity of Subject with its <u>own</u> past S_1 S_2 S_3 etc.

S_2 inherits from --- S_1 a modification of its Essence as it is for itself – its own intrinsic reality simply with aspect of being past. – It is the fullest inheritance you can have. So inheritance is of previous duration. Subject inheriting itself. It inherits its own unity.

Every enduring Subject inherits ∧or anticipates∧ itself under aspect of {different/other} circumstances. That is how you get the <u>same</u> thing <u>enduring</u>. <u>My unity</u> is not a mere pleasing convention. But a real and portentous fact.

This brings up view of what <u>Physics</u> is. Physics is account of how this[1] modifies essence of this[2] – taken simply as account of how that modification of essence affects Spatio-temporal character of temporal inheritance.

How subject passes on into a new phase – so as to place itself in a new present.

Thus  and not thus

It has to place itself <u>focally</u> in the new present.

There is a <u>choice</u> of conditions however, ~~Physics~~ is simply how the reality here affects conditions <u>there</u> so that Subject passes to <u>this</u> event rather than <u>that</u> one. But we have obviously changed the <u>meaning</u> of Subject here. To get that which is permanently real you have to make Subject the whole strip S_1 can only pass into S_2 by having some anticipation into which it issues. Embodiment of Envisagment. The Subject holds out before it the conditions into which it is entering. It is how → conditions are dealt with which is permanent. There are two subjects there

⟨The following lines are written up the left-hand margin of |132|.⟩

That permanent selective character – Permanent purpose of Being which, taken by itself, is a pure abstraction. But is same in series of different realities. What it <u>is</u> can never be disjoined from its being itself <u>in</u> a whole community. The Envisagement is not an <u>independent</u> one – but part of general taking account of whole order of Reality. So <u>Spinoza</u> <u>left out main</u> attribute of underlying Substance. Instead of "modes" as "affections" of Substance. Rather is realization of it. It is the Substance <u>being itself</u> <u>in a limitation</u>.

⟨The following lines are written at an angle in the upper the left-hand corner of |132|.⟩

What is really odd is not that we're in a universe but that we're not in an infinite number of them. Peculiar fact is that there is a <u>definite</u> course of Events. It is not that this is a <u>convention</u>. There is a <u>definite</u> achievement and underlying activity to <u>be</u> itself has to bring in the definite and exclusive atomism. That is where Freedom comes in.

1. Bell leaves a long curving line connected to the diagram above, attached to the line coming in from the centre right.
2. Bell also leaves a long curving line connected to the diagram on the previous page, attached to the arrow coming up from the centre bottom.

Hocking's notes

April 9 –1

1. <u>Is the monadic view contradictory to continuity</u>?

Berkeley & Leibniz introduce a *deus ex machina* to overcome the difficulty of atomism & continuity.

Subjectivism & objectivism reach the same impasse

Starting with events, & bringing the future & past into it, don't give enough differentiation. Had to introduce 'reality' as 'real togetherness', bringing in the time-idea. If you take time as merely generating the event, Zeno gets at you. There is no such thing as a moment. What must be real is the togetherness of the content of the event.

2.– Cognition is self-cognition. What is an event for itself? The subject of Descartes' *Meditation*. If you differ from a philosopher you differ from him before he has begun, in his presuppositions.

Descartes' cogitation involves '*inspectio*' of what is other than himself, raising the epistemological question in its most acute form. *Cogitata* – things cogitated. The first thing he finds in asking what he is is that the ideas of substance & quality doesn't apply. *Cogitata* cannot be represented as qualities of the mind.

Self consciousness is simply self-knowledge.

The true way of putting it starts from idea of internal relations. The relation of subjectivity to objectivity, modifies its own essence. In asking what it is in itself you can't avoid taking account of what the other real things are. Any real subject is one among other things, constituting a community of internal relationship. Better than substance-quality idea, in Aristotle.

3. How represent these internal relationships?

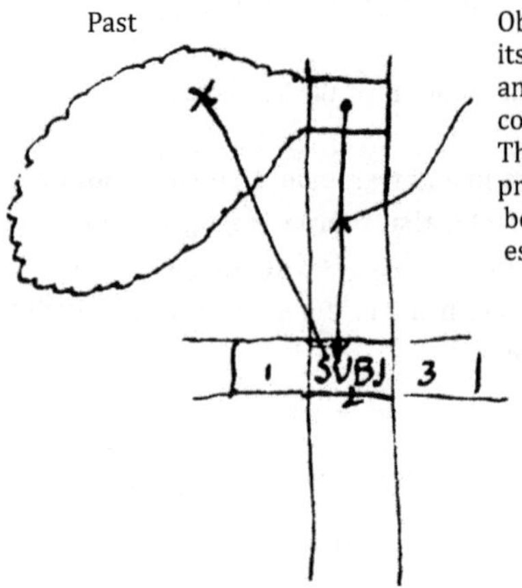

Object event as lending its aspects to Eternal Objects. and vice versa. "That is colored," or "color is there." The object of each is to prevent the other from being simpliciter in the essence of subject.

The mind is not something standing behind events but something realized in events. It is of the essence of what is being realized.

Taking the subject as real is to take it atomically as having in itself the flux of Time, spatio-temporal. Realization is limitation, embodies the whole in limitation. How does the subject achieve this?

1. The whole is as NOW, a duration.

2. The beyond-the-now. (PAST)

The past is for the subject qua realized.

What the subject is for itself includes the present as issuing from the past. Internal relation with the past.[1]

April 9 –2

Schoolboy catching cricket ball

Perceptual object – data appearing as issuing from a strain of control.

Not so far from what Kemp Smith has been saying in his recent book.

Which everybody has been saying is so inadequate. Would have done better if he had drawn a diagram.

Continuity of the subject.

Inheriting a pattern. [2] inherits from [1] the totality of the subject as being past. The subject inherits its own unity, its own grasping-together, as the same process only with the aspect of being behind.

"Every <u>enduring subject</u> inherits [or anticipates] itself under <u>the aspect of other circumstances</u>" (matter on board)

What is it that I am the same person as lectured to you last time – seems portentous to me –

Physics is the account of how <u>this</u> (past) modifies the essence of this subject. How the passage is accomplished. How this subject can find itself under different circumstance. How the reality of the past affects the conditions so that the subject passes from event 1 to event 2.

What Spinoza left out is that the underlying substance is always in its limitations, is always itself in a limitation. What is odd is that there is one universe – instead of an infinite number of universes –

Imagine a solipsist deity enjoying every world at once, & he is realizing nothing – can't be discriminated. There is a definite achievement. Here freedom comes in. The underlying activity has to be definite and exclusive.

1. The 10 lines of text at the top of this page appear in Hocking's notes above and to the right of the diagram at the foot of the previous page.

Lecture 69

Saturday, 11 April 1925[1]

Bell's notes

|133|
Whitehead: 11. April. 1925[2]

Endeavoring to consider Coming of Entities and Spread of things in Space or development of internal relations. So that Totality is to be discarded, not *simpliciter* but as related, in each individual thing. That is what you mean by internal relations – Thing is what it is in and because of its relations. So if you know what each thing is you have some knowledge of the totality. Whitehead holds this is only way to escape skepticism of Hume. This at once brings you to point of basing your Metaphysics on a Monism. So you are, then, with Spinoza. But you take a step which Spinoza ought to have taken: – that the attribute [the one attribute – all others are merely unfolding, explication of this] is realisation in a pluralistic actuality. It is then in each individualisation because of limitation of the Substance. Whole is there qua aspects of the pluralistic actuality.

Space and Time exhibit that general metaphysical position [This is what Whitehead has been showing] – they are our general scheme of adjustment of internal relations of realised entities. To get this clear:– driven gradually to this position:– What comes to us under aspect of continuity in Nature – where you get Space-Time – things as wholes and parts – you are dealing with realm of possibility. Representation of the monistic totality under Abstractions. Generalization of --- ⟨blank⟩

Abstraction of Eternal Objects, which are there for ingression. Those Eternal Objects – among them you get whole realm of possible relationship. Realisation is grasping together of whole into a Unity. There is a whole realm of possible relationships. Whole actual scheme of actual relationships is the scheme of mutual extension which is to be actualized into the particular emergent entities. It is that scheme of possible relatedness which is the Past, Present, and Future. It is by reason of its possibility that you always get that ending -ible or -able. Reference to Potentiality – "Divisible" etc. Space-Time relation is way in which plurality of individuals – actual realisation – is related to whole via each other – but also via its possibilities. To have a particular position

1. What might be the Radcliffe version of this lecture, from notes taken by L. R. Heath earlier the same day, begins on |77| of her notes, pages 508–9.
2. Although Bell simply records in his diary 'Lectures etc.' for that morning, later that day: 'Dinner at Phil. Dept. & Visiting Com of Overseers' & one or two others at 7:30 in Union. Very pleasant. Harry James presided. Palmer & Whitehead as chief speakers. Byrne (of Corporation) spoke excellently also Morton Prince, Prof. Clifford Moore, et al. variously Broke up rather late.'

with regard to possibilities of Whole. Scheme of Relatedness here is the Space-Time schema.

Then we have the process of Becoming. You have here the realisation of a pattern – Something detached as a possibility and realised as a fact. An enduring Entity is source of differentiation of Space and Time. In *Concept of Nature* (Chapter on Time) – discussion as to how, in some sense or other, Time is more fundamental than Space. When you try to conceive to yourself what an enduring individuality is in terms of this monism – To express what you mean by an enduring entity: – You have to make two cross-cuts to get --- ⟨blank⟩

Conceive on one hand abstraction of underlying activity as individualized and yet retaining its universal position as out of Time and out of Space. – The ultimate Ego of any Entity. The individual essence as realising itself in that individualized activity. As individualized it is both itself now, and the same as in the Past. As "same as in past" its realisations exhibit, as "Becoming" in itself. Every realisation reproduces in itself every character of the Whole – only under limitation. That gives you Space-Time in two cross-cuts.

Individualisation at any instant.

Whole life history of monistic individual.

That again to be real, individualises itself.

This, too, not independent. Only expressible via its relations to every other entity. So you get:–

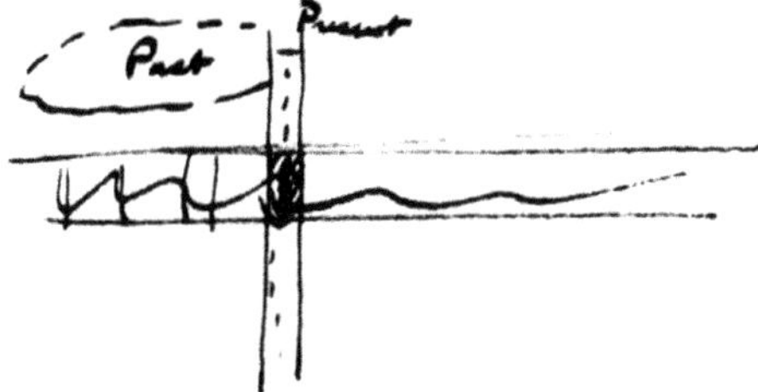

This is what it is in its immediate realisation.

This expresses totality of things in two ways.

Its own past in its immediate Self. Its essence consisted by two systems of relationships – as totality of its own past, as in its immediate Self, and as Totality of Otherness as in its immediate Self.

A selection of the Spatio-temporal Positions under the Aspects of Eternal Objects. Possibility is always possibility of ingression of Eternal Objects (under aspect of Event, are grasped together in this That is how the Totality is related to that particular individualisation.

But you haven't these individualisations separate and independent.

The past is going in like this:

The internal relations of these in their immediacy gives you "actual potentiality". Potentiality isn't nothing. Your totality of possibilities – There they are, and

|134|

whatever is has got to take account of them. Relation of Past to what is individually and immediately real is antecedent condition under which latter becomes real. Form under which this relationship appears to you is essentially limitation of possibilities. Relationship of actual potentiality is Extensive – it is extracting rather than adding something. You have got to look on possibility in the abstract as complete totality of things. Actual potentiality is what is left as result of Past – not something added.

This turns up even in Logic. A priori as sheer possibility one of two contradictions cut out. ["Logic defined as shadow of truth falling"]. That is how Present has its relevance to the past.

Own past is, as it is modified by whole

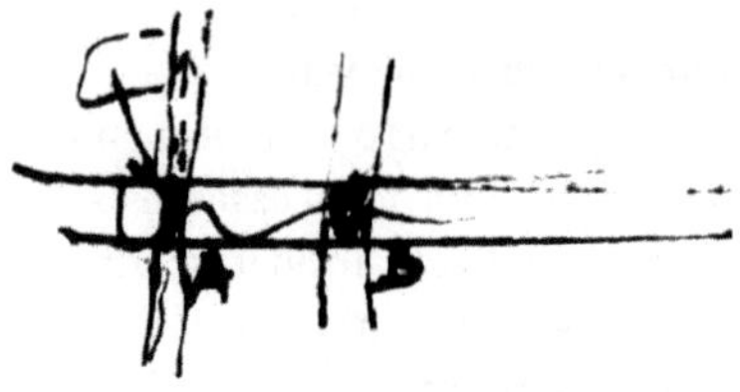

Merely expressing here diagrammatically what is for sense perception etc. the immediate display.

Also A is immediately in B, but as recollection. It is your recollection – only recollect what happened to yourself (to anyone else only via communication of it.)

Otherwise your knowledge is of something in present.

Whitehead has left out whole future and whole aspect of potentiality, so far in diagram. Diagram silly. But equally silly to go and agitate air with noises and tell people to understand these. Make another noise and say "that's what I mean".

Advantage of diagram:– It does preserve that scheme of relationships which are relationships among the internal relationships – which gets set out in aspect of each thing being for us in Time and Space. Other questions turn up. – The idea of the "any". There you are obviously dealing with Possibilities. – Realising is connection throughout whole region of Space and Time.

Your present as enjoyed is outcome of Past. – Conditions under which you are yourself, are laid down by Past. Past as issuing into the present. That is exactly what Science reproduces. – Reproduce influence traveling from Sun, e.g. and your present (8 seconds later) is for you under aspect of Sun. Actual Potentiality is of the Past issuing into Present.

Here we come to question of relation of physical to psychological field [!careful!]. We think of physical field as causing it. But physical field, unless related to past --- ?

People who over-clarify their ideas and their danger. – Then you'll at once call the physical field "purely conceptual". But what on earth the conceptual physical field is about – wherein it differs from 1001 Arabian Nights, Whitehead can never see.

My immediate experiences set me meditating on ⟨blank⟩

If just conceptual field – then as thought

past isn't cause of present but present of past!

You must, *au contraire,* retain some immediate knowledge of that as causing that.

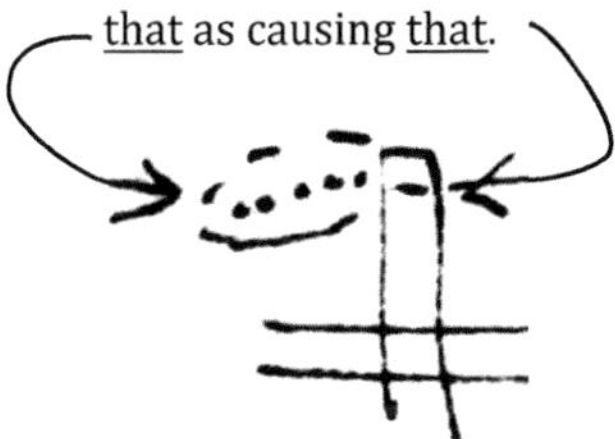

Boy catching ball – Past as issuing into present is what is grasped.
"Strain of control" – the physical object.
Question of Alternative Time-system next.

Hocking's notes

April 11, 1925

1. Internal relations.

Means that the essence of the individual is what it is by virtue of the relatedness. This brings you to monism. MONISM

"You never get away from the totality."

"There you are with Spinoza"

But you take a step which I think Spinoza ought to have taken: that the one attribute of your monistic substance is that it is realized in a pluralistic actuality – That is in each individual realization because it is a limitation of the substance qua aspects –

2. Wherever you get the aspect of continuity – in space-time relation, you are dealing with possibilities. Whenever you talk of space & time you always get the termination "-able".

We have the process of becoming – the realization of a pattern. See *Concept of Nature*. Ch. on Time. On why Time seems more fundamental than Space. The ego of any entity is out of space – the monistic total as exhibiting itself in that individual activity of realization.

Individual as in time

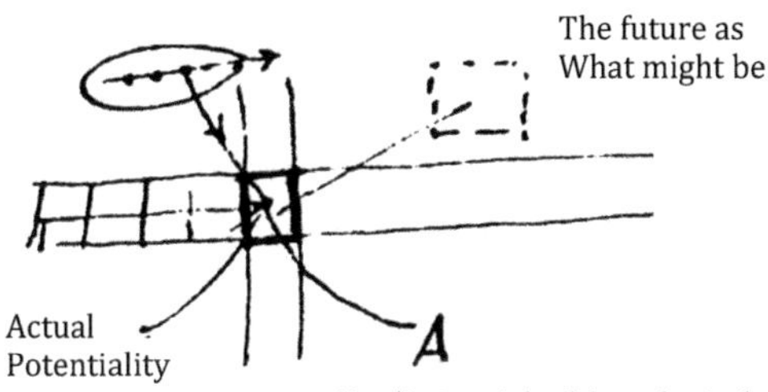

(As distinguished from logical possibility, only certain things can be realized in view of the past)

Its life history

This again, in order to be real individualizes itself.

[Drawing cross-time]

Must express its relations to every other part of the whole.

A is what it is in its immediate realization. Expresses the totality in two ways. The Present as in its immediate self. And also as its own past

The burden of the Platonic & Pythagorean message is that possibility cannot be non-entity.

You don't recollect what happened to somebody else, you see.

Lecture 70

Tuesday, 14 April 1925[1]

Bell's notes

|135|
Whitehead: 14. April, 1925[2]

Associated symbols vs. Substitute Symbols

Essence of Language and its enormous power from fact that it roams over all field and is at home there. Like human being among animals. They can do everything better than men. General inefficiency as animal. But enormous adaptability.

But in mathematical symbols special adaption to a particular region of thought. As soon as you move one step beyond that field – no significance or ⟨blank⟩

Wittgenstein pointed out well – Also Stout already in *Psychology*[3] of substitute symbols. In the Symbol, viewed merely as physical fact before you – you have relations which embody in general character same character as embodied in subject matter. You see in the symbols same characters as in object. Then, if Symbols easier to reason about than the more complex objects symbolized --- etc.

Connection of our Experience ("Perception" when we are aware of it) with the Spatio-Temporal Continuum. Advantage of a diagram is that it is itself Spatial. Some of the characteristics you are speaking of are presented in the diagram. Diagram must be used and kept in its special significance. Philosophy has lost there a valuable help.

How experience involves Spatio-temporal Relationships:– Assuming alternative Spatio-temporal relationships (modern Relativity). Take event A_0 (has definite past and definite future) and then there is the "vague present". (That which will fall within some duration of Simultaneity according to some Spatio-temporal system)[4]

(Really a 4-dimensional diagram wanted, instead of 2)

Now consider an enduring entity [A man, e.g.] --- not as discreet but as atomic.

Same pattern reiterating itself in the life-history. Present duration for ⟨blank⟩

[The "one" Substance sounds like one thing among others,

1. What might be the Radcliffe version of this lecture, from notes taken by L. R. Heath earlier the same day, begins somewhere on |78| of her notes, page 509.
2. Bell's notes in his diary simply: 'Lectures & consultation hour as usual also tutee'.
3. This must be George Frederick Stout (1860–1944), who was named a Fellow at Cambridge the same year as Whitehead, and who published his *Analytic Psychology* in 1896.
4. Closing parenthesis provided.

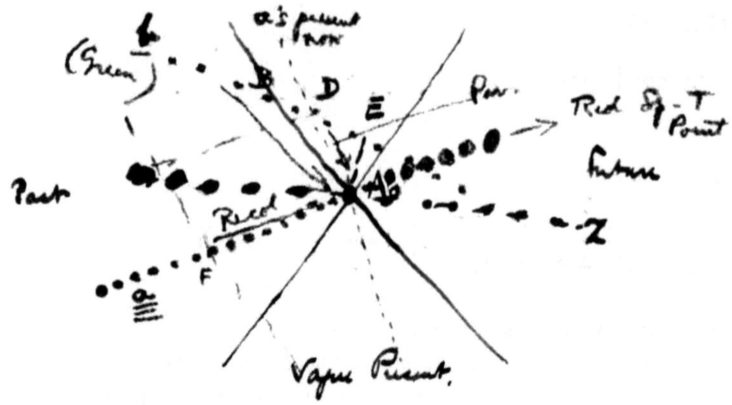

but:– underlying Activity whose essence is individualization of what is all in the different things. "Aspects of things in A_0". – These have a dual form – essentially related to one of two types

– (1) Aspect of the whole – the other things – as now – under aspect of being Now – Limitation of Whole into the Now. ([1]) Aspect of what the whole other than a is for a at A_0 now. Unity of World depends on there being a unity of Scheme of Limitation.

Take aspect of another entity (green entity b) – How is its aspect in A_0 (life-history course) Take the more direct aspect that comes with velocity of light [anything coming with less is to be looked on as indirect – to be looked on infinite influence of this on this, on this ---and then there].

The main abstraction comes in your body. Bergson, rightly; – Body rather as a method for keeping out – it is only what it selects that really is for one the aspect of things. Finally the effective part of one's body --- ? ? – "Direct aspects" – Where we have what we call normal or true perception. Owing to abstractions of body it is often very vaguely located Spatio-Temporally. – It is at D for you.

Under aspect of "now" you get B at D.

Then aspect of the Past. Grasping whole essence of a together in F. So "this" ["Recollection"] will be of the type of Recollection and this ["Perception"] of type of direct perception.

Now Whitehead puts on a purple entity just whizzing by. Z_0 enduring pattern or entity at Z_0 (that individualisation of the ultimate activity)[2].

Z's now What is happening at Z_0 for z.

What is for z at Z_0 under aspect of being at E and What is for a at A_0 under aspect of being at D is entirely a representation of essence of a_0 or z_0 as outcome of same event: b at B.

You have got a 3-term relation for each perceptive experience [knowledge of a [3] term relation between A_0 B and D.

1. This might be where the '(2)' implied by there being a '(1)' should go.
2. Closing parenthesis provided.

for a at $\{A_0/Z_0\}$
For z at Z_0 – 3-term relation between $\begin{cases} Z_0 \text{ B, E} \\ A_0 \end{cases}$

You are getting account of same fact in both. So this difference of Space-Time relatedness
|136|
does not mean that the two entities are giving an account of a different physical events. But they are talking of the same thing in so far as try to talk of it in an impartial manner – divested of its particular aspect for then. That divests idea of there being different Space-Time systems of one man thinking of a thing before it's happened. Both ⟨blank⟩

This is overspatialised. Everything always has its aspect in everything else. It is not simply that as there that one knows – but always that as issuing from its past and going towards its future. It's as an element in the process that has its aspect at A_0.

We have an awareness of actual potentiality under its true aspect. A_0 as embodying ingression of Eternal Objects into world takes its aspect from that effective potentiality, but is limited to this "present" in this. That ∧effective potentiality∧ as a modification determines, so to speak, how the A_0 selects its eternal objects.

This in abstract language is what put in concrete language in story of small boy catching the ball. He is not intent on any "class of sense-data". He's not catching the brownness – he's intent on the determinative element of the universe with reference to these aspects. That's what is really interesting him. He is grasping it there as a strain of Control issuing into Future. But that Strain of Control is the intrinsic reality of that event in itself. Here Whitehead disagrees entirely with the idea of private worlds. If there is no apprehension of the thing itself you cannot get along at all.

What's embodied in any individual entity is set of relations with the whole. How the present modifies essence of A_0 is by lending its aspects to the eternal objects. That's in A_0 systematically, the other "effectively."

B, D, E ordinarily[1] indistinguishable.

Now Whitehead obviously in danger of "Bifurcation" against which he fulminated in *Concept of Nature*. – A picture of psychological Space-Time as against physical Space-Time.

Next time....

1. This assumes 'ordinarily' is the correct expansion of 'ord'ly', although it could be 'orderly'.

Hocking's notes

April 14, 1925

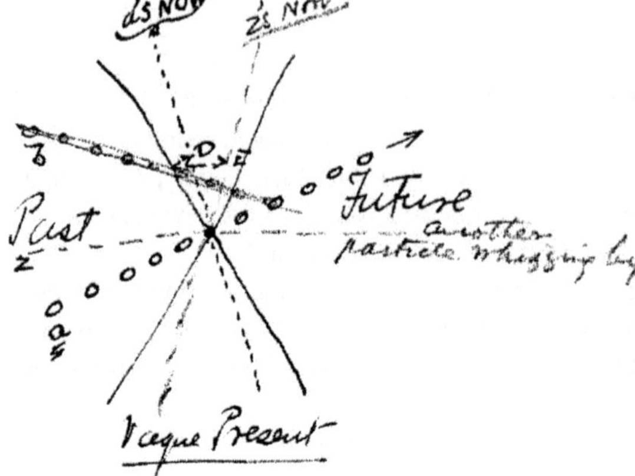

Reiteration of a pattern. Metaphysical view I am drawing is in a sense a monism – not one among others. But it is the essence of the activity that it individualizes itself in a plurality of real things

---

The body is to be looked on as a method of keeping out – letting in. It selects what aspects shall be considered.

a at a_0. Perceptive experience in knowledge of –
3-termed relation between a_0, B & D

z_0
z at z_0 – { Perceptive Experience is a idea of a 3-termed relation } z_0, B & E

The small boy catching a ball, is not catching roundness but the determination of the universe in respect to the ball being there.

Not sense-data. A strain of control issuing into the future. Very much disagree with the idea of private worlds. If there is no apprehension of what the event in itself is 'there' you can't get on at all. The essence of A_0 is being modified by the aspects of B, etc.

Then in danger of bifurcation – psychological space-time and the actual space-time –

Lecture 71

Thursday, 16 April 1925[1]

Bell's notes

|137|
Whitehead: 16. April. 1925[2]:-

Metaphysical question as ca Point of View and my points there[3]

Event taken in its essence; what it is in itself – How it is what it is by reason of its relation to rest of World. Three-fold relation always turning up:-

(a) The World as now (Reality limited ---)

(b) World as recollected – inherited – under limitation of antecedent objects as inherited.

(c) World as real – and thereby as selected from abstract possibility – World as in thought or concept as extending beyond the real world.

Three limitations. Two or three root problems you always come up against:- Question as to whether really you ought not to invert whole idea and start with a thorough-going Subjectivism? Am I not analyzing psychological field of this Entity? And whether that's not really the final word on Metaphysics.

– Here's the Subject which is in enjoyment of its own Experience and that to be described just as Experience qualifying the Subject. (People would not go that far, but ---) If you take Descartes he really has this attitude. In modern times people a little shy of talking that way. – Tell you Subject-Object relation is the fundamental. Whitehead thinks that's just the ghost of the Cartesian method. You are frightened of the Subject-Predicate relation because it doesn't quite fit. – So you put it in the Subject-Object phrasing. Descartes does talk of what mind enjoys, as its Attribute. Moderns simply have a new way of trying to "get away with" old idea. Whitehead prefers latter in its nakedness; because there is some truth there.

Psychological point of view (Subjectivist) is of something there to be analyzed. But the mere predicate of the mentality won't do. Psychology today into hands of Doctors (Neurologists etc.) – goes beyond that. The older Psychology (derivative from Hume, rather than from William James and [Wundt][4] is really psychology acknowledging that pure subjective

1. What might be the Radcliffe version of this lecture, from notes taken by L. R. Heath earlier the same day, begins somewhere on |80| of her notes, page 511.
2. Bell's diary indicates 'Lectures etc. as usual'.
3. This comment is written in Bell's hand in the upper right-hand corner, although it seems to be in pencil (written later?), while the lecture notes are in ink.
4. In his lifelong career at Harvard, William James (1842–1910) made significant impacts on both philosophy and psychology. Whitehead's comment is reflected in James also being trained as a physician, and his first appointments at Harvard were in fact as instructor in physiology and anatomy. Wilhelm

attitude leaves out something [Vice versa!]. Subject–Object point of view is a compromise – Endeavoring to express something more complete than Subject–Attribute relationship and yet keeping the "Subject". "Object" then becomes or remains ghost of Predicate.

Whitehead walking into lecture room. Experience was not of himself ["Subject", if it means anything, is of something that underlies] as underlying anything. The room was not entering into Whitehead in any way; he was entering the room. There are rather vague essential situation which is universal, and because universal is never expressed in language. Language expresses what may or may not be. Where you get something essential it is hard to express it. Experience is always of entry into reality. – But not even of World as entering into Experience – You're always entering into the World. ([1]All these formulae are mere metaphors)

Primarily:– An Ego-object amid Objects. My experience is not now of this room; but of myself in this room. Subjectivism gets rid of that primary factor in all Experience. Any analysis of Room as myself – dependent, is an abstraction. The aggregate which is for my abstract knowledge is limited by myself as in this room and is to be recognized as such. But all that is realized by abstraction from "myself as amid this room". Analysis of how my relations to other Entities enters into my constitution now. --- exhibits modality of --- as me now.[2]

This Modality always exhibits me now as in relation to Entities which are not me. Modality of limitation has that 3-fold character:–

The present:– (a) It has transition within it. There is otherness within it. The otherness is not me and the other things, but the otherness.

(b.) It has transition from otherness which lies without it – from an otherness which is past. The present doesn't represent itself as a complete entity.

Also transition to otherness which is future. So the present exhibits itself as with internal relations to that which is beyond it.

Every other Entity exhibits itself as complex of Relations of which I'm one relatum. It all enters into our knowledge via perceptual object. The perceptual object is really the permanent – enduring – object as issuing into the present. It is this determining what is vague into what is precise. – there there's always an element of hypothesis. But differs with James, Dewey, Schiller[3] in holding that there's something definite there

Wundt (1832–1920), also trained as a physician, established a psychological laboratory within his own Philosophy Department at the University of Leipzig, and together with James is credited with founding experimental psychology. It is not clear why his name is included in brackets, unless it might be Bell's own addition, as in all likelihood is the addition in brackets at the end of this same sentence.

1. Opening parenthesis provided.
2. Although this looks like a parallel rephrasing of the previous sentence, such a construction is usually shown using ditto marks, whereas here Bell uses dashes (or series of hyphens), which often indicates that he has not fully captured what was said.
3. Sometimes seen as aligned with the pragmatism of James and Dewey, Ferdinand Canning Scott Schiller (1864–1937) spent most of his career at Corpus Christi, Oxford. Whitehead had earlier acknowledged his obligations to Schiller in the Preface to *Principles of Natural Knowledge*.

|138| All excellent[1]
to be known – Some intrinsic reality in itself. My hypothetical determining of its status may require correction. Every Metaphysics must retain element of hypothesis because without it no Error. Without Error no ---

Problem is to get a neat theory which allows for Error. Great service of James:– Dragging out Error, and patting it on the back. [W. James was always on the side of the underdog.]

Recollection takes form of aspect of me then as modifying essence of me now. Permanent event is individualized activity of realization.

The general character of Enjoyment as exhibited in Cognitive Experience gives you an objective World.

But all this general talk ought to be buttressed up by showing that detailed items of experience as they force themselves on our attention, bear out this interpretation. (Way written out for Lowell Lectures– Criticism of Subjectivist theory ⟨blank⟩ Common world of Thought; but no common world to think about. Type then in Equations of higher Mathematics.)

– A position which Eddington and other extreme relativists are tending to take up.

[Whitehead reads a lot, excellent, from Lowell Lectures.] [2]

Broad reasons for rejecting Subjectivism.

(1.) Arises from direct interpretation of our perceptive Experience [already given more or less] – we as elements of the World. In our sense-experience we know away from and beyond our own personality. Even intermediate Subjectivists (Kant) place our Personality between Subjective Experience and common world.

(2.) Based on particular Content of Experience. Our historical knowledge tells us of past ages when no life existed on Earth – Of Star-systems whose light takes 1,000,000 light-years to come to us. Far side of Moon and interior of Earth:– What's happening there.

In face of the Content, it is hard to believe --- ⟨blank⟩

(3.) Based on instinct for action – Seems to issue in an instinct for self – [Here Kantian. But Kant should have thought of his "Practical" before writing his Theoretical Critique. It is not activity directed to veiled world of the intermediate Subject. Activity within the known world and yet directed by us. Humpty-Dumpty and great principle of technical terms.]

But "Subject" as a term begs the question

And is ghost of a philosophical theory of Subjectivism][3] Everybody

1. Written at top left of page|138|, in pencil, and seems to have been added later.
2. This bracketed comment is clearly Bell's, and it seems that Whitehead does in fact read from the script of his Lowell lectures, since the following three points and the last lines from this lecture can be found in chapter V of *Science and the Modern World*, pp. 89–91.
3. There is a confusion here of partial bracketed comments, in this case with a closing bracket but no opening bracket. These seem to be Whitehead's comments, aside from his reading from the Lowell lectures. Hereafter (once past the Russell anecdote) he picks up the reading again, to the end of this Emerson Hall lecture.

wants to struggle back to an objectivist position again:– B. Russell's experience with the Solipsist surprised that there are so few of them[1].

Whitehead doesn't see how the "common world of thought" can be established independent of the common world of things.

Any community must suppose somewhere or other the objective position of world of immediate experience.

Distinction between Realism and Idealism is not that between Subjectivism and Objectivism. But Objective Idealist, when comes to analyze reality of World finds Cognition involved in nature of everything. That the Realist denies.

So the Two have quite a bit of road possibly in common. Objectivism distorted in past because supposed bound up with "scientific" view, then implying primary and secondary qualities etc. – a position which falls easy prey to Subjectivist criticism.

Hocking's notes

April 16. –1

The world as real, selected from abstract possibility –

The mind as subject and the object as predicate.

Modern psychology with physiology as its guide shows that Subject–Object view won't do, the nerves are something not just predicates of the mind. Subject–Object view is a compromise – more complete than Subject–Predicate view, or Subject–Attribute of Descartes.

The object is obviously the ghost of a predicate – Not the fact of experience. I walk into the room. I enter the room. The room doesn't enter me in any sense at all. Our experience is always one of entering into the world, not into the world as supported by oneself in any sense at all

Apr. 16 –2

Sub-ject – lying under – wrong.

We are always entering into things –

I-amid-objects is the truth of things. Myself within this room.

Subjectivism gets rid of & abstracts from that

My relations to the other entities enter into my constitution now

2. Modality of limitation has a 3-fold character –

a. as of present

b. as of recollection

c. as of thought

We perceive the past as issuing into the present as in the cricket ball. "Definite knowledge that there is something definite to be known always involved in hypothesis" hence disagreement with James & Schiller

1. In Hocking's notes this 'solipsist' is named; see footnote 2 on the following page.

Any sound metaphysics must preserve hypothesis else there would be no error. It would be easy to get a neat theory without error. "James' service to metaphysics, in bringing out error and putting it on the back. Always on the side of the underdog – one side of it"

3. Recollection. Me-then as modifying Me-now –
Remnant Me as an individualized activity of realization.

4 – Eddington tends to take up extreme subjectivism.
Halfway house. Kant –
Objectivism – Whitehead. Elements perceived by senses are the elements of a common world – Things are to be distinguished from our knowledge of them. Enjoyment of experience discounts – from cognition – Bacon.

Things enter each subject on equal terms – hence subject spread[1] to three objections to Subjectivism

1. Direct perception. I am <u>within</u> space seem to be elements of world – Instead of making the world depend on us – Holds with naive experience.

2. Particular content of experiences. Historical knowledge ages in the past. Star systems beyond our ken – Distances. Interior of earth, or far side of moon

3. Action. Instinct for self-transcendence. Activity passes beyond self – Kant. Should have put his three critiques together. But activity goes to determinate ends – Known world transcends self.

<u>April 16 –3</u>

Everybody wants to struggle back to an objectivist position – Russell & Miss Ladd Franklin[2] –

I do not see how a common world of thought can be established apart from a common world of sense. Can't remedy it by saying "anyhow we have a common world of thought" – Any community must presuppose the immediate objective position of the world of immediate experience –

Realists & Idealism division does not coincide with this. Objective Idealist finds the common world, but finds it involves mentality in every detail. Here realism departs. Should start with provisional realism.

Next thought & abstraction.

1. This term is uncertain and the reading is made more difficult by the sentence not being completed.
2. Christine Ladd-Franklin (1847–1930) wrote a thesis entitled 'The Algebra of Logic', supervised by Charles Sanders Peirce. Bertrand Russell is often quoted as reporting that she had written to him, saying she considered herself a solipsist, and was surprised there were no others. Being a logician, Russell was surprised at her surprise. This story is usually cited from Russell's *Human Knowledge*, published in the 1940s, but as Whitehead's comment in this lecture shows, Russell must have told it decades earlier.

Lecture 72

Saturday, 18 April 1925[1]

Bell's notes

|139|
Whitehead. 18. April, 1925:–

Idea of Subject–Object. Subject means underlying something.[2]

So far as meaning of words entered into and controlled signification of Metaphysical view you're starting from, phrase is an unfortunate one. Whitehead suggests that Subject–Object ∧idea∧ really has in background: Subject–Predicate concept.

Discussion started with individual substances either mentalities or extensions. Taking this idea of Mind and simply viewing it under guise of ~~someth~~ ∧a Subject∧ whose essence is to be determined by predicates – and the awareness of objects are its primary determinants. Whitehead objects ⟨blank⟩

Prime Experience is of an Ego-object amid Objects – An apprehension of a parity. An Object-Subject amid Objects. But Whitehead does want to retain what he holds to be the immediate apprehension – of a Subject in a World. If you once drop that, of course you can never recover it. It is a question of what immediate point of view accounts when developed for all that which you immediately assume and can't help assuming.

Analysis of the "What" of the Subject. If you start in the Cartesian way you can't ---[3]

– little dig at Aristotle (greatest of all Philosophers all the same) – Correcting an unfortunate derivative of bias he gave to Logic.

Whitehead:– Internal relation is fundamental metaphysical idea. As soon as you are dealing with what is real. Ego-Object in a World of Objects; and that world of Objects enters into (modifies) essence of each object in that World. That is why the Ego-Object in its self-knowledge (self-recognition of what it is in itself) knows itself as amid other objects. Whitehead looks on Predicate-Subject point of view as being expression of what Ego-Object is in itself, whereby the other objects are abstracted from as forming a complex and are conceived merely under guise of entering into relationship with the Ego-Object. – In so far as they can be expressed as constituting essence of Ego-Object ("Subject").

1. What might be the Radcliffe version of this lecture, from notes taken by L. R. Heath earlier the same day, begins somewhere on |82| of her notes, page 511.
2. These phrases are written in the margin above the text, at a slight angle.
3. Bell usually leaves a series of dashes, perhaps question marks, and blank space that we interpret as his having failed to capture what Whitehead said, but in this case he leaves dashes directly beneath this phrase, and another row beneath.

Predicates express how relations modify nature of Subject. You are expressing Relationship in terms of the Essence. The Synthesis of the Complex of Relationships is the real Entity – and what is expressed by "Essence" – which is an abstraction from the other Relata. Thus notion of Predicate and Subject covers a variety of Relationships. ---

You always come to Relation of "Ideas" "Concepts" ("Eternal Objects") to what is actual. Again you come to that difference of Emphasis which marks Platonic and Aristotelian school. Plato was a mathematician, Aristotle son of a doctor – and that is essence of difference between the two. Plato was considering complex abstractions – and how actual presupposed the idea. Aristotle was classifying – brought up to see the ideal after the example. Plato studied (equivalent of Euclid) and then looked about and saw – "Here and here they find their realization". For Aristotle it seemed that Universals sprang out of the actual – the redness of plumage of stuffed bird, etc. A mathematician is bound to be Platonic. (Whitehead:–) The potential underlies actual and transcends it. Can't get his Metaphysics started without it. But ideas are held by the actualizing process before the realization as potential for it. – Can't tear the two apart.

If you insist on the potential as being in particular individual realization (realized entity) you will get into a muddle:– because in that case how are you ever to have anything new in the world. But Whitehead does feel that the potential (the "Ideas") are always in a general relation to process of Reality. And insofar as every individual instance of what is real by theory of internal relations embodies within itself under limitation the Totality, thereby ~~it is held up~~ the realm of Ideas is in a general relation to every particular entity. But insofar as the particular Entity is essentially a finite limited individualization the potential is there merely in its various grades of Emphasis, Selectiveness.

It is there that thought comes in – an exhibition of the Selectiveness. If you once admit that anything that is is entirely out of relation to the Actual, then it is nothing with relation --- and you can't get --- The realization can only be the realization of what is already related as potential. When Aristotle comes to realization/ΛchangeΛ he has to drag in Potentiality. But Change is only an account of how Realisation takes place. So the real instance can only be accounted for by bringing in the idea of potentiality and potentiality Λcan only be account for asΛ realm of Eternal Objects as transcending what is real and yet in relation to it.

|140|

Now when you come to realm of Eternal Objects relations you have are external relations and that is why they are to be looked on as logically antecedent to individual realizations

And it is this external relationship which gives content to the relationship of identity. [Going back here to his very beginning.][1] You get relation of Identity by comparing one this with another this – difference

1. Closing bracket provided.

of circumstances etc. – (External relations). You have therefore your Metaphysical situation:– in endeavoring to get the analysis of it you have really a 3-fold analysis as being the presupposed fact of it. – You have ideas or eternal objects; you have Activity; and you have the Actual;– and you have in conceiving the "Situation" [Spinozistic "Substance"] in combining ΛthoseΛ two concepts of activity and ideas you get to account of Process as being an emergent individualization – viz. that the Substance is emerging into a plurality of individuals. Each individual being what it is by reason of its relationships to realm of Ideas, and to totality of Individuals. And then whole essence of Activity (Process) – this individualization then, in its turn, divides itself into the immediate and actual and the potential. Each individual immediate entity goes back on that Plurality and is a synthesis – a unity – expressing itself as an all-inclusive unity whose essence is its relationship with the plurality of individualizations and with the plurality of ideas – So that you never get away from either the Plurality or the Unity. Real ultimate point being the notion of the complex of relationships which are external insofar as the Relata are "Ideas" and are internal so far as each relation is itself an expression under limitation of the totality.

Each Subject [of predicates?] ΛEmergent IndividualΛexpresses itself as a
{ totality / [unitary phase] ?? } of the total of what is actual.
This is an endeavor to take seriously:

(1.) (a) Theory of internal relationships (b) fact that we are one among entities. Primary Fact for Consciousness is an interrelated world – in some extent a synthesis.

(2.) (a) That it is essentially a plural world of "obstinate" and irreducible facts (b) That there are these universal ideas to which concept of identity applies and for which the relationships are external.

Furthermore, even within the emergent entities you have really to discriminate between the immediate Self which is simply here-now and the Enduring Object which is the individualization of the Activity as the embodiment of Ideas. – It is giving the individualised Activity [off on a "Myth"] as the immediate entity as embodied already in eternalities[1] of past and in the "immediate" entity. In addition to this there is an expression of embodiment of the underlying activity. This is itself individualized with a character related to realm of Ideas. Insofar as the general activity does reproduce itself under same character of embodiment of Ideas in addition to its essence of immediate realization, there you have an enduring object.

In case of higher type of enduring objects (as You and I) Whitehead regards this as dissociating itself [in thought] into two more abstract parts:– that which exhibits itself (amid loss of molecules and such changes, and is yet the same – the same man, although the body all different) – that permanent is that which underlies the activity of the Synthesis; and you can separate that, somewhat vaguely, from that which is the immediate

1. Bell actually has 'aetnalities', but seems more likely to have meant 'eternalities'.

realization of the Synthesis. ~~The~~ ∧Our perception of an∧ enduring object is how our essence is modified by its relationship ---?

Whitehead cannot make out how Hume, talking of Impressions on our mind, says that we perceive nothing which connects impressions of Past with those of Future. Hume says its first importation is entirely Arbitrary. Seems to Whitehead perfectly obvious that in perceiving the piece of chalk as an object you are perceiving something on same level as yourself – and as issuing into the Future, and, in being what it is, determining what your future's going to be for you too. You don't know what is going to happen to piece of chalk in future but you perceive that the future is going to issue from that object. The "Control" is in the things which you perceive.

Descartes sees that your apprehension of any object is to be discriminated from having of Sense-data [Calls it an *Inspectio*]. Whitehead:– That *Inspectio* is apprehension of Control-relationship between the various individual examples of realization. Here Whitehead disagrees with Descartes and whole Materialism of Science – This all conceived under guise of Space-Time – in external relations. Then
|141|
Idea of "Simple location"
you have to endow the Substance with another field of Force. – That is what Newton had to do. If you once admit External Object just passive thing in itself to be looked at, there is no getting away from Hume. – If you once admit idea of Simple Location – you have seen all you have
? seen. You have then left the perceptual object no function whatever. But this contradicts immediate apprehension. – Future will be what it is by reason of that object. So far is Whitehead from preaching a paradox, he is preaching what physical Science always assumes. When Physics talks of electricity the charge is ⟨blank⟩

Faraday, and J.J. Thomson[1] [utterly unmetaphysical] say: – Only use of ideas as "Charge" etc is as marking distribution of a field of force – Something spreading throughout all Space. – So far from Descartes' idea of "lump of matter" as just there being scientifically convenient – it is inconvenient for Physics as well as for Biology and Psychology. Scientists regard you as talking metaphysical nonsense when you are just phrasing what is implicit in all their own statements.

– Whitehead doesn't regard Eternal Objects as something there which the Actual has ∨somehow∨ found out – you can't have one without the other. etc. Metaphysical Situation is essentially dualism.

We now can give an account of the Future, too. The Future is realm of Abstraction. Ideas in its particular relation to the Actual – Some ideas extruded by latter (as possibilities etc).

1. There was a passing reference only to Michael Faraday (1791–1867) back in lecture 3. Despite little formal education, Faraday was well known for his experimental work in electromagnetism in both physics and chemistry. Sir Joseph John Thompson (1856–1940) – whose lectures Whitehead attended while still a student – became Professor of Experimental Physics at the Cavendish Laboratory at Cambridge, where he is credited with discovering the electron through his experimental work.

Hocking's notes

April 18 –1

Subject–Object – has in its background the Subject–Predicate conception.

The essence of the mind – the holding-before it of objects. One objects to this as the prime expression of the metaphysical situation.

The relation is one of parity.

Immediate apprehension of oneself in a world the axiom.

If you once drop this you can never recover it. & it is an immediate question what point of view you can hold.

The internal relation is the fundamental metaphysical idea. As soon as you are dealing with what is real. The ego-object in a world of objects. The world of objects modifies the essence of each object in that world.

The ego object knows itself as amid other objects. The predicate-subject point of view. The idea of the essence is an abstraction from the other relata.

April 18 – 2

Relation of ideas to what is actual.

Difference of emphasis – Plato – Aristotle – Plato, mathematician, Aristotle, Doctor – Plato considers abstraction: Actual presupposed ideal

Aristotle first saw the stuffed bird & then found the idea.

Potential underlies the actual – any mathematician agrees to this

But these ideas are held before the realizing process as potential for it. Can't tear the two apart.

"I speak here with immense ignorance of what I am talking about. So please understand that."

If you insist that all your universals are in the particular realized entities, you get into a muddle. Can't see how the world then is to get onto anything new.

But they are always in a general relation to what is real.

Thought is a manifestation of the selectiveness

As for the eternal objects. The relations they enter into are external. Hence logically antecedent to realization. This gives content to the relation of identity (see beginning lectures of this session)

It's nonsense if you have not got two instances of the same thing.

Activity ≡ process. Underlying activity.

Our perception of the enduring object is how our essence is modified by its aspect

Can't make out how Hume says of impression that we perceive nothing that connects impression of past with future.

Its first importation must be entirely arbitrary.

Obvious however that in saying that anything – this chalk – is an object, you are perceiving something on the same level as yourself & as issuing into the future & thereby determining what the future is to be to you too. In perceiving an object you perceive that the future is to issue from that object, though you can't know what the future is to be. Your apprehension of the object is thus as Descartes said, to be discriminated from the sense datum. Called it *inspectio~~n~~*

On this ground we object to the whole materialism of science Passive matter in space & time & in external relationships – Then have to endow it with a field of force, as Newton did. Once admit that the perceptual object is merely where it is & what it is. Hume has you, for you have exhausted all you have seen. Faraday 1849 said an electric charge is nothing but its tubes of force.

Lecture 73

Tuesday, 28 April 1925[1]

Bell's notes

Continuing on |141|
Whitehead – 28th April. 1925[2]

To avoid muddle now need a point kept in background so far.

Title of Whitehead's notes now: – "A Cognitive Experience" Bacon's phrase.

Between Cognition and Experience to intermediate Concrete Unity of individualised Experience as an "Event". But how Cognition looks sticking close to point of view developed so far in this course.

All Cognitive Experience is to be analyzed into and conceived as being a cognitive apprehension of how essence of an immediate occasion is modified by (particularly constituted by) its relationships with other entities. What is it we Know? – We know how our own essence is being particularly constituted by our relationships with other Entities. – That is our immediate knowledge of immediate occasions (not knowledge of our permanent nature). There Whitehead believes himself in complete agreement with William James in his "Does Consciousness Exist" – In complete disagreement therefore with Descartes for whom existence of the knowing stuff is fundamental.

Thus cognitive apprehension is realisation of reflective reflection between an Entity and its own Essence. Firstly – note the Subjective Side of Cognitive Experience. It is essentially Self-knowledge. Any account of cognitive Experience which doesn't emphasize the Subjectivity must be wrong. One of the difficulties of all Philosophy is to escape complete Subjectivity (Solipsism). Whitehead avoids it by his internal relations – Not looking at it as a Stuff with its predicates. The otherness of the Entities is included in essence of the Emergent reality.

The Essence qua composite is essence in its --- ? ? ⟨blank⟩

A paradox which arises – Cognitive Apprehension presupposes both cognisant Entity and its Essence qua Composite which is the potentiality out of which the occasion emerges into Reality. Entity has to be there to be cognised and if its nature is only by being Cognised then --- etc. But it presupposes the Essence as less than itself – emerging from a potentiality antecedent to Consciousness. Potentiality which enters into --- is admissive

1. What might be the Radcliffe version of this lecture, from notes taken by L. R. Heath earlier the same day, begins on |84| of her notes, page 512.
2. Bell's diary records that he had been ill during the break week, and on Monday, 27 April, notes: 'Didn't start own lectures till Thursday (cough as a result of grippe etc.)' but records nothing about the Whitehead lecture, for which he did take notes.

of Cognisance but not inclusive of it. Entity qua cognisant is other than the Entity in response to whose essence the Cognisance has regard. We distinguish therefore between the Occasion as inclusive of Cognisance and Occasion as exclusive of Cognisance. Names for these two: – Latter occasion is "The Standpoint" Occasion; former is "the Cognisant" Occasion. Here only utilizing sharp differentiation between unity of Experience and Consciousness of it. It stands in nature of cognitive apprehension that these two occasions emerge into real facts. When Standpoint and Cognisant are
|142|
not so distinct, there is no Consciousness. Translucence of Cognisance arises from this (Cognisance doesn't alter facts – presupposes same occasions exclusive of Consciousness). This is the central belief of Realism. Either one's own psychological Experience is or is not dependent on one's Consciousness of it. Central point of Realism is that it is not so dependent.

Certain Alternatives along which you can try to make up Philosophy: –

(1) That Cognition is somehow formative of *Cognita* (Kant etc.)

(2) Can look on Cognition as distorting *Cognita* in some way

(3) Cognition apprehending something derived from deeper sources – manifestations of that (H. Spencer [Kant]).

(4) Cognition as finding the *Cognita* (Realism). Distinction between "Realism" and "Idealist" depends on what University you have been brought up in. Hoernlé[1] (Glasgow) – "Idealist" – If Cantab "Realist".

(Story: Bosanquet and Russell agreeing on Symbolic Logic. Agreeing on a fact and disagreeing only in emotional reactions to fact. [Same re differences of philosophical terminology (-isms)]

As long as a man admits (4) above --- But "absolute Idealist" is harping on (1.) Cognition not formative of immediate facts [But difficult to know how to Classify some people].

So far immediate occasions. But parallel duality arises re permanent objects. The permanent "Self" (reiterated emergence of Same achieved entity [Value] for its own sake) vs its manifestations. We have got to discriminate the Self of the flux of Cognisant occasions from Self of the flux of Standpoint occasions. Former is the Cognisant Self. Latter is the Experient Self. [Derivation: – The Industrious, Active [Experient] Self].

Cognisance is fitful, broken, and selective – and variable in its partial analysis of immediate essence of any standpoint occasion. The Experient Self is, then, the important thing. Properly speaking there are many Cognisant Selves. This Multiplicity gains a derivate unity by reference to Stability of Experient Self. The Standpoint Occasion is therefore immediate

1. R. F. Alfred Hoernlé was definitely an idealist and educated at Oxford, and was mentioned previously by Whitehead in lecture 21 (see footnote 4 on page 79). The reference to Glasgow is puzzling, however: he was at St Andrews University for a time with Bosanquet, and at a college within the University of Durham (both northern), but not at Glasgow. In Hocking's version of this contrast, it is Oxford (idealist) versus Cambridge (realist), 'Cantab.' being the often-used abbreviation of the Latin name for Cambridge.

experience of the Experient Self. Experient Self is the Identity which connects the succession of Standpoint Occasions. This brings in the way in which your enduring Identity can change. Enduring Self is more abstract than immediate occasions.

The Cognisant Occasion is the immediate "Image" which constitutes immediate knowledge of Cognisant Self. Therefore it is the Image ∧in∧ which is an element the immediate Conscious arises. In traditional phraseology as to being within or without the Mind. Standpoint Occasion might be said to be within the Cognisant Occasion. Experient Self is within the Standpoint Occasion. Furthermore, when we talk generally of "Self" we must mean the Experient self [The one that was born so and so many years ago]. The Self within the Images and not vice versa: Images within the mind.

Instance of saying the images are mental; ought to say: Mind is Imaginal.

Idea of Images as modifications of Mind is misstatement of ontological status here. Whitehead here against Aristotle and the Subject-Predicate (Substance and predicate-quality) obsession. That which is permanent is not cognitively mental. Permanent Mentality as abstraction from the Images. Here diverges from Descartes, where he tacitly identifies the "I" which endures with the immediate cognitive occasion and also with the

Point of View I

Cognitive Mentality which is an abstraction from that Occasion. This Cognitiveness is a complete Universal – just one for all cognitiveness. The Self which has endured from birth therefore is not Cognitive. There have been many of these. Cognitive Self is not totally aware of whole wealth of experience of the Experient Self. [Psychologists know this]

("So and so doesn't know it but his whole judgement is warped by "Envy" e.g.).

Plato's Theory of Knowledge as reminiscence, and only fitful and dim awareness of its inheritance. Whitehead finds identity in transmission of pattern, and presence in identity of "route".

Whitehead thinks this required by any theory based on "translucence" of Cognition. But the translucence doesn't mean Consciousness merely looking on and never affected! Every occasion as an unification of aspects of entire Universe. Has Aspect in next Occasion – Cognisance has its effect in steering the course of realisation. Doesn't make the things we know now, but those we are to know tomorrow etc.

Hocking's notes

April 28. 1925.

Cognitive Experience
A subject hitherto kept in the background –
Amplifying Bacon – as Bacon would be surprised to follow
What are we drive at by the point of view of course?

Cognitive Experience – How the essence of an immediate occasion is modified by, or partly constituted by – its relations with other entities –

Believes is in complete agreement with James – in His Essay, "Does Consciousness Exist?"[1] objecting with the idea of a stuff of mind, & in complete disagreement with Descartes –

Subjective side of Cognitive Experience – It is essentially self-knowledge – (requiring a discussion of 'self')

Everybody knows the difficulty of getting away from solipsist "Your reason ought to spring out of your philosophy, & not out of the fact that you don't believe your philosophy"

"Consciousness is a relation between an emergent entity and the composite potentiality from which it emerges"

A paradox. Cognizance presupposes the entity as less than itself.

An occasion may permit cognizance without requiring it or including it. The ∨standpoint∨ occasion & the cognizant occasion.

The translucence of cognizance. It does not alter the facts – An occasion inclusive of cognizance presupposes the same facts exclusive of cognizance. The essential belief of realism.

You may look on cognition as distorting the *cognita* in some way, or as apprehending representatives of reality. (Spencer). or as finding the *cognita* (realism)

Here absolute idealism comes close to realism –

If you have been brought up in Oxford, you call yourself an idealist. Cambridge a realist –

A matter of emotional reaction. Russell & Bosanquet agreed on value of symbolic logic – for Bosanquet's kind of philosophy.

As long as one admits that cognition does not alter the facts the absolutists are harping on the importance of consciousness[2] in the universe, – indubitable – the realist harps on the translucence of consciousness

April 28 –2

A parallel distinction in re the enduring entity

The self – the reiterated emergence of the same achieved value for its own sake

The self of cognizant occasions must be discriminated from the self of standpoint occasion.

Cognizant self & experient–self. (industrious active). Cognizant self – fitful & variable – in its analysis of the essence of any occasion – In relation

---

1. James' 'Does "consciousness" exist?' appeared in the *Journal of Philosophy: Psychology and Scientific Methods*, vol. 1 (18) (1 September 1904).
2. In these three lines Hocking uses the abbreviation 'c.' twice, which could represent either 'cognition' or 'consciousness'. We have tried to rely upon context.

to endurance, the important self– is the experient self– There are many cognizant selves – sleeping, waking –

"The cognizant occasion is the immediate image which constitutes the immediate knowledge of the cognizant self."

When we talk generally of the self we mean the experient self. It is within the images. q. the images within the self.

Mind is imaginal q.[1] images mental.

Images as modification of the universal mind distorts the cosmic situation – terms of substance & attribute or predicate – That which is permanent is not cognitive qua mental – associated rather ~~than~~ with image.

Descartes tacitly identifies the self which endures with the immediate cognitive occasion

The self which has endured since birth is not cognitive. "The man's judgment is warped by envy" – much that has not entered into his cognitive self at all –

1. Both here and in the line above Hocking uses an unfamiliar abbreviation, 'q.', that we have interpreted (from context and comparing with Bell) as 'instead of' or 'rather than'.

Lecture 74

Thursday, 30 April 1925[1]

Bell's notes

|143|
Whitehead: 30 April 1925[2]:-

Mind imaginal, rather than Images mental. Whitehead looks on that grammatical form as, generally, a cloak for a well-known situation you are pointing at, where Subject is some wider view and Predicate is some narrower ingredient. Viewed as habitual location it is defensible, but it is not (in either way) a very happy way of expressing situation.

Two questions from Demos[3]:- (1) How can Consciousness be translucent in view of doctrine of Internal Relations (i.e. that every object modifies every other one).

(2) Is it correct to throw Consciousness into the flux. Knowledge seems to have an eternal quality and be neither transient nor enduring etc.

Whitehead welcomes these questions – In enthusiasm of trying to make theory neat and effective one constantly overstates situation.

There is an essential difference between Past, Present and Future. Getting in that essential difference is all the difficulty with Internal Relations. – What caused Bradley to discard all that we really care about. Unless you get essential difference between Past, Present and Future, you're landed here:- That total content of experiential facts is given and then you can only assign Consciousness role of turning over the leaves of a book.

Cain killing Abel – has then got to be in eternal nature of things, and we come on it in a certain order and that is all. Freedom, Moral responsibility, etc. all go. You get a logically neat Metaphysics; but wholly opposed to Whitehead's direct apprehension of the ontological situation. If try to avoid this point, no alternative except to describe a Universe in which there is room for freedom.

---

1. What might be the Radcliffe version of this lecture, from notes taken by L. R. Heath earlier the same day, begins somewhere on |87| of her notes, page 513.
2. On this day Bell records in his diary: 'Lectures etc. In evening to supper at Douglas's (Whitehead's there). Thence to Harvard Club. – Canadian Club meeting as Guests of Mr. Warner. Whitehead spoke on the British Connection & its Values. Later after driving out to Cambr. Tea with Whitehead's. Thence home & prepared for Phil. A.'
3. Raphael Demos (1891–1968) came to study at Harvard in 1913, where he earned his PhD. He immediately joined the faculty, and taught there until 1962. During 1924–5 Demos was, like Winthrop Bell and Ralph Eaton, an instructor and tutor. All three served as tutors for the basic Phil A course, in which several of their senior colleagues lectured, and each of them also taught a course of his own. So far as we can tell, Demos attended Whitehead's lecture course regularly and also his Friday evening seminar. Whether the questions Whitehead is addressing in this lecture are ones Demos raised in this class, in the seminar, or outside of class, we do not know. Back on 15 November, Whitehead spent his lecture 23 discussing questions that had been raised the previous evening at his seminar.

First step is to hold that what is not actual does not modify essence of what is actual – Except systematically via the Spatio-temporal scheme of relational essences. It is only Future as being in the system of relational essences in which Actual is. Via essence of Actual as including in itself aspect of conditioning what is not actual (the Future). Thus the internal relationship as between an occasion in Past and occasion in Future is two-fold:– (1) The actual occasion directly modifies essence of non-actual (2) Non-actual indirectly modifies essence of Actual – viz: by reflexion from Actual's modification of it.

Now relation of Present to Present. Isn't going to worry about Alternative Time-systems. – Doesn't essentially modify explanation, really. Suppose a stand-point A, – definite occasion; Now – Then in A the Universe is synthesizes doubly (even trebly, [later]): – (1) Under modal limitation of being included in the immediate present. (2) Further limited as being the present from standpoint A.

Under (1.) Endurances from Past are limited by their relationship to present – as issuing into the Present. ∧Past as issuing into present via its various permanencies of achievement.∧

Under (2.) Illustration of the Present as an achievement for the standpoint A

Now: –

(1.) Our apprehension of Endurances of Past of issuing into Present is Descartes' "*Inspectio*" [in his wax example]. That is the perceptual object. Line Whitehead took from beginning in talking e.g. of "Strain of Control" – "The table" as: What happens issuing because of a certain enduring pattern.

Under (2.) – there you get Sense-awareness – Awareness of Present as exhibiting a Pattern of Sensum. You have to steer close to erroneous distinction between Primary and Secondary Qualities – Galileo, Descartes, Locke, etc. wouldn't be so happy unless there were something.

Philosophy as a series of extraordinary insights to be harmonized somehow (Not a lot of silly things all different).

Whitehead doesn't look on Mind as Consciousness of its Mental images. – Consciousness of Relations in the actual universe. [Now illustrate this by one of Whitehead's metaphorical diagrams.][1]

Actual occasion B at same time as A. What B is in itself doesn't modify A except via its systematic Spatio-temporality. But systematic relation of B to its immediate past, modifies A. Immediate past as issuing into B modifies A; and for essence of A there is B as a situation of Eternal Objects (Sensa e.g.) as at B, and for A.

1. Closing bracket provided.

|144|

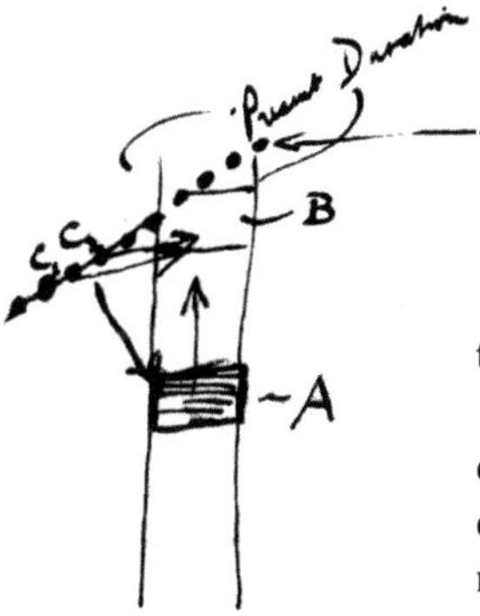

An object (Trolley car e.g.) as it issues into present [Each line here is an instantaneous 3-dimensional Space, of course]

There A takes Past as though issuing, as if at rest, into the present.

Relation is a complex one of the Sensa as arising out of C ... Cμ , issuing into B, modifying essence of A etc. – Result = "Eternal Objects" there (B) for that [A.] As matter of fact that is a complex relationship as in essence of A. – Cannot be resolved into a simple relationship.

If you simply say "The Bus is red" and is there; you entirely misrepresent what is nature of our Experience (which is of A with these other entities) Whole Spatio-temporal Experience is really --- ⟨blank⟩

You have got to explain Sense-awareness and correlate it to "Inspection" of Sense-object; but also have to allow for Error.

Think of yourself sitting opposite an extremely good mirror and pressing one eye so you see double. If your theory of Perception survives that, it's pretty good! We always associate the illustration which is associated with the permanence issuing from immediate Past into Present, as being actual in B. But while this is habitually so in every day life, it will often not fit in to our systematic reconstructions of things.

Essence of B in itself (as actualized) in certain independence of essence of A in itself. There is an immediate Inspection of the Past as issuing into and conditioning B. But no immediate Inspection of whether the endurances which are illustrated in B by the pattern of Sensa (as illustrating B and enduring there) correspond to ? ?

There is a distinction there too, and it is in the identifications there that error arises. We have a certain atomic independence of immediate (simultaneous) occasions, except via their relations to the Past. That atomic independence of what is immediately present, is absolutely necessary if you are to have any Freedom or any Error. And these are in the Universe.

Because of this atomic independence; that Universe has for us an immediate aspect of radical Pluralism.

Only role allowable to creative energy of realization is the making of it real. Effectual potentiality in A.

Individualization of realising activity not wholly featureless and ⟨blank⟩[1]

Underlying Activity – Creature of the Activity as created by the Past. It is not there and then finding perchance, what it creates. Underlying creative activity is not purely colorless.

1. Partially left blank, and partially filled with indecipherable lines and squiggles.

What is embodied must be looked on as something selective – something in nature of purpose. Whitehead thinks purpose is self-preservation[1] of the ? ?

Now come to Demos's question. Whitehead has been showing it inherent to his scheme that in the Present he would hold to a disconnect – an atomic structure. Cognizant Occasion (Image associated with it) is also in past.

Translucence of Cognition then is example of Atomic independence of all occasions in same present type. Only type of explanation is exhibiting it as illustration; as ⟨blank⟩

Then second question: How do you get immediate fact of Knowing something into the flux.

Where do I want to do so? I want it in the bit outlined in pencil above. Must decide whether that square is red or white.

The grey square is equally an aspect of the whole Duration. You have got to have the image realisation of Whole [Realization in time of the relationship between grey square as a realized entity]

Something which stands outside and beyond the flux of Experience. But is in itself an actual fact then and has its immediate relations to that. And is in fact because can ⟨blank⟩[2]

That I'm thinking in the "Now" is even clearer than that I'm thinking in my head.

Read chapter on Time in *Concept of Nature* [Had not full theory of Mentality then. Talking there only of the Experient Occasion – not the Cognizant one] Time enters into Thought in a clearer way than Space does.[3]

Sleepy[4]

Hocking's notes

April 30 – 1

Demos.[5]

1. How can consciousness be translucent in view of the doctrine of internal relations? Should not consciousness modify its object?

2. Is it correct to throw consciousness into the flux – seems to be neither transient nor enduring – as when I say the sun is shining.

---

1. Given the abbreviation, this could also be 'self-presentation'.
2. Bell's handwriting in much of these last several lines is unusually sloppy and cryptic.
3. These last three lines are written up the left-hand margin.
4. This comment has been added by Bell in pencil, perhaps later.
5. See footnote 3 on page 351 of this lecture.

1. In re internal relations we must admit an essential difference between past, present, & future.

The total content of the time-process is given, unless you do that, relates time-process to consciousness, & turning over leaves of a book.

We must have a universe in which there is room for freedom.

Anything which is not now – in the future – can modify the present only "systematically" – via the essence of the actual, as containing in itself the condition of how it can affect the future. It stands within the essence of the actual occasion how it is modifying the future.

Now we come to the Demos' question, "how the present is modified by the present."

Trolley issuing into the present"

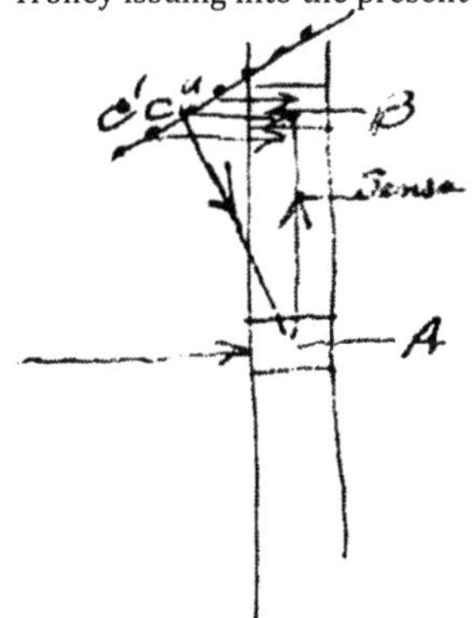

Standpoint a, which is now, suppose, – in this the universe is synthesized doubly (or trebly)

1. Under the modal limitation of being included in the immediate present.

2 Further limited as being the present from standpoint a.

A takes the past as issuing, c', c'', into the present perceives sensa (eternal objects we called them before) A's experience or essence is of A as modified by other essences, – this is true without consciousness.

But now we have to provide for sense-data.

There is an immediate *inspectio* of B, but not of whether the endurances which are illustrated in B. – of what the present is in itself.

The atomic independence of --- is necessary if you are to have a breath of freedom or of error

Radical pluralism. I am here & that chair is there.

Translucence of consciousness is only an illustration of the atomic independence of all occasions in the present time

2. How did get A into the flux? By *inspectio* it doesn't seem to be there. But the awareness does take place in time.

Read Time in *Concept of Nature*[1]

1. This phrase is added, at an angle, in the bottom right-hand corner.

Lecture 75

Saturday, 2 May 1925[1]

Bell's notes

|145|
Whitehead: 2. May. 1925[2]:-

Experient Self. Question: What is meant by Experient Self not so easily determined. That ˅pattern˅ which is reiterated through different occasions. – Not very definite Principle of the Irishman's Sock[3] comes in. The enduring with respect to my bodily life this momentously[4] different from: if you mark it from some great crisis in my life. There is an indefinite variety of meaning we can give to the Self. One's entirely in agreement with common ways of stating things, though perhaps in <u>dis</u>agreement with views which look on mental part as definite Substance. Take religious Conversion:- We say there is a <u>different</u> Self:- Person. In <u>some</u> sense, though, the <u>same</u>.

Experient Self, Whitehead looks on as embodiment of Creative Activity whose law of being is Self-creative. But not an independent Substance but <u>emergent</u> upon occasions from <u>possibilities of</u> relationship. – It emerges by reason of the niche which is there for it in the Universe. <u>Translation</u> from potential to actual. Here idea of "embodiment" of creative energy is required. Role Aristotle attributes to his "Forms". Whitehead's individual "embodiment" is: Active principle involved in formation of chain of particular instances – and it gains its identity and essence through its particular route.

What is universal in – compared[5] to – the particular instances – and gets its individual identity from being that particular route. Active principle of formation as defined in particularity by its definiteness of Achievement. – To be understood by reference to:

(1) The systematic relationship to the general scheme, by which niche <u>systematically</u> related to that scheme, is provided.

(2) Particularity arising from <u>past occasion</u> and their --- ⟨blank⟩

(3) The character of the individual embodiment of creative energy.

---

1. The Radcliffe version of this lecture, from notes taken by L. R. Heath earlier the same day, begins somewhere on |88| of her notes, page 514.
2. In his diary Bell recorded: 'Lectures, etc. as usual. In P.M. marked some Phil. A. papers.'
3. What 'principle' Whitehead has in mind is uncertain, but it might be a sock so often darned that it is questionable whether it is the same sock.
4. This is one of a number of possible expansions of abbreviation 'mom^{y}'.
5. The abbreviation in this instance is simply 'com.'

But this creative energy is somewhat arbitrary, in this realisation in the creature, as to how you assign the creative energy its character. There comes in question as to whether character of creative energy is not entirely determined by the antecedent potentiality. Whitehead thinks not, himself.

The character "the realised standpoint occasion which is the creature"

The reiterated unity which maintains itself within successive diversity of potentialities issues from the --- ?

What is stable is different under short-time and under long-time view. Thereby you get different determinations of "Experient Self". Most concrete Experient Self comes when you take very short time. – Concreteness of immediate experience. A certain stability in what is immediate – You've got more content there. Most concrete is just what is common to just past and immediate standpoint occasion. There you get the "immediate" Experient Self – the Experient Self of the moment.

Other extreme:– Same organism from birth to death. "Organic Experient Self" arises from life history of the dominant organic scheme whose identity characterizes succession of all antecedent standpoint occasions. Life history bound together by a thread of Self-preservation. You don't get a Self unless you get a dominance of --- ⟨blank⟩

[Mustn't be a chance permanence due to Environment. – There you lack the organic nature].

Organic Experient Self is the schematic unity of Organism. Immediate Experient Self is more concrete but Organic Experient Self is the complete embodiment of a limited unity of creative purpose; and may therefore be more significant as instance of general translation of purpose into actuality. – The "process of realisation".

When it arises to be a very major affair you get realisation of cognitive Experience. This, with its images, is in some sense return of Actuality into underlying creative activity. This underlying relationship of "taking account of" – "envisagement".

Partiality of creative energy [it's not impartial with reference to World of Ideas. Definite relation to past] becomes a fresh realized fact. Cognisant occasion, therefore – the Image – is the emergence into Actuality of how the Standpoint occasion is modifying (conditioning) the Actuality. This Conditioning is then emerging from factor into fact. This is occasion of Knowing. And this occasion is what mediates between standpoint occasion and what is Known. This is function of the Image – mediation between Knower and Known. – It is not the bare fact but fact in process of being Known. And this is a new occasion.

Now further – Cognitive Experience is always a selection from complete Experience which is synthesized in the standpoint occasion. [Delphic oracle and difficulty of Knowing oneself].

|146|

Standpoint occasion involves relations with entire universe. Cognitive Experience only an Abstraction from this. Former there as providing only a systematic background of the Cognitive Experience.

This systematic relationship enters into relationship by things enjoyed in Cognitive Experience. Thus, partial knowledge is possible within a system of internally related ⟨blank⟩

Cognitive Experience always involves Abstraction. And Abstraction without distortion is thus possible, because what is needed is only the other relations in their systematic character and not in detail. The abstract character of Cognitive Experience enters into its very character in this way:– Complete Image is for itself as analyzable into component Images, and each presents itself as possible as a complete Image. There is only a historical reason why in each of us the component image isn't the complete Image. Thus the abstractive character of Cognitive Experience depends (1) on fact that relations are external as regards eternal objects in relation to concrete occasions (internal as respect to actual occasions) (2) fact that all eternal objects and actual emergent objects are in systematic relationship to each other. (3) that there are in actual occasions definite relationships between eternal objects, which only refer to these definite occasions.

Whitehead is here preparing for a correspondence theory of truth. The necessity that this systematic background and empirical facts that Spatio-temporal scheme provides is why Whitehead is reluctant to abandon systematic character of Space-Time. Whitehead very doubtful whether Einsteinian doesn't make Space-Time relationships unsystematic. There is a systematic relationship entering into character of all things. That seems to be this Spatio-temporal relationship. Every eternal object "pervades all Space-Time".

Another reason for reluctance to abandon this systematic character of Space-Time is that the immediately actual Space-Time of our experience is the Space-Time of the immediate present; and this Space-Time (as pointed out last time) is as an Experience antecedent to any experience of the intrinsic character of its specific occasions.

We see the Sun over there:– But that is the nature of 8 minutes ago as issuing into present. It is not the intrinsic character of the occasion we Know but this as illustrated by --- ? ? ⟨blank⟩
The intrinsic realities of the present are veiled for us (and of future).

Space-Time in which we locate things is the Space-Time as determined by succession of immediate occasions. We know it.[1] And If what it is depends on what is realized in successive moments of experience ---? ⟨blank⟩

We do know it – but qua a niche as illustrated in our own intrinsic reality. But our Knowledge of its own ∧intrinsic∧ reality is on a par with our Knowledge of Future. – Except that we do Know the past as emerging into it. Here: is possibility of Error. You get some abnormal passing of Past into Present. You may be deceived. There are always hypothetical elements in issuing of Past into Present.

Now further analysis of Cognitive Experience in itself.

1. This phrase is inserted from the left-hand margin.

Hocking's notes

May 2 - 1925

"The creature emerges by virtue of the niche that is there for it in the universe"

Cognitive experience is always a selection from the Whole experience.

Only requires the Whole in its systematic aspects -

Abstraction without distortion is possible because relations to the whole may omit detail.

Correspondence theory of truth -

Intrinsic realities of the present are veiled from us. We have only the intrinsic sun of eight minutes ago.

Lecture 76

Tuesday, 5 May 1925[1]

Bell's notes

|147|
Whitehead: 5. May. 1925[2]:–

Test of self-consistency of ∨an∨ ontology is in its possibility of giving a place to Epistemology. Whitehead objects to point of view that first step in Philosophy is an investigation into powers of human mind (How Knowledge is possible). You can't express yourself until you have a certain minimum of ontological doctrine. – I.e. a certain <u>general</u> description of things known (<u>perfectly</u> general). <u>That's</u> ultimately based on your apprehension of the immediate occasion. What you immediately find before you includes your recollection. Character of things known shared by you and the man on the trolley-car. Unless you have got that, you've got no terms to discuss question How knowledge is possible? at all. You start with hasty ontological doctrine uncriticized. Then unless careful you get an Epistemology based on this hasty Ontology, and <u>then</u> you get your Ontology again revised to suit this Epistemology!

E.g. you cannot help thinking that both Locke and Hume <u>assume</u> Mind and <u>Impressions</u>. All Cognitive Experience has form of cognitive apprehension of how Essence of an immediate occasion of realisation is modified by its relation to other Entities. To be modified by is to be partially constituted by.

Cognitive apprehension is therefore realisation of reflexive relationship between an Entity (i.e. an Immediate Occasion) and its own Essence. And this reflexive relation is ground for <u>Subjective</u> interpretation of Cognitive phenomenon. If you add to this description in terms of Subject-Predicate categories you get hopelessly entangled in Solipsist difficulty. Whitehead thinks only road out is by way of Internal Relations.

Direct apprehension is not of images in my mind or anything like that. Whitehead's language is abstract and difficult simply because putting in most general form what is most immediate and familiar to us.

The Experient Self requires or refers to or: <u>is</u> the embodiment of Creative Activity whose law of Being is self-perpetuation. The creature emerges by reason of the niche which there is for it in Universe, and is

---

1. It is not clear whether there is a Radcliffe version of this lecture, as the notes taken by L. R. Heath seem to indicate she had to miss this class; nonetheless, compare Bell's and Hocking's notes with what appears on |90| of her notes, page 515.
2. In his diary Bell recorded: 'Lecture etc. as usual.' He was then out for the day, but later: '... Home, & to work on Iredell's Ph.D. thesis.' This was the first of several examples of Bell working on doctoral theses, as he had done the previous year, including the thesis by Charles Hartshorne.

translating the niche qua Potentiality to niche qua Actuality. Here is where concept of Creative Energy is required. Whitehead applies to individual ^Embodiment^ form of this energy what Aristotle to his Form. – No case not involved in formation of a Chain of Particular Instances. Gains its individual identity from that particular Route of its ? ?

Defined by its definiteness of Achievement. What an extraordinarily odd fact definiteness of achievement is. Such an odd fact that Bradley could not believe in it. Vulgarity of Wonder is man who begins to wonder at immensity of heavens etc etc. The real wonder here is assumed and jumped – That is why so many people say they don't see what there is in Metaphysics. Metaphysics tries to set forth what is involved in it.

In this definiteness you can distinguish three roots or sources: –

(1) We discern it as arising from systematic relationship within a general scheme, whereby a niche systematically related to the Actual is provided.

(2) That in respect to this niche there is a potentiality arising from particular relationships to past occasions – (viewed as issuing into present niche).

(3) There is a definite character of the individual embodiment of creative energy.

This question of the Creative Character is somewhat arbitrary. – In making the two abstractions: Creature and Creative Activity, must remember that character of latter is realized in former.

Evidently concept of individuation of Creative Energy maintaining its essence through different occasions must be derived from identity of pattern --- ⟨blank⟩

Must be grasped by distinction between realized Standpoint occasion and its potential essence as defined by relation to other occasions outside itself.

Reiterated unity which maintains itself amidst diverse occasions issues from identity procuring thread of uniformity among ⟨blank⟩

This identity is permitted by potentialities derived by antecedent occasions and is achieved by the creative self-perpetuating energy.

Whitehead looks on general ontological situation as being the March of Events, and this has the general character of being progressive realisation of particular ^matters^ of fact from infinite possibilities.

|148|

In expressing character of particular Matter of Fact there is mistake in ignoring its relation to perfectly general welter of Possibility. When we try to analyze what we find. (seven headings and three sub-headings):

(1) Activity of Determinism – Has got clouded by Scientific idea of matter as passive Stuff which "of course" endures. Inertia of matter – only later that field of force recognized. There wasn't assumption that these substances were eternal, though. But the miracle of the "going-on" ^and of the determinate going-on^ that is inherent in Process. You can't go on into Nonentity.

(2) There is the Actual– the actually determined Matters of Fact – Caesar did cross the Rubicon.

(3) There is the general impartial implication of Possibilities in a determinate System of Relationships which also implicate matters of fact. – Not the sensible possibilities only, but any of them. Aladdin might be here standing on his head. Must not put aside Possibility as Poor Relative to be shoved aside seems to Whitehead to be cloaking what is is very Essence of things.

(4) Actual things exhibit themselves as determinate selection from these possibilities.

(5) The actual as preferentially conditioning possibilities. Conditions germinate these possibilities by Exclusion. Future, because already included by limitation there. – The Future when realized has already a realized Past. The Future can only be as patient of its own History. The Actual is the History of the Future.

(6) The plurality of particular Matters of Fact. These are correlated by the Self-perception of particular Entities.

(7) That this self-perpetuation of the particular is to be expressed as:–

(a.) A particular embodiment of an activity of self-perpetuation. There is that purpose of Self-perpetuation [Otherwise why doesn't that Radiator turn into a Dinosaur?]

(b.) There is the emergence of a determinate Character whose inclusion of the Future conditions the Future so as to condition Future with a patience for reiteration of that Character. ([1]Whitehead simply expressing that idea of a Substance as Causing itself. "*Causa Sui*")

(c.) The Cause permits the Effect and Excludes what is incompatible with fact that it itself has happened. Thus:– Alternative between Freedom or Determinism.

Does this patience, and accumulation of Conditions – Does that completely determine what is to be the realisation in that niche. There you have alternative between Freedom or Determinism.

Either we divest our activity of everything except activity – so that there is activity in bringing to birth a non-ambiguous opportunity or ? ? ⟨blank⟩

We must ask Physicists, Psychologists, Theologians etc to make that up between them. So far as we observe, the Molecule seems to behave as though it emerged from a non-ambiguous opportunity of potentiality. On other hand, great divergence of opinion re ourselves. But Ontology seems to Whitehead to leave it open. Personally, Whitehead believes that this law of Self-perpetuation seems to be natural to ascribe it to very inmost character of the activity of realisation and not to any casual conditions. This law of Self-perpetuation – Everybody is assuming it. E.g. Hume assumes it:– He's got his Association. Says: Minds will go on associating as before. Dome of St. Paul's in danger:– Assumption is that Dome goes on or that it is the Dome that is falling. – Elements will have

1. Opening parenthesis provided.

Self-perpetuation. You will never have Dome suddenly gone and no issue from it.

Now to another point:– Cognitive Apprehension is return of Actuality into the active creative purpose. Underlying relationship (is taking account of) – That what is Emergent is being conditioned by the Potentialities, thereby emerges itself as a realized fact.

Relationship of ∧Past∧ Actual to the immediate Creation – the immediately realized Creature – here emerges into an additional fact within the realized process. Selective Partiality of ⟨blank⟩ becomes a fresh realized fact. – Emerges from a factor into a fact and this we call Knowledge – or rather:– it is an occasion of Knowing. And this is what mediates between the permanent ~~things~~ ∧subject∧ and the things known. My private psychological field viewed as immediate standpoint occasion is what it is because you are over there, known by me. That is why my knowledge goes beyond myself.

Now come back to Demos's question.[1] How Consciousness can be translucent in view of doctrine of internal relations. [Then to consider how process of Actuality differentiates itself into Past, Present and Future] (Must be this differentiation or it is all eternal, and Consciousness must timelessly turn over leaves of already printed book.)[2] Whitehead's objection to this simple doctrine is that it is wholly against immediate apprehension.

[3]If we are to make Metaphysics agree with direct apprehension, we have got to describe ontological occasion leaving room for freedom and leaving to the individual Sciences to tell what Freedom there is there if any. Logical Incompatibility is first hint you get to solve wonder of the ∧determinateness of∧ particular matter of fact. Other (Spatio-Temporal) limitations The Actual doesn't issue into any Future – but into one related to the Actual. Mutual relations of Future occasion to each other is indirect relations to same past facts which condition them. The future is in the present.... What has a history is conversion of Potentiality into Actuality.

Hocking's notes

May 5. – 1.

Back to content of experience

1. An ontology that finds us room for Knowledge stands self-condemned. But it is not true that the first step in philosophy is an inquiry how knowledge is possible.

---

1. Demos' question was first mentioned on 30 April in lecture 74. See footnote 3 on page 351 of that lecture.
2. Closing parenthesis provided.
3. Bell's notes from here to the end of lecture are written up left-hand margin.

2. All cognitive experience has form immediate occasion modified by its relations with other entities. Modified by is partly constituted by.

3. We apprehend not images in our mind but entities there. Our expressions only a restatement in general form of simple experience.

4. The experient self embodies creative activity whose law is self-perpetuation. A creature emerges because of the niche there is for it in the universe. Individual embodiment is the active principle involved in the formation of the chain of particular instances.

"Definiteness of achievement is such an odd fact. Such an odd fact that Bradley couldn't believe in it, as over against an Absoluteness which is everything at once. The supreme wonder that a particular individual should be sitting here.

5. 3 Antecedent sources of Definiteness

a. Systematic relations in a general scheme whereby a niche is provided

b. In respect to this niche there is a potentiality arising from

c. A definite character of the individual embodiment of creative energy.

This question of creative character is somewhat arbitrary.

Lecture 77

Thursday, 7 May 1925[1]

Bell's notes

|149|
Whitehead:– 7. May 1925[2] –

[late][3] --- There is the alternative theory according to which you can't know anything at all, because you would have to Know everything – then you get theory of "Appearance and Reality".

Another difficulty:– Must describe Knowledge so that it can be wrong. For God metaphysics is awfully easy. Distrust any trenchant simplification because difficulties you face are intrinsically complex.

Having regard to history of this question in Philosophy --- ⟨blank⟩

First thing:– Whether Error merely begins with conscious experience – Is it merely that when you begin to think you think wrongly, or do you think wrongly because you are a wrong individual. Natural thing [Bergson's line – Whitehead dissents] is to think of animals going along beautifully – thus exalting intuition and throwing all error into intellect. In a sense that is true. Effect of Error when consciously believed is greatly enhanced. Appalling effects when you first had conscious thoughts about Religion – All errors inherent in thought is enormously enhanced – and they sacrificed children to Moloch. But that is a different matter – importance of error. But question of Error itself is quite different. When you look on animals it is not true that they avoid Error. On contrary their actions are most ill-judged. In long run Consciousness is a mode of enabling you to avoid error. Köhler judging of intelligence, etc. – More thought the more error [?]. Bergson's glorification of non-thoughtful side Whitehead has great suspicion of. We think wrongly because we are already erroneous. First result of thought then is to magnify effects of error. In long run consciousness is a method of avoiding Error. But action of thought in doing this is a long business. When you haven't time you suppress thought (In war, Shipwreck, etc.).

Spinoza contra

We have got really to make an analysis of the complete Experience so that it can be abstracted without distortion – ∧intrinsically∧ separable into parts (capable of analysis) ∧and also so that it∧ includes some elements

1. The Radcliffe version of this lecture, from notes taken by L. R. Heath earlier the same day, begins on |91| of her notes, page 515.
2. In his diary for this day Bell recorded: 'Lectures as usual. Iredell's thesis.' And then, later: 'In evening to the Hocking – Pound Seminar on Philos. of Law. On return finished Iredell's Thesis, marked Sec. B.4 Phil. A Papers & prepared for sections.'
3. This is of course Bell's own acknowledgement that he arrived late.

which are erroneous, in some sense. Whitehead trying to do this – not altogether satisfied with his analysis though.

All this without any reference to thought.

(1.) If able to separate, you must have Experience in itself as in some sense Experience of External Relations, otherwise everything dependent on everything else. Where does external relationship come in? – Comes in over that principle that the present is only known to us as what it is in itself via its systematic relationships.

Now schema not in diagram but words:– (1) we are dealing with Experience – what present occasion A is in itself. Further – Any event is in itself a mere triviality of Experience unless there is embodied in it an enduring object. – An enduring pattern which is being self-perpetuated. A is present state of enduring emergent object ā. ā means that what A is in itself embodies a pattern of experience $\breve{\alpha}$ which is qua pattern a universal. ā is $\breve{\alpha}$ as actualized in a sequence.

What this experience is in itself is:– $\{A_3, A_2, A_1, A\}$ $\breve{\alpha} \rightarrow$ ā   ā as an emergent entity is that sequence.   "is life history of"

How are we to analyze up the essence of A. Eight different types of things we have got to put down:

Essence of A is the analysis of the experience which is A =

["Conditionally realized" – Term to be explained. Aspect of y in x – that x is realization of whole universe under limitations expressed by relations of x to other things. Essential difference of Past and Future is that y as in x is conditionally realized in x. It is in x as it is, is already conditioned ["*a priori* Possibilities"] y in its realization has to realize itself subject to condition that it is already presaged in x.]

= {A is conditionally realized in $(A_3, A_2, A_1)_{\breve{\alpha}}$ [Self inheritance of life history of object]}

How A is essence derives from past of ā.

Various Sources of Modification of the Essence

+ {A as conditionally realized in $(X_3, X_2, X_1)_\xi$ [Any other enduring objects in past] etc.}

Also in all the Events whether in Enduring Objects or not.

But their effects really all included in the X_3, X_2, X_1

+ {A as conditionally realized in (U,V, W)}   [The detached antecedent events.]

+ {A as standpoint for the impression of Eternal Objects, β, e.g. in the sequence of occasions $(B_1, B_2,\ldots)$ in which B is always embodied in the present.

And of γ in $(C_1, C_2\ldots)$ in the present etc.}   [That is what is called "appearance".

Red is there in certain shape, for me, here, etc.]

|150|

[Going from last one to next is where the erroneous individual comes in.
Looking at Moon, e.g. – I not only see patch of light; I see a moon.
I see something there. But time of passage of light is 1⅓ sec.
Plenty of time for moon to have vanished. When I say there is something there, I'm associating endurance of patch of colour with endurance of an emergent object of type ā in shape of moon. m̄. We've got to run idea of "Selfhood" qua idea is eternal object just as much as anything else. If I'm to run point of view which Whitehead is attempting, I'm bound to bring in notion of ingression of Selfhood in experience of A.] therefore –[1]

+ { A as standpoint for ingression of selfhood in occasion $(B_1, B_2,...)_\beta$ in the present }....}

[Descartes' extraordinary good sense in not sticking to case where he was right. But case where he was wrong. Hence his Subjectivism – with "*Inspectio*" connecting Substance and Subject.]

So here there may be What B_1, B_2 are in themselves is in A only systematically not constructively. Relationship of "Ingression" is where you pass from internal to external relations. Relation of Ingression is Internal as far as A is concerned, but external so far as b is concerned. And so far as B_1, B_2 , B_3 etc. concerned.

+ {A as conditionally realising the future via system}
[Future issuing from its aspect in A]

[Partly general, partly particular]

+ {A as conditionally realised from the past via system}

+ {A as with preferential ~~envisag~~ abstractive envisagement of eternal objects.

[Point here is: – I'm looking at whole process in general, you have to conceive it as having all possibility before it. Relationship between what is already achieved and what is possible, is mediated by Space-Time relationship. What is actual stands as a selection from amid all possibility and what its Future is. Abstractedly there is possibility of anything anywhere. But Actuality gives you definite matter of fact:– definite relationships of any immediate occasion to --- This is what Whitehead calls "preferential envisagement". Calls it "abstract envisagement" because it's not the envisagement as exemplified but as possibles, pervading Space-Time is also preferential

(Taking account of things.)

[Lastly, including nearly everything we are interested in:– Psychologists at least]

+ {A as intrinsically emotional and purposive [what A is for itself] etc.}

1. These notes with which Bell begins his |150| are all written in this unusual configuration and all tilted up on right side. Also, it begins with a bracket, and it is not clear if end bracket belongs at end of first sentence or at end of this whole section.

Some of this emerges as immanent [?] in conscious occasion. Selection of this in the individual occasion. The Erroneous individual turns up at 4 and 5. He has a Knowledge of himself in past. Only it's vague and not definite.

At 4, we have what Descartes called "Apprehending with his senses"

At 5 we have what Descartes called "*Inspectio*"

Then you begin to correlate (2) and (5)

There you get Error in the Crude Experience; and that is what emerges as Error in Thought.

Hocking's notes

May 7 –1

Idea of the Self. last time

Today. Relation to Abstraction

Don't want to run into solipsism –

Have to describe Knowledge so it can be wrong.

1. Does error begin with conscious experience –

Do you think wrongly because you are an erroneous individual – Bergson rather exalts intuition, & tends to assume that error begins with thought. When man begins to think, he thinks wrongly.

With self-consciousness error grows in importance. Children are sacrificed to Moloch

But the real question is whether instinct is inerrant. The actions of animals are most ill-judged. And in the long run, consciousness is a mode of enabling you to avoid error.

Kohler was judging the presence of thought by the ability of apes to overcome an obstacle.[1]

Whitehead believes we think wrongly is because we are already erroneous – then the error is magnified in importance.

2. We must therefore analyze the whole of experience without reference to thought.

First you must have experience as of external relations, otherwise everything depends on everything else.

Where does the external relationship come in? –

Over that principle that the present is only known to us through its systematic relationships.

Any event in itself is a mere triviality unless there is embodied in it enduring pattern, self-inheritance

1. Wolfgang Köhler (1887–1967), well known as a proponent of Gestalt theory (see the quotation from Bell's diary at the outset of lecture 78), had done experimental work reported in his book *Mentality of Apes* (1917).

A, ā, $\bar{\alpha}$ {A_3, A_2, A_1, A} $\bar{\alpha} \equiv$ ā
A = present experience in itself
ā ≡ enduring entity
$\bar{\alpha}$ ≡ universal pattern
the unity running through As in series
How are we to analyze up the essence of A?
8 different things.

May 7 –2

more or less adept
Essence of A. (Various Sources of modification Needed)

1. A as conditionally realized in the sequence $A_3A_2A_1$ etc $\bar{\alpha}$
+ A as conditionally realized in $X_3X_2X_1$ any other enduring object
+ A as conditionally realized in $U_1V_1W_1$ – antecedent events
+ A as standpoint for ingression of eternal objects, β
in sequence of occasions $\{B_1B_2\text{-}\}_\beta$ in present
and of $(C_1C_2\text{–})_\gamma$ in present, etc.
"appearance" – Red is there for me here, etc
+ (and here is where error comes in)

Moon could have vanished in the 1⅓ seconds for light to come over. I do say I see the moon, not merely the patch of color – an enduring entity. The idea of self, qua idea, is enduring object as much as any –

I am bound as running this point of view to bringing in the ingression of selfhood

Error sense *inspectio*

+ A as standpoint for ingression of selfhood in occasions $(B_1B_2\quad)_\beta$ in present
+ A as conditionally realizing the future via system
+ A as conditionally realized from the past via system
+ A as with preferential ~~env~~ abstract envisagement of eternal objects
+ A as intrinsically emotional & purposive – (character of the underlying creative activity)

Lecture 78

Saturday, 9 May 1925[1]

Bell's notes

|151|
Whitehead: 9. May: 1925[2]:-

Actual world (not totality of all Experience) is a multiplicity of all occasions. Each is something for itself. – A centre of Experience. In more Kantian language:- Each occasion in itself is a synthesis of Experience. Analyzable as detached items – Each item representing how something which is not that occasion is in that occasion via the modification introduced by its relation to that occasion. This is internal so far as that occasion is concerned. So for each occasion we have to distinguish its value for itself and its --- ? ? ⟨blank⟩

Another point:- There are grades of actuality. Fuller actuality is attained by an occasion which inherits a pattern of intrinsic coherence from an antecedent historical route. Such inheritance involves two types of unity:

(1) A unity of achieved experience

(2) A unity of creative purpose of Self-perpetuation

Grades of actuality according to grades of Selfhood which are perpetuated.

In schema:- A is an actual occasion inheriting an experienced pattern $\bar{\alpha}$ from antecedent historical route ... A_1, A_2 ... as actualising $\bar{\alpha}$ is history of the Self which we will call $\bar{a}$.

Two analogous but distinct ideas here. You can't really tear apart world of Eternal Objects from world of Actualities. The actual enduring self $\bar{a}$, as actual, is paralleled in Eternal Realm by a corresponding ideal pattern: $\bar{\alpha}$.

As typical of an indefinite number of others – another self $\bar{x}$ – Perpetuation of pattern ξ but historical route .... X_3, X_2, X_1. ... X_3, X_2, X_1.... as realizing $\bar{\xi}$ is $\bar{x}$.

Now take (U, ...) as meaning empty events in Past, if there are any (V1 ...) in future. Not, i.e., empty when they arrive – but not yet determined.

$(B_3, B_2, B_1...)_\beta$ as ∧historical route of∧ Events contemporary with A's which is route of situations of some eternal object β which is an aspect for A etc. for other objects.

Point:- Whitehead's mere trifle of 10 headings last day. But could go on to 999. There are the interconnections of the interconnections etc.

1. For reasons unknown, Hocking provided no notes for this lecture, nor for the next two Saturdays. Neither (as in the past) did he arrange for someone else to take notes. The Radcliffe version of this lecture, from notes taken by L. R. Heath earlier the same day, begins somewhere on |93| of her notes, page 516.
2. In his diary for this day, Bell recorded: 'Lectures.... Dinner in event at Harvard Club. Later to At Home at McDougall's – Kohler there & talked on Gestalttheorie.'

All things related to all others. But question is type of thing in result of analysis ⟨blank⟩

"Analyze essence of Actual Occasion A is Synthesis Experience of A (which is A) as in antecedent Analysis ∨[crude Experience] ∨

= (i) {Aspect of $(A_3, A_2, A_1)_{\bar{\alpha}}$ as from A, and as conditionally realizing in A}
[Basis of natural determination of things][1]

+ (ii) {Aspect of $(X_3, X_2, X_1)_{\bar{\xi}}$ as from A, and as conditionally realizing in A}

+ etc.

[In principle (ii) is same as (i). Difference extraordinarily greater intimacy of (i). – Identity of pattern and continuous historical route. My own past in myself now is in principle same as you for yourselves as in me. Whitehead entirely denies the private world view altogether. But such an overwhelming difference in concrete manifestation as to bring a difference in quality. But your past is for me just as mine is.][2]

+ (iii) {A as the standpoint of the Eternal object β as illustrating route $(B_1, B_2 ...)_\beta$ i.e. the ingression of β in the duration as from the standpoint A

+ ... [β as one for many].

[This is a very complex set of relations]

[Blank fact of "green" as there for me now. Brings before me that occasion there as enduring through my presents – as potential succession of occasions]

+ (iv) {A as standpoint for ingression of [enduring] Self-hood in sequence $(B_1, B_2 ...)_\beta$ etc.} [Double reference to a standpoint and an occasion. – Spherical shape there for one here etc]

+ (v) {Aspect of past occasions, e.g. U... as systematically related actual occasions conditionally realizing A}

[The general conception of Past as in definite Spatio-temporal relation to oneself and as in itself what it is conditionally laying down its future – but not determining them

+ (vi) {aspect of future occasions ... etc. and as conditionally realized in A}

[Future is realized as from present, whereas Past realized as A being from it].

[Knowledge having another character – a selection from Experience realized in another way. So I've got to find the "basis of thought" in Experience, otherwise whole Theory of Knowledge goes to pieces.][3]

+ (vii) {A as standpoint for ingression of preferential abstract envisagement...

[Possibility is definitely something – in some sense actual. – Not a mere pendant

---

1. The bracketed comment is added by Bell at angle in the right-hand margin.
2. Closing bracket provided.
3. Closing bracket provided.

|152|
to matters of fact – but <u>in</u> them. "Eternal objects <u>pervade</u> <u>all</u> Space-time". – In any <u>one</u> occasion not simply <u>impartially</u> there– but there is a <u>selection</u> of this taking account of.]

So continuing the double bracket:–

i.e. eternal objects (selected) as in (selected)(possible) Spatio-temporal rolationship)

+ (viii) {A as emotional, and feeling, and valuing, purposing. I.e. – A as realizing a character of individualised creative activity.}

[<u>This</u> I can only find in something that is actual]

+ (ix) {Aspects of correlation of (i), (ii), and (iv) – I.e. the past as <u>issuing</u> into the Present}

+ (x) {Aspects of other correlations between (i)....(ix)}

[General waste paper basket for remainder].

Point – Any neat schema of this sort is entirely and absolutely foolish. Point is that language is made only casually in contemplation of Metaphysics. Things which people have repeated long enough don't <u>appear</u> silly. [Like talking of "Ideas in the Mind" etc....] of course phraseology wouldn't have been used if there weren't <u>something</u> in it. Your language will <u>always</u> be <u>metaphorical</u> here. "Aspect" "Standpoint: etc, are <u>here</u> <u>Metaphors</u>.

In (i.) above, where you have "A_3" it is <u>all</u> the schema as for A_3 melted down into an aspect. What is <u>in</u> one at present includes one's immediate past in Memory etc. Separation between case in (i) and in (ii) is not quite as definite in one's mind as might consider; because any part of your body (a molecule in your toes e.g.) is a thing on its own, <u>only</u> so intrinsically connected with pattern of total that it is of a superior <u>grade</u> of <u>intimacy</u>. – Involving your <u>direct</u> Knowledge of your body as a thing in itself.

Again:– If you use an instrument – a pen – you feel in the pen point:– The type of intimacy you have for your body can be extended (very <u>vaguely</u>) <u>beyond</u> it. That is why Whitehead maintained no essential difference between (i) and (ii).

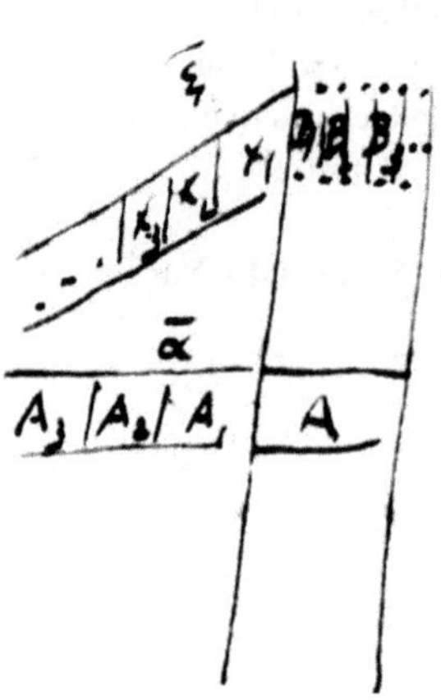

But X_3, X_2, X_1 etc.... <u>also</u> involve <u>antecedent</u> <u>cognitive</u> occasions involved in A_3, A_2, A_1. So there are the cognitive occasions <u>as in</u> crude experience!

If you construe the X's as cognitive occasions <u>or</u> as crude experience you have --- in --- exactly as you have A_3, A_2, A_1 in A.

Whitehead would appeal here to obvious experience:– We have a vague – very vague but quite decisive – experience of a tone of feeling or mentality as distinct from specific details. Whitehead entirely disbelieves reigning and orthodox view that our construction of other people as mutual and feeling is entirely derived from extraordinarily acute inference from physical analogy to myself.

Infant and mother obviously sharing in Emotions. But that infant is arguing from analogy!!!

What ingression of Eternal Objects does do is to present the occasion to me as definite when I look across to Moon, intervening space not presented to me as definitely marked out occasions:– Whereas --- ⟨blank⟩

Aspect of an "illustrated" occasion ⟨blank⟩

Illustration of X, in A, is immediately in A, and that enhances the aspect of what X_1 is in itself. X_1 is illustrated in A, and as immediately antecedent for A, has then in A, a more efficient aspect of what it is in itself, and there, Whitehead maintains, that Knowledge of what A is in itself.

– I look out and see President's house and have realization of that as something in itself, and not merely for me. That is basis of all realism.

There is in one's own essence an aspect of what the other thing is in itself. This is vague and always theoretically behind the illustration. The immediate illustration gives me apprehension of object as being something in itself [I see something 10 feet behind the mirror – and that is something for itself. This is what Whitehead's talking about in (iv). The β of events behind mirror for me is a very trivial fact.][1]

Error[2] comes in where something definite in X_3, X_2, X_1 issues – passes over in something indefinite. B_1, B_2, B_3 You ascribe something important in B_1, B_2, B_3 – apprehend a self there where it isn't. Here is where you've an erroneous individual. General apprehension of past as strains of creative purpose issuing into present is very insistent but vague in detail. Thus you are generally right and occasionally wrong. Past may have for you aspect of an antecedent self as issuing into B_1, B_2, B_3.

1. Closing bracket provided.
2. From here to the end of the lecture, Bell's notes are written up the left-hand margin, presumably to avoid beginning a new page.

Lecture 79

Tuesday, 12 May 1925[1]

Bell's notes

|153|
Whitehead: 12. May. 1925[2]:–

Couple of lectures not immediately following up details of preceding lecture

(1) Rather abstract, on theory of Mathematics as illustrated by its history

(2) Chat on recent development of Michelson-Morley work.

But at present hammering away at point: How, if you start out with a through-going Realism and hold translucency of Cognition (Knowing doesn't alter fact) you have got to make fact as distinct from Knowing [the image which arises out of fact as selection from it] – what is Known as a synthesis of all the Knowable things in that occasion. – So "crude" experience (the standpoint-occasion) must be explained in terms which include everything which can then be Known in image. Synthesis of Experience which may be and will be Experience of other Cognizant moments. But its own Knowledge of itself is also another Cognizant Experience. Here we come to Thought and the "Private Worlds".

In a sense essential for this view that there should be no private worlds – Everything is synthesized in every actual occasion. You can find everything in it. That we have to apply to thoughts. There is in every standpoint occasion its relatedness to whole realm of "Eternal Objects". That relatedness is in nature of case, whether Known or not (whether there is a realized image of it or not) – and that is not a purely abstract relatedness but a "preferential" one. Relation to some eternal objects different from that to others – and image of this is "the thought".

Usual to look upon "Thought" as in nature of things private. But Whitehead doesn't see how this can work. God Knowing all our thoughts, e.g. (Whitehead objects to hauling God in to perform metaphysical impossibilities, but –).

---

1. It is not clear whether there is a Radcliffe version of this lecture, as the notes taken by L. R. Heath seem to indicate she had to miss this class; nonetheless, compare Bell's and Hocking's notes with what appears at the end of |94| of her notes, page 517.
2. In his diary for this day, Bell recorded: 'Lectures etc. as usual. Finished K's thesis at 3.55. Took it over to Perry at library.... Got to work on Buchanan's Thesis.' On the previous Sunday, he had noted: 'Did some careful reading on Krikorian's thesis.... Then around for a few minutes to the Whitehead's. Later read further in Krikorian & prepared to-morrow's lecture.' And then on the Monday in between: 'Lecture, reading K's thesis, getting up lecture etc. all day & until late at night.... Department meeting 4.30–6.30, theses etc.'

If you are to take certain activity in simply creating the thoughts – a sort of synthetic activity – what you make in your minds. But if to allow an element of thought in perception, difficult to get away from, – Whitehead doesn't see how you would get rid of Kantian conclusion that phenomenal world is outcome of synthesis of Categories essentially ? ?

? ? in connection with a given element. If feel that this leads to a Subjectivism which is not accordant with your immediate experience, you have (to avoid this) got to put the Synthesis into the non-cognitive part of the occasion. – An active emergence of an occasion of Experience, synthesizing its relations to other things. – And this must enter into everything else – and that is your public world. But privacy is an infection, very difficult to escape. Seems, you only have to have bit of it, and you land in theories that don't allow what we all believe.

We have to allow for Error. Experience is what is. Then if Cognition is simply a selection from what is, and therefore you can't Know wrongly! A difficulty here, therefore – not insuperable but must see how it is a difficulty here. You've got to retrace to Descartes. I experience what I do experience. So you have got to show how, in world of experience, there are discordances which are those we term as Error. To explore what you mean by Error is to explain what you mean by Error. But to explain Error is more important – Many people in accounting for Truth get theories which would render Error unintelligible.

To do this:– You must bring in grades of Experience in some way or other; and our old friend: Primary and Secondary Qualities, must turn up somehow. But: are you going to shield the secondary from External world, or are they to be all public?

That is reason why Whitehead emphasized that our Experience now of ourselves in the Past is in principle same as our experience of any other Self in the past. – More vivid and immediate etc. (Different quality therein).

There's got to be a general publicity of Past in every present occasion of concrete Experience. If you haven't got this you'll find yourself sooner or later in great difficulties.

(2)[1] You have got to have some facts (in order that partial Knowledge shall be possible) which can be isolated in some way. That is a limitation to the notion of internal relations. Whitehead thinks that limitation comes in in two ways: – (a.) In particular in relations of present occasions to each other and of future occasions ~~related~~ to them. Idea of Internal Relations is intrinsically bound up with Causation on one hand and "systematic" nature of Space-Time on other. Causality relates a detail of Past with a detail of Present; and spatio-temporal relationship is the general internal relationship which relates every existent, actual or eternal to one another. [As Whitehead said: Eternal Objects pervade all Space-Time] – [Whitehead comes down on idea of relational essence].

1. This is (2) and Bell seems to have missed (1).

Expressed this last day by stilted and formal Analysis of Experience under 10 headings.

|154|

But we can get a X-classification which is quite necessary. Another set of headings equally required

If there were not different Universes you couldn't say: it is this sort of thing.

A is Standpoint Occasion – Its analyzed occasion is[1]

(Imaginable occasion) It's something in itself first. Then, analyzing its essence, there are items you'll find which can call "Memory Experience". Simplest form of ? ? ⟨blank⟩

A is Standpoint occasion –

(1) Memory experience [Past as in A].

Simplest type of relation to past:– How A itself includes the past. If there is no intrusion of Past, you can not admit of a direct Energy. On other hand, take my direct memories. I remember myself as in Room B. – It is the memory of another occasion. But experience of that now is ? ? ? ⟨blank⟩

Vivid relations of certain attributes of him in past. But that is not in essential relations any more than my Present as general outcome of past. No difference in principle.

[Primary] (2) Anticipating Experience [Future as in A.] This is very different on different sides. Very detailed in some respects, vague in others.

(3) Ingression Experience I. [Present as in A.]

(4) There is your abstract Experience ["Eternal Objects"]

Drowsy

There have to have some form of Emotional Experience which is some form of A, so to speak, as in itself.

(5) Emotional Sequence [A as in itself]

You can't find any role[2] for Error in (1)-(5) taken simply.

You are bound to come down to some form of secondary Experience – not private – which is of nature of Ingression.

Ingression is a complicated relationship in which Eternal Object is complicatedly related to a Standpoint and a situation [Also as an outcome]

(6) Secondary Experience [Ingression. Eternal Objects].

Outcome of (1)–(5) as re-envisagement ∨more abstract∨ of relationships of Eternal Objects in relation to the primary Experience, which is represented by (1)-(5)

(6)[3] Secondary Experience [Ingression: more abstract/ more definite [?] Eternal Objects: to primary Experience not to events as for A]

In abstract all Eternal Objects are in relation. This abstract Experience then takes a preferential Character:– to be taken as Ingression though secondary Experience.

1. It is hard to tell from way the notes are arranged on this page whether this carries on to the next line or whether Bell is leaving a blank.
2. This term could also be 'rule'.
3. It is not clear whether Whitehead meant to repeat (6) – the two are closely related in this scheme – or whether he simply misnumbered the point.

The ingression from secondary Experience also takes the form of ? ?

Last day's list too special:– Selfhood – Ingression of B there, for A

A is realisation of occasion with its transition.

But that also gives you an ingression of a permanent self for A as in B. – Which is not what B is in itself – but as ingression of a self in B as for you [A].

You have an ingression of Eternal Objects as in other occasions, arising out of your primary Experience! This ingression is particularly definite. It includes past and future.

In respect to secondary Experience you have a wider ingression than in respect to primary one.

Hume begins with "Impressions" and "faint copies" of them. He immediately defines immediate experience into a primary and secondary one. Hume brings in word "Copy" here. Hume says fainter. Whitehead doesn't think it always is. – Faintness is very weak just on this difference. Only point that is quite different from Hume – and from Descartes – is that Whitehead has put (4) – what is in nature of thought. – in primary Experience. Also because it is essential to Universe as ~~conxxg~~[1]

(4) is spatio-temporal relationship, as bringing all Eternal Objects into A, which then the Ingression in (6) is an Ingression of Eternal Objects via Past not Past as it is in itself but as it is in reference to total Experience of A. Eternal Objects = X_1, X_2, X_3 as they are in A. Here is an opportunity for accordance or discordance. Further: (6) not comparable with ?   You will get truth in as you get your Secondary Experience in consistency with the primary ones.

Finally[2]:– (7) Selective Experience [Character of the Creative Self]

That presupposes all the rest [especially (1)]. Cognitive Occasion would then be not only Knowing wrongly but Knowing something which includes intrinsic discordance. Intrinsically accordant or discordant individuals.

Hocking's notes

May 12. – 1

Thought & Private Worlds.

Essential to Realism that there be no private world.

Everything there is, is entering into the character of every thing.

Natural to look on thought as in the nature of the case private. Yet they say God knows all our thoughts. How can he, if they are in the nature of the case private? Object to God being brought in to do impossibilities.

1. It is not clear what this word is, nor whether Bell intended to strike it out or to underline it.
2. These last lines are all written up the left-hand margin.

What if the mind shapes in synthesizing? "If you run any privacy, you will find it an infection difficult to get rid of"

To explain what you mean by error is another way of explaining what you mean by truth – But it is more important to attend to the explaining error. Only one ~~say~~ way to do that – Grades of Experience. And your old friend primary & secondary must turn up in some way.

Hold that our experience of ourselves in the past is the same in principle as of anything else in the past General publicity of the past in every present occasion of concrete experience. If you don't have that publicity, you will sooner or later find yourself in great difficulty.

Secondly, you have got to have facts that can in some way be isolated from internal relations.

Necessary to get a cross-classification of experience to bring out the meaning of experience.

A ≡ standpoint occasion.

[Primary]
- Memory Experience [Past as in A]
- Anticipatory Experience [Future as in A]
- Ingression Experience [Present as in A]
- Abstract Experience [Eternal Objects as in A]
- Emotional Experience

No room for error above this line.

[A as in itself, its own value]

We are bound to come down to some form of "Not private!" →

Secondary Experience
[Ingression: more abstract/more definite]
An outcome of all the above as a reenvisagement, an ingression of eternal objects into events, – a relation of eternal objects to the primary experience.
Hume. You have your impression. Then you have faint copies of your impressions.
Copy = secondary –
Differs from Hume & Descartes in putting Abstract into the Primary Experience.

Self-hood

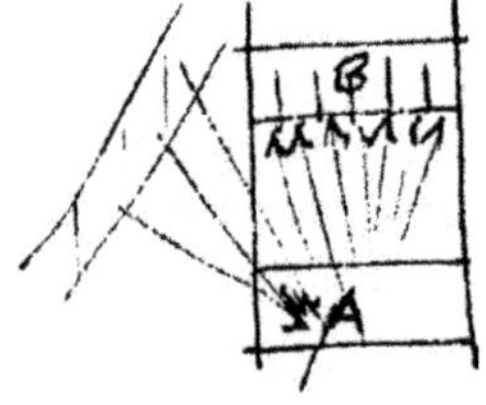

Lecture 80

Thursday, 14 May 1925[1]

Bell's notes

|155|
Whitehead: 14 May. 1925[2]:-

How details of a metaphysical system grow more incredible as you go on. Stop and confront yourself with your general principle. The difficulty in relation to that is: difficulty of expressing it otherwise than in details. But you are more certain of principle than of your handling of the details – and principle gets overlaid with the difficulties of the detail. [The mixed metaphors of the old parson – Gospel preached by my imperfect hands].

Principles of Natural Knowledge – obviously by someone immersed in mathematical physics. Getting relations of Time and Space straight. Equally obvious that Whitehead's principle is same with Morgan's *Emergent Evolution* and Alexander's *Space, Time, and Deity*.[3] Whitehead feels a little that Alexander has carried through a system with a general genius that is admirable, but dissatisfied with this:- (1) Alexander rather in a Muddle over early general muddle re Space and Time in early chapters. (2) In handling idea of Emergence he doesn't seem to Whitehead to give sufficient emphasis on the provision of the niche for Emergence. Take emergence of Cognitive Experience where you have got a sufficiently complicated state of affairs there is Cognisance. Then he proceeds to treat Cognitive Experience in common-sense way according to proper terms etc.

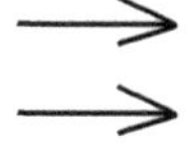

But Whitehead feels you ought to have a principle of Metaphysics that given one state of affairs that shows you in analysis that there is thereby provided a niche for emergence of something else. Primary object of Metaphysical Theory is not so much to describe emergent thing (that is task of a special science) but to demonstrate [not as logical proof but[4] as medical lecturer "demonstrates" something –] – to point out the "niche". By "niche" mean this: – Relationship between elements (in fact) which is a possibility for synthesis in a real togetherness. That distinction between a

1. The Radcliffe version of this lecture, from notes taken by L. R. Heath earlier the same day, begins on |95| of her notes, page 517.
2. In his diary for this day, Bell recorded: 'Lectures etc. as usual. Finished Buchanan's thesis. Got Phil. A ready etc.' And on the following day: 'Phil. A sections.... Dep't meeting 4.30–6.15 (Reports on Prelims, Theses, etc.).'
3. These two works are first mentioned, together, in lecture 13. Among more than a dozen other books, Cowy Lloyd Morgan (1852–1936) published *Emergent Evolution* in 1923. Samuel Alexander (1859–1938) published *Space, Time and Deity* in two volumes (1920) developed from his Gifford lectures of 1917 and 1918.
4. Bell has 'by', but from the context it seems that this should be 'but'.

relationship qua possibility and what emerges into real synthetic actuality is really a distinction of a fundamental character.

Suppose colors Red and Green and a being with Knowledge of Reds etc and other with Knowledge simply of Greens. or: One who had never experienced Redness in any synthetic relation with greenness. Concrete actual relation of contrast which arises for anybody experiencing redness there and greenness there (for me here) – it is not experience merely of greenness and redness but of greenness and redness as swearing at each other or harmonizing with each other. There is a specific actuality of the unity of the two in the same actual experience. There is emergence there of something beyond the two. What is emerged is a relationship.

What Metaphysics demonstrates: – When you've properly described how the Red is entering into experience and how the Green – you've left a niche for emergence of concrete entity:– the red and green in that relationship.

You can describe everything about it except what it is. [Strong point of Empiricism]. Can describe the niche. But the specific way in which they swear at one another will evade all description. This specific way in which they swear at one another might have occurred in an Experience in Chaldean times. Therefore that is an eternal object just as much as any other. Therefore there are eternal objects which require a niche provided by realisation of "lower grades" of Eternal Objects.

A Metaphysical theory of Emergence is essentially --- ? ? --- starting with lowest grade of Eternal Objects and endeavoring to explain their true relations within the Actual as to demonstrate the niches for the emergence of a higher grade of Eternal Objects within the actual.

This procedure is then obviously unending.– For whatever emerges establishes new relations. Merely restating old truism that you can't exhaust concrete fact by analysis in terms of universals.

When you have described a concrete occasion as synthetic unity of the potentialities which arise from relationships which are internal in itself to rest of things, you have then got that occasion with that essence. There is then the relationship of that occasion to that essence not as a synthesized unity, but as an antecedent multiplicity of potentialities. That essence to what there was for it to realize. Another actual occasion which requires actual occasion and its essence. Whitehead calls it "image". That occasion is [Whitehead] cognisance.

Most Whitehead did was to demonstrate a niche. And to say that the Cognisance was that was to go beyond anything he demonstrated. Whitehead's metaphysical principle led to that – led to see that there was that relationship for actualization. Then
|156|
there is the re-appeal to the empirical fact that Cognisance is exactly what emerges as an actualization of that possible relationship. Just as you cannot know by any metaphysical principle what specific swearingness of red and green, so you cannot demonstrate what Cognisance is in itself.

Then there is the *creative* side of the Synthesis:– The universe is such that given that possibility of an occasion: that definite occasion which is real with its complete actualization of eternal objects in definite niches:– universe is such that emergent entities – that eternal objects are exactly what they are. Leads to view:– However you have analyzed the matter of fact as being eternal objects which are enjoying those relationships; how those relationships lead to emergence of *further* eternal objects *what* emerges always belongs to what you've left out – so you've always left out the most interesting part of the story. You've provided paper, type, even words; but *story* and *author* left out.

So you can never get away from creative activity with creative character. Relations with possibilities *qua* possibilities play a dual role – remain alternative to what is actual but also are possibility for what is creative. When you've got totality of universe into matter of fact and *productivity* of matter of fact, your eternal objects have noble role:– (1) as possibility for what is creative. How creativity is oriented with respect to eternal objects: – (2) actuality as in what it is oriented in respect to eternal objects, and furthermore you get that hybrid relationship of Space-Time which can be looked on from point of view of actuality – what is alternative to past. Julius Caesar at Rubicon *might* have been in *any* costume [possibility as alternative] – Possibility of past – *cf* that of future. There *is* however the matter of fact process. – You get the eternal object as [?] attribute of *substantial* creativity.

Further: – There is not merely creativity in general, but in respect to each particular occasion. Particular character of Creativity in respect to *that* synthesis – and in respect to a *succession* of occasions. In handling account of the relationships of things at any stage in respect to which relationships you can see a niche, you can't help bungling it. – Only possible way of expressing it is in metaphors [because of conditions of creation and development of language]

A description is *good* in Metaphysics which takes you along for a good way without imperfection of its metaphors breaking down. Then it discloses that *cannot* analyze more in *those* terms. Then need description which doesn't completely throw away the old but involves it in wider form. This suggests Hegel. [Hegel represents genius and certain amount of arrogance.][1] Main principle of dialectic is that where Hegel is involved in contradiction it can't be because *Hegel* is at fault – must be in the things. *If* he'd said: whenever you've got to point where an *inclusive* description breaks down you've got to go back and get subtler vocabulary!

You can usually express error there is as one of Misplaced Concreteness. Oversimplified relations which *work* – because there *are* higher abstractions of which they *do* hold. In *language* we express very *high abstractions* of the real concrete fact but having definite one-to-one relationships with the concrete facts; so we can go a long way without

1. Closing bracket provided.

realizing any divergence. Attribution of quality to object as <u>though</u> a simple two-term relationship carries us along pretty well. But on high level of abstraction. – You'd have to take in whole Universe! There is the pink cloud. But for <u>me</u> <u>here</u>; if ⟨blank⟩

→ The more concrete you are the more complex someone elsewhere, it'll your relationships. <u>Therefore</u> have to go back over your work. Find not be pink to him in order to provide niches of real concrete level, have to have more complex description.[1]

Two ways to study History of Philosophy – Both necessary:– (1) [natural and scholarly way] to try to reproduce in yourself Plato's, Aristotle's state of mind and by scholarly discussion of states of mind <u>open</u> to men at that time etc.

(2) By way of knowledge you have at present (all your rubbish heap of Knowledge) you both correct and improve your own description and verify it, too, by seeing how <u>your</u> description is a generalization of the descriptions which other people had before. Ought to be able to fit it in as a correction and consummation etc. You see where Descartes ought to |157| have altered <u>his</u> description etc. Reason for retaining <u>hold</u> of <u>History</u> of Philosophy is the astounding difficulty of pushing back your descriptions to a concrete level. <u>Last</u> success of intellect is to describe the <u>perfectly general</u> relations of things – so that it will be in relation with "concrete state of mind" of any epoch.

The Philosopher ought to know a lot – Ought to have very concrete hold of what experience really <u>is</u>. – Ought to play baseball or the like e.g.

Now back to <u>previous</u> <u>point</u> (next lecture)

Another point:– Curious fact that you never can get back to the <u>absolutely</u> <u>general</u>. When dealing with Actuality. (matter of fact spatio-temporal relationship) – It has 4 dimensions. – Why not 444? There is always something which <u>is</u> so and <u>might</u> have been otherwise. Whitehead feels even so with Logic. Logic couldn't be otherwise if Reality is to be what it is. But very suspicious fact that we can vehemently and passionately believe contradictory propositions.

Then we are haunted [Very controversial ground] in some general sense by Knowledge that Bach beautifully played by Symphony Orchestra <u>is</u> finer than Barrel Organ playing "Wearing of Green". There <u>is</u> an external standard of Aesthetics and Ethics – hard to define, but there. Vision of something beyond you and <u>yet</u> particular. Get beginnings of that mysterious flair from which you develop both Metaphysics and Religion and that whole field of generality which <u>must</u> be otherwise and <u>yet</u> is <u>that</u>. (and is interesting for that reason). Only reason why you can <u>state</u> Laws of Thought is because <u>might</u> be otherwise and are yet that. Things which <u>couldn't</u> be otherwise you can't even state. When people say something doesn't <u>exist</u> – means they have placed it in wrong set of relationships.

1. The notes for these lines are oddly broken up and the phrases do not seem to connect.

Nonentity you can't even talk about. – You talk of something in other grades of relationships.

Hocking's notes

Thursday. May 14, 1925
Emergence of Thought

Having arrived at a maximum of incredibility, we turn to reconsider the general principles. Details always reach the incredible.

Very much the same general line as Lloyd Morgan and Alexander (Whitehead was prior) in *Emergent Evolution*.

Dissatisfaction with Alexander

1. In a muddle in his early chapters on *Space, Time* –
Never found anybody who did.

2. In handling the idea of emergence, not sufficient emphasis on the provisions of the niche for emergence. He says that where you have a sufficiently complicated state of affairs there is cognitive experience. He ought to have a principle such ~~that~~ as shows you in an analysis that there is provided a niche for the emergence of something else –

The niche is the relationship between elements which is the possibility of a synthesis.

Get red & green together, you have left a niche for the particular togetherness, as swearing at each other , or harmonizing – eluding all empirical description. Can describe everything except what it is

Procedure evolves because whatever emerges has new relationships. You can never exhaust the analysis of concrete fact by means of universals.

The most Whitehead has done has been to demonstrate a niche. "To say this is cognizance is to go beyond anything I have demonstrated" – What one described led to seeing that there was that relationship for realization – Cognizance is precisely what emerges as the realization of that possibility. Can't foretell the specific character of cognizance[1] any more than of the swearingness of red & green

---

3. The creative side of the synthesis.

"The universe is such that given this occasion that the emergent entities are exactly what they are". What emerges is always what your analysis has left out. You have provided the type & the words, but the story has been left out. You can never get away from a creative activity with a creative character.

1. Abbreviation for this term was simply 'c.'

May 14 – 2

In respect to what situations do you recognize a niche? Can't tell it without bungling it. "Any metaphysics is a good metaphysics which takes you a good long way without its metaphors breaking down".

Hegel a certain arrogance. "When Hegel finds himself bound in a contradiction, that must be the character of things in themselves, & not of Hegel's thought." Doesn't go back to revise his own metaphor, or change his Error of misplaced concreteness, & oversimplified relations. Attribution of redness to the cloud carries you a long way, or the sixpence size to a star.

Two ways of metaphysics.

1. Try to reproduce in yourself Plato's state of mind Aristotle etc., try to see how things were for them

2. Describe things as well as you can, & revise it. See that your description is a generalization of what people have said before.

Keep philosophy in connection with the concrete state of mind of your own epoch.

You can never get back to the absolutely general. There is always something matter of fact about it. Why 3 dimensions, or 4? Always get something which is so & might be otherwise, even in logic.

Standards of music & ethics are <u>there</u>. Matters of fact. Something beyond you. Metaphysics always has to begin with that. Religion concerned here.

|3 ?|[1]

Change terms–

Laws of thought can only be stated only because they might be otherwise and are yet that. You can talk of anything that might not be otherwise.

---

1. A loose page from Hocking's folder of notes contains these lines in Hocking's hand. It is unidentified and undated, but the content matches the point Whitehead is making at the end of this lecture, as we find in Bell's notes, so they have been included here.

Lecture 81

Saturday, 16 May 1925[1]

Bell's notes

|Continuing on 157|
Whitehead: 16 May, 1925[2]:-

Question of abstraction. Comes back to Mathematics (as most complete Abstraction you can have). General lecture on Mathematics as factor in History of Thought [written out as more formal lecture than usual].[3]

Mathematics in modern developments is the most original creation of human spirit. Originally consists in fact that in Mathematics relations exhibited which, apart from human intellect, are not at all obvious, (in perception) ^except a perception already guided by Mathematical Knowledge.^

Suppose project our imagination backwards. What an extraordinarily odd creation Mathematics is. Matters obvious to us, to even greatest in

1. Hocking does not seem to have taken notes for this Saturday lecture (nor for the previous Saturday, nor the next Saturday). Neither (as in the past) did he arrange for someone else to take notes, but on the following Tuesday he does begin those notes by saying: 'Absent in N.Y. Saturday May 16. Reading a paper on History of Mathematics, to be published.' It is puzzling that the same topic should be presented by Hocking in New York but, clearly, Whitehead was lecturing in Emerson Hall on Saturday, 16 May. Also, for these two lectures (81 and 82) on the history of mathematics, there are no notes taken by L. R. Heath at Radcliffe; see explanation on |97| of her notes, page 518, and the footnotes below.
2. On this day Bell recorded in his diary: 'Lectures etc. as usual. Took tea down to Mrs. Whitehead.... Then in town buying things for my dinner party to-morrow.' On the next day, Sunday, he noted: '... to "Old Grey House" – Pleasant little dinner:– Outerbridges, Hockings, Coles, Levinson's, Miss Allen.' And then on the Monday: 'Reading Farber's thesis.... At 4.30 PM Miss Mull's Ph.D. exam... Dep't meeting – Shimer's Ph.D. exam. He made a very poor showing (Relativity).'
3. In a letter from Whitehead to his son, North, dated 12 April 1925, Whitehead says that he will be giving a lecture before the Mathematical Society at Brown University entitled: 'Mathematics as an Element in the History of Thought.' They anticipated driving down to Rhodes Island ('not an island', he explains to North, 'but a State with a small island attached') on Tuesday, 14 April – see letter transcribed in Victor Lowe's *Alfred North Whitehead: The Man and His Work, Volume 2* (Johns Hopkins University Press, 1990), p. 310. He already knew at that time this paper would be included, along with the Lowell lectures delivered in February, in a volume called *Science and the Modern World*. This is repeated in Whitehead's Preface to that volume, which is dated 29 June. Bell's notes on lecture 81 at Emerson Hall follows the first half of chapter II of *Science and the Modern World* quite closely, which prompts three comments. First, Bell seems to be struggling to keep up; his notes are riddled with missed sentences and endings. This suggests Whitehead was reading from a prepared text, and either going quite quickly, or even reading 'into' his text. There is something about the pace and delivery of this lecture that seems different. (The further implication is that Whitehead did not usually 'read' his lectures, but talked from notes.) Secondly, chapter II of *Science and the Modern World* follows this lecture (and presumably his presentation at Brown University) quite closely, until he introduces Pythagoras. Lecture 82 at Emerson Hall, on the following Tuesday, picks it up from this point. And thirdly, this lecture, as far as Bell manages to capture it, shows all the signs of being an earlier draft, which Whitehead presumably corrected, clarified and added to for publication in *Science and the Modern World*.

primitive[1] ages are matters of only dimmest apprehension. Relationships between groups entirely independent of essences of the things. First man who noticed analogy of relations between groups of fishes and groups of days. How little of results of all this could have been envisaged. Erroneous literary tradition confining love of Mathematics as prerogative of few strange individuals.

History of thought without Mathematics is like Hamlet without Ophelia (she a little mad, too – Mathematics a divine madness of human mind).

Number, Quantity and Geometry and (today) more abstract – Order and purely logical relations. In Mathematics, point is we have got rid of any particular entities. (in pure Mathematics. – In realm of complete and absolute abstraction.) All insist on is that if any entities satisfy certain purely abstract relations, they must satisfy others.

Not even now generally recognized. Supposed that our Knowledge of Space in this world is certainly of Math (Geometry) utterly wrong. The relational forms of our Spatial world. So far as observation is concerned we are not quite accurate enough to be sure. By slight hypothetical extension we can correlate this with one of those systems of unspecified entities in the pure Mathematics. In pure Mathematics the "point" may be a turkey or anything.

But when you come to deal with Space of the Universe – ~~Say~~ etc.

"Abstract geometrical conditions".

|158|

When we come to physical [?] Space we say that some ∨definite∨ group of Entities does satisfy certain of these conditions and therefore must etc. ⟨blank⟩

Certainty of Math depends ⟨blank⟩

But can have no *a priori* certainty that any observed entities will satisfy any one of these formal relational schemata.

In criticizing an argument based on application of Math to matters of fact, there are three processes we must have in mind. (a.) Be sure no errors in your Mathematics itself. But when a piece of Mathematics revised and "before expert world for some time" – the chances --- ⟨blank⟩

(b.) Determination of abstract premises from which Mathematics proceed. – Matter of considerable difficulty. (Idea that all continuous functions possessed a differential coefficient.) Chief danger is that of oversight. – two opposite oversights – another doesn't lead to error but only to lack of simplicity – may think some postulate necessary which can be proved from others.

(c.) Verification – That our postulates etc. hold for the particular case. Here is where all the trouble arises. Counting 40 apples – arrive at pretty good certainty. But in more complex cases --- ⟨blank⟩

1. The abbreviation for this term is simply 'prim.'

Two things involved:– (a.) there are particular definite things involved and we have got to be sure what direct matters of fact are. Error always possible here. Scientific method is to cut it out here.

(b.) Things directly observed are almost always only samples. Want to know that hold for all of what appear same kind. Process of reasoning by induction is here and all difficulties of its theory – Despair of philosophy. [margin: cf. my[1] attempt →]

Not true that verification for anything like gravitation depends on independent verification of independent postulates. You verify entirely along whole line – that within region of Experience to which you are configuring your pure Mathematics, all the truths which are deductively deduced from your whole bundle of postulates are verified within limits within which verification is possible or in that region of experience and that everything that happens appears as instance of your pure mathematical deductive science. Then you inductively infer high probability for applicability for your deductive abstract mathematical theory to your particular case. If it isn't it is quite obvious that you've got your whole system there for verification.

Ordinarily you will feel there is one suspicious member of your family. Possibility for Science to be very seriously misled by taking police record of one of your family and overlooking ∧hidden∧ double-life of one of your most respectable members – usually something that has drifted uncriticized down the centuries.

E.g. Sort of way Kant insisted on taking formal logic of his time as an absolutely fixed and finished Science.

Thus often happens that in criticizing a book, whole trouble is with first chapter or first page – where author may slip in his assumptions. Trouble is not what Author Knows he has assumed but what he has assumed unconsciously. Every generation criticizes the unconscious assumptions of its parents.

History of Language is History of progressive analysis of ideas. Inflected vs prepositional and auxiliary language. ["Compact absorption [margin: cf. Toronto Logic →] of ⟨blank⟩ into English is a very highly analyzed language and is more complete exhibition of --- Pure Math is resolute attempt to go whole way in direction of complete analysis. Emphasizes direct aesthetic appreciation of ~~aesthetic~~ direct content of experience – what it is in own essence.

Question of direct experience dependent upon ∨sensitive∨ appreciation subtlety ⟨blank⟩ Also emphasizes ? ? ⟨blank⟩

|159|

Lastly is further apprehension of absolutely general conditions satisfied by those particular entities as in that experience.

Such general conditions [e.g. that of one-to-one – each person here, so far as observable, sitting on one chair. That is not abstract and necessary thing. Might loll on two chairs – the one-to-one-ness is a perfectly general thing but its presence --- ⟨blank⟩

1. The use here of 'my' (since Bell does not insert 'Whitehead') seems like Bell is referring to himself.

These conditions must hold for indefinite variety of other occasions. Exercise of logical reason is always concerned with these absolutely general conditions.

Discovery of Mathematics is discovery that totality of absolutely general conditions are themselves interconnected in a pattern of relationship with a key to it. This pattern of relationship among abstract conditions is imposed on - - - with assumption – necessarily that each thing is itself and nothing else.

Any interrelated actuality we can find [question whether in realm of unknown might be entities related "illogically"?]

Key to pattern is from select set of these general conditions a pattern involving infinite variety --- can be developed from --- ⟨blank⟩

Any such select set is "Premises" or "Postulates". Reasoning is exhibiting whole pattern from --- ⟨blank⟩

Wherever there is a unity of occasion there is established aesthetic unity between general conditions involved in that occasion.

Whatever falls within this relationship --- ? When key exemplified all the pattern is exemplified. Complete pattern thus determined is determinable from any one of number of sets of equivalent ⟨blank⟩

Reasonable harmony of being required for unity of a real occasion is Primary article of Metaphysical doctrine. – Involves that: for things to be together they are reasonably together. Provided we know something perfectly general about anything perfectly definite about any occasion we can then know --- ⟨blank⟩

Reasonable harmony etc. has two sides – inclusion and exclusion. Pythagoras is first who had any appreciation of general sweep of this sort of thing. [Burnet[1]] Pythagoras asked: What is status of Mathematical entities in Universe – Starting a controversy lasting ever since.

In Shape we have got to a relative concreteness that Pythagoras didn't notice – Shapiness of ⊙ e.g. doesn't come in in pure analysis. Aren't there perhaps particles of such general order that when you have said --- you've said all --- You've said almost all you can about ⊙ when you've stated Mathematical properties. Exactly mistake Einstein people make. Tend to identify abstract equations with the particular Shapiness which is exemplified in the abstract equations.

Seems silly not to have recognized (Pythagoras that numbers are material of world). Athanasian creed[2] and Hegel. Importance of definite number in divine nature etc.

1. Likely, Whitehead is here referring to J. Burnet's *Early Greek Philosophy* (1892).
2. The Athanasian creed was used by Christian churches from the sixth century; it was once thought to have been authored by Athanasius, and was the first to state explicitly the equality of the three persons in the Trinity.

Lecture 82

Tuesday, 19 May 1925[1]

Bell's notes

|160|
Whitehead: 19 May 1925[2]:–

Stopped at Pythagoras.[3] Limitations to be made as to generality of Mathematics – or of any generality. Qualification applicable to all statement. "How a general statement is possible?" [Kant] etc. Whitehead holds no statement but one can be made re any remote occasion ∧except for connections∧ so that it forms a constitutive element of immediate occasion.

By "immediate" occasion means one which involves as ingredient the judgment in question. The one:– If anything out of relationship then complete ignorance as to it. H. Spencer "Knowing more than was proper to Know of the unknowable."[4] Either we Know something of the remote occasion in the cognition which is ingredient of immediate occasion or we simply can say nothing about it.

Universe is one in which every detail is such as enters into its proper relation to immediate occasion. Generality of Mathematics is most complete one consonant with continuity of occasion cons.[5] ⟨blank⟩ Ground for entity into Mathematics and Mathematical Logic of notion of the "Variable". Irrelevance of the particular entities not usually properly understood there. Roundness of ⊙, Shapiness of Shape etc. don't enter into mathematical reasoning. Pythagoras: Number in its relationship to shape, had more of physical assumption than he was probably quite aware of.

Now continuance –
Importance of individual thinker owes something to chance. Here Pythagoras fortunate – for his ideas reach us through Plato. Owing to Greek notion of representing numbers by dots, divorce not so much as by us.

Einsteinians following pure Pythagorean tradition – nearer to Plato than to Aristotle. Practical advice of Pythagoras was to measure and

---

1. For these two lectures on the history of mathematics, there are no notes taken by L. R. Heath at Radcliffe; see explanation on |97| of her notes, page 518, and footnotes below.
2. On this Tuesday Bell simply recorded in his diary: 'Lectures etc. as usual.... Farber's thesis.' On the next day: '4.30 PM. Dept. meeting with Miss Mull's Ph.D. exam.... In with Demos to dinner at "Athens" afterwards. Prepared lecture etc. In P.M. Dow's divisional oral (Hocking, Ford, myself) – He did very well.'
3. Whitehead is continuing to read from his paper on the history of mathematics, prepared for presentation at Brown University in April, and intended for inclusion in *Science and the Modern World*. See Hocking's somewhat confusing comments at the outset of his notes, below, and footnote 1 on page 385, lecture 81.
4. In his *First Principles* (1862) Herbert Spencer devotes part I to 'The Unknowable' and part II to 'The Knowable'.
5. It is unclear how this abbreviation should be expanded.

express qualities by number. But biological sciences overwhelmingly classificatory – Hence Aristotle's form. Classification is halfway house between given thing and complete abstraction of Mathematics. Unless you can progress from Classification to Mathematics your reasoning won't take you very far.

2,000 years from Pythagoras & Plato to 17th century Mathematics.

Geometry gained conic sections and trigonometry. Method of Exhaustion had anticipated Calculus.

Arabic arithmetical notation and algebra. Progress was on technical lines. As formative for philosophy Mathematics hadn't recovered from deposition by Aristotle. You don't find the new aspects of Mathematics entering into the Sciences.

17th century is age of great physicists and mathematicians and philosophers and philosophers were mathematicians (except Locke etc.). Locke also influenced by the "Anti-rationalism" – revolt against Rationalism of Middle Ages. Even Locke living with mathematicians (Newton etc.) But Galileo, Descartes, Spinoza, Newton, Leibniz – they all remind us of typical uses of Mathematics.

But Mathematics which now emerges is very different from early Mathematics. Modern career – piling subtlety of generalization on subtlety of generalization and finding in each stage some application either to ? ? or philosophical thought.

The forming of new "niches" which discover themselves. Where ever you see a particular kind of generalization realized you therewith find a new niche emerging. Relations of relations of relations etc. Whatever has emerged into actual generalization gives new opportunity.

Relief from struggle with mathematical detail (Arabic system) gave root for development already faintly foreshadowed in Greek time. Algebra just a further generalization of Arithmetic first employed in Equations (Asking complex arithmetical questions).

Then suggestion of a "function" of one or more ⟨blank⟩
Letters are "arguments" or "variables"

Trigonometry absorbed into this new Algebra. Grows into general Science of Analysis in which we study various --- of functions.

Then generalization to "any function". Fruitful conclusion, *aber*[1], is generality with some happy exemplification (e.g. any continuous function). Descartes' Analytic Geometry. The Calculus, etc.

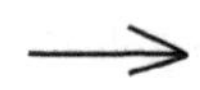

The point to make is Dominance of idea of Functionality found itself mirrored in nature in form of mathematically formulable laws of Nature. Galilei, Huyghens, Newton etc. produced formulae. But this
|161|
type of formulae wasn't in the mind of Greeks at all. Crop of formulae here was result of men whose minds were full of process of formalization.

1. It looks like Bell slipped in a German term again; *aber* can be translated as 'but' or 'however'.

Consider notion of Periodicity in nature – All sorts of things which recur. Apart from recurrence Memory and Knowledge impossible. – Without regularity of Recurrence no measurement.

So Galilei and periodic measurement of pendulum and stars. Mersenne, Newton, Huyghens etc. and what they did. – Application of abstract idea of periodicity to groups of phenomena.

Simple abstract periodic function exemplified by sides and angles of right-angled triangle became general and by becoming so it became useful. Illuminated analysis between sets of utterly different sets of physical phenomena and therefore showed how these could have their relational forms exhibited.

History of 17th century Science reads as if vivid dream of Pythagoras or Plato. [Reason of demand for autonomy of conduct of centers of intellectual thought. Facts that couldn't have been divined before. That apparently most extremely diverging lines of thought from the concrete are going to come back as fundamentally most important practically]. [Those soaked in intellectual appreciation of things should have complete freedom of keeping that culture alive].

Paradox now fully established:– Most utter abstraction is best tools for --- ⟨blank⟩

18th century thoroughly mathematically minded (except for, perhaps, English Empiricism from Locke) – especially in France. Newton's direct influence seen in Kant.

In 19th century the direct influence of Mathematics declined. (Romantic Movement in Literature and Post-Kantian Idealism). History of Kantianism due to foundation by mathematically minded man and carried on by men not mathematically minded. Need Physics to understand Kant, but not to understand Caird on Kant.

Even in Science:– growth of Zoology, Geology etc. Chief scientific excitement of Century was Darwinian Evolution.

But Mathematics not neglected nor uninfluential. Pure Mathematics made almost as much progress as in all the previous centuries. Period of discovery of Elements of the Sciences stretches from Pythagoras to Descartes, Newton, Leibniz.

Developed Science is matter of last 250 years.

Applications in Physical, Engineering etc.etc. sciences and therewith on popular life. Therefore two great periods of general influence of Mathematics on human thought.

(1.) Pythagoras – Plato

(2.) 17th and 18th centuries

Both periods had certain common characteristics. Interesting to see how philosophy rooted in general thought characteristics of age. Both ages are ages of general upset of thought. In Religion two movements in both cases. Mysteries and cool rationalist critical scientific interest. ∧Puritan and Catholic reformation∧ Both mean greater curiosity.

Periods of rising prosperity and new opportunities in both cases. Interests of a population which is unhappy (2nd and 3rd century when Christianity arising) are just as important as those of a population which is happy, but they'll be different.

Where you get "curiosity" awakened it is rather a happy epoch. Christianity wasn't the product of a happy epoch!

Characteristics of interesting life centering around New England and Boston etc. in last 50 years. Whenever Mathematics becomes relevant to philosophy you are near Platonism.

Temporary submergence of Mathematical mind from Rousseau onward now passing. Philosophy must take full account of those ultimate abstractions ʌthe relationsʌ of which it is the business of Mathematics to explore.

Then example of relation to Physics and encouragement to it to take daringly abstract form. [Quantum Theory].

Hocking's notes

May 19 –1

Absent in N.Y. Saturday May 16
Reading a paper on History Mathematics. to be published.[1]

Pythagoreanism

Classifying on Aristotelian ground tended in Middle Ages to displace measuring. How much we might have learned had this not been so.

Modern Mathematics – Shows the emergence of new formulae by further generalization, Relations of relations, etc., being equivalent to new niches.

In becoming abstract, trigonometry became useful – broadened out into diversity of possible applications. The utmost abstractions are the true weapons for controlling concrete fact.

Christianity was not the product of a happy epoch. It is only in a happy epoch that the age can undertake the revision of fundamental concepts. Then mathematics becomes relevant to philosophy. & you are then very near a Platonic view of philosophy.

1. To be clear about these two comments: it was Hocking (in a rare moment of referring to himself) who seems to have been away to New York on the previous Saturday. He may have been reading a paper there, but this is not to be confused with Whitehead, who presented a paper on the history of mathematics, at Brown University, in April. For this lecture at Emerson Hall, Whitehead is reading from the second half of his paper on the history of mathematics.

Lecture 83

Thursday, 21 May 1925[1]

Bell's notes

|162|
Whitehead: 21 May, 1925[2]:-

Theory of Abstraction:–[3]
Oddity that way your thoughts go corresponds to way things go. A metaphysical theory should show structure of things in a way that lays bare root of this business of abstraction. (Work out system of numbers and then try it on apples and it works). Pythagoras's finding the Abstract in the fact, as that which built it up. Get this in its most beautiful form in Plato – In effective form in 17th century and after, in way Mathematical Thought governed Physical theory ∧and practice∧.

Queer fact that the more completely abstract you become, in Mathematics, the more useful it is. In more fundamental way: epistemological problem: – how is Knowledge possible? How is abstraction possible? Theory of abstraction is Study of relationship starting from point of view of Eternal Objects.

We can ask:

(1) How are actual events indicated in relationships?
(2) How are emergent objects indicated in relationships?
(3) How are eternal objects indicated in relationships?

As we proceed along this course the relations gradually lose their internality. In actual occasions you can't drop the relations. But the emergent objects do run about on their own.

Are eternal objects merely implicated as being in relations external to themselves? – In a sense, yes, in another: no. I.e: not quite.

---

1. The Radcliffe version of this lecture, from notes taken by L. R. Heath on the previous day and then continuing on the same day, begins on |98| of her notes, page 518.
2. On this Thursday Bell recorded in his diary: 'Up early. 8.45–10.00 Farber's "topicals" (Hume, Schelling, Metaphysics, Epistemology) (Hocking, Demos, myself). Then my lecture – later Whitehead's. In P.M. Feldmann's divisional orals. In evening last meeting of Pound's Seminary. Also finished Farber's thesis – Did all Phil. A papers & prepared for to-morrow's sections. Got to bed very late. Only about 4½ hrs. sleep per night, Wed, Thurs., Fri.!' And then on the Friday: 'Phil. A sections as usual.... Then Dep't meeting. Farber's Ph.D. exam. (Too glib altogether!)'
3. Issues as Whitehead is working them through in these lectures have echoes throughout his later work and especially *Science and the Modern World*. The previous two lectures are especially concrete examples of material he had written up for presentation at Brown University, read out for lectures 81 and 82, and which then became chapter II of that book. Chapter X, 'Abstraction', seems to be another example, perhaps not so explicit as chapter II, but clearly the repository for the thoughts he works out in these two lectures, 83 and 84, which he introduces as his 'Theory of Abstraction', and at the outset of lecture 84, 'from the point of view of Eternal Objects'. Page 159 of *Science and the Modern World*, for example, leads into 'two principles', which also find expression here, beginning on the second page of this transcription.

Consider an eternal object A. If we ask how A finds itself in universe of things, we must answer that this status of A enters into essence of A. What A is in itself involves A's status in the universe. It doesn't just chance to bump up against a metaphysical situation.

In essence of A there stands a definite determinateness as to relation to other eternal objects and indeterminateness as to relationships of A to actualities. [Eternal objects "pervading all Space and all Time"] In what A is in itself there is an indeterminateness as to actuality.

Think of a definite shade of red. Question whether a definite bull somewhere in Massachusetts tomorrow is going to see red isn't contained. It may see red and it may not. This indeterminateness is in A's essence and is "A's patience to external relations". External relations involving ---

Two principles:

(1) An entity cannot stand in external relations unless in its essence there stands an indeterminateness which is its patience for those relations.

(2) An entity which stands in internal relations has no being apart from these. I.e. Internal relations are constitutive relations.

Thus when in essence of an Entity there stands a definite determinateness of relationships, these relationships are internal. – Where indefiniteness they're external.

But in no case can an Entity stand in any relationship which isn't already in its essence internal or external.

Spatio-temporal Continuum as representing when viewed in its completeness – future and past and what might have happened in past [Trevelyan's prize on History of 19th Century if Napoleon had won at Waterloo[1]].

An eternal object is to be analyzable as to (1.) what it is in itself [back to beginning of Session] ("individual essence") and (2.) in its determinate relations ("relational essence") and (3.) in its patiences. – It is in a definite status with relation to actuality. This turns up everywhere in philosophy. Where people say: "The thought is entirely in your thinking of it". If were so, how could people think of same thing? But status in respect to actuality is in essence of an eternal object.

Whole dispute between Plato and Aristotle turns out that point – exact description of that status. Thus eternal object A analyzed into (1.) A-ness in itself, (2.) A in all the determinate relationships of A to every other eternal → object ("relational essence" of A) – Realm of possibility, (3.) A as patient of relationships – Relational essence of A as patient of realizations. When you talk of Realm of Possibilities you are talking of patience of realisation there.

1. In July 1907 the *Westminster Gazette* offered a prize for an essay on this subject. The winner was George Macaulay Trevelyan with the essay 'If Napoleon had won the Battle of Waterloo'. This was during the period when Trevelyan (1876–1962) had left his fellowship at Cambridge to take up writing full time, and before he later returned as Regius Professor of Modern History .

|163|

Then three aspects of A cannot be apart. Can ∨be∨ themselves looked upon as eternal objects of a very high and abstract grade. The determinate relationship of A to every other eternal object is how A is systematically related to every other eternal object. This system of relations is A.

Space-time is the relationship of every eternal object to every other one. Stands in nature of all eternal objects to be thus related. Thus stands in ∧essence of∧ A and is internal to A.

"A *simpliciter*" here.

But in respect to Actualities one indeterminateness stands in Essence of A. An actual occasion synthesizes in itself every eternal object, including its determinate relation to other objects.

Every particular occasion in a sense includes everything. This synthesis involves limitation. Grades of entering into this synthesis are ⟨blank⟩ Any particular Relationship of A to any other particular realisation of any eternal object is realized in A under different grades of actuality. These can only be displayed as relevance [?] of Value.

What A is in itself is relevant. Relevance of Value varies as you compare different occasions from inclusion of individual essence of A in aesthetic synthesis (some ∧grade of∧ inclusion) to lowest grade which is exclusion of A as client in the aesthetic synthesis. In this lowest grade every determination of A is an ingredient of occasion merely as a possibility. – What A is in itself fails to make any contribution.

Nothing new here! – Repeating what Aristotle says: nonmusical man becomes musical. Music included in that realisation.

Thus A conceived as merely ∧and barely[1]∧ in its determinate relationships to other eternal objects and in no other way, is A conceived as not-being. And A conceived as in its patience for its ingression in actual occasions, is A as a possibility. Thus A as standing in synthesis of a definite occasion is A as including or excluding as an effective element in the synthesized --- ⟨blank⟩ Bergson, etc:– What does happen is an exclusion of what doesn't happen. In this sense [perhaps misquoting Hegel], every occasion is a synthesis of Being and not-being. Further: – though some eternal objects are synthesized merely as not-being. Each eternal object which is synthesized qua being, is also synthesized qua not-being. – Under its inclusion under certain relationships [me as sitting in this chair[2]] and excluded under other [me not in that other one] [I am, of course not an eternal object][3] Excluding what is not actualized.

Here getting to whole question of actuality being a selection and therewith an abstraction from ? ? ? ⟨blank⟩

---

1. This might be 'rarely'.
2. This seems to suggest that (unless at the blackboard) Whitehead sat through this lecture, and perhaps others.
3. Closing bracket provided. The joke would be extended if the closing bracket were placed at end of next phrase.

Terms here rather Hegelian than Whitehead's own.

Also there are different meanings of Being corresponding to different grades of Ingression. "Being" here means effective in the ⟨blank⟩

Experient occasion viewed as self-created

The general fact of the prehension of all eternal objects in every occasion: – this is the indeterminate relationship of all eternal objects to every occasion.

Now analysis of eternal-world-makes[1]-to one another: Consider a set of three eternal objects A, B, C. Call it the "set" of eternal objects. Relations of these to one another always expressible by some relation to Spatio-temorality: This:–

(1.) Allow themselves to be analyzed into set of relations of A,B,C among into finite statements i.e. you don't have to know everything at once

(2.) Into all other relatednesses (including[2]

That you have particular truths arises from real of Possibility

R(A,B,C)

For this specific relationship of ABC mentioned above under "(1.)" – Under "(2.)" we may find ~~an~~other determinate relationships to

|164|

D_1 D_2 ... D_n. Suppose it involves set A,B,C, D_1 D_2:–

exclusiveness. Then perhaps second determinate relationship

D_1 D_2 ... D_n and R(A,B,C).

I.e. since our A,B,C is after all only another eternal object (more complex than A,B,C) the new relationship is --- ⟨blank⟩

Eternal object may be complex and be in fact a relatedness among eternal objects of a lower grade.

Analyzability of general relatedness of eternal objects means that complex eternal objects like A,B,C exist.

We think of it the other way on. A as possibility, however, is A with its complete schema of possibility

If you bring in remaining "idea" you emphasize relative to consciousness of it. Considering realm of possibility we start with simple eternal objects.

A complex eternal object may be looked on as an abstract situation. – Abstract in two senses:–

(1.) If abstracted from the occurrence.

(2.) R(A,B,C) abstracts A from "A *simpliciter*" and you have the complex eternal object [R(A,B,C)]

A there is A in which there has been --- is in limitation. A in that relationship has abstracted from total possibility of relationships.

---

1. This term in the hyphenated set is difficult to make out.
2. It is not clear whether or how either (1) or (2) were concluded. Perhaps the 'R(A,B,C)' is the conclusion to (2) but that is not obvious in the handwritten notes.

When think of <u>possibility</u> of A just in that relation you are thereby thinking of the R(A,B,C) actually involved in ⟨blank⟩

What is actual is an <u>abstract</u> from realm of possibility – and is itself an analysis of it. A is not in actual situation *simpliciter* but under abstraction (or "limitation") of <u>that</u> eternal object.

Thus very foundation of abstraction is abstraction from complete realm of possibility. "A" is <u>itself</u> an abstraction. An abstraction, but in all its relationships.

A in R(A,B,C) is more abstract than A simpliciter. – Its "A under a limitation". Thus an actual occasion (α) synthesized in itself every complex eternal object – either as <u>included</u> or as <u>excluded</u>. If R(A,B,C) is synthesized in α as being then R(   ) synthesized there as <u>not</u>-being. A as standing in definite relationship ρ̇ (ṙ ḃ c)[1]

The synthesis of A in α, both as being and not being, is reason why every <u>actual</u> occasion ⟨blank⟩

Hocking's notes

May 22, 1925[2]

Pythagoras' problem: How is abstraction possible?

More fundamental than Kant

How are actual events implicated in relationships? emergent objects eternal objects.

Relations progressively drop their character of internality in this series. It is the essence of an eternal object that it does run about on its own.[3]

Are they merely external? Not quite.

Any eternal object A has determinate relations to other eternal objects, & indeterminate relations to actuality.

It doesn't lie in the nature of reds whether a given bull somewhere in Massachusetts will see red tomorrow.

Red may or may not have that relation & yet be the same thing

Principles

I. An entity cannot stand in external relations unless it is indeterminate in its relations to others which it is patience.

II. An entity which stands in <u>internal</u> relations has no being as an entity <u>not</u> in these relations

---

1. It is not possible to be sure of the transcription of these symbols.
2. While Hocking definitely dates both pages of notes for this lecture as 22 May, it was undoubtedly 21 May.
3. Here we seem to have a rare occasion when Hocking and Bell do not agree on what was said: Bell has 'emergent objects' run about; Hocking has 'eternal objects' run about.

Individual essence. Relational essences, Patiences.

Analyze an eternal object A –

1. Its A-ness
2. Its determinate relations to every other eternal object
3. A as patient of relationships which may be realized

A *simpliciter* is (1) & (2)

Bergson & Aristotle said it: What happens is an exclusion of what doesn't happen.

Every occasion is a synthesis of being & not being

Every eternal object which is synthesized qua-being is also synthesized as not-being – (not Hegel)

I am not in that chair but in this one –

I should not have said "I" – I am not an eternal object.

May 22 – 2

R(A,B,C) is another more complex eternal object.

$ABCD_1\ D_2 \ldots D_n$ These exist.

What is actual is an abstract from the world of possibility

Every actual occasion synthesizes every eternal object, including the complex eternal objects, either as existing or as not existing. This is the reason why every actual occasion is a limitation.

Lecture 84

Saturday, 23 May 1925[1]

Bell's notes

|165|
Whitehead:– 23 May 1925[2]:–

Half way through a topic last day. Considering things from point of view of Eternal Objects rather from that of actualities. Determination of relationship of ∧(not of actual occasions but of)∧ "eternal objects" [neutral phrase]   R(A,B,C) as complex eternal object. In laying down that there are such complex eternal objects, are laying down ground of possibility of finite Knowledge. How is finite Knowledge possible?

Because general determinate relationships of eternal objects *inter se* is ⟶ capable of analysis. If not so, there couldn't be any ~~fin~~ Knowledge without Knowing all.

R(ABC) is a relationship exclusive of all other eternal objects. Those others there merely as systematic background. This determinate relatedness is itself a complex eternal object. Analyzability of general relationships of eternal objects means that complex eternal objects exist. (Hume's Simple and Complex impression)

Complex eternal object is a complex situation. Foundation of Abstraction is that there is an abstraction from the complete realm of possibilities (Analyzability of this realm)   (ordinarily think of abstraction as from Actualities only!) *Mais non*!

– A in itself must mean A in all its relationships. So A in R(ABC) is more abstract in some sense than A *simpliciter*. What corresponds to Concreteness in an actual occasion ~~is~~ corresponds to simplicity when you are dealing with eternal objects.

Thus an actual occasion α synthesizes in itself every complex eternal object such as R(ABC) either as included or excluded. If R(ABC) is synthesized in α under lower grade of exclusion then R is synthesized in α as not-being.

Every occasion is a synthesis of everything.

---

1. For reasons unknown, Hocking provided no notes for this Saturday lecture, nor for the previous two Saturdays. Neither (as in the past) did he arrange for someone else to take notes. What might be the Radcliffe version of this lecture, or portions of it, from notes taken by L. R. Heath on what was likely earlier the same day, begins on |103| of her notes, page 520.
2. On this Saturday Bell recorded in his diary: 'Lectures as usual. Then busy clearing up papers – …'. And then on the Sunday: 'Finished cleaning up papers, etc. In evening to supper at Whitehead's & then took them in to "Symphonic Programme" at the "Pops"…. Looked after ticket applications for the W's. Later – tea & chat with them. Then home & lecture preparation! Also questions for Phil. A. paper.' On Monday: 'At 4.30 Dep't. meeting – Miss Amen's Ph.D. exam. (Experimental study of Consciousness of "Self").'

Also A as standing in determinate relationship R(ABC) is then synthesized in α as not being even if A in some other relationship is synthesized in α as being. This synthesis of A in α in both ways is reason why every actual occasion is a matter of limitation. A is in α under an aspect– not *simpliciter* but under limitation expressed by how it's not in α. This limitation depends on fact of unanalysability of total determinate relationship of A into complex eternal objects with A as ingredient element in them.

Only way to get over difficulty is to enunciate a principle which will enable you to say it isn't a difficulty. Introduce terminology – then it is all right. Whitehead:– Principle of "Translucency of Actuality". Translucency of Cognition is only a particular case of it. Correspondence theory of truth depends on this same general principle. Actual Experience is not a predicate, i.e. ⟨blank⟩

By Translucency of Actuality mean: R(ABC) as in actual occasion α, as excluded or in whatever grade of inclusion, has no other content than just R(ABC). It is just that principle by which finite Knowledge is possible. Being actual is not an addition to the content, but is emergence of that content as informing the value emergent in that occasion. Aristotle would say: then "value is your word for my ὕλη". Whitehead thinks "Value" has a better suggestiveness than "wood". Thus R(ABC) in actuality is just --- in which A-ness of A, B-ness of B --- are in α as conditioned by this determinate relationship R(ABC). It abstracts its value from realm of possibility by specific way in which formed by relationship R(ABC). But type of their relevance to each other depends on grade of the actuality. Thus relationship R(ABC) in its determined realisation in the actual occasion α is condition which limits ingressions of A,B, and C into α. Grade of realisation it is which becomes determined then. The form is already determined as possibilities.

What is positive is exclusive of what is negative. [ABC as how they are not in the occasion]

Thus possibility is analyzable – i.e. is a self-developed realm of ascending objects. (Difficulty of possible *Verstrickung*[1] in infinity[2] required unless you have a first grade in this realm.

1st grade:– A B ... ... Z
simpliciter

2nd grade: R1(ABC), R2(ABC), ... in which you get A not *simpliciter* but as abstracted from its simplicity. But do get R, *simpliciter* etc.

3rd grade: S{ R(ABC) R'(DEF) } S1....[3]

1. The German inserted at this point, *Verstrickung*, could be understood as 'entanglement'.
2. It is not clear whether this is the correct expansion for 'inf.'
3. There two are additional marks in Bell's handwritten version, but both seem to have been crossed out, and we have left them out.

|166|

It is the grade of complexity which is simple in that grade. (Extremely
→ unlikely that it runs up so very symmetrically. It may, but there is nothing in Whitehead's conception that presupposes that it must.)[1]

Getting to higher and higher grades of Abstractions. Each representing a higher abstraction from simplicity of simple objects you started with. Reality is all of them. But never can think of them as if all one class [Because always the stage beyond, and the stage beyond .... are there].

Presupposition that metaphysical situation could be talked of as though it were an entity within itself. α includes every eternal object which has ingression in it in a form which is a complete abstraction from A *simpliciter* – of every eternal object however complex.

So any actual occasion deals preferentially with whole realm of possibility. But since A in R(ABC) is still A, in A-ness of individual essence (also has its relational essence) it is totality of A as a possibility as excluded. Since A is there you can return to the complete simplicity [Starting from complete concrete occasion, α, in which A is there abstractly.][2] α is in its nature an abstraction from reality.

(1.) by abstracting R(ABC) from α and then A from R(ABC). Abstraction when you are abstracting from the actual you are then returning to simplicity in the possible. Contrasting Simplicity in the Possible with

Leibniz! Concreteness in the Actual. The actual embodies – is informed by – the possible, in which each object is there with minimum of simplicity.

Exclusion from a common occasion is exclusion. What is thought in cognitive occasion is therefore in some grades of inclusion in the standpoint occasion (Translucency of cognitive occasion, remember!). When you are thinking of R(ABC) – It is realized as a real togetherness via the Standpoint occasion. The simplicity of R(ABC) has, however, not been abstracted from in this realization. R(ABC) as a whole, as one entity, enters into the occasion as itself an entity, not conditioned by relationships with other entities as a whole, but as conditioning entity of A, B, and C into Standpoint occasions by its own internal relationships. – When you are thinking of any situation the object in their Spatial-temporal relationships (in that limitation) are entering into your experient occasion. But that object as a whole is not entering into any including relationships – It is its internal relations which do so. As a whole it is excluded from adding to Value of your ?   ?   Internal relations as possibilities of that thing anywhere, – that relationship is ---   ⟨blank⟩

Relationship of that [partial] relationship to anything else is not included, but excluded, from Experient occasion. So, if R, turns up in second grade there is no realization of third grade object S, etc. You get it as "inset" so to speak. Looking on realization   ⟨blank⟩

1. Closing parenthesis provided.
2. Closing bracket provided.

Accordingly R(ABC) as a whole is not set in determinate realized relationship to the eternal objects in higher grades of realization. Accordingly realization is a process of abstraction from the simplicities of Eternal Objects, and this process of Abstraction is merely an ascent from more abstract grades of eternal objects, investing them with that grade of --- ? ? ⟨blank⟩

Thus a thought (they thought of) is an image – an element in a conscious occasion – of an eternal object: R(ABC) which as ingredient in standpoint occasion is (not completely) abstracted from --- ⟨blank⟩

Note:– Whitehead not very far away from Hume's "faint copies".

Logic stresses the general relations of Exclusions to Inclusions, which are inherent in the Patience of Eternal Objects. Represents how actuality is patient of ingression of Eternal Object or:– How Eternal Objects are patience of abstraction into actuality. This gives you as sort of structure of the patience – the possibilities.

You cannot separate Form from Matter of Eternal Objects. All Forms are in every occasion, and the How of Ingression is the grades of determinate Actuality of the Content of the occasion, but does not add to the intrinsic character of the Content. Whitehead not quite certain but equally disagreed with here by Aristotle and Plato.

Realisation both Spatializes and Temporalizes (gives you your succession) – coloured language here suitable. Two plane grades of Realisation – Experient and Cognisant. Cognisant has unique relevance to special experient occasion but except for that has no Spatialisation. – So it is an epoch – a "hold-up" – in creative --- of Substance.

Lecture 85

Tuesday, 26 May 1925[1]

Bell's notes

|167|

Whitehead:– 26. May. 1925[2]. –

Just recently statement of Prof. Miller (Mt. Wilson) repeating Michelson experiment;[3] What Miller got hold of. Repeated Experiment at A –

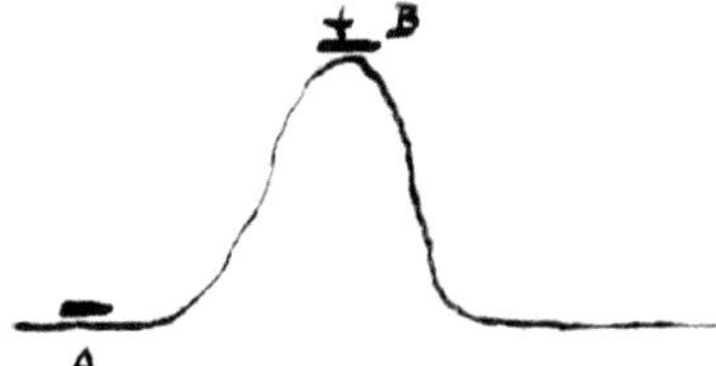

(confirming Michelson) and B (getting a positive result!) (Confirmed null result) Michelson went over experiment three times (always with differences). But Michelson always did it on ground.

---

1. There are no notes taken by L. R. Heath that correspond to what was the last lecture presented at Emerson Hall.
2. Bell noted in his diary on the day of Whitehead's (and his own) last lecture: 'My lecture as usual. Was so exhausted & tired that I simply couldn't keep awake in Whitehead's lecture (Really suffered with nervous exhaustion).' The next day, Wednesday: 'Eaton woke me up about noon (!) & we did final work on Phil. A paper. H.C. Green drove Demos & me home. Maurice came in later with his portrait of Whitehead. Still later George. I went home with him & we talked Till very late.' And, finally, on Thursday, 28 May: '3 P.M. – Dep't meeting. Roelof's Ph.D. exam. ("Authority") (very good exam.).' In early June, there were the last occasions of socialising for this academic year. The diary records for Monday, 1 June: 'Up early. Cloudy, warm morning. At 7.15 to Hocking's car needed repairs. We finally got away from the Whitehead's door shortly before 9 o'clock. Drove via Mattapan, Stoughton, Taunton, Fall River, to Newport, R.I. & thence out to Berkeley's old house at "Whitehall". – Charming! Then to "Hanging Rock" (B's cave) where some of B's letters were read aloud. Then in to Newport where we had lunch at a very nice Eating Club of Hockings. Then H. & I changed wheels on the car (flat tire).... Then drove around Shore Drive, & started back for Boston. Ran through heavy thunder-storm from Taunton on, after which everything was very beautiful (Double rainbow). Arrived at the Whitehead's about 8.30, where a fine dinner awaited us....' On Tuesday, 2 June: 'Finished marking Phil. 25b theses. After dinner (Col. Club) took photos around to W. James's (nice chat re: Germanic Museum etc.) delivered Phil. 25b essays. Took H.O. Taylor's book around to Whitehead's.' And, finally, on Wednesday, 3 June: 'Phil. 25b exam. At 10 A.M. meeting of Board of Tutors in Phil. & Psych. re requirements for distinction.... At 4 conference with Woods re Bandler's reading for distinction in Ancient Phil.... At 7 o'clock with Dow to dinner at Whitehead's (W. gave D. letters of introduction for Cantab.) Then I tried to get things for W. to Pratts (sailing for England to-morrow).'
3. Dayton Clarence Miller (1866–1941) was an American physicist known for holding onto the ether theory and resisting Einstein's theory of relativity. Using apparatus similar to that used in the Michelson–Morley experiments, Miller worked with Morley to improve its sensitivity, but in 1904 still acknowledged null results. Miller himself continued to refine these techniques for many years, and claimed he could obtain results indicating ether drift. Whitehead's last lecture begins with an examination of results only recently released of measurements made by Miller on the top of Mt Wilson, where he claimed the effect of the 'ether wind' would be greater. In 1926 in *Physical Review* he published the results of experiments conducted during 1925, which may well have been announced earlier. These are likely what Whitehead is responding to in a recent statement by Professor Miller. The consensus of opinion drifted away from Miller. But the problem with his results seems to have been less with his experimental measurements and more with the methods he used to analyse the data.

There's obviously point that on sheer pure Relativity idea of Einstein view (that there is a peculiar – almost metaphysical quality in light – always moving with same velocity) you ought to get same effect everywhere.

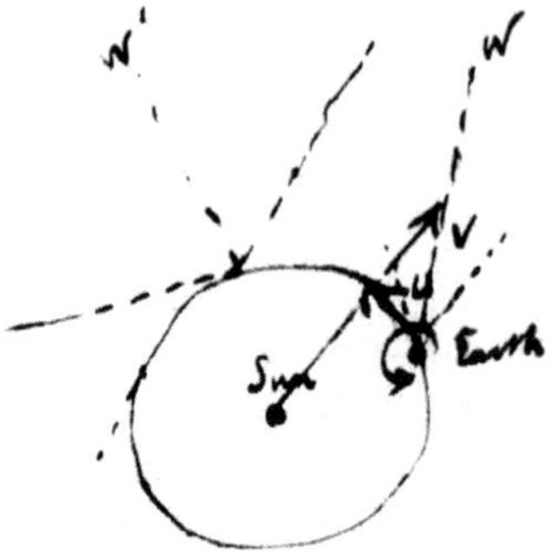

Question in such cases: – What, precisely, has Miller got? We don't know yet. Movement of Earth relatively to fixed stars = W. (Not quite sure re possible movement of stars) W, W' near V. Velocity, will vary from V + U at one time of year to V - U at another.

Abstractly metaphysically might be at rest (Earth) all time. But would expect a seasonal effect (undiscoverable according to Einstein theory).

Velocity of earth (and therefore of instrument) on a certain day when would exactly reverse effect of "fringes" because relation of W to apparatus exactly reversed! (on any one day).

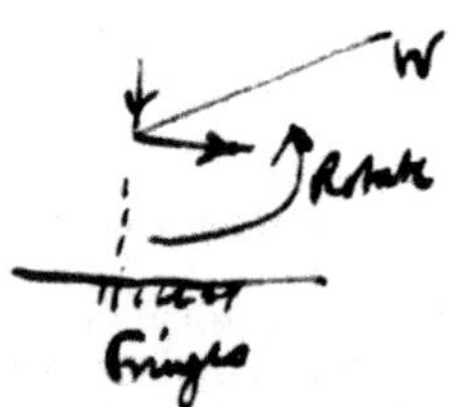

Might get W so. (Revolving apparatus). Magnitude of displacement when turned one way and other way for days of year, ought to get Seasonal fluctuation. You'd expect the ordinate to be higher than the variations.

But that effect should be at B and A, except on one hypothesis:– that Earth drags along Ether with it. Then at A you would get fuller dragging than at B. Very nice explanation – but has its difficulties!

Sir George Stokes[1] (Cambridge – in 1880s) pointed out that, if Ether-drag, difficulties re aberration.

One man getting some result in place like this ⟨blank⟩
Over questions of Seasonal effect, it is difficult to know that they are due to (Temperature etc.) (of course, Miller's greatest care)

Nap

No sensible person would ever have grounded their emphasis on Relativity in this one type of Experiment. It is only your point of departure. Michelson-Morley experiment is only one of a nice number of Experiments faulting the principle. This is the point from which Einstein etc. went out.

Whitehead would expect Troutman experiment to go by board if Michelson-Morley demurs one day. Wouldn't expect the Mercury Experiment to do so, not Spectral lines from atom.

Then St. John, here, – Shifting of spectral lines.[2] (Easy to observe. Still bigger reason.)[3]

1. Sir George Gabriel Stokes (1819–1903) was Lucasian Professor of Mathematics at Cambridge through all of Whitehead's years there, and a colleague for a time of Maxwell.
2. Charles Edward St John (1857–1935) was on the staff of the Mt Wilson Observatory, and announced in 1923 that his measurements of solar spectral lines convinced him of gravitational redshift.
3. Closing parenthesis provided.

Newton took view: could say of <u>one</u> of them absolutely at rest in Space and other moving. But nothing <u>in</u> Newton <u>using</u> this.

$x_\beta = x_\alpha -$ [illegible]

Newton said Velocity in shape of direct[1]

Newton showed in opposition to Aristotle:– that not velocity but uniform acceleration which is the important thing.

If this isn't way to look at formula of Space and Time the velocity doesn't come in, at all except[2] by way of reason[3] ⟨blank⟩

So long as Newton's gravitation equations were looked on as trying...pics[4]

"Mechanical" explanation ⟨blank⟩

What Clerk <u>Maxwell</u> did? Maxwell died thinking this one step to toward getting the ⟨?⟩ It was Herz who pointed out what Maxwell[5] ⟨blank⟩

|168|

Herz's way[6] was new one entirely. Since that time: Electro-magnetic Theory of Matter has turned up. – That molecule is to be looked on as an electro-magnetic <u>happening</u> (as a thunder storm) Therefore all derivative from general equations which express Electro-magnetic World.

You get entirely inverted <u>there</u> relation between Newton's equation and Maxwell's. The dynamic equations of Newton come to be derivatives of the Maxwell equations.

Now come to difficulty much[7] older than Einstein. Einstein did not work out the equations – Larmor did that (Lorentz, too).[8]

<u>Maxwell</u> equations aren't <u>invariant</u> as Newtonian (so that it doesn't matter whether you took coordinates A-system or B-system. (x, y, z or x', y', z'). But with thorough-going <u>Relativity</u>, no reason why one set of axioms preferable to any others.

Then suggestion that, whatever the fundamental and proper view of Spaciness – important question for <u>Maxwell</u> is whether you are at rest in reference to the Ether.

If you get "<u>Ether-drag</u>", then you get a great muddle.

Question: what transformations, if any, for Maxwell's formulas which will reproduce them in the new form. That's what Larmor found out.

1. Perhaps because of the 'nap' mentioned by Bell, his writing is very unsteady at this point, such that these terms and the equation above (part of which has been reproduced as an image here) are barely readable. Fortunately, it looks like the details of what Whitehead was presenting are captured in Hocking's second page of notes, below.
2. Almost undecipherable.
3. Almost undecipherable.
4. These terms are undecipherable.
5. Bell's notes barely readable.
6. Heinrich Rudolf Hertz (1857–94) was the German physicist credited with experimentally proving Maxwell's electromagnetic theory of light.
7. Bell definitely has 'must' in his notes, but from context Whitehead may have intended 'much'.
8. Lamor and Lorentz were both mentioned previously on this same topic; see footnote 4, page 295, lecture 63.

Larmor said: suppose $V_{\alpha\beta}$ [Velocity of β system in terms of a system][1]

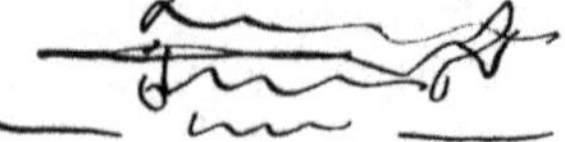

Metaphysical System of Whitehead:– Space and Time differentiated from one another by Endurance. Question whether Endurance utilized one and same Space-Time. No reason for Metaphysics to turn down Einsteinian. Doesn't exclude fact, either, that there should be one determination of Space and Time.

P. 163 in Whitehead's Book.[2] Whitehead cannot see one reason more than another. But if you do assume doctrine of Relativity, you get a simpler and nicer doctrine by which you can shorten up what is rather vague in Metaphysics to concise ? ? ⟨blank⟩

E.g.:– from idea of Extension only impossible (Whitehead) to get to "point-event" [minimum of simplicity]. If you get alternative systems, – three conceptions relevant to a point which is accounted for. Also get account of the specialization of Spatiality into Planes, Lines etc.

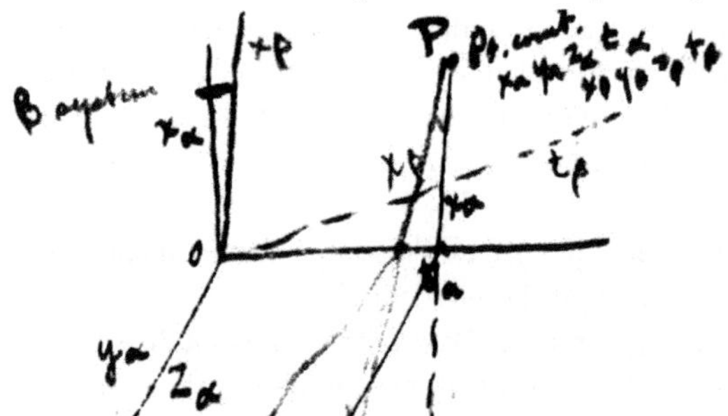

Everything at time t_a is to be looked on as simultaneous with P in x_α Or in x_β

That's got its paradox. Give different Simultaneities.

~~All I have to do is~~

Reduction of Geometry (as hypothesis) from these. You have then got to bring in extra principles:–

(a) How to get a point-event. [Must bring in notion of physical convergences][3]

Two abstractive Classes have same convergence if all possible physical properties derivative from these convergences are identical.

Brings in idea of position. So related as to have identity of position. Identified "position" with moments of time. Idea of position derivable from bundle of alternative Now's in which it lies. If any point-event lies in any one moment of time, it may be anywhere in it. Position must then depend on something beyond Time (which isn't so nice). Position is primarily, the relation of all possible things.

Idea of Space-Time whole of possibility must be correlated with idea of Position. Two abstract classes are convergent to same position. When the physical limits to which they thereby approximate [qua mere abstractive class it converges to nothing] – Physical quantities involved do converge.

1. At this point Bell records only the blurry diagram below, but fortunately much of the lecture material here is captured in Hocking's notes.
2. The book, in this case (as we see in Hocking's notes) is Whitehead's *Principles of Natural Knowledge*.
3. Closing bracket provided.

Hocking's notes

Whitehead – last lecture May 26. 1925

Miller's positive result.

You don't quite know what to expect because you don't know the resultant motion

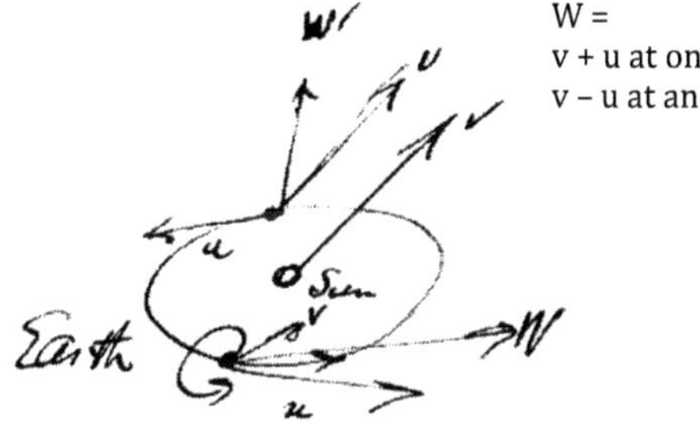

W =
v + u at one period
v – u at another period

If the earth drags the ether along with it – then the result at the top of Mt Wilson might be more than on the ground.

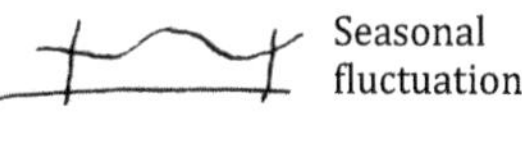

Seasonal fluctuation

Newton's laws. $m\frac{d^2(x_\alpha, y_\alpha, z_\alpha)}{dt^2} = (F_1, F_2, F_3)$

$$\frac{d^2x_\alpha}{dt^2} = \frac{d^2x_\beta}{dt^2}$$

1. Naturalism

18th & large part of 19th Century: You have explained a thing when you have brought it under these equations.

Clerk Maxwell produced another set of equations – and died thinking he had made a step to bringing electricity under his equations –

Hertz pointed out that "Maxwell's theory is Maxwell's Equations"

Since that time the electrical theory of matter has turned up –

Now the dynamic equations are derivative from the Maxwell equations. Relativity diverted.

Maxwell's equations however are not invariant – for different axes – According to him the important thing is whether you are at rest with reference to the ether. Would be very muddled if you have your ether dragged about. So you are "up against a very well documented theory"

Larmor & Lorenz found out the new equations which would be invariant.

5/26 – 2

Larmor's Equations.

Now refer this to a system of axes moving along x axis with velocity $V_{\alpha\beta}$

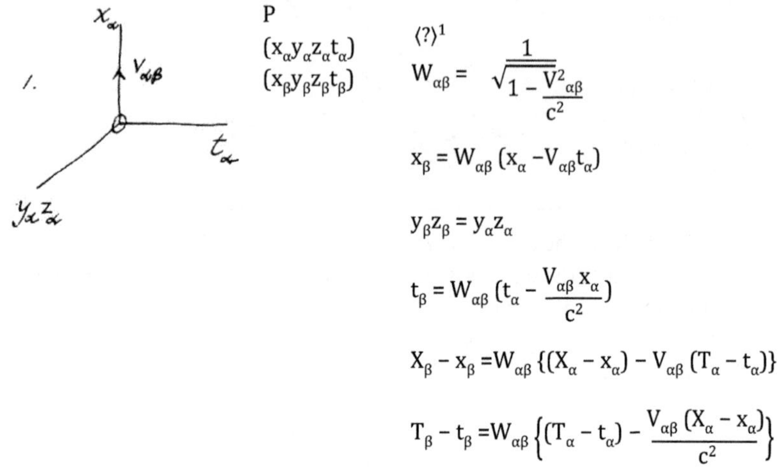

P

$(x_\alpha y_\alpha z_\alpha t_\alpha)$

$(x_\beta y_\beta z_\beta t_\beta)$

⟨?⟩[1]

$$W_{\alpha\beta} = \frac{1}{\sqrt{1 - \frac{V^2_{\alpha\beta}}{c^2}}}$$

$$x_\beta = W_{\alpha\beta}\,(x_\alpha - V_{\alpha\beta} t_\alpha)$$

$$y_\beta z_\beta = y_\alpha z_\alpha$$

$$t_\beta = W_{\alpha\beta}\,(t_\alpha - \frac{V_{\alpha\beta}\, x_\alpha}{c^2})$$

$$X_\beta - x_\beta = W_{\alpha\beta}\,\{(X_\alpha - x_\alpha) - V_{\alpha\beta}\,(T_\alpha - t_\alpha)\}$$

$$T_\beta - t_\beta = W_{\alpha\beta}\left\{(T_\alpha - t_\alpha) - \frac{V_{\alpha\beta}\,(X_\alpha - x_\alpha)}{c^2}\right\}$$

The ordinary Newtonian system.

$x_\beta = x_\alpha - V_{\alpha\beta} t_\alpha$
$t_\beta = t_\alpha$
$X_\beta - x_\beta = (X_\alpha - x_\alpha) - V_{\alpha\beta}(T_\alpha - t_\alpha)$
$T_\beta - t_\beta = T_\alpha - t_\alpha$

Whitehead regards it as a question of fact whether any given body is using any special space-time system. Neither view excludes the possibility that there shall be one unique determination of time. See *Natural Knowledge* p. 163. If you assume relativity you get a nicer doctrine. Connect your metaphysical doctrine with your physical entities –

Get a nice explanation of the different conceptions of a point & of the 3 dimensions of space

1. Hocking seems to have identified someone or something here, within parentheses, which might be 'Mega' but we cannot be sure.

Radcliffe College lectures, 1924–1925

Notes taken by Louise R. Heath on Phil 3b, 'Philosophical Presuppositions of Science', delivered by Alfred North Whitehead

First semester

Sept 27[1]

Whitehead – Philosophical Presuppositions of Science. Phil 3b[2]

General Introductions.

Each age has its own philosophy. All systems incomplete & characteristic of age. Revealing & rationalism[3] human life.

The problem of our day is that of Science as Ethics was that of Greeks.

A peculiar problem for each age of philosophy

Our immediate difficulties are

1. What is the real nature of the Scientific movement which has unexpectedly arisen in our day?

a. Might begin with chief motives to scientific thinking? (Psychology).

b. What is there in nature of things which necessitates that science should have its peculiar characteristic.

Our starting point. Start from experience.

General relations of theology metaphysics & scientific.

1600. Theology died
1700 Metaphysics died } as far

as influencing scientific thought.

Happened because of need of complete freedom of scientific

Method of Least Action Dedekindian Sophus Lie[4]

1. Louise Heath in fact often fails to note the date, and there are few instances where it is clear where one lecture ends and another begins; consequently no attempt is made here to divide the material by lecture. This, though, does seem to be the same introductory lecture given at Emerson Hall from notes taken by Bell on 25 September. Here, the date is written diagonally in the upper-left corner. The dates of each lecture are written in same ink and hand as the notes, throughout, but are such an important marker we have put them in bold to help reader locate them. Throughout this semester, except for a period in December, the content of lectures presented at Emerson Hall seems to have been presented at Radcliffe on the next class day, or later.
2. Although course numbers at Harvard often distinguished 'a' from 'b', depending on the semester in which it was taught, Whitehead's course of lectures at both Radcliffe and in Emerson Hall was 'Phil 3b' and was offered through both semesters.
3. Heath often provides only partial underlines, and we have opted to keep them as we found them.
4. This box is drawn diagonally along the lower left side of the page. It is not clear whether this was an aside within this first lecture. The principle of least action was of interest to Whitehead and of the two

|2|[1]

thought for freedom. We must resign ourselves to rough edges of thought.

We are seeking a rational synthesis – therefore cannot be autonomous entities or groups of entities. i.e. not separate fields of knowledge.

Therefore Philosophy & even theology are capable of serving science in so far as they have arrived at sufficiently true formulations.

Even medieval theology may have fostered science.

Romance or wonder everywhere – how does it pass into science.

Why did it happen in 16th & 17th centuries in Europe – not in 1500 yrs of La Arintine[2] prosperity or in China.

Weakness of Hume's philosophy is no justification for scientific generalities

Scientific spirit must include belief in fundamental rationality. ~~This~~ 16-17th centuries inherited scholastic theology belief in

|3|

rationality of God. Therefore + method of observation → modern science.

Our metaphysic must support this belief in rationality of universal.

Kant – we only know ~~on~~ly edition de-luxe which has this

Forms turning point in metaphysic.

Whitehead's idea of what

Process of cognition is one type of relationship between things which occurs in things of reality.

How is any entity possible in view of its relation to other things?

Another class[3] for same problem is that of inductive logic.? Why are general methods of inductive logic sensible proceedings.

How can knowledge of a give knowledge of b not included in a?

Hume's answer is no knowledge from however large a number of particulars.

We must first establish the question of how one fact

|4|

can be related to another.

Once admit one independent fact & science collapses.

3. Do not get rid of difficulty by saying that in practice nobody doubts it?

4. No help by basing trust on past experience. According to Hume you must assume that If present is irrelevant to future its past is irrelevant to present.

Answer is

1. Becomingness of reality is the process of realizing modes of the object therefore any one entity which is realized reveals a beyondness.

---

named mathematicians he mentions Dedekind elsewhere. Why these all appear in this box together, here, is not known.

1. While Heath provides dates only intermittently (and not at all for weeks at a time), where she goes on to a new page in her notes is of course always clear, and we will mark each page change in this manner, as with the Emerson Hall lectures. The pages may well have been numbered by someone else at a later date, but we will rely upon them. They will become the key by which we can link her notes with the lectures recorded by Bell and Hocking in Emerson Hall.
2. It is not clear whether this is meant to be Argentine, or what it might be.
3. This term is uncertain.

|5|
First step in philosophy of science is the question of togetherness of things.
Can be no horse out of time & space – i.e. togetherness with other things. Its ~~definite~~ one of many relations may be ~~different~~ but must be definite.
This essential togetherness of things has been in past obscured by Aristotelian classification & its success.
Difficulties of science result from[4] exaggerated emphasis – abstract entity, apart from its relations.
Every entity must be studied in its environment.
This to-getherness takes form of a process of realization. (or of becoming)
In spite of this there has been a tendency to believe in static reality back of changing appearance. This has separated science & metaphysics, for science is concerned with this flux.
|6|
Whitehead – Nothing behind the veil of becoming.
Reality is process of becoming, open to consciousness, but our individual consciousness is aware only of small factor. Reality applies to connections & only relatively to the things connected. a is real for b & b is real for a but not absolutely real independent of each other.
Realization is making real of value.
True reality is achievement of reactive significance
A doctrine of complete relativity of reality – expressed from point of view of realist who finds Spinoza most significant of modern philosophers.
|7|
~~Relativity~~
Guard against impression that metaphysical philosophy is merely handmaid of science. Metaphysics is critical appreciation of whole intellectual background of man's life. As near to poetry as science
Concerned with what can be, what can not be, what is, was, will be & will never be.

Ref – Whitehead – *Principle of Natural Knowledge.*
Whitehead || *Concept of Nature.*
Whitehead *Principle of Relativity.*

constructive
Spinoza —
Destructive

Hume *Inquiry, Principle of Human Understanding.*[5]
Last section he brings out his scepticism.
C. D. Broad – *Scientific Thought* 110–12[6] Scientific Method.

---

4. Throughout her notes, Heath seems to use 'fr' as an abbreviation for either 'for' or 'from' (and sometimes 'fm', though it can be difficult to distinguish); in all cases we have expanded the abbreviation according to context, despite it not always being clear.
5. This is presumably Hume's *Enquiry* and his *Treatise on Human Understanding.*
6. The numbers are written in pencil.

|8|[1]

Sept 30.

Even in revolt of modern philosophy from medieval and ancient it has retained its taint – i.e. due to language & subject-predicate nature of propositions & division of matter & its properties. Essence of material not so much as process as static.

Better if had never been a theory of substance. Result has been that in physics, i.e. when they found something going on they looked for a subject which is going – i.e. so came about "ether"

We shall consider processes fundamental. Shall attempt study of Great Ideas of Continuity, cf Atomic Entities. Are places where divisibility does not apply (Inapplicability of divisibility) (Sheep cut in half is mutton not half a sheep). Looks as though question is Is Nature continuous or discontinuous? This is a mishandling. Will show it in the next two lectures. Not in any sense of philosophy lecture but an attempt to give the scientific outlook, of which we are to study the presuppositions.

|9|

Process involves a basic idea of continuity. At present science is leading toward atomicity. i.e. Quantum theory suggest that not only there is a discontinuity as of above sheep but that either Space or Time is discontinuous.

According to Whitehead – space & time must be continuous. Unless he has pointed out how some of requirements of Quantum theory can be satisfied without discontinuity, his theory would be hopelessly out of touch with modern science. While he has not a neat little theory, he thinks he has shown ∨the∨ above.

Fundamental Point of View of all physical science is that the electro-magnetical phenomena on the whole give one the fundamental elements on which universe is built up. Whole theory is built upon group of formulae discovered by both men Clerk Maxwell,[2] at Cambridge

|10|

Continuity of space – when anything is going on anywhere there is something going on everywhere

P = density of electric charge

$(f_1 g_1 h)$ electric force at P

Think of relation of above four forces at P. at definite time t

Consider Temporal rate of variation at P.^{fixed} at time

E.$^{\text{jumping point}}$ = $T_1 v$ of things at P. at time t.

or you may ask how electric force distributes its force all over space at time T.

i.e. of Spatial rate of variation at P.$^{\text{jumping point}}$ at $t^{(\text{fixed})}$

S.V

1. This is the beginning of lecture 2 in the notes taken by Bell in Emerson Hall on 27 September.
2. James Clerk Maxwell (1831–79), Scottish mathematical physicist, and the topic of Whitehead's own dissertation.

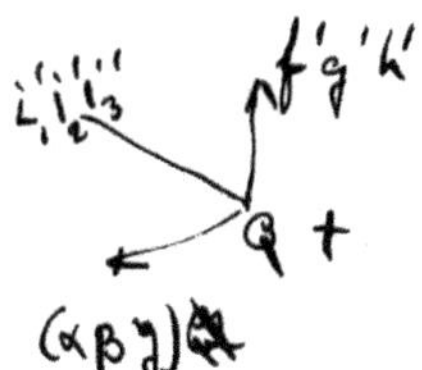

more complicated as T is linear & S is tri-dimensional.

On this matter C. Maxwell studied group of <u>8</u> equations, hold anywhere.

Set I. Quality concerning 1. S.V of (f.g.h.) ~~with~~ = ρ

Two facts[1] 2. Quality concerning | S.ν of × (αργ) = 0

~~Set II~~ When dealing with vector quantities are very apt to have three equations to

|11|

state one fact.

Set II. State one fact relating in a definite way the S.V. of (αργ) with T.V. of (f.g.h.) with $(i_1 i_2 i_3)$.[2]

Set III. Relates in same definite way the ~~spatial variation of~~ S.V. of (f.g.h) with T.V of (αβγ) (with no analogue to current). Note similarities in above.

This is based obviously on idea of c<u>ontinuity</u>.

Then C. Maxwell searched about for an ether to explain this. Whitehead only presupposes an "ether of events."

Next the electrons turned up – are little bundles of atomic charges – all same charge – Protons are also made up of charges.$^{+}$

These two have all charges.

In an electron current $i_1\ i_2\ i_3$ = ~~of element~~ (u v w) electric force times ρ the density = uρ, vρ, wρ.

Next physicists studied the structure of atom. Electro-magnetic theory of matter is

· = – electrons
● = + proton.

|12|

Way to study an atom is either to bombard it or to analyze the light it gives when agitated.

Every atom emits vibration (spectral analysis). Every vibration has a frequency.

Frequency ν.

ν = Greek <u>nu</u> T=period <u>ν T=1.</u> =

Inverse of period is frequency.

A certain amount of light energy

Theory of radiation. comes out & amount is S = integer H = constant ν = frequency[3]

h = Planck's constant.

Can only give out quantities of energy which can be expressed by formulae Shν: not ½Shν

E=energy E = Shν

ET.=action. ET = Sh.

ν Seems to be a sort of ~~animosity~~ atomicity of action.

Dynamics.

1. This appears at an angle in the left-hand margin.
2. Heath began to write '$(i^1 i^2 i^3)$' with superscripts, but crossed them out and changed them to subscripts.
3. These three were stacked up vertically by Heath.

$\frac{1}{2} M v^2$ = kinetic energy.

Seems to be incapable of any atomic motion since v is continuous.

But ∨since∨ ET = Sh

Therefore it is held that Quantum shows discontinuity of space & time.

|13|

Oct 2.[1] ~~But~~

General Remarks – Is prevalent idea that when nature has been observed under one set of circumstances, ∨concepts∨ can be used under other circumstances. Totally erroneous & not witness by science. New theories always turning up which don't fit in. To say earlier forms are wrong is not as good as to say think they are lacking in richness of conception.

Always some connections, ~~but~~ modification rather than overturning. Real change is a gain in richness.

Example of Galileo & Inquisition – both shown to be wrong by the richer theory of Relativity.

Return to last lecture

Prophecy –1: Biology youngest of sciences & there is every reason to believe that the modern study of biology will lay bare characteristics of matter never could be discovered by physics or chemistry, which do not deal with nature as organisms. Study of physiological process as good a ~~g~~ chance as any other for discovery of new physical facts.

2. Intimate structure of matter is essentially rhythmic; an interweaving of occurrences. (a prophecy.)

Will later point out that fundamental fact is process, or transition & something else + retention of value. "Passing on" supplies

|14|

"change" & retention amid "change" necessitates either "identity" (will explain in his theory of objects) or "recurrence"

Accordingly Whitehead should expect nature atomic structures of vibrations to be established within the electro-magnetic field (Last lecture.) Discover vector forces to be factors in ~~rec~~ atomic structures. & essence is a certain vibration or recurrence.

In *Concept of Nature* & in first lectures. Every entity expresses itself in whole of reality & rest of reality is patient of it. Therefore should expect every structure to impress its significance on whole of physical field so that every electron is everywhere. Most fundamental point of view is that of something rhythmic, something recurrent, a focal region (where electron is). & the essential nature of an electron would acquire tentacles stretching out throughout all space & all time.

Enunciated by Faraday.[2] "Tubes of force" are really essential parts of electrons.

1. This seems to record to be something like lecture 3 from notes taken by Bell in Emerson Hall on 30 September, but is the first of many instances where the match is not exact.
2. Michael Faraday (1791–1867), British physicist and chemist. In *Concept of Nature* (p. 146) Whitehead notes: 'As long ago as 1847 Faraday in a paper in the *Philosophical Magazine* remarked that his theory of tubes of force implies that in a sense an electric charge is everywhere'.

|15|
Doctrine of rhythms with regard to living organisms

Stated in Last Chapter of *Principles of Natural Knowledge*.

With respect to Life. A final suggestion. Essence of life is the extraordinary delicacy of its processes.

This delicacy of process depends on a delicacy of ~~rhythm~~ timing. Makes a great difference as to the phase.

Depends on adjustment of the phases of the organisms.

And in this way minute disturbances may accomplish a great deal if they readjust the harmony or disharmony of phases.

Summary of philosophy which we've been discussing.

1. Reality is a flux, a process, a continuity of becoming. Can't call an isolated, simple entity real in itself. There is a process of realization & this involves fact that A breaks in on B & B breaks in on A. i.e. breaking in of essences on each other, ~~constitut a~~

There is realized in social entity is value of various ingredients for each other.

Furthermore, this process can be analyzed into two detached
|16|
abstractions

a. Character of display or sense-data. Greenness of tree.

b. Character of control or of determination of process.

a. Display represents abstraction of realized factor of achievement.

b. Control is abstract character directive both of its own flux & of flux of display. Have not concrete fact until have relations between physical field (of controls) & display.

3. Continuity of flux exhibits atomic structures as imbedded in itself. These represent permanent character impressed in that flux, which exhibits itself as the changing relations of these entities. Permanent & also rhythmic characters.

4. No finite entities (Spinoza's mode) are in themselves independently real, but ∨thought∨ that every such entity is an atomic entity character is a mere
|17|
abstraction.

Oct 4. Lecture on Scientific Method.[1]

Concrete illustration necessary. Much of literature is vague & abstract because it lacks that.

Present situation in science.

Strong reasons to believe that Clerk Maxwell's equations, if not true, are tending toward the truth. Therefore will be our starting point.

1. This still includes material similar to that within lecture 3 from notes taken by Bell in Emerson Hall on 30 September.

Then there are these atomic structures, the electrons & protons. Electrons have never been divided at all. Protons are to some extent breakable, but never seem to get down below nucleus of hydrogen. Except there is some evidence of little bundles being shot off as light, i.e. some evidence for corpuscular as well as wave theory of light.

Then Quantum theory & here energy again seems to come out in little lumps & dissolves into light

If nu ν is ~~measure~~ unit of energy, the energy always come out in

S Integer.

h Planck's constant

ν unit.[1]

|18|

Furthermore there is rather vague evidence that when you get either within nucleus or pretty near it the field seems to go much more in ridges. Evidence of a complicated field inside of regular field of force where ~~ener~~ attraction varies as inverse of square of distance.

Quantum theory has led people to say Space & Time are discrete.

This Whitehead considers nonsense.

Then another point, when you have an electron or proton at rest there is no magnetic effect whatsoever.

Owing to ridge business it is suspected that Maxwell's equations do not apply within the ridges of field.

~~They~~

Whitehead will try to make C. Maxwell's equations apply within this field.

All above considerations must be in back of mind – [Only things that know so well that you forget it that is of any use].

1 ρ = density

2 (f,g,h) = electric force

3 $(\alpha\beta\gamma)$ – magnetic force

4 $(i_1 i_2 i_3)$ current.

|19|

Ordinarily think of (i i i)[2] is only electrons coming thru wires, but Whitehead will hold theory that there is a current at every point whatever & motion of electrons gives us this effect everywhere.

Essential basis of physics is a structure of elements. 1, 2, 3, 4 merely ~~describe~~ define the structure of energy.

Density of energy at a point defined in terms of 2 & 3.

[First must preserve as much as possible the well attested conclusions & conceptions of science.]

Current is vector essentially associated with motion, expect it to be associated with flow of energy.

1. 'S h ν' is written horizontally, with the identifiers written vertically below the characters.
2. These should, perhaps, have contained subscripts, as above.

<u>Now</u> will ~~consider~~ run the theory "That both electrons & protons are built up of ultimate corpuscles which he will call primates. A vibratory structure of one definite frequency.

A great thing if can find two sets of primates, one made up electrons & other protons.

Although electron has never been
|20|
knocked to pieces it may be made up of several primates.

Protons have been knocked to pieces, & then aren't going off = S h ν. Therefore must find something we can identify with h.

If we could do this we would show that knocking energy in b<u>undl</u>es out of proton does not spell di<u>scontin</u>uity although it does spell atomicity.

In a sense proton & electron with its ~~fi~~ influence extends through all space. But it is still correct to put it in definite point of space.

Probably therefore find that must have arbitrary <u>for</u>m of primate. Common-sense to make it a s<u>phere</u>. [Cf. Aristotle & Greeks]

1. Simple law is always better than a complex law.
2. Can reason from it.

|21|
will in first place consider all vibratory. Then 3 goes out because we will consider primate at rest.

Let us put

$\rho = \bar{\rho}\ e^{1.2\pi\gamma t}$
$(f,g,h) = (\bar{\iota}\ \bar{g}\ \bar{h}\ )e^{2.2\pi\gamma t}$
$(i_1 i_2 i_3) = (T_{1,2,3})e^{\iota,2\pi\gamma t}$

$e^{\iota,2\pi\gamma t}$ defines vibration & other factor is amplitude.

When you have done that (f,g,h) is derivable from a potential function. Everything goes in surfaces, spheres ~~of~~ everything at same stage of vibration & f,g,h. is normal (⊥. to surface).

A form which is not being considered & depends on taking $i_1 i_2 i_3$ as all through field.

Then you find that it follows from C. Maxwell's equations that you cannot have your ρ density 0 everywhere if you are to have any structure. Thus
|22|[1]
gives a focal region for primate where ρ is not zero & all rest where ρ is zero

Then notice that inside focal region M. equations are not sufficient to show how ρ is really varying.

---

1. This seems to be roughly the beginning of lecture 4 in the notes taken by Bell in Emerson Hall on 2 October.

Taking simplest conditions – assume that ρ satisfies the wave condition in focal field. Its velocity is electro-magnetic velocity. Then you will find

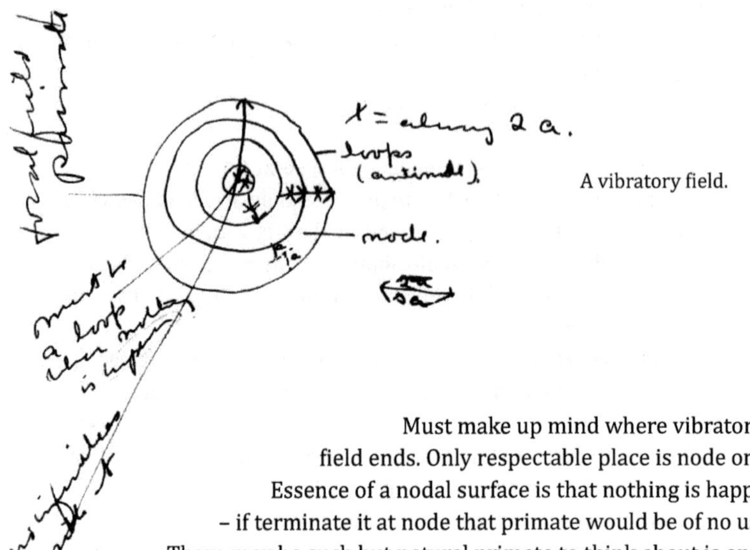

A vibratory field.

Must make up mind where vibratory field ends. Only respectable place is node or loop. Essence of a nodal surface is that nothing is happening. – if terminate it at node that primate would be of no use: There may be such but natural primate to think about is one that ends at loop & communicates itself to all external region.

|23|

Certain conditions necessary therefore for transmission of energy from internal to external regions. Physicists assume that there is no sudden transformation.

Find energy outside focal region etc energy enters by flowing in streams in directions of & determined by current. No destruction or creation of energy outside. Inside where ρ is not zero the rate of change of ρ,(in addition to flow of energy caused by current) remainder of energy is proportional to rate of change of ρ. Therefore wherever have ρ have vibratory creation & destruction of energy.

Oct 7, 1924.[1]

In modern physics mass is looked on as a concentration in an atomic structure. Total energy = e

C = velocity of light in vacuum.

Mass = e/c^2

Two alternative ways of constructing a primate in general way

|24|

described above.

Call extension <$\overline{s\ \ a}$> corresponds to an electronic primate.

Call extension <$\overline{g\ \ a}$> corpuscular primate.

s and g = any integer.

1. Lecture 4 from notes taken in Emerson Hall by Bell on 2 October seems to continue.

Each primate has a definite frequency.
$2a\nu = c$ [From Maxwell's equation]
Therefore when have ν have selected a
for c is a universal constant.

sa = s-fold primate
ga = g-fold primate

Period = T
Frequency = number of vibrations per second = ν
$\nu T = 1$.

What does energy come to
Av. energy of a s-fold primate = $\frac{3}{4}\frac{E^2}{sc}\nu$
E is intensity of vibration
a constant.

Bigger primate the smaller its energy.

Av. Energy g-fold primate. = $G\frac{\pi^2E^2}{16c}\nu$

very nearly = $G\frac{3E^2}{5c}\nu$

According to Quantum Theory energy = $Sh\nu$
S = integer G = integer

$\therefore\ h = \frac{?E^2}{5e}$

we have found h as we sought.

This shows us that E has got to same for all primates whatever their frequency.

If you look on proton as mash up of
|25|
primates you either have to shake out
1 – 2 – 3 etc $\frac{3E^2}{5c}\nu$ account for Quantum theory without any worry about alleged discontinuity of time & space.

Must suppose some conditions of permanence. Whitehead has not worked this out.

Not ordinary point of view, but satisfies Quantum Theory & proves that Quantum Theory does not necessitate discontinuity of time & space. Illustrative of way of scientific thinking.

Must presuppose perfect timing for the steady attractions. Will get into difficulty with theory of relativity; but one mustn't be too worried by Like difficulty about shadows for wave-theory of light, which were gotten over by better math & more sensitive tests etc. etc. –

Recapitulation[1]
Two fundamentally different ways of looking at nature of things. (Both in Aristotle.)

1. (Note his Logic). Predicate–Subject type of proposition (starting point)→ Quality–substance (as foundation philosophy).

1. Heath begins here to present the content of lecture 5 in notes taken by Bell in Emerson Hall on 4 October.

With this process & time tend to be unrealities

Configuration at an instant is fundamental. Or

2. Start with Becoming or Process as fundamental.

Process at an instant is nonsense

|26|

Process is essentially transition to otherness

Idea of "at an instant" is pushed into background as derivative.

A slab or duration of time becomes the most concrete fact.

Transition during a duration is event & is fundamental.

Whitehead's view is that Aristotle had both of these but Aristotelian tradition emphasized 1 as in Logic, more than 2 as in the metaphysics. This has colored or tinged ideas of modern philosophy.

On 2 must describe "an instant" as a higher degree of abstraction. (see *Concept of Nature*) explained as a root of approximation.

On 1. duration becomes a series of instants & that is more abstract than instant.

Because Russell's difficulties with causation very largely depend on assumption of 1.

1. Has advantage of simplicity.

On 2. For simplicity's sake make a sort of approximation & get idea of velocity or rate of change at an instant.

For Process must have foundation of continuity. On Process must embroider

|27|

some idea of atomic structure. [We've done that in our previous lectures]

On 1. must embroider an apparent continuity on top of atomicity.

Two types of scale wanted temporal is ν or velocity & spatial or quantitative is E or energy.

Sort of considerations necessary for forming all necessary concepts

1. Best empirical concepts which already exist.

2. On dealing with these generalizations, you must search for generality. For example, held today that Maxwell's equations will not apply inside a molecule, this is a inadequacy. & an argument for Whitehead's way of running Maxwell's equations through all space.

3. Utilize minor suggestions for empirical knowledge. Ex. $(\alpha\beta\nu) = 0$ when primate is at rest. Suspicion that corpuscular theory had something in it.

4. Use of simplicity.

5. Curious interweaving of induction & deduction.

6. Get up against rough edges of thought.

|28|

Oct 9, 1924[1]

Keynote – What is it that we do observe? In what sense, if any, do we observe the physical field? Answer is absolutely essential.

Space-Time-Motion: An Historical Introduction to the General Theory of Relativity. By Prof A. V. Vasiliev.[2] Translated by H.M. Lucas & C.P. Sanger with introduction by B. Russell. B. Russell's Introduction[3]

"The philosophical principle that distinctions which make no difference to observable phenomena display no part in physics, has had great effect on philosophy. Called phenomenalism & very influential in modern philosophy. Although it might seem self evident to any empiricist, it cannot be said that even the general theory of relativity implies it. ? if it is possible to construct any physics in accordance it? Assumption of other objects, or persons is impossible. Contrary ∨to∨ it to assume that observed is wider than observer. Phenomenalism may then be regarded as an unattainable ideal.

note – There is a contrary principle to the effect that the general body of physics must be obtained. Physics tries to dodge between two principles."

|29|

Whitehead's comment on above.

B. Russell has taken phenomenalism as derivative from Hume,. as observable in sense data & relations (space, time, resemblance, contrast) Anyone would admit that in general unobservable has no influence, but then there is slipped in the Hume definition of what is observable.

Also in last paragraph slips in Hume's method of getting out of difficulties. i.e. Hume instead of saying that either premises or reasoning is wrong – ∧principle∧ says that "in practice act on things which philosophy does not justify. [Note cited reading Not as a practical principle for a moment, ~~but~~ until can get philosophical principle right, but as if philosophy were right & would permanently have practice different.]

Again Whitehead is skeptical as to obviousness of a great principle with regard to science, if we find that we can't apply them now & never can. Furthermore it is an audacious statement to imply that Einstein's work represents a step toward the adoption of phenomenalism. Especially as it kicks out idea of field of force which seems to be resolvable into sense-data & puts into its place the 16 tensor equation as a condition of space-time. Weightiness is removed as fact & there seems to be no advance toward phenomenalism.

|30|

Take it more generally. What does principle of phenomenalism mean to exclude? Does it exclude any evidence, for any field, which if true, would be relevant in physics? Certainly not – no watertight compartments in

1. This seems to be the beginning of lecture 6 in the notes taken in Emerson Hall by Bell on 7 October.
2. See footnote 2, page 23, lecture 6, in Bell's notes.
3. The source of the following passage is in Bertrand Russell, *Collected Papers*, vol. 9, pp. 285–91, although what Heath has attempted to transcribe does not match Russell's original introduction in some particulars.

knowledge. The relations between different fields may be vague & therefore not much useful help, but if there were a relation it should be considered. Only a practical Warning ∨not general principle∨, that inaccuracies present in all thought are of great importance where fields are related.

B. Russell & phenomenalism mean that we are not to introduce into physics any entities of which we have no direct evidence of their existence or of their relevance to physical phenomena. ∨This principle∨ wants limiting very much on lines that Russell points out. i.e. we would not rule out molecules in middle of earth just because we can't get them on the surface.

To return. What is it that we do observe?

1. Phenomenalism is right in that it excludes purely hypothetical entities?

a. Means "sorts of" not minor differences. i.e. Cardinal Newman's angels pushing around planets. Not because there are not angels, but because there is no direct evidence of the habitual interference of that type of entity in physical phenomena.

|31|

b. Superadds the premise drawn from Hume are sense-data & certain relations of sense-data.

2. How did science get itself going. Obviously by considering those ordinary entities ∨(watch)∨ which in common life we presuppose, & the qualities of those entities (sounds, etc.) & bodily feelings. Then now it presupposes electrons, protons, electro-magnetic fields, (19th century the ether) energy-quanta & all are bricks of the general structure of things. ~~Expressed in terms of~~ Bodies, such as stones etc. are close-packed aggregations of these fundamental things. Things which in ordinary life are predicated of entities (colors) are secondary qualities or mental reactions to the stages of ∨spatio-temporal∨ adventure which our bodies (molecules) are going through. [Note *Concept of Nature* for Whitehead dissent.] Whitehead & phenomenalism agree that above point of view is untenable. Whitehead feels that idealism involves ~~subje~~ solipsism.

Scientific view is that only way secondary qualities interest science is that thereby we gain knowledge of spatio-temporal adventure of the entities. This phenomenalism wants to sweep away.

|32|

Quite obvious that can't rest here.

1. On scientific theory what has become of stones, chalk etc. Do we observe them?

2. What are the we who are doing the observation? Our relation to the electrons? i.e. Really are three types of things presupposed in naive point of view of science.

1, Observers. 2, Things observed. 3, Entities.
condition observation.

3 interests science. & the vague idea of ordinary scientist is the 3 affects 1 to observe 2. That Whitehead in his books calls 3 perceptual objects & scientific objects (electrons.)

Kant makes observer{1} construct for himself, perceptual objects{3} & adds Things-in-themselves i.e. the ghost of entities{3}.[1]

Kant rescues observer from skepticism by making him construct phenomena out of things-in-themselves (which are unknown).

How does he rescue him from solipsism. Great influence of mathematical physics– Kant & also logic.[2]

Hegel & Absolute idealism did this by putting center of gravity outside of observer.

|33|

Oct 11[3]

Concrete fact is not also 1, 2, 3, but entities which by abstractions are observers{1}, display{2}, physical field{3}.

One of critical dividing points of various philosophies are the proper point of abstraction.

Wrong abstraction is fallacy of misplaced concreteness. Usually taking things to be more concrete than they really are.

Also the converse.

First question. What do we mean when we talk about electrons & electro-magnetic fields?

1. Electrons are mere myth. A way of stating things, useful but mythical. [K. Pearson[4] calls it a description of routine.]

a. Sheer mythical theory, pushed to an extreme, seems to be nonsense. A mere myth not only does not throw light, but is intrinsically incapable of throwing light on anything.

b. What they mean is that the electrons, etc. have certain VpurelyV formal relations among themselves, of spatio-temporal character which are the same as entities which we can discern as arising from the sense-data. Therefore When we have reasoned about the consequences of these spatio-temporal-relations of electrons & merely stick to those relations we have equally deduced truths concerning the

|34|

spatio-temporal relations of these entities we discern as arising from the sense-data.

Why not call entities arising from sense-data electrons? There is a point; that the entities we discern are of a very highly abstract character, requiring much logical intuition in order to discern that they are of classes, (abstract & complex) & that they are very difficult to think about, whereas the electron is thought of as more concrete mental sensibilities for above mentioned entities.

1. Here and on the next sheet of her notes (|33|, below), Heath places a total of six superscript numbers directly above their respective terms. These have been placed in brackets here to avoid confusion with our series of footnotes.
2. These two lines are written vertically up the left margin next to the two paragraphs on Kant.
3. This seems like the beginning of lecture 7 in the notes taken in Emerson Hall by Bell on 9 October
4. Karl Pearson (1857–1936) was an English mathematician. Pearson is referred to only in the first of the Emerson Hall lectures.

A great deal in above explanation. But – if you are going to knock out all conditioning entities you raise Hume's problem. ~~If.~~ Therefore Essential for rational philosophy of science that there should be some status for conditioning entities.

When you come back to the concrete fact, the most concrete fact you can think about is an event, part of its essential nature comprises transition. From one point of view, if you abstract from that event, you obtain observers, from another you obtain display, from another you obtain conditions. The event itself is only relatively concrete. Absolutely concrete is in whole totality of process from beginning to end.

(Concrete universal?)

|35|

Assuming there is a physical field, science considers the goings on in physical field in themselves & display as evidence of physical field. If you analyze what is meant by "physical field in itself" you find it is merely in terms of what influence it exerts on itself. (in future) Display indicates rather conjecturally the physical field.

viewed in present it indicates display & display somewhat conjecturally brings one back to physical field. If as Hume supposes there is no ~~condi~~ connection of display & physical field the physical field → physical field is mere myth, & all is nonsense.

Back to main point.

Where are we to look in the present in display for any ~~evidence~~ conditioning entities. From conjectural character of scientific objects (physical field) ∨it is clear that∨ what we do observe must have lack of precision. But if we are to be at all sure it must be insistent, although imprecise. In perception (to Whitehead) the object is more insistent than the color, shape etc. i.e. than the sense data.

Three theories as to status of object.

1. B. Russell is Class theory. We deal simply with a class of sense

|36|

data associated with a space-time root.

2. Substance theory – Subject underlying all predicates observed. Predicates become secondary qualities.

Whitehead.

3.∨Perceptual∨Object is really your apprehension of the controlling aspect of the concrete character of things, & that controlling aspect can be discriminated as associated with this definite spatio-temporal location.

Object under very differing conditions (including the conditions of your brain) in conditioning the display of sense-data. That is very insistent but somewhat vague. Our primary apprehension is apprehension of display generally as conditioned & that we further discriminate that vague apprehension as being condition by spatio-temporal-locality & these are objects we perceive. That localization is very apt to go astray.

Objections to 1. Class theory.

a. We do apprehend our sensa as conditioned. i.e. appeal to immediate consciousness.

b. Vagueness of class theory can change sensa. What shall we associate in order to classify. Bound to think that there are only sensa & their appearance in time & space.
|37|
Cannot then bring in "lump of chalk" to explain how you classify the sensa. This must be done simply by continuity in time and space.

c. Never completed. Always different from instant to instant. Class of sense-data of chalk yesterday are entirely different from now.

Whitehead – Idea that as we perceive sense-data, they quicken our apprehension of their own display as controls. In a more general sense we have in addition to sense-data we have strains of sense-data. That the Berkeley criticism of the whole externality of things really applies to a very over-simplified conception of the way sense-data are interconnected with the control. This over-simplification springs from the subject–predicate analysis of fundamental logical nature of things (originated with Aristotle) & that therefore Berkeley was considering yet a third theory of sense object that it was the substance of which we perceived the qualities. Berkeley comes to realize that everyone sees a different set of qualities. Quite fatal to the quality–subject point of view but does not touch the theory of Whitehead i.e. the ingression of
|38|
sensa, of a very highly complex character, involving irreducible, many-sided relations which cannot be separated, one from another.

Philosophers here tended to ignore these complex relations because they are difficult to think about, or some philosophers have ~~infl~~ insisted that relations are essentially two-termed.

This Whitehead cannot understand. Further a tendency to look on all relations as reducible to the predicates.

Whitehead thinks this a linguistic cook-up.

In dealing with control theory think of it by considering some vivid experience V(sense-perception)V. i.e. school-boy catching a cricket-ball in an important game; & see whether one apprehends a "class of sense-data" of that ball since the beginning of the match. No worry about that if there is "the ball" underlying it. but otherwise there is no way of deciding on continuity of sense-data. But boy thinks of it as a definite entity, not as class of observable phenomena which may have change, i.e. in color, in game. True explanation is that ingression of the controlled display involves very complex relations, & what science does is to take apprehended strains of control & make them more precise. Called in *Concept of Nature* – scientific objects.

Physical field enters by way of perception of objects & . Therefore electron is not a "mere myth" but an

~~III. Time.~~[1]

|39|

attempt to make more definite (conjecturally) the classifications of entities belonging to this observed genus, but conjecturally as to species.

2 ————————————

Explanation.

In discussing perceptual objects, we are at one of points where philosophers diverge. Here Kant involves conceptual machinery where by order replaces the unrelated sensa which he had received from Hume. His center of gravity is the knowing subject, so that for him time & space are forms, while objects in causation are outcome of conceptual activity. He is using a control theory, i.e. control of sensa for our apprehension only. Instead of Kant's question How is knowledge possible? Whitehead asks How are things known possible including my knowledge of relation between observer & things known? He finds things known imbedded in a process of realization. What is analysis of the process, including entities which maintain their self-identity amidst the transitions inherent in the process & also including as it does cognizers,

|40|

cognizing the *cognita*. Can we not define this in terms of something we can discuss vividly in cognizant   Wants to think of it as something which represents fundamental character of realization. (Cognizance is most real & complicated form of realization) His machinery is far below the conceptual level. (Kant over intellectualized the experience.) Accordingly it is Behaviorism in which the end to be attained is in a sense already

Like Platonic Idea.

there conditioning its own attainment, so that end itself emerges from existence into realization. But the togetherness of End & Behaviors is not effective in way of concept in mind of knower but is a fact. Ultimate principle is identity of efficient & final cause in concrete fact. Process of realization is a process of achieving mutual VreactiveV significance, which has stages:– faintest stage is merely having a relevance generally for the process as a whole; but that the higher stage is where it attains a particular relevance, under limited conditions for other entities. Limitation

|41|

being supplied by time-space which characterizes process. Thus process of achievement is envisagement of concrete fact. Achieved value where entity has been drawn into some complete reaction. is Display. Then included

1. This is written upside down.
2. Heath does not indicate the start of a new lecture, other than perhaps with this line, but this is now likely 14 October and seems like the outset of lecture 8 in the notes taken in Emerson Hall by Bell on 11 October.

partly in display, but abstraction, this includes an urge to transition, i.e. condition the process. This envisagement of existents where a concrete fact, is the efficient cause conditioning the process. Whereas the existent quā envisaged, as a candidate for realization is the urge to transition is final cause. Only final cause & efficient cause are some thing in concrete fact of envisagement of existent.

And therefore to substantiate this analysis it is essential that we should be able to discover among our observed things, a functioning of conditions: i.e. entities which self-condition the process. In order to do this point to conceptual objects as strains of self-conditioning process. Very vividly but vaguely observed. Lack of precision comes in, i.e.
|42|
1. limits of perceptual objects 2. i.e. object seen behind a mirror, i.e. location. spatially.

3. stars are indefinite. temporally

4. Everything depends on community of space-time. Whole point is unity of process. [Psychological procedure has very adversely affected the metaphysical view of space-time. We have measurements of space-time which represent real space-time facts but of interest to us alone. But entirely dissents from being psychological space-time & the scientific space-time] Takes very seriously fundamental premise of phenomenalism – that something ought be brought into science which is really unobserved. Nor can it be

[1]A sketch of metaphysical standpoint we are approaching.

1. Rejects question How is knowledge possible? but discusses act of knowing. But when ask what is the character of process of being realized it is natural to ask what is general character of the process of reality which involves ourselves. Then what do we mean by ourselves being real? In some
|43|
general sense our reality consists in our knowing things i.e. taking account of other things. Taking account of involves two or three more abstract factors. 1. Sheer display. [appreciation of what entity is in itself – as display. a strain of control.] 2. Satisfaction or dissatisfaction. i.e. ∨Display is there as∨ conditioning the transition.

Must we then assume that these entities have about the same kind of knowledge as I have. But notion that knowledge is definite, sharp, all or none process is all wrong. We know dimly, more or less, & we can discern in ourselves a taking account of which is not really cognition i.e. is below consciousness.

Cognition is a reflexive process. ~~Taking account of is~~ Realization is a reactive significance between things. Thing in itself becomes significant

1. Heath's notes might from here have been taken on 16 October (she does not indicate a date), as the content is something like that for the beginning of lecture 9 in the notes taken by Bell in Emerson Hall on 14 October.

for the other. – in a sense this is a fusion of two individual essence. But cognition Suppose that fusion, A envisaging B, is reenvisaged by A, so that A takes account of its own envisagement of B, then we have
|44|
conscious knowledge. A taking account to B is A's image of B. Entirely deny "private worlds" which Russell is so fond of. Metaphor of idea within the mind is absolutely wrong. Rather the mind is in idea. i.e. A envisaging B is idea. A envisaging of A envisaging B is conscious knowledge.

An idea is a social entity, i.e. involves both A & B.
Spinoza's message.

If you are going to look on reality as a fusion of the individual essences, ~~if~~ V& haveV you start with a plurality of existents, it is a little difficult to understand how you keep your existents separate from each other. Obviously something is left out. i.e. You must get that fusion with limitation i.e. must take the form of a limited fusion, controlled or expressed by definite relationship. & it's there
|45|
that Whitehead brings in idea of the relational essence of things. i.e. how a thing can enter into VlimitedV reactive significance with other things. How is significance – patience relation.

This system of relationships both limits & makes possible relationship. This is the field of Spinoza's Modes. [Difficulty about Spinoza is having got his Substance going there is no reason for anything happening]. so ... shifts center of gravity to modes. Modes always in reference to space-time. i.e. Space-time is field of possibility & also of actual realization. A primate is a structure in space time. These structures give structural limitation under which the fusion takes time & at same time a field for the exhibition of fusion. ~~Essence of real~~

We've got to describe not something that is partial, with something else back of it, but the totality – think of fact as a totality – amid presuppositions such as assuming space- & time. Describe this process, with its conditions which characterize it, – leaving
|46|
nothing out. Basis must not be another process, needing another basis – but rather its standing character.

Mere being – this is compatible with complete isolation & therefore is equivalent to a nonentity for everything else. Can't prove that there are no mere-being but for us is nonsensical.

Existent – ~~as~~ in its simplest relation, involved in process of realization. [Like Platonic Ideas.] The relation of the totality of existents – how related as implicated in process which forms the standing character of the process. That standing character is really what we call possibility. This must be found from the envisagement of the totality of things. [Spinoza's God.] Taking account by which procedure determines elements for realization in process [Spinoza's modes]

Modes – Reactive significance in some stage of the process. But share in limitations & envisagement of whole.

|47|

Whole in one sense is static, but view as shared by parts is process. Partitionment of envisagement is always at a new stage of reactive significance.

Process of realization involves.

Display = achievement

Valuation = Selection

Directive passing on to otherness of modes.

Seems to have explained whole thing is terms of final causes. But the concrete fact is both a final & efficient cause. i.e. is VaIsoV *a tergo* behind in the Existents, & is reflected in science as the efficient cause.

But from point of view of existent to be envisage it is a final cause. Final & efficient cause are identical except that final cause is an abstraction.

|48|[1]

|49|

Oct 18, '24.[2]

Beyond: What do we mean by beyond:

What do we observe corresponding to it? It is obviously relative – beyond something that is immediate for us now. Apprehension is not limited to the actual sense-data in this room ~~now~~, and their display. Beyond sense data is obviously other than sense data.

1. Mere otherness tells us nothing; it may be related to the immediate fact of my sense data, VorV experienced in one way – namely it is other than. Mere otherness does not exclude complete isolation – leaving each entity by itself alone. Viewed as sheer isolated plurality you have sterility.

Whitehead – . Whatever logic is satisfied with: in fact mere otherness or mere identity is an abstraction founded on specific distinction, embodied in ~~a~~ some mode of realization in one relationship; as thus envisaged there is a specific contrast value – as a specific identity value thus exhibited – Realization is establishment of definite relationship. Identity establishes identities as mutually relevant in their individual essence & so capable of joint envisagement in their relevance. There status distinguishes them as expressible in that relationship with a contrast or identity value, which will be specific. The foundation of every observation of identity or diversity is a more fundamental specific & relational diversity between modes of

|50|

realization. Such an observation requires a diversity of instances which is a relational fact. These diversities are within instances.

Relation of beyondness is relation ~~of~~ which leads to observation of diversity or identity as between individual values. Diversity means

1. This page has been left entirely blank.
2. This material is like that at the beginning of lecture 10 in the notes taken by Bell in Emerson Hall on 16 October.

difference in way they occupy space-time. – lies wholly in individual essence. Relational essences – very complicated. – relation to space-time very complex. Individual essences are difference. relational essences are same. Redness & greenness are individual essences of red & green. Red & green differ from a kind deed. Different in a more fundamental way. – different in way are in space-time.

In talking of beyondness there is always a possible identity. Space-time relation if then of beyondness if relation is such that relata occupy regions of space-time in same sense of word occupy.

Motion of beyondness – relation of beyondness. Allows this relation to appear as relative in different appearances – is conferred by the station in conjunction with individual essence. Individuality of green patch is conferred by patch which is green as well as by part green which is patch.

We get idea of beyond – always a certain identity of ~~re~~ essences. Duration of ignorance – Ignoring is the basis of abstraction. How come in?

|51|

Two distinct factors in knowledge:

1. Where – Space-time relation Where & When – but do not know inside.

2.

Cognizance means awareness of happenings as well as of individual essences. [worked out as happening is knowledge of ingredient of essence].

Cognizance of relatedness is cognizance of a happening merely as (space-time) related to a happening of which I am cognizant by adjective. Where I am cognizant by adjective is full perception.

Cognizant by adjective includes cognizant by relatedness not essence. Something beyond springs from bare possibility of bare cognizance by relatedness. Anchored in the immediate full perception. When abstracting you are reducing. Every finite entity is an abstraction.

Whitehead – theory of variables – namely use of term any – some (abstract & general terms) there is not a definite entity which is any man. What is there in fact, what is definitive of fact which any man refers to. You are always dealing with an indication – some method which indicates a definite entity but fails to indicate full individual essence of that entity. You can think of some one entity – you can indicate entity without its individual essence.

|52|

Oct 23.[1]

Idea in back of Kant's mind – in introduction to *Critique*.

"Experience tells us what is but not necessarily as it is. Never gives us

1. Heath jumps from 18 October to 23 October. The sequence in Emerson Hall, found in Bell, would suggest there should have been one on 21 October. In Emerson Hall, Whitehead spent that Saturday morning responding to questions, and so he may have not repeated that at Radcliffe. Therefore, 23 October at Radcliffe still echoed Whitehead's lecture 12 in Emerson Hall on 21 October.

general truths & our Reason therefore is aroused rather than satisfied. General truths are independent of experience & have necessary truth. For even if remove from experience everything belongs to senses these remain. Draws distinction between organized experience & mere experience, or the given.

Somewhat blurred distinction between experience enshrining of generality ~~experience~~ & distinction. Our final form experience depends on general proposition.

Kant had been reading Hume i.e. Two principles

~~Here K. &~~ 1. No objects imply beyondness.
~~H agree~~ 2. Constant conjunction does not give reason for induction.

All ideas from impression therefore search for idea → necessity.

Kant & Hume agree that any occasion of VbareV experience never gives us really general truths beyond. Hume says nothing but bare experience [Whitehead agrees]. Kant says

|53|

is a generality in complete experience provided by activity of cognizant. – A general activity, applicable to all exp & therefore goes beyond occasion & gives general truth.

Whitehead disagrees with Hume & Kant in that ~~occ~~ VexperienceV gives no beyondness. VButV If are any truths, they are to be deduced by analysis of that occasion & not by any procedure irrelevant to that occasion. Believes that Aristotle would agree with him.

Read Ross' *Aristotle*.[1]

Ways of getting out of Hume's dilemma.

[1. An occasion does imply a beyond. [Hume is saying no beyond – really think is no beyond via causation.]

1. ~~2~~. Hume & B. Russell – Appeal to Practice. Whitehead says Practice lies outside their Philosophy.

2 ~~3~~. W. James with Pragmatism. by taking practice into his philosophy. [Whitehead – not wholly successful but very valuable.]

|54|

3 ~~4~~. Kant's way. Deny that we have no idea not from an impression. Some ideas are not from without.

4 ~~5~~. Some agree with both of Hume's points & then are always making postulates. [Whitehead If you have no knowledge, a postulate is no use.]

"Because Heaven lies about us in our infancy, I don't see why we should lie about it in our old age." J.K.S.[2]

5 ~~6~~. Disagree with first. Can get beyondness.

6. ~~7~~. Might disagree with both. (Maybe Aristotle would??). Whitehead hopes not.

---

1. This is probably referring to Sir David Ross' book *Aristotle*, which was first published in 1923 and remains in print today.
2. James Kenneth Stephen (1859–92) made this rejoinder to Wordsworth.

Where Whitehead diverges from Kant.

Caird[1]–Kant: "Whenever therefore we can make any universal assertions as to objects presented through sense ... our assertions must be based on nature of our own sensibility*, & not on nature of object "affected."[2]

~~*. Would agree down to.~~

*. Does disagree on negative statement after.

|55|

Down to *. based on presuppositions which render it ambiguous.

Whitehead differs on whole question of how far you can get beyond immediate senses. Can get universal statement? True in most complete sense cannot have any. But are grades of universality relevant to particular object you are dealing with. What do ~~yo~~ we mean by universal assertions & objects presented through sense.

Take "that particular green patch". In considering it Caird–Kant is guilty of V"fallacy ofV misplaced concreteness"

Says we do the "patchiness" & outside world does "greenness". Whitehead says concrete cognizant, stands in respective otherness before cognizant & more concrete universal than greenness or patchiness but is green patchiness. Any dissent is product of conceptual activity of mind. – abstraction ignoring equally important factors.

|56|

Greenness VaioneV confesses its abstract character by its inability of realization – Exists for thought but not reality.

Greenness therefore is not a sense-data but apprehended as result of conceptual process & immediately perceptual data is green-patchiness. Greenness is individual essence. Patchiness is relational essence. Patchiness is modes of spatio-temporal limitation essential for realization of greenness.

Difference Caird–Kant – Greenness is naked data. Whitehead – Green Patchiness is naked data. i.e. standing up in front of cognizant in its objective otherness.

Whitehead goes further & thinks green patchiness is also a conceptual abstraction. The real naked-data is that particular mode of green patchiness as triangular, voluminous etc.

|57|

Particular "mode-iness" of sense data is most concrete universal of sense data type. Universalize it throughout its beyond.

Admits has made too rash a jump all through. Has ignored relativity which is essential to whole procedure of abstraction. Namely started from

---

1. Edward Caird (1835–1908), Scottish philosopher. Bell records Whitehead quoting Caird at the end of lecture 10 (16 October); see footnote 3 on page 38.
2. From Caird's *A Critical Account of the Philosophy of Kant: With an Historical Introduction* (Macmillan, 1877), p. 233.

VaV definite occasion of realization of green patchiness. Least abstraction "particular mode-iness of green patchiness" can be universalized in relation to spatio-temporal beyondness which is required by & Vthat "particular mode-iness of green patchiness"V in which it is set. Abstraction is thinking of it as a universal experience, equally possible through this VparticularV relevant space-time.

Can go also to a higher abstraction any extension etc. in any space time. But particular space-time has its
|58|
characteristics assigned by the particular mode-iness of ---.[1]

Must not confuse degrees of abstraction. One degree is relevant to each particular experience. It is this procedure of abstraction which lies at base of all "beyond" & "beyondness" of that patch is not "beyondness" *in vacuo* but as disclosed by that particular occasion. As long as keep your relevance to a particular occasion, have abstraction in 1st degree is different from particular anchorage in particular occasion in 2nd degree.

Occasional abstraction
Praetor (is beyond) occasional

Usual form of fallacy in misplaced abstraction is mistaking 2 & 1 or 1 & 2.

Logicians talk of "universe of discourse" (James) is that manifold of entities including relations receiving attention in logical thought. According to Whitehead
|59|
is that there is still more fundamental type of universe – universe of perception. Manifold of perceptions included in occasion of beyondness derived from some definite mode of realization. Furthermore a universe of perception is closed in the sense that it is delivered by any one of occasions within it. An occasion of realization bears in itself "knowledge by relatedness" of happenings in that "universe of perceptions"
~~Requires a further consideration~~[2]
But going much too far to say off hand that I shall never have any experience of another universe of perception. It's untrue. i.e. dreams are in a different space-time continuum. A vague relationship but obviously not the usual simple one.

Agrees with Hume etc. that all our knowledge comes from observation. But Whitehead says any limited
|60|
entity we observe always confesses itself as an abstraction, by exhibiting as essential to itself a beyondness Vwhere contentsV described in terms of "logical variable" – devoid of content but indicating definite relation to starting point. No entity is self-sufficient.
"Logical universals" are outcome of this fact.

1. It seems that Heath could not catch these words.
2. It is not completely clear whether this line is meant to be a strikeout or an underline of the text above.

———————————————[1]

Back to Caird–Kant quote

Can make no ∨universal∨ assertions in praeter-occasional sense.

Therefore rewrites. "Whenever therefore we can make any occasionally-universal assertions, as to objects presented through sense as being in the universe of perception derived from any definite occasion of realization, whenever we can say of such objects that they must be such & such our assertions must it seems be based on the nature of the occasion from which the universe of perception is derived."

Our own sensibilities are part of
|61|
the occasion.

———————————————

Grades of abstractness.

1. This modiness of green patchiness.
2. Green patchiness.
3. Greenness.

or

1. This modiness of green patchiness.
2 This modiness of color patchiness.
3. color
4. coloration.

Many other ways. But there is a lowest species of existents, i.e. taking up green chair.

This modiness of green patchiness

exact spatio-temporal limitation – & perceptual world ||

"an existent informs & represents the greatest particularity the simplest thought can obtain. It is equally relevant throughout its space-time but carries with itself the structural modality (limitation) of its
|62|
relational essence. It relates this ~~limitation~~ structure to anywhere at any time & requires spatio-temporal geometry by fact of this modality. Its modality (is form etc.) has an inherent limitation (spatio-temporal-structure) & an inherent universality. (anywhere)

is spatio temporal pervasion

In this respect is an essential difference between space & time [in spite of great truth in modern relativity.] If chop up time have time spectacle case – but if chop it up in space you destroy it. i.e. the modality you really think of has more spatial relativity to display while time is relative to endurance.

1. Heath's line here may indicate the transition to next lecture, on 25 October, as the following content is like that at the outset of lecture 13 in the notes taken in Emerson Hall by Bell on 23 October.

Lurks in the math under the $\sqrt{-1}$

Is there an existent greenness etc. an entity for each degree of abstractions. Agrees with Aristotle, as far as higher abstractions) that these abstractions
|63|
are only factors discernable in lower abstractions. [i.e. No greenness Idea.] Differs from Aristotle in saying that lowest existents are to be taken as having Being & relevance to Process of Realization apart from any particular occasion, but their relevance in addition to particular occasion are the static elements out of which the character of the whole process is built. i.e. determine the character of real world. Lowest existents are Pure Forms [like Aristotle's God]. And Aristotelian ὕλη would take to be <u>value</u>. [like Hume's term impression]. No such thing as value by itself.   always an emergence or a becoming of these forms. The "Ideas" never are real but what is real is the emergent entity, their relations etc. Value as shaped is ingression of these forms into events. Objects to subject–predicate point of view & substance-qualities because it
|64|[1]
has come through simplification of language & does not do justice to the baffling complexity of reality. Three sides in perception

? ? &

occasional particularity.

Aristotle on "Becoming": As he fits into modern thought. – Authority – Ross[2] on Aristotle.

Aristotle's Logic – extra-acute technical device has been misleading in philosophy.

R.A. 65–66.[3] Two sorts of things as coming to be.

Man becomes musical &

A becomes B

unmusical becomes musical.

Not A becomes A.

But what always happens is <u>a-not-b becomes ab</u>

Two elements substance & form.

Third element "privation-form" is presupposed by change.

Matter-Form-Privation. ---

Explanation. 1. Thing comes into being from its privation: but it comes into being incidentally: not from bare privation, but from privation in a substratum. othing comes into being *simpliciter* from being – from what incidentally is & not as it is mere being.

2. Grades of Being – Potentiality & Actuality.

---

1. A few lines down Heath is recording what seems like the beginning of lecture 14 in the notes taken in Emerson Hall by Bell on 25 October.
2. This is likely referring to Sir David Ross; see footnote 2 on page 48, lecture 14, recorded in Emerson Hall on 25 October.
3. Referring to Ross's *Aristotle's Metaphysics* (Oxford University Press, 1924), pp. 65–6.

|65|

Whitehead – two inconsistent sides of Aristotle are here. Starts from subject–predicate point of view. This presentation of change somewhat excludes transition. Becomingness has been left out.

Therefore writes rest of paragraph to explain first part. 1. Slips in Matter & Form. They are not so worried by Law of Excluded Middle the b & not-b. Finally in last clause get to Potentiality & Actuality. a-not-b now seems to assert two things: A is not B, but A is potentially B. Seems to upset subject-predicate point of view.

"*Simpliciter* – incidentally & privation is a substratum" can only be explained as admitting that subject-predicate is only a cloak or shorthand device.

p. 176–177. Different meanings of δύναμισ mainly interested in it as "potentiality in a single thing of passing from one to another." Potentiality is indefinable. Megaric school had denied existence of potentiality. Aristotle says that is not all. Ross – point is that change is not catastrophic – A has certain conditions of B'ness else it would never become B.

According to Whitehead – ὕλη is a value, not bare value but <u>superject</u>.is product of ingression of

|66|

existentia infima into process of realization. Process is taking account of existence as potentiality or envisagement.

<u>a</u> part of process <u>not b</u>

[sharing in & conditioned by

C envisagement of forms b, c, d, e etc.]

Envisagement is a mild form of reality

Indicates the complex meaning of our propositions. Berkeley's recognition of futility of simplification led to his complicated meaning with ideas in mind of God. These ideas etc are not unlike Whitehead's envisagement in a not b of b c etc.

whole event contains active transition

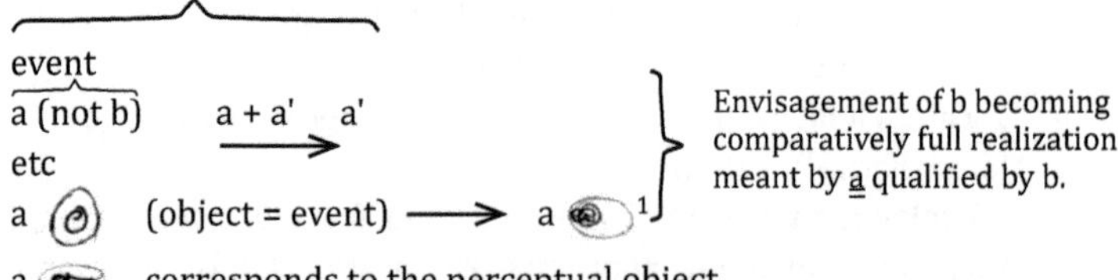

is our apprehension of the mutual significance of the realization localized. This value also conditions (control) the event.

~~Object is a realization~~ Observation of

|67|

an external idea.

1. We have no idea what Heath is trying to capture with these three symbols.

Idea = M yself ↔ T hing
M → · M ↔ T
. T
M ↔ M ↔ ↕
. T

Oct 30.[1]

Another side of Aristotle, i.e. that idea that what we are dealing with in metaphysics is not something with background left out but is all-inclusive. Never refer to a background outside your subject matter, but that your subject matter must ~~exclude~~ include its own background, i.e. must include all that is.

Ross 93–94. In physics Aristotle is rather muddled between ether? ∨Why are things moving?∨ & "ground of general becomingness?" "Circular motion" i.e. get repetition of vibratory motion of Whitehead & modern physics.

Prime Mover on circumference. i.e. Movement must originate either at center, or circumference & movement ground by Prime Mover must be fastest because gradually dies away.

Now considered significant, but from Galileo to 1905 would have been considered the silliest.[2]

[Modern Relativity – A standard motion is maximum velocity, & all other motion is derivation from that.]

|68|

& because of supposedly observed fact that fastest movement is of fixed stars. (Now we say light.)

Physics leads to questions.

1. Incorporeal Mover at circumference.
2. Incorporeal Mover give movement.

Aristotle recognizes pushing & pulling as only cause of movement. (cf. Lord Kelvin,[3] 19th century). This is ether side of his question. ~~While~~ Explain ether by stresses & strains etc. ∨& ether as like matter. ∨ Electro- magnetic theory has turned up recently – matter is our appreciation of certain properties of electro-magnetic field. & therefore instead of ether are left with field of force. Cf. also Faraday, 1847.[4]

Attempts answer in *Metaphysics.*

Ross 179. (149?) First Mover causes movement by desire. Here Whitehead agrees.

"Unconscious teleology of nature" as cause of movement. Whitehead agrees.

1. This material seems to be like that at the beginning of lecture 15 in the notes taken in Emerson Hall by Bell on 28 October, but would represent only a portion of that lecture.
2. This sentence is written vertically as a marginal note on the left side of the page, with an arrow pointing to it from the 'Prime Mover…' paragraph.
3. William Thomson, Lord Kelvin (1824–1907).
4. See footnote 4 on page 11, lecture 3.

Ross: God is the efficient cause by being the final cause & in no other way.

|69|

|70|

|71|[1]

~~The Perceptual Object~~

Nov. 1. What is Reality?

Aristotle's distinction between ὕλη and εἴδος or μορφή

~~Under~~. This is much better than his subject–predicate logic.

Any answer must give ground by which things are what they are. Only possible in a generalization from all the questions as interrelated.

If not we must take up the fundamentally irrational point of view of practice – no forecast possible.

This ground in West has unfortunately been identified with a personal God. The East has kept it impersonal. In Western thought there have been three approaches. 1. Theology – eastern medieval largely Semitic influence.

2. Physical Science – Greek approach

3. Our Subjectivism – psychological point of view – analysis of process.

Each approach has its characteristic danger. Theological is impatient of true width of metaphysical demand. Theology demands the Best & definition – of aspect

|72|

of universe in terms of Best. In West has ended with personal god.

The Best to the exclusion of the Whole. Metaphysics demands the Whole – The Eternal does not take sides.

2. Scientific approach is too abstract. Leaves out aesthetic & spiritual side. Commences with abstractions & ends with problems essentially insoluble due to gaps in human thought.

3. Psychological approach leaves out whole problem. Presupposes organism in an environment. Always assumes the problem.

Nov. 4.[2]

Idea that there is a ground of change has always haunted mankind. The "eternal in the particular" is what we are interested in. Always difficult to get Eternal into particular. Tendency to make Eternal is the Real & the particular is the Appearance. Aristotle – The Eternal "God" only thought. – Whitehead says envisagement is thought. – Thinks only of self. Danger in thinking of Eternal as a Being.

|73|

Course of History of Philosophy. Always three stages (more or less Hegelian) – thought.

1. Heath's pages 67 and 68 capture only the first half of what Bell records as lecture 15. Heath's next page, 71, matches the beginning of Bell's lecture 16, delivered in Emerson Hall on 30 October. We can only conclude that pages 69 and 70 have gone missing.
2. Despite Heath recording that it was 4 November, her notes continue with the second half of lecture 16 at Emerson Hall, given on 30 October.

1. Primitive 2. Culmination. Almost the beginning.

3. Developed but also "sense of running down" Because any mode of procedure, however happy & fruitful, extrudes a great deal which genius of primitives had seen to be relevant. [Then comes a return to some primitive notion & the cycle again.]

Burnet's *Greek Philosophy*[1] Read Herodotus.

Primitive Western philosophy starts in Ionia with scientific impulse. Pythagoras with numbers had Got hold of something very important but in a queer way.

Democritus etc. etc. Culmination

Plato & Aristotle

Always two sides to a culmination – Side looking to the primitives (Plato) & looking toward a long development (Aristotle) Development

1. Begin ethical development.

2. New Platonism.

3. Added interest due to Christian theology, in early stages was Platonic.

4. Arabs & Middle Ages

|74|

transformed itself to Aristotelian alive first to logic. gradually to general metaphysical standpoint Culmination in Thomas Aquinas With a fortunate & unfortunate (accuracy) emphasis on logic. Dies out.

Renaissance.

Throw back first to Plato in own dialogues & in one sense to Primitives.

Christianity there enormous high light on individual human soul – eternal value of, etc. A strange influence of religion ∨(individuality.)∨ on an age not fundamentally religious. & then in Descartes

Cogito ergo Sum

I I.

Thus modern philosophy gets its subjectivism.

Whitehead – that one sided view is being worked out & we must go back to the primitive to get our balance.

Spinoza as last expression of time Aristotelian – non-personal view. characteristically was a Jew for it was Christianity which added to Judaism vivid personal ideas.

His God is last edition of Aristotle's God.

|75|

Great difficulty in getting his God in relation to particulars (modes).

The Cartesian tendency culminated in Hume & Kant. Hume showing subjectivism or skeptical side; Kant trying to develop ordered world out of subjectivism. Rather difficult to rescue phenomenal world from solipsism. & then Hegel came in with quite right idea of consciously going back to more impersonal, objective Greek standpoint. Not quite enough. His self-development idea is really very magnificent human mind.

1. John Burnet (1863–1928), Scottish classicist.

Stace[1] Must go back & balance objectivism & subjectivism

Hegel *Cognovi ergo cognita.*

I have ascertained therefore it " [2]

I know & therefore

Supremely known cognition is mind as home of cognizing process according to Descartes etc. Whitehead does not agree.

Start with experience of individual & therefore are objects. Individual is one of objects.

|76|[3]

Problem to describe in general terms the *cognita* – occasions of realization.

Can't get beyond relevance to the occasion – but other things are fitted on to the occasion. Something general which is discerned in what is observed.

Fact is everything fittable on to the immediacy of the observed occasion

Fact is Determinate, Processional, Realization.

The Eternal

Stands outside the particularity of any one occasion.

is VTheV Determination

is Ground exhibited on every sort of occasion of realization.

is That which is true respecting every occasion because it is an occasion.

is The Processional Character in itself. but itself is static.

is The Substance of which the occasion is a limited affection (Spinoza.)

is The Substance requiring Processional occasions.

is The Ground of Generality.

is The self-contained Reason within fact.

If there be no Eternal there are no generalities, fixed principles or possibility of science.

|77|

(But even so how can an Eternal whose "prime attribute is procession" be ground for fixed principles.) Without an Eternal Hume's analysis of all impressions & ideas (first sentence) is not valid.

So Hume & Aristotle alike presuppose an Eternal. For Whitehead the non-processional side of fact. Every statement about the Eternal is perfectly general – on other hand it must find its verification in each particular occasion of realization. Therefore there must be particular relevance to that particular & therefore there must be a side about which one can say particular things. The Eternal is the static in Fact – the binding force within Fact. Only with the Eternal can there be Fact & not mere isolated facts. [Cf. above with Leibniz theory of monads.] All

Cf. Aristotle

reference to possibility Vor impossibilitiesV is reference to Eternal as

1. Walter Terence Stace (1886–1967), *The Philosophy of Hegel: A Systematic Exposition* (1924).
2. It is not clear in this instance what the ditto marks are meant to repeat, unless it be the *cognita* above.
3. What Heath marked previously as 'Nov 4' completed the lecture given in Emerson Hall on 30 October. So this juncture seems to begin the material from lecture 17 given at Emerson Hall on 1 November.

something general. May look on the Eternal as the How of limitation – itself transcending the limitation. How is there monistic Fact (organic unity) when is (amid pluralistic diversity)

|78|

The Attributes of the Eternal – express how the Eternal is itself in its function of conditioning processional realization. Conditioning etc is an abstraction. The idea of conditioning & conditioned (both abstractions) are together just It-in-itself. Limitation always means some reference to alternatives inherent in every occasion. ~~Bears~~ Every occasion of realization bears impress of Eternal both generally & occasionally (by reason of particular shape by which the occasion exhibits the eternal character of thing.)

Plato as Aristotle

The comparison of the general with the particular relevance. The finite occasions are of the essence of the eternal

Nov 8[1]

Now will turn to idea of passage, procession. Can only describe it by many phrases. Way things flow by. Transition to something else in transition. Always otherness in transition. But always element of retention. Becomingness of retention & omission.

Cf. Bergson & next lecture.

|79|

More limited idea is an occasion of procession [Amid event as a slightly more limited]. An occasion of procession is itself in transition & includes transition beyond itself. Things emerging & expanding but with certain directive linearities. [Humanity has thought of only one linear ~~view~~ ∨flow∨ & Whitehead says that is too limited.] In our minds we break up this procession into occasions. Therefore may consider it a transition within occasions, from occasion to occasion.

Restriction to one linear progression. ~~First~~ difficulties.

1. The[2]

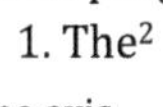

time axis

[How say that b has moved since we have only time ∨approximately∨ instantaneous view. [Illustration for Milton & Dante].

Absolute idealists are right in saying there must be self-reference.]Sometimes woven into the eternal, timeless complex space of the physicist. Old classical theory only one time as above.

1. This Radcliffe lecture begins with material from the latter part of lecture 17 delivered in Emerson Hall on 1 November.
2. We have eliminated a double 'The the'; it is not clear what Heath's intention was.

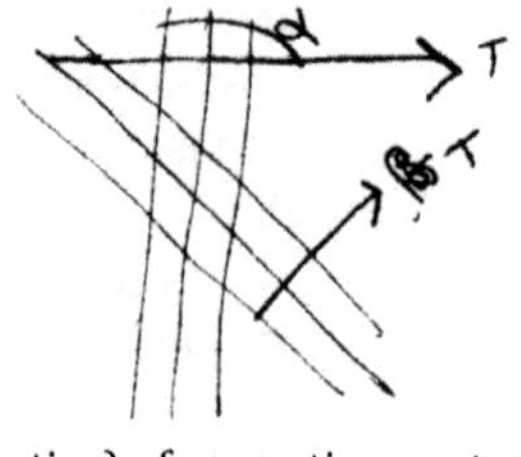

New is below α is whole cluster.

Anything at rest in β is in motion in α.

etc. etc. This is much more self-explanatory system. This

|80|[1]

system has ground in immediate perception by science [Einstein.]

Are really talking of different meanings (alternative) of space-time, not merely of time. These are all real – being real is togetherness of degrees of reality. Equally good, but different groupings. But as soon as get idea of instantaneously & sharply real can have only one system. This is not what we have. What is happening is expansive development.

[Digression:– One might object that we are muddling the conceptual space-time of science with psychological space of immediate experience. Whitehead says conceptual is merely a cloak to hide our failure to deal with problem. What we want to know is the relation between cognition & thought in sense perception.][2]

Realist holds in every procedure that in procedure the cognition imposes concepts on thought. Where Kantian holds that thought imposes concepts on the cognition.

So to say that that space of physics is conceptual may mean one of 2 things.

1. Object of mere abstract thought in that case the objective thought is only an indirect ingredient of

|81|

actual occasion of which thinking process is direct ingredient.

Indirect ingredient via ~~ingr~~ envisagement. Need never have been realized. If space of science is conceptual it has no more relevance to actuality than Old Mother Hubbard – is mere envisagement of an idea.

2. May mean that actual occasions of sense awareness impose this concept on the cognizant (knower.) Because you are aware you discern the concept of time & space as immediately relevant, as abstract from the actual occasion. Objective thought is of real fact. Whitehead says that space of physical science is conceptual in this sense, & then in philosophy is used in other sense.

Another reason for doubting adequacy of single time-~~space~~ system. How explain dreams & day dreams. Even in "real" world there is also a passage of mind & a passage of nature. Yet both are equally objective & real.

Concept of Nature 66–70

There are realizations of value whose modality of spatiotemporal type. But we can get realizations which cannot be adequately apprehended with actual spatio-temporal

1. Heath does not mark the date, but on this page Whitehead is presenting the same material as at beginning lecture 18 in the notes taken by Bell in Emerson Hall on 4 November.
2. Closing bracket provided.

|82|
system although bear close relations. Even multiple space-times will not adequately relate these.

Bergson insists that his durée is not divisible. Now can always divide a linear system. Here Whitehead is with Bergson. While Bergson says spatialization of time is distortion – Whitehead says it is merely high degree of abstraction & O.K. if you bear this in mind.

Nov 11[1]

Realization

Foundation of all thought: *Cognovi ergo cognita*.

Things that we know are the relations of ourselves & other things. Are Hume's impressions ideas. It is the impressions that are realized. What we know as immediately real is relation. Therefore Cognition is a further relation. The primary relation is the thing known. – Whitehead indicates this by "impress." Therefore Most concrete immediate *cognitum* is an impress is togetherness of self & other.

Criticism of Hume for subject–predicate point of view. Hume treats the impressions as predicates of the mind. In Whitehead's usage an impress is expression of relation of in between terms. Can be predicate either of subject or object. Has no
|83|
essential connection either with humanity, mental, cognition. An impress is the togetherness cognized in cognition, but is distinct from such, & does not require it. Cognition requires that cognizer should be one of relata of impress.

Nature of an impress considered in itself. An impress is transitory & limited. Transitoriness is expressed by character of limitation & vice versa

~~Passage.~~

Impress is internally transitory & transitory with regard to external. Finite in Spinoza's sense.

A continually increasing impress. Impress from which & to which are together layer impress; *ad infinitum*.

Though impresses are limited there are no natural boundaries which exhibit them as units.

The limitation is not therefore as merely transitory sufficiently individual. Achieves individuality by a certain unity of values or contrast which expresses another aspect of the essential limitation.
|84|
Therefore An impress is a limitation of something which could not be itself apart from that unity of impress.

---

1. Heath's date in the left-hand margin could be read as 'Nov 9', but that would have been a Sunday, and so unlikely. The next normal class day would have been 11 November. The content of the notes from this point roughly match lecture 19 in the notes taken by Bell in Emerson Hall on 6 November. Clearly, the Radcliffe and Emerson Hall versions of Whitehead's lectures have slipped further apart.

A passage of what? Is the breaking down of isolation of individual essences into partial fusion. That which becomes is the basic stuff of value. This presupposes that value can be fitly applied to the general ground of Becoming is The Eternal, because as viewed as abstraction from impresses it is nothing but that general value which is in process of realization in particular value. It is ὕλη. Also value as characterized as what is generally true is eternal. As shaped in the Eternal as captured within the finite. The impress is itself a substance in so far as it is itself a substance. Limitation is of the essence of value.

Will consider Aesthetic Value.

A commonplace that art is selection & thus is limitation. Every work of art depends on significance of positive content being separated from the irrelevant. Insistent individuality of great work of art, achieved by structural
|85|
limitation amid unbounded absolute. Read Newman's *Apologia*

Though properly speaking there is only one substance (Spinoza) there is an emergence of that substance in every mode of realization. That is why finite truth is possible to us.

The intense value of this capture of the eternal by some passing circumstance is illustrated by intense pathos in literature of reference to intense passions of the past, which have thus become eternal facts. The irrevocable past is the eternal relevant past.

This capturing of the eternal in one sense enriches the eternal, i.e. that fact has happened. i.e. Outside of this ~~rela~~ realization the envisagement of the eternal is merely into the possible & impossible, the logical & illogical. Now one further shadow the question of truth. (actuality.) So that the Eternal itself has to be in a sense comprehended in a gradation of limitations. [Neo-Platonism] The eternal is general envisagement & also relevant to every particular occasion. It is thus that realization requires solution, omission – particularly – extrusion of irrelevant detail & being realized it stands as eternal enrichment of eternal ground of becoming. Every realization points beyond itself. The ground for distinction
|86|
of past & future. Insistent individuality of a work of art or any event.[1]

Will dwell on emergent substantial individualities as superjects. A real enrichment of the Eternal. The irrevocable past is the relevant past. The Eternal has to be comprehended by a gradation of limitations. (Neo-Platonism)

Wherever get limitation get Change & this Change in the Eternal is the Shadow of Truth in distinction from the eternal logical compatibility. [Cf. Pure Math.] Requires the extrusion of irrelevant detail. Therefore stands

1. Heath gives no indication, but this paragraph matches notes taken by Bell and Hocking in Emerson Hall at the outset of lecture 20, on 8 November – but only roughly.

as a permanent enrichment of the eternal ground of Becoming. This is ground of distinction of past & future.

Cf. Dean Inge. "On Eternal Life."[1]

Because what has been – states of realization now – is relevant to very ground of being, therefore the very laws of nature change

This eternality of realized value is very ground of memory. The memory belongs to "this occasion" now & therefore if hold that it is memory of past it is because ground of this occasion contains the past. [cf. Bergson] Apart from that the memory is simply a silly little illusion.

Endurance is entirely different from eternality. It is the relation of the individual value as emergent & quality

|87|

in definite spatio-temporal reason. Exhibition of a derivative substance, with limitation expressed in space-time terms. A structure may extend throughout all space & time, but some nucleus is portion of space & time which are thoroughly characteristic & all the rest is its influence. And this derivative substance (superject) is activity of substantial value necessitates more complete analysis of idea of substantial value. Then points,

Display[2]

1. Fusion of individual essences.

~~2.~~ Realization of fusion, a process

Determination.

2. The emergence of the determination as part of the eternal. The exhibition of that ground of being as relevant to that structural limitation.

This endurance through a period of time in a volume of space. The eternal is nothing but the eternal how of its limitations. There is not one eternal substance – its unity is the organic unity of its limitations. The plurality of individuals is an eternal either[3] but which is of its very essence. Best marked by speaking of eternal is sheer generality as Substance (Spinoza) in the particularized eternal as enduring as Superject Concrete fact is the eternal Subject emerging with a

|88|

plurality of Superjects. That which emerges is a realization of particularized fusion & valuation from determination.

Above has generalized two problems.

1. What is ∨meant by∨ perceptual object } into
2. What is ∨meant by∨ mind }

What you mean by limited individualities which emerge is process of determination.

1. This line is written vertically along the left margin. It extends from the line 'Wherever get limitation' down to the bottom of paragraph. Sir William Ralph Inge (1860–1954) had moved from Oxford to Cambridge as Professor of Divinity while Whitehead was still there, and thereafter was appointed Dean of St Paul's Cathedral in London.
2. This word, and the one below, are added at an angle in the left margin.
3. Neither this term nor 'ether' seems likely, but we cannot suggest another.

In some sense they've got substantial independence & in another sense they are merely aspects of the stream of things so as finites they are abstract. In abstracting a finite individual from the Whole we make all the rest the systematic space-time-iness. On Bradley's philosophy cannot say anything true except of whole. Whitehead provides for this.

Relation of eternal ground to the particular occasion. The One & the Many.

The Actual is an Eternal fact in ground of being. An outcome which does not pass. Holds together the passage of things as an enduring utility. Illustration of time. Thus it is integrated tension of the whole in which there is endurance of past in immediate realization. Thus realization dawns & expands.

|89|

Ultimate Meaning of our Knowledge

There is a Becoming of Reality. We are in this Becoming or Being Real. Being Real does not touch a pure atomic undifferentiated entity. It is the Becoming of a certain mutuality What is real is structure & shape of this mutuality.

Paper clip must be abstracted from all spatio-temporality.

General systematic character which must be discerned in & beyond every entity. – its relational essence, of which space-time is most prominent & obvious relationship.

What are the relata (terms) we don't know. Only know the relations. These entities therefore give the emergent pulling-together of a contrast.

Where does this stop – What are the terms? & Who is presiding over this imposition? i.e. must be ground of how there is a Becoming? This is itself inherent in nature of realization. Must be general taking account of conditions & possibilities, which as a whole are taken account of.

Is there any thing suggested as permanent. – in its becoming but not itself that which becomes. We find objects which stand out of change as being just itself. example: sheer yellow.

The concepts or ∨Platonic∨ Ideas. The how

|90|

of the realization is determined by the Objects for realization. Not quite same sense as Plato's Ideas.

Cf. Aristotle's Potentiality. These Ideas Objects are there forming, the structure of the contrast.

How does this general envisagement get itself going? It is nothing but a Totality being itself by its emergence into a subordinate & dependent plurality of units.

Consideration of a Criticism on Whitehead's recent Lectures.[1] See Intro to Broad *Scientific Thought*, 18–23. Contrast of Critical & Speculative

1. Heath does not indicate a new lecture or date, but this juncture seems to match the beginning of lecture 21 from notes taken by Bell and Hocking in Emerson Hall on 11 November – but only roughly.

Philosophy. Broad suggests that Whitehead has deserted the safe Critical Philosophy & embarked in Speculative Philosophy.

Whitehead questions whether this distinction is so sharp. Broad discusses Memory, Induction, Possibility & Probability. Whitehead says – either there is or is not in every occasion a taking account of what has been, what will be, may be & can be. If there is no such taking account of the above for merely stand for a baseless habit of mind at the instant. If there is any "beyondness" this can only be expressed as relating to nature of the eternal

|91|

process itself. If take point of view of English Empiricists intrinsically there is nothing in Philosophy

Only three general lines in philosophy

1. A priori ideas. Rather a hopeless point of view when you are to think of what "mind" is & why it posits it. Only man of Genius who followed this is William James.

2. Kantian – subjective activity as giving form.

3. Realist & discern in each occasion of experience elements in that occasion & reaching beyond. The universal procedure in ordering life.

Agrees with Broad that can't go beyond our immediate knowledge but must not arbitrarily limit this with ∨mere∨ sense impression. (cf. Broad.)

Whitehead – very task of critical philosophy that must not arbitrarily lay down what it is we directly observe, & then say ~~all~~ any thing else is futile.

Speculative philosophy – Broad defines – Object take results of sciences & results of religion & ethical experience – Then reflect on this. Broad's criticism is based on uncertain character of results obtained.

Whitehead – a certain ambiguity lurking in word results. The broad & final results are its foundations.

|92|

It is really science itself which is the speculative philosophy that excites Broad's suspicion, i.e. the getting at these results by detached reasoning. He objects to putting these things together. (Hoernlé[1] in "synoptic vision" says one must get beyond abstractions.)

Cambridge tends to suspect ethics religion ∨aesthetics∨ & be bored by them. Of course they are knowledge where discovery of fundamental principles are difficult.

Natural science is obvious in its reliance on sense data. In other sciences must call in critical philosophy in scrutinizing their fundamentals. No matter how much we distrust them, still they remain as facts of man's interest. Discarding them makes critical philosophy impossible.

Another tacit fallacy is his outlook. Assumes that what you observe is given/ fixed (sort of "wages find.") A trained mind actually sees more than the untutored rustic. Therefore above is quite erroneous. So critical

1. Reinhold Friedrich Alfred Hoernlé (1880–1943); see footnote 4 on page 79, lecture 21.

philosophy when it has done its work, & when Science has done its work, the observational premises which are directly apprehended are increased. Mistakes in past are no reason against our reasoning from them.
|93|
Three lines of philosophical procedure (above)

2. Kantian theory is designed to meet obvious objection to theory number 1; . i.e. ~~that~~ why postulate? Kant answers that ~~it~~ postulating is element in foundation of our subjective experience. It is what experience has to be in order to get to objective realization.

1 & 2. start from conception of experience as in its subject. Whereas last procedure starts from consciousness of subject as in things experienced. A fundamental difference. Any admission of "private worlds" (Russell & others) seems to be contrary to realism. For this reason also Aristotle's quality & substance has been dangerous, because subjectivist conceives experience as qualifying mind while naïve realist conceives mentality as sentiency among objects of which it is one. Must find our mind in impress & not impress in our minds. Really a natural way of putting it – you're immersed in philosophy; the actors in the scene, & not vice versa. Impresses is a good word. ~~but~~ – Modern philosophical tradition tends to invert the immediate delivery of common sense.

Our starting place was "Every finite entity is abstraction" Can't get away
|94|
from totality of process in which it was embedded. Whitehead has recently been partly other side – i.e. that very fact of being one individual emergent entity (like perceptions objects, minds), makes them in some sense or other the capture of substantial individuality, which is reproduced in each entity.

Therefore The emergent entity requires its environment it is within itself in some sense an entity, an individual process. Systematic character of beyondness therefore it can be discerned from standpoint of the particular relation to the definite entity.

Nov 20.[1]
General Make-up of Real things that we perceive and know. In passing are entities not exactly passing. Very idea of identity & sameness. These Whitehead calls Objects. Has taken the view that these objects transcend realization in so far as they stand in very character of the ground of realization. The plurality of objects is an essential attribute of character of eternal ground. That is always in a particular realization.

Division of objects into primary & emergent (actual) object.

1. Primary Objects How are they implicated in what is passing? Belong
|95|
to the essence of the eternal as such. Perfectly general statements which

1. Heath indicates the beginning of a new lecture, but instead of the next in the sequence given in Emerson Hall, this one skips lectures 22 and 23 and launches into lecture 24, recorded by Bell in Emerson Hall on 18 November, which means Whitehead is, once again, just one step behind.

have any content (not purely logical) must be expressed in terms of these primary objects. Any particular relevance of a primary object to a particular occasion is as it were accidental. – accidental of the immediacy – can never get back of the accidental in case of particular relation of that which is in its nature general. Primary object in itself is perfectly neutral as between all ~~essences.~~ occasions. ~~The community of relation essences. The assemblage of individual essences represents~~ Every particular occasion stands in its own particular adjustment to manifold of objects. It all fits together: It belongs to the relational essence of every primary object to be in some relation to every particular occasion. – either relevance of potentiality or actuality. The self-conditioning side is taking account of itself – or is emergent. The actual particular relevance is not in the relational essence; the potentiality is the reference to the relational essence.

Primary objects are two kinds

a. Pure objects Only relevance to an occasion is as to ingredience or potentiality of ingredience.

b. Social Object – The Concept itself

|96|

of an occasion – a pattern of primary objects. Nothing else than the housing of every possible Social object in every possible way in Space & Time. Pervades its whole relational essence – the potentiality of everything anywhere. This structural relation is there merely as a potentiality In Social object is qua mere social – objects are not related except by their relational essences.

[Whitehead In talking of relational essences talks as if Space & Time finish it up. He thinks that is enough to try to deal with although he sees no reason why it might not have begin & end.]

Can't describe a pure object. A social object can be described because you can get some cognitive perception of some other social object in course of realization which has relational structure identical with that of social object. Can get a model for the structure. Having done that you have to get hold of the pure objects by pointing out a set of emergent facts where the pure objects are to be found.

Another difference between pure & social, is that it's nonsense to talk of a pure object being realized by itself, because it is always a togetherness that is being realized.

|97|

And therefore that togetherness might be realized i.e. a social object might be realized.

Hume recognized this (beginning of *Treatise*). Impressions vs. ideas – Perceptions into simple & complex impressions/perceptions or ideas

2. Emergent Objects. Its essence is that it is the fact of something that remains – i.e. an eternal fact. "There was ancient Rome." It has both an

endurance in time and an eternal actuality. The Shadow of Truth falling on the eternal principle. Now question of truth in pure & social objects. Fairly easy to conceive of emergent objects of high type – i.e. our- selves. Can't always try to explain in terms of lower object – else infinite regress – finally have to come to a basic object which we will describe simply in terms of social objects. i.e. in terms of patterns of pure objects which are eventually realized.

a. Object as it is known in perception. In a sense I perceive it "out there" – also "in my head." or "from me, here." [This latter Berkeley is always pointing out]. The basis of idealism.

b. An electron. defined in terms of what its effect is everywhere else.

c. Memory (essential to knowledge)

|98|

Must be a sense in which past is directly here now.

d. Induction ∨Anticipation∨ (another aspect of memory). is really knowledge of the present as in the future.

All these bring out fact that the relevance of a pure object to time & space is not nearly so simple as "It's there ∨now∨ & that's an end of it."

Can't[1] avoid dualism. Takes Platonic Form. General Ground of Realization gives the monistic basis. Different forms gives the pluralistic basis. Each entity has individuality & yet doesn't exist apart from the general entity.

Basic objects. Mass production leads to inferior types but the higher types only exist by reason of the mass production of lower.

Mass production in itself. Parallel to American civilization. Basic Objects – lowest type. is the emergent object which ~~has no~~ is ~~oute~~ not referred to as outcome of any other – up against the union of the potentiality with the realm of realizing fact.

Four aspects of basic objects which we know.[2]

1. Object is perceived ∨here∨ where you are, but it is there

2. In physical field we talk about electron & again there is

|101|[3]

here where it is & there which is its field of activity.

3. With respect to time, we again have here & there, of immediate memory. The living past. The past is both here & there.

4 We know the present in anticipation as having its reference to the future. Whole of induction & anticipation is based on that. Has been assimilating space & time as is the tendency of modern thought.

Difficulties.

1. Union of Ideality & actuality. has been age old ⟨?⟩[4] of philosophy All arise from neglect of essential aspects of what it means for an existent to

---

1. From this point the notes shift from blue to black ink.
2. This content from here matches that at the outset of lecture 25 as Bell recorded it on 20 November. There are similarities in content, as well, in some passages in previous pages of Heath's notes.
3. Either pages 99 and 100 are missing, or the pages were wrongly numbered. The content from page 98 does seem to continue without interruption on page 101.
4. There is a short word here that we cannot decipher.

be realized. ~~There~~ The relevance to space-time of a pure object, is not of a simple character which it is usually assumed. Not sufficient to say "color is here."

In what sense is it here & in what sense is it there. Relevance of ingredients to space-time is complex. Pure object has a diversity of function through its structural relation if a basic object. Broad's discussion is wrecked by this.

Four categories are applicable to an emergent object.

Actuality, Selection, System/Pattern, Unification

|102|

1. Actuality means that in object we find ~~the~~ delegation of eternal underlying principle ground of the process. Stands in very essence of ground; is only through the emergent object that it exists. Very reason of time is to be found in actuality. What is subsequent to an object is to the general becomingness is what it is as qualified by actuality of that object. Getting idea of field of action. Whole idea of subsequent bears with it the notion of eternal as stamped with limitation subsequent to the stamping. Irrevocable past.

The subsequence of a basic object b. is that space-time region throughout which eternal becoming is limited by actuality of b. In calling this region space-time has restricted meaning of its subsequence. Becomingness is rather expansive than linear. A certain regional divisiveness exists however, & therefore in a more complete sense, by the subsequence of b. means region within expansion of becomingness is such that throughout it the ∨expansion∨ of eternal principle it is limited by b.

2. Basic object is primordial example of derivation of actuality

|103|

from substance & therefore exhibits its self as solution. Essentially something is happening in a systematic way to elements selected from the potentiality.

3. Patterned. Every part of region receives qua itself. definite qualities & thus quality issues from very nature of basic object. The selected entities concerns primarily pure objects. There is an indefinite range of potentiality & the selection is simply out of a definite range of entities within that potentiality.

4. Unification.

~~a~~. Structural. ~~as a~~

~~b~~. Selection manifests itself in restriction to that structure.

a~~c~~. The Structure as a whole 1.[1] its distribution. even as divided there is a unification through subsequent field. In every portion of the field there is something going on in nature of a unification of selected range of potentialities into an immediate realization. Togetherness, relatedness

1. The '1' is circled. There is some confusion in numbering/lettering of these sections; Hocking and Bell each provide somewhat different attempts.

exact transcript of scientific description That fusion of individual essences is in field – modality
|104|
of essence of object & modality of other position.

Object is both <u>there</u> & <u>here</u>. Importance of multiple-term relation has been neglected by philosophy. This is an essential case of ~~at~~ a three term relation.

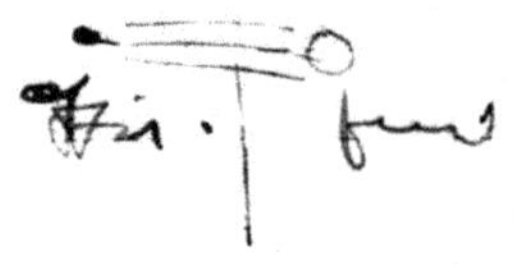

Confer Aristotle on no presence *simpliciter*.

For standpoint P there are a range of Objects O_1 O_2 O_3 which are in a peculiar sense formed at P. Relevant with modality they get from their location in various other regions. L_1 L_2 L_3

Call aspective set of objects all objects from standpoint P. or the P-objects.

L_1 L_2 L_3 = actual ~~positions~~ aspective locations for P-objects

L_1 to P = modality of L_1 for P

P' objects have systematic relation to the P-objects.

P' = another standpoint, whose objects are (O_1' O_2' O_m')

Close analogues O_1' O_2' etc to O_1 O_2 & if L_1' L_2' etc to L_1 L_2.

O_1 is known at P only with modality of being at L_1 = O_{L_1}

All aspects of the pure object end in a pure object – if simple this is basic object B. = $O_{L_1} + O_{L_2}$ etc.
|105|
and is called aspect of set at P or the P objects. B might be molecule & the $O_{L_1} + O_{L_2}$ the contributions in magnetic field due to B. From another standpoint P' get P' objects from B.

O = green, L_1 = modiness } cf. Theory of abstraction.

Basal object <u>B</u> grows out of a <u>structural pattern</u>., in some sense a systematic correlation between all these general possibilities. 1. Different O_1 O_2 etc. of B. coalesce. 2. Definite correlation between aspective sets.. 3. Actual locations from different standpoint coalesce into coherent actual location.

Three categories which apply to the fusion or the Becoming Reality.

1. Quantitative Cumulation
2. Qualitative Cumulation
3. Qualitative Contrast

1 is Aggregate of effects O_1 at L_1 & almost same O_2 at L_2 etc. & aggregate up into a changing intensity
2 is Qualitative Fusion 1+2 Cumulation of Likeness & Likeness
3. Cumulation of Difference ~~time~~, etc.

Being together because of their difference.

1 - Identity of Quality & Difference of modality.
2. Kinship of Quality.
3. Contrast of Quality.

|106|[1]
Objects first to saying have mere class of objects O_1 O_2 etc at P. Where does contrast & unity come in then? There is a real unity. B described as patterned way in which B qualifies its subsequence. Therefore unity of P. is P aspect of B.

Two sides to this – display & control – but not really a bifurcation, only in relation to a further unity. i.e. the totality is the more concrete fact. This unity has its intrinsic & extrinsic reality. Realization is a togetherness of differents or of the discrete. Therefore in one sense this emergence must reject the self-enjoyment of the unity. It is integrating itself by reason of its logical formulation. Is not saying object is formula, but in itself has formulation. If you hold that B is unreal & is merely a formula then have to start out & find reality of the something else which has formula.

[2]

|107|
Nov 29.[3]
Although there is a general monistic process of becoming, yet the realization is a pluralistic fact, from another point of view. i.e. Realization is realization of individual entities, which, to that extent have independent of each their own being, yet this realization of the entities is only as within the environment & as conditioned by the relevance of the environment. So that full concreteness of reality is pluralistic at same time that each individual entity is only concrete as in its environment. Without environment it is a mere formula; in it, it is a formulation. By this concreteness means that individual entity is the achieved & achieving or controlling, also it is in that particular limitation – also embodiment of general ground of whole becoming.

Actual Location }
Modal Location } For every object.

Sense qualities are at actual location with modality of the modal location or standpoint.

Degrees of realization.

1. The following page of notes touches on topics in both lectures 25 and 26 in the notes taken by Bell in Emerson Hall on 22 November, but even if from lecture 26 there are these topics only, and then Whitehead moves on to the next lecture.
2. Here, five lines of notes were written at the bottom of the page in ink of a different colour. It seems likely these were from one of Heath's other classes, and she subsequently struck them out.
3. The following material is at least roughly similar to the outset of lecture 27 recorded by Bell in Emerson Hall on 24 November.

|108|

Let us take any 4-dimensional region of space-time and ask what is happening at event E.

Qualified by modal presence of pure objects is Actual Potentiality of E

To be distinguished from Ideal Potentiality is qualification of regions as being within the relational essence of objects.

Every pure object pervades space-time in every region by reason of ideal potentiality i.e. it is one side of their general envisagement. i.e. Space-Time is how the objects are relevant to realization.

Actual Potentiality is modal presence.

[Cf. Aristotle

So far as E is within system which includes former modes of realization E' & E'' what happens with respect to E (i.e. its emergent reality) is prejudiced by the former occasions, i.e. modal limitations.

Prejudice of E from E'& E'' or Aspect of E'& E'' from E. Can be analyzed into the presence of various pure objects.

|109|[1]

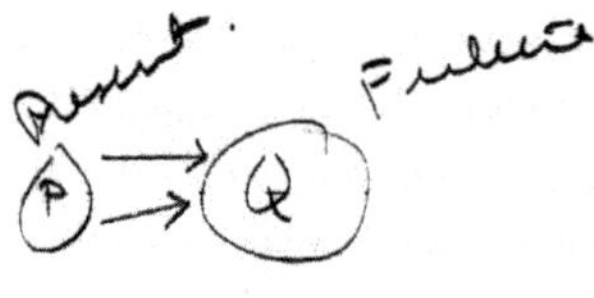

Part of the nature of occasion P that the spatio-temporal relation to P of Q should be what it is. Furthermore that aspect of P from Q should be what it is. There is no aspect of Q from P VYetV but there is in the nature of P the aspect of P from Q i.e. return to idea of beyondness as essential of induction.

Future does not qualify the Present, but present qualifies the Future.

Total Actual Potentiality as it arrives at Q is not a mere assemblage but is analyzable in terms of categories of qualities & qualitative cumulation & Qualitative contrast.

Space-time etc are isolation & realization is achievement of unity under limitation.

Present integrates as actual potentiality what preceded it & what is subsequent.

Exap.[2]

Metaphysics has been too largely descriptive in modern times.

Criteria.

1. From particular sciences.
2. From logic.
3. From epistemology.

---

1. The diagram that appears at the top of page 109 of Heath's notes appears following the first paragraph of lecture 28 in the notes taken by Bell in Emerson Hall on 29 November.
2. It is unclear whether this is 'explanation' or 'excerpt'; what is clear is that Heath begins in a different colour of ink, and though she does not otherwise indicate this, it is very likely the beginning of a lecture – the one Bell had recorded as lecture 29 in Emerson Hall on 29 November.

|110|
Must have self-consistent metaphysics & therefore main criticism of metaphysics comes from epistemology but should not start with epistemology. Must first have metaphysical description & ask if it allows for an adequate theory of knowledge, to show how you can know everything you describe.

Return[1]

We have been considering the most general ordering of things. under space-time with a query in mind as to what was beyond space-time.

Can't consider one part of space-time separated from another.

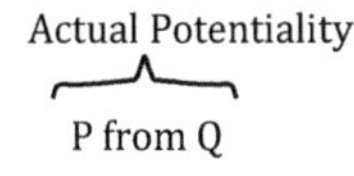

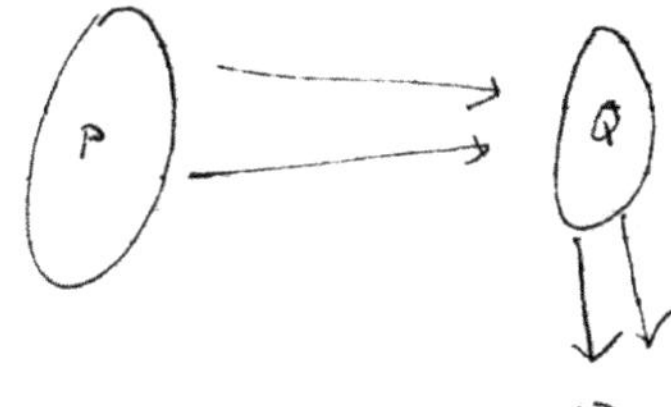

Pure objects are for you at Q an aspect of being from P.

[R with Q from P = actual effectiveness]

Also relevant to any region in the subsequence. i.e. how Q is going to impress itself on R.

Q is in some ~~mere~~ unity, either this or effective, but more than mere assemblage of P's from Q & R with Q from R's. Q is a real fact as conditioned by P & bearing future R. it emerges as in some way a unification of Past & Future

|111|
together in actuality of the Present.

Becoming real is really the coming of always something more – analysis gives plurality but real fact is always essentially unity – i.e. a process of unifying, essential~~ly to~~ two things, i.e.

a. Pluralism VgivesV materials

b. Monism gives realization, always a becoming of a unity, & what gives the standing character is always expressed as a pluralism.

Coming of Unity is not an abolition of the Pluralism because it is a limited unity, i.e. a mode of unity. – limited by that which is beyond itself. & therefore cannot be abstracted from what is more than itself without becoming an abstract formula.[2]

Any division of region is arbitrary & gets one into a vicious circle which we attempt to escape by concept of points which we do not experience.

---

1. The import of this term is uncertain, but Heath's notes do skip about a page of Bell's notes on this lecture, and may pair up with the reading as 'excerpt' of the term noted above.
2. Whereas the content presented just above seems to echo parts of lecture 26 and even lecture 28, that presented immediately below seems to coincide with outset of lecture 30 in the notes taken by Bell in Emerson Hall on 4 December.

What we experience is the region as a whole. Unity of (a) comes from the reality. If you are dealing with space-time as a mere relational essence it has
|112|
no regions. Space-time is not built up from relationships but is abstracted from individual character of reality.

Any region (a) is not only ⟨left blank⟩ but is something for itself – is a mode of realization in all that there is fulfilling itself under that particular mode of limitation – this mode involves taking account of what is beyond itself & what is in itself. Whatever can be said generally of Ground of being can be said of that mode.

The emergent matter ὕλη is value & the ~~for~~ method of VemergenceV is its form.

No region is self-sufficient & yet in a sense is a cause of itself as it is.

What freedom there is in the universe is in emergence. Here is secret of determinism vs. freedom.

E = event Q = region.

E_Q is an immediate substantial novelty, is only itself in its environment. Has subjective apprehension as feeling, emotion desire. i.e. Within the limitation Vof environmentV it is self- creation.

The emergence of E_Q is ~~relati~~ matter of degrees. How far has it that character of relevance which permits of effective emergence. The effectiveness of the character depends on the further question of endurance. (will ⟨left blank⟩
|113|
An event is an achievement in itself & for itself. For itself is subjective apprehension – which does not always rise to consciousness but is embryonic consciousness.

[My own note[1] – Leibniz & Spinoza]

The dependence of the entity on its environment issues in the reflexiveness of experience. Each event is from its environment because it is a modality of totality – which is formed[2] & emerges as contents of its own experience. These contents include their own fusion in that unity. This experience that an entity has of being within its own experience is Alexander's enjoyment.[3]

Reason for reflexiveness's enjoyment is understood by noting that the selection of modal contents of fact is merely due to the fact of its fusion in the experience of an emergent entity. Act of emergence of the entity is act of selection. The modal contents have ingression into reality merely by virtue of its emergence in reality. ~~Their experience merely as found content~~

The emergent entity accepts the past. (cf. Kant) & affects the future.

1. Unusually, Heath seems to be making a note to herself.
2. This word could also be 'found'.
3. It was Samuel Alexander's contention that we are conscious of the process of our perceiving the world, directly, which he articulates as 'enjoyment'. See footnote 3 on page 130 lecture 30, in Bell's notes for 4 December.

|114|
What the entity is which accepts is the actual event in Q whose emergence is E_Q. Q in itself – as being actual is the emergence of E_Q.

All above is making point that these are entities-in-themselves- (not mere relations).

Question of Endurance.

Have actually nothing unless some endurance – something going on the same – is essence of an achievement – of anything that has any reality about it.

Tunnel for endurance

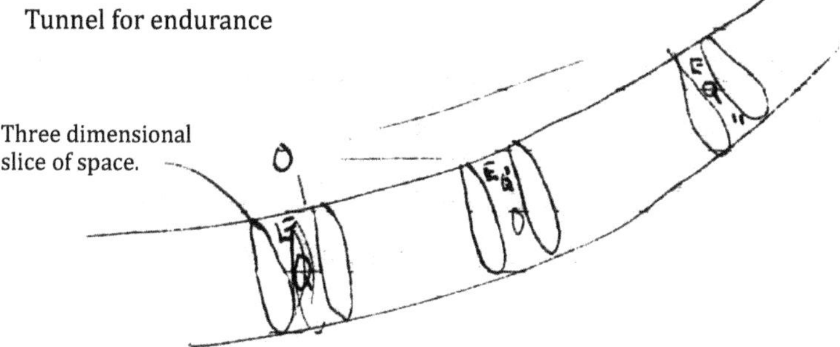

Object O is equally $E_{Q'}$, E_Q, $E_{Q''}$

Object endures when in any slice of event it is the same. Fundamentally an identity of pattern. Actual event is most VconcreteV entity; what endures is more abstract than the event.

The individuality of anything depends on its concept (pattern) & that particular
|115|
historical route. The pattern is not a causal thing but must be such as to involve its own endurance. i.e. its Actual Effectiveness; i.e. way in which pattern at one point affects immediate future. This disengagement from the environment is only ideal. In any actual case, the survival value is dependent on the environment. Vi.e.V so as it is actually shaped by environment.

Two sides to survival character of things – Only one is usually active.

1. Adaptation of organism to environment.
2. Adaptation of the environment to the organism.

Electrons & protons represent (in abstract) the evolution of entities whose very existence forms the production[1] of entities of same sort, or of one or two contrasting types of entities of same sort.

Very laws of nature may be considered the product of a gradual evolution. Subordinate basic values have to VbeV realized or made actual before you can go on to the new values & realize more elaborate values. Have to have a stability in basic value so that emergent entities can take their place along with abstract as

1. This word could also be 'products'.

|116|[1]
entities or ingredients in organisms of higher & higher types.

Has been warning us against mere idea of a linear advance. Under some aspects it is linear but other alternative types are alternative linear development & therefore an expansion.

Speculation

Would like to suggest that general aspect of things rather suggests that process of realization of value takes an indefinite no. of forms & shapes, always under modal limitation; & that stage presupposes stage & just in the same way as here therefore you may have stages even in the general type of mode of realization. Possibility of stage is always there, but may require other stages for realization. We are realized in a spatio-temporal stage of realization, ~~but not~~ not only stage – our stage presupposes other stages which are a priori to it & therefore the given set of possibilities from other stage, the sensibles from our stage, are perhaps the achievements of presupposed stages.

|117|[2]
Amplification & Recapitulation of Preceding lectures. Metaphysical position as a philosophy of evolution. Has been constantly insisting that concrete fact is an event – cannot take change out of things & leave the things.

Change & evolution are obviously the same.

Older philosophy starts with the permanences – Cf. Plato & Aristotle. Perfect reality is timeless.

Influence of mathematics (note Plato as summation of antecedent philosophy), was for the timeless. Also Spinoza.

In a philosophy of evolution change is the concrete fact – then you have to find the permanences to be evolved. Must find also motive of change –

Whitehead – "the emergence of that which in itself has value for itself" – Value is essentially limited, i.e. in modes – no vague value. Then you have to get the evolution of the enduring value.

Whitehead started from a ground of being – the existents which were within the envisagement of the process. There the old static idea reproduces itself, ~~cer~~ here there are certain definite matters of fact "out of space & out of time" in sense of being eternal.

A philosophy of evolution to be thorough going has got to obtain the blank matters-
|118|
of-fact. It ought never to find "the eternal" anywhere. Should be no "primates"

In order to get around the notion that in modern science linear quality of time was rather thrown over & get expansiveness. Then whole

1. The content of this page matches up to that of lecture 31 from notes taken by Bell in Emerson Hall on 10 December, but it may have already shifted to this lecture on the previous page.
2. The following material appears to be the start of lecture 32 in the notes taken by Bell in Emerson Hall on 12 December.

spatio-temporal point of view may be considered an evolution, i.e. other forms of issue, or transition lying behind space-time. Wants here to guard himself – namely what you have, however far you go back, nothing can happen which is not in the nature of things – any detail by itself must be considered the purest of pure abstractions already present in possibilities of things.

For a particular mode of past green there is a space-timeiness to it from which it can be abstracted.

Space-time is field of action (already the field of possibilities) & as such has been evolved as an actuality. Cosmology is itself an issue from something else – mitigates the Sheer matter-of-factness of 3-dimensional space & polarity of time.

All matters-of-fact are a stage that you are passing through. Each stage adds something to actuality & thus you get a new
|119|
situation. The past is modally in the present.

That endurance is essentially the evolution of societies. i.e. evolution of entities which create a favorable environment for similar entities. Not of one species but of an association of 2 or 3 species. Also differentiation within species. Suggests electron & proton – difference of sex – that is one reason he suspects that real space-time is an evolution i.e. a certain differentiation between fundamental unity. Social side of evolution is important to keep in mind – natural selection as ruthless competition has been very much run, Other aspect put by Prince Kropotkin[1] – called "Mutual Aid" – evolution impresses idea of mutual aid quite as much as competition. VPhilosophicV Basis of "mutual aid" is that it is creation of environment. But also is fact that there is a great deal of environment that you can't alter, ~~i.e.~~ ex. – Absolute untameableness of the sea.

Difference between stable progress & rapid progress.

Rapid – founded on competition   Aristocratic (Athens)

Stable – founded on fundamentally altering the environment. Then should alternate. Applies just as well electrons as men.
|120|
Can't be certain you won't evolve (stably) beyond maximum number & then get slow degradation by fact that begin to evolve feebler & ~~yet feb~~ feebler types. Then get ruthless aspect turning up. (Religions that have survived have always taken account of the ruthless) Must be some way of evolving higher type & arranging environment which will automatically discourage weaker type.

The evolution of a unity of realization, which is the value, shaped in the particular modes of togetherness. Emergence of an entity which is finite and limited. Value of abstract possibilities in their togetherness. An eternal difficulty in metaphysics with regard to matters of fact. In nature of things viewed in abstract there is every sort of possibility – but in realization

1. For Prince Kropotkin, see footnote 4 on page 141, lecture 32, given in Emerson Hall on 12 December.

of things the essential meaning is that it is realization of abstract possibilities – naming through the particular modes – mode is value, etc. Order of nature has certain peculiar qualities – i.e. why do Clerk Maxwell's equations & not others apply throughout all nature.

Community of Space-Time & community of physical laws is to be noted.
|121|[1]
Return to Abstraction.

Difference between Occasional Abstraction is that ∨particular mode of∨ green patchiness there ∨in the space-time relation is∨ proper to this occasion

Preteroccasional – that particular mode of green patchiness in any space-time is just greenness in any dimensional space.

The order of nature is essentially an order within the community of alternative occasional abstractions in which the occasions are interchangeable, i.e. all give same community.

This community of occasions is field of realization for our space-time cosmological order. That particular community of occasions in relation to which there are the particular ~~the~~ truths such as 3 dimensions for space & other particulars[2] of nature.

If look at preteroccasional realm of existences you have none of this particularity, because every possibility is there. If you are going to run a thoroughgoing philosophy in evolutionary, expansive terms, it is fairly obvious that every thing that is definite in respect to realization must be looked on as itself an emergent entity – a mode being realized. Otherwise you get back to the static type of philosophy.
|122|
Therefore We must look to a hierarchy of cosmological orders – each is in itself a super-mode, a getting things into a to-getherness which – itself as a whole has a value. Realization has degrees. With the utmost stretch of praeteroccasional abstraction is all that is left of a mere static machinery. ~~Wh~~ Then get question of whether within our knowledge there is any knowledge of cosmological order beyond our system of nature (nature is having a definite community of Space-Time & laws). In philosophy & literature thought is baffling duality of that which is in & that which is outside ∨nature∨ & that which is Nature. Cf. Bergson's "Spirit" which seems in some way not to be spatio-temporal – Also Reason etc.

[Every mode is fusion of all that there is in one particular aspect.]

Cosmological order in that process of realization is the issue of the finite entity gathering its unity & therefore its effectiveness in value in one general mode, or cosmological order & this finding a particular place in

1. The following material seems to match the start of lecture 33 in the notes taken by Bell in Emerson Hall on 16 December.
2. The letters 'icu' are completely missing from this word (possibly a pen malfunction?), but 'particular' seems to make sense in the context.

a super-mode – a wider cosmological order. Therefore you can trace the emergence of a definite unity through the different cosmological orders.

Philosophy must go to antecedent, especially since the consequent is sertid.[1]

|123|

The inherent retentiveness of things is only superficially expressed in any definite cosmological VorderV. Therefore must not think of other cosmological orders as antecedent to this. There is a more fundamental expansiveness which is superficially & modally expressed in spatio-temporal form. That is where Whitehead looks on time (Bergson's *durée*) as more fundamental than space. Should conceive this order as time spatializing itself & other modes would be relative to this cosmological order. Modern physicists generally concerned with measurement of time: time as measurable is so within this particular order.

Cosmological order of the Spirit – out of which this Space-Time cosmological order is emergent is known to us but known as something more than mere Space-Time fact.

Dec 18.[2]

Francis Bacon "All bodies whatsoever, whether they have sense, have perception" etc. Is a certain experience of individual essence of Body. Perception is often far more subtle than sense. (Cf. Bergson). Is sometimes at a distance as well as at touch.

Whitehead starts from that. Bacon, though he haunted ideas of 18th century, never gripped the

|124|

practical men of science as he might have been expected to. Entire absence of any quantitative ideas. Contrast with Newton.

Suppose electron comes into existence at moment.[3] Its influence through space & time is matter of transmission expressed in terms

$12^{h} - 12^{h}\cdot 0^{m}\cdot 1^{s}$ φ (.[4] & $(a_1\, a_2\, a_3)$ directive vector.

φ' $(a'_1\, a'_2\, a'_3)$ is its contribution,

Distributed through space.

Decreases inversely as square of distance. Is moving outwards with velocity of light. 3×10^{10} cm/sec

---

Pattern of aspects must be taken as a whole. Functioning of an organism is fundamental fact & detailed static part is derivative. Works in with that idea in modern science which Whitehead thinks is biological age.

1. While this is how this term seems to be spelled, it makes no sense; perhaps 'sorted' was intended.
2. This material matches that at the start of lecture 34 in the notes taken by Bell in Emerson Hall on the same day, 18 December, although the sequence seems somewhat altered.
3. In the Emerson Hall version of this lecture, this passage about the scalar and vector quantities of modifications in the electromagnetic field comes later in the lecture.
4. Although we have retained the '(.' here, it is not clear what was intended, if anything. It is also not clear how the numbers (which seem to indicate hours, minutes and seconds) are related.

Organism requires that Whole be considered previous to the part. Biology is analysing & physics is additive.

Problem of vitalism & mechanism.

Whitehead – at very basis of our idea of things is idea of the organism.

It is a structure functioning – if you consider both space & time. If you try to tear apart space & time, so as to gain an eternal space & something absolute

|125|

Read Jack Haldane (nephew of Lord Haldane)
B. Russell "Icarus & Daedalus"[1]

happening in it. You are dealing with high abstractions. Must describe the pattern of functioning in ~~spac~~ timeless space & in time – in terms of transmission. [Cf. Newton.]

Beauty of timeless space is that it enables us to connect the theory of the medium. Has been considering the electron as a pattern of all space & time. Has been considering it by itself. but really there are a whole lot of electrons.

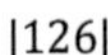

P = standpoint.

Electron E_2 will change due to E_1's contribution to E_2's electro-magnetic field. So if you notice the structural pattern of a electron viewed as a persistent entity, takes the form of a description of inwardness of E_2 + form of electro-magnetic field caused by E_2. So pattern of E_2 at P is that of E_2 as modified by the contributions of E_1 etc.

|126|

And so as a matter of fact you get a most highly complex effect of the medium. Then E_2 as thus modified produces modifications all around therefore you have also its <u>indirec</u>t effect at P. i.e. what E_2 will be at P then E_3 etc.

That problem is as matter of fact quite solvable in terms of effect of medium – as a transmission generally through the medium.

Any finite entity is only completely concrete as in its environment. When try to take it out of its environment get a very highly abstract formula.

1. This reference to the 1923 debate between Haldane and Russell seems to have been fitted within Heath's notes on this lecture. For background on the Haldanes and the debate, see footnote 1 on page 149, lecture 34, as recorded on 18 December by Bell at Emerson Hall.

|127|

January[1]

The enduring object is essentially an organism – the endurance of a plan of recurrency. That brings us up against the idea of nature as through & through a system of organisms as versus a materialism. Revolt first led by Driesh[2] vs. materialism – Some element entering into an organism which issued from the organism as a whole. Could not build up the functioning of an organism simply by considering the physical forces acting & that aggregate was simply way body does act. Closely allied to point, are final causes (brought up by Francis Bacon.) Revolt vs. scholastics was revolt to efficient as vs. final causes. That is why scientists have been so prejudiced against philosophy – has been an anti-rationalistic bias in science.

As between vitalists & mechanists of late 19th finds with mechanists. Real difference between them.

Materialism starts with idea of some stuff. is simply location just there now & then this stuff has more or less accidental changes arising from locomotion. What you

|128|

know are the qualities of stuff or else the qualities of mind which the relation of the stuff to your mind makes you apprehend. Then if you have laid stress on the divisive side of space & time, you come down to the point, the instant & density. That holds rather you take the relational or the absolute view of space. The movement of each particle is regulated by the laws of motion etc. electromagnetic field. Then when you build up an aggregation the aggregate is simply the sum of the parts. Each element is completely defined in itself. This absolutely & blankly contradicts our immediate experience of trying to ~~control our~~ preserve our existence as an organism. The idea of the organism is undoubtedly something in which instead of constructing the whole of parts the parts cannot be fully determined without reference to the whole.

If you once start with materialism you can never get back to organism. Whitehead does not believe in one view for stones & another for animals, etc. etc. But question is whether there is any mechanism anywhere. If you are going to sweep out materialism – must consider how it arise – Categories of Aristotle presuppose

|129|

materialism. Where, when, etc. etc. Are uncommonly useful ideas.

Also must make up clearly your mind whether you are going to start from a subjectivist or objectivist point of view.

No simple location from sense data or electrons etc. Simply the organization of a field of force.

---

1. The material here matches that at the start of lecture 35 in the notes taken by Bell at Emerson Hall on 20 December, the last day of classes before Christmas. Heath notes it was now January, and we can assume that was 6 January, but hereafter she provides few dates for several weeks.
2. Hans Driesh (1864–1941), a leading vitalist, was especially known for his *The Science and Philosophy of the Organism: The Gifford Lectures…* (1907 and 1908).

Ultimate entity is an organization of all space-time in which different locations are connected by their diverse structural relations to eternal objects.

[1]How do we start unflinchingly from the point of view of the organism – argue from whole to the parts – the concrete unit of reality in something which is capable of analysis – but what you get by analysis is not more concrete, but may be less concrete than the totality we started from.

Question – What is the unit (W. James – "Does Consciousness ∨Exist∨?"[2])

Eternal objects represent the ingredients in the organism. The unit of realization is an event. Major events can be analysed into minor events without losing concreteness. Or can be analysed into ingredients which are not events. In realization there is an experience of "prehension" (not cognition which is raw experience).[3] An event includes
|130|
within itself aspects of other events.

A unit which under certain limits includes in itself an experience of all other units of the same community. This is our machinery.

The "eternal object" is not simply here or there but ~~the~~ ∨its∨ function in the organism is the interconnection of separate events. It is in one event with the aspect of the other events. (Cf. W. James on intersection.) (Cf. Leibniz – Monads mirroring.) ∨Reciprocity∨ Correlative aspects are the basis of spatio-temporal relations between events. Space & time must not be looked on first as merely the loci of simple location, but reciprocity of aspects is the fundamental fact about space & time, & simple location is high degree of logical abstraction. From that point of view one is sticking quite closely to what space & time are in our own cognitive experience. There is the experience, which we cognize, but world is not constructed on lines of universal self-cognition. [How does that relate to Bosanquet.[4]] The organism is the "prehensive" unit of a pattern, which by its nature it distributes into patterns of same kind.
|131|
Question as to the enduring entity.

Admits he has been muddled & thinks it is due to lack of consistency. Has not been sufficiently dominated by idea of the organism. Has been thinking of parts → whole (pattern promulgated along a route.) Should not do this. Should think in terms of patterns.

1. A change in ink seems to mark the start of a new lecture, matching lecture 37 in the notes taken by Bell in Emerson Hall on 8 January.
2. For details of James' publication, see footnote 5 on page 161 lecture 37, in Bell's notes at Emerson Hall.
3. Closing parenthesis provided.
4. Bernard Bosanquet (1848–1923), English philosopher and political theorist.

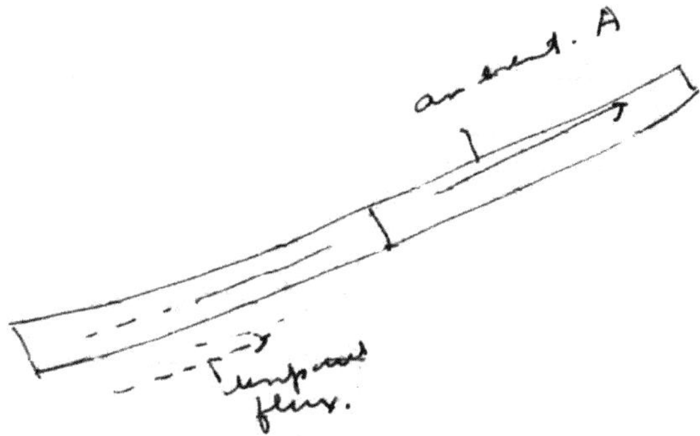

Whole idea of organism has to do with essential difference between space & time. The event A is pervaded by an enduring organism when the whole or any temporal slice exhibits in its patterns a certain identity. Then it is pervaded by an enduring object or an emergent object which endures – Quite different from the eternal object.

The ∨identity of the∨ enduring object may be most trivial. i.e. persistence of hearing noise of trolley car. Also may be aspects simply of environment, i.e. extrinsic. & in a sense accidental or casual.

|132|

Or it may have intrinsic endurance if it inherits from itself its identity. This intrinsic endurance will refer to a certain particularity of character in the prehensive activity which lies behind. – i.e. the "*élan vital*" in this limited realization of itself.

But this is not "in vacuo" so that it is not merely inheritance from there, but in connection with the environment. An enduring object & its relative importance presuppose the favorable environment.

Evolution as the gradual ~~evoluti~~ emergence of spatio-temporal modality into emergent objects. Achieving definite types of value. & thus also limiting itself. To be unlimited is to be nothing.

Whole philosophy problem.[1]

1. Start with Flux. Need motion. Must deal with problem of identical entities; i.e. beings which have existence apart from the flux. – called "eternal entities/objects" or "existents." ∨also "primary object"∨ We only have to deal with them as part of the machinery of realization.

Every occasion has a significance beyond itself. – That is 1. our habitual, immediate experience, 2. If flux is really a system, any finite entity has to have its environment adjusted to it. 3. We always

|133|

aware of a "beyond" in space & time. i.e. its a sensible question always – "What is outside?".

This he called knowledge by relatedness as distinct from knowledge by adjective.

1. The following material seems to match that at the start of lecture 38 as recorded in Emerson Hall by Bell on 10 January.

Then Whitehead pointed out that in dealing with these abstractions – eternal objects like greenness etc, that there were grades of abstraction. What we perceive is analyzable into an eternal entity in a particular mode. This in turn is a more concrete eternal entity – again more concrete as that mode as seen from here. Qua existences, the eternal objects are to be looked on as pervading all space & time.

Eternal objects are divided into

pure objects – greenness.

social objects – like concept of a situation.

Pure object is not ingredient in space-time by way of simple location. It is from my body here under the mode of being there; it's an "aspect" of green that we deal with.

The very ingredients that are pulled together "prehensively" into the unit of my consciousness has these aspects, the essence of which is pulling together of time events. Happens also with regard to shape – cf. Broad on shape of pennies, in which he rather muddles the "bare shape" & the "aspect of shape."

|134|

The event beyond is always an ingredient in my bodily event. Then you get the conception VofV the eventV as a pattern of aspects. Then you have pattern ~~as a~~ VquaV concept is social object, i.e. a situation.

If you choose Your event carefully you have an enduring pattern & will see that it is an important endurance. The endurance of a social object. Just here with endurance comes in the distinction between space & time. The endurance of pattern in an event discriminates the transition in an event into spatiality & temporality. Can divide up total event into partial events with definite temporality as to total event, & every part exhibits an identity with pattern of event. The endurance of any pattern is rather trivial. And that at once brings in fact that pattern of what lies outside this particular event, i.e. the pattern of the environment i.e. endurance arises from external aspects. What is important is pattern arising from its own endurance, inherited from its own past. But even here the environment enters as part of the inheritance. In self-endurance of an individuality there are two points to be considered.

|135|[1]

1. There is a underlying activity that realizes itself into individuality under limited modes.

Read Bosanquet – System[2]
a Vitalist controversy.

1. This material could be that presented at the start of lecture 39 in the notes taken by Bell at Emerson Hall on 13 January.
2. It is not clear to what source material Whitehead is directing his students, but the *Stanford Encyclopedia of Philosophy* (http://plato.stanford.edu) on Bosanquet says: 'In the first series of Gifford lectures, *The Principle of Individuality and Value*, Bosanquet holds that when we speak of "the real" or "truth", we have in mind a "whole" (i.e., a system of connected members), and it is by seeing a thing in its relation to others that we can say not only that we have a better knowledge of that thing, but that it is, as Bosanquet writes, "more complete", more true, and more real.'

You see at once that the essence of the idea of organism & you see that the whole is to some extent determinative of the whole.

Idea of a basic object, apart from the influence of the environment, is that it is just that type of pattern that it is.

As a matter of fact – it is what it is in its environment.

I Basic Objects.

With regard to electrons.

Lorentz[1] a. Rigid electron. is always a spherical distribution of electricity. i.e. always same under different environments.

Abraham[2] b. Plastic electron modifies itself. → Electron becomes a spheriod under field of force.

What it is depends on influence of whole electro-magnetic environment. Thus in defining what electron is you have already allowed for the environment.

I. Enduring objects – organism of organisms.

In the case of electron within human body you have two environments –

|136|

(a) general influence

b. particular influence of human body.

(c).[3] Does not aspect of total pattern of body influence the particular electron.

Exactly same idea as for basic physical endurance of scientific matter. Only point is that in ~~org~~ ordinary physical patterns – like a stone, the pattern is very blurred & therefore its pattern VinfluenceV is negligible.

Also is question of "a crystal". Here there is plenty of pattern.

In order to get life – there must be a very decided pattern & also great power of adaptability to circumstance.

["Crystal" has no adaptability. Stone has little pattern.]

Calls pattern of life a rhythm. Have whole pattern determining the part & an immense responsiveness to the outside.

On mechanist side, Suppose you are investigating how matter is acting. Investigate particular molecule. & will find physical laws determining the molecule

|137|

but already dependent on influence of the whole. So the way to investigate is usual scientific one.

Can't work with final causes.

This, our method, will result in ~~That~~ finding laws of molecular action which applies only to living body.

Cf. Eddington's[4] dealing with increased densities in stars which destroy ordinary molecules & get them confused & marred. A physicist inside such

---

1. Hendrik Lorentz (1853–1928), Dutch physicist.
2. Max Abraham (1875–1922), German physicist.
3. '(c).' may well not be transcribed correctly.
4. Arthur Eddington (1882–1944), British astrophysicist. See footnote 3 in page 143 lecture 33 (16 December) and footnote 1 on page 174, lecture 39 (13 January), in Bell's notes.

a system would not ever conceive of the organization of our molecular matter.

[1]Whitehead – suggests get types of organization of molecules in living body which have no exemplification outside living body.

Notice immense difference between "basic object" which has as its ingredients "eternal objects" & one of higher types of enduring objects which has as its ingredients "basic object". This type may not be important, its pattern may be confused. (Stone) or it may have an important type of unity like molecule, crystal, or living body. i.e. has marked pattern of a higher type of unity.

|138|[2]

Question of the locomotion of a physical object.

1. Has assumed a definite route, in four dimensions.

Question of rest, or motion throughout time, is one of locomotion. Basic question for physical science. Is question of dynamics

Question for mechanics is can property of ∨body∨ part be expressed as property of route. Yes.

F = force in certain direction v = velocity

T = intrinsic (kinetic) energy of body at time t.

dt = energy during time ending

[Tdt + T'd't' .... + Fv.dt + F'v'.d't' .....] = total action.

~~Then the law~~

Then assume that any slight variation will not make difference.[3]

Ask about. Principle of Relativity.

Space & time are exhibitions of relatedness between events therefore has character of systematic uniform relation – reject Einstein's heterogeneity in space-time.

Atomicity of physical field.

|139|[4]

Obstructive action of the medium.

[Started from realist point of view that cognition is superadded emergence & is in a sense self-cognition. Content of psychical field is what cognitive event is in itself.

Realized unity of an event is a limited mode of the totality of nature. By limitation the value emerges. An event must be conceived from two points of view – intrinsic reality & extrinsic reality.

Intrinsic – prehension of appearance of aspects.

Extrinsic – its own aspects in other events. It is a factor by reason of which other events are what they are.

---

1. At this point there might be a transition to the next lecture. The content from an Emerson Hall lecture seems to be missing.
2. There seems to be a break here to the following content, which appears near the end of lecture 40 (15 January) in the notes taken by Bell in Emerson Hall.
3. There is possible another break at this point.
4. There seems to be yet another break here. Also, the next lines match the material at the beginning of lecture 40 in the notes taken by Bell in Emerson Hall on 15 January.

Science is the study of the ~~aspect~~ Vpoint of view (side)V of control, abstracting from the ~~aspect~~ Vpoint of viewV of appearance. Each event taken as a whole. This side of control & side of appearance is momentary except so far as you have endurance or retention of value. It is the endurance of a pattern, so that you take a whole event, & it is significant & important because pattern of chair is there for whole of two minutes. Here again by limitation you get clearness & value.

In higher types of organisms
|140|
the pattern of the event has in itself the pattern of lower types of organism.[1]

Pattern responds to the environment.

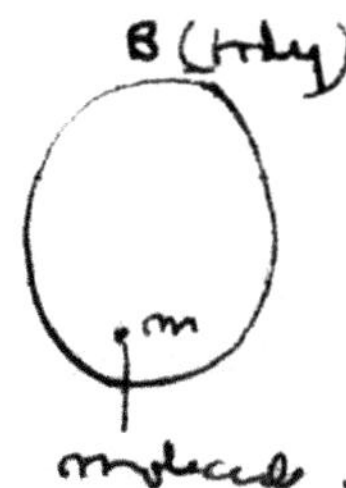

B is part of environment for m.
m is part of environment for B.
Their aspects are particularly important for each other.

No gap in nature.

m changes to ~~cha~~ shield B from changes of wider environment.
Cf. springs on change.

Environment is favorable by being obstructive. Principle of organism is principle of obstructing, i.e. shielding from influence of outside environment.

Principle of quantitative limitation in nature. A quantitative limitation in its aspect of control. Generally expressed by intrinsic energy. When get molecule from neutral into stressed environment the intrinsic energy is increased just as molecule is contorted.

Whole of nature is made up on the spring principle. Owing to principle of quantitative limitation the energy is borrowed from the environment
|141|
and turned to new purposes. That is how you get intensifying action. Principle of a dynamo & way of directing energies by machine.

General principle of whole of nature is also principle of living body.

[2]Discrimination of time from space.

General relativistic reduction of time to space.

Whitehead discriminates on basis of "endurance" which brings in the <u>time</u> element. Insistency of pattern in endurance is due to taking definite kind of succession, i.e. temporal. i.e. a definite temporality. Must be something inherent in general pattern of events.

---

1. The issue addressed here, and in this image, arises in the middle of lecture 39 and again within lecture 40 in the Bell's notes, but those Emerson Hall lectures cover a number of other considerations not present here, and in both lectures Bell and Hocking record diagrams not found here, while this sketch is not in theirs.
2. There is a definite change here in Heath's hand and ink. There seems to be a transition to the next lecture, which matches lecture 41 in the notes taken by Bell in Emerson Hall on 17 January. No other indication is left by Heath, but this is likely recorded at Radcliffe on 20 January and is the last class of the first semester. Whitehead's lecture in Emerson Hall on this same day presents content not found in any of the Radcliffe lectures.

Events in Space Time

The space-time continuum is significant of endurance of events. Endurance of events expresses the patience of space-time continuum.

Does not require that actual enduring objects should have developed, but when developed, it is patient of them. But very meaning of endurance requires temporality, i.e. is significant of them.

|142|

Spatialization of nature, or reality, i.e. there is a display standing for itself, connected with just one unit, i.e. unity of the event. Therefore is spatialized. (This Bergson calls the spacialization of the intellect)

Spatialization is display of event as a unit & temporality gives the event dissolving into something beyond itself.

Question of different time ~~systems~~. VsuccessionV in order of nature.

Comes into philosophy as soon as you interpret psychological time as your knowledge of the world. No question as long as had private worlds. Thirty years ago one would have been so certain that there is only one time succession that one would have assumed agreements among psychological time. Now due to ~~psy~~ physical theories it is different.

But does not deny that in our own lives there is just one order of time succession.

Spatiality will be dependent on your temporality.

|143|

Difference in time succession is only manifested in more or less abnormal experiences – not in our consciousness, only manifest in ultra fine physical experiments that observations have been made that manifest different time systems. Michelson.[1]

If space-time system of A is same as That of B get rest; if different get relative motion. If A is now at rest & now in relative motion with respect to B have to assume that one or both have altered their space-time system. Question of principle of dynamics is wholly concerned with conditions under which objects change their space-time systems – (i.e. laws of motion). Then it is question of fitting together experiences together to find out the proper ~~relation~~VcorrelationV of measurement relation either in one or various space-time systems. To be conjecturally decided by scientific evidence, i.e. purely from empirical evidence. Metaphysics does not decide them but makes them VsensibleV questions; i.e. are natural if start with view of organism.

1. This is likely referring to American physicist Albert Abraham Michelson (1852–1931), who was referred to in Emerson Hall lecture 41.

Second semester

|1|[1]
Phil 3b
Geometry

How it starts.

A carrying to ∨ideal∨extreme limits of relations we do perceive. Involves a belief in an ideal of exactitude to which we may approach.

By surface & point etc. We use high abstractions. Here comes in danger of Fallacy of Misplaced Abstractness. Are, however, natural abstractions from our perceptions. Similarly we all in a muddled way know what 5 is but if we try to make it exact it becomes very abstract. These natural abstractions of our muddled thought to which there are definite entities corresponding are very important abstractions

The concrete is an event, a prehension of aspects of the whole universe. Not exactly Leibniz monad which was self-subsistent & enduring through time but still an event is the most concrete.
|2|
In this concept of event we

1. Argue from our own psychological field. a. Our cognition is self-cognition of our body, not in detail, but as it is for itself – as a totality.

~~2~~ b. Also holds that there is no break in nature.

~~3~~ c. Obviously a variable element in cognition, ex. memory.

d. And on principle of no break we apply c to any event. Some events, however, are more patterned off & important than others.

When you go to the abstractions of physical science – as electrons – these are simply enduring objects looked from the point of view of the interrelations of their life histories, you find the psychological field reproducing itself. An electron is simply its contribution to the electro-magnetic field here & there. Same idea of aspects prehended into a unity. A formula of the pattern in itself from which all the aspects can be derived & therefore you get what can

1. Heath's notes for the second semester seem to have been kept in a separate batch, and the page numbers reflect this by beginning again from 1. We are restarting our inserted page numbers from |1| as well. This is, then, the first lecture of the second semester, most likely on 10 February, and the content matches that of lecture 43 in the notes taken by Bell in Emerson Hall on the same day.

|3|
be said about space & time apart from any particular aspect & in terms of which all the aspects can be expressed. Thus come space & time as a simple set of locations. A high-degree of abstraction.

Voluminousness is most concrete fact of Space.

First Quality of Volume in simple ~~concepti~~ location is that of Euler's diagrams.[1]

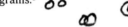

A contains B. <u>A K B</u> but explain that it expresses a relation of volume without any reference to <u>points</u>. Not the relation of <u>All & some</u> but obviously has many analogous qualities & when you get to points can then explain volume & tens of points & all & some. But must not begin here.

2/12/25[2]

aKb . bKc . ⊃ . aKc
a,b,c) : aKb . bKc . ⊃ . aKc
for all values of a, b, c
p ⊃ q   either q is true, or p is not true

|4|
Reread *Introduction to Mathematics*
Concept of Nature
Principles of Natural Knowledge

The dot means that we're dealing with relations.[3]

aK~~b~~ . ~~b~~Kc
aK^2c

aKb . aKc . ⊃ . aKc
$K^2 \subset K$
included in
This we call transitiveness

Other type of assumption is of <u>existence</u> theorems.[4]

1. These diagrams are closely related to Venn diagrams, and Whitehead is here attributing them, as they still are attributed, to Swiss mathematician Leonhard Euler (1707–83). Euler's diagrams are also mentioned in the notes taken by Bell for Whitehead's lecture 43 at Emerson Hall, but there are no examples sketched as there are here. (There are no Hocking notes for this lecture.)
2. Assuming this date is correct and positioned correctly, then some material from lecture 43 given in Emerson Hall on 10 February carried over to 12 February at Radcliffe. And as noted below, the content of Whitehead's lectures at Radcliffe on 12 and 14 February seem to be a somewhat different presentation of similar material compared with lectures 44 and 45 in Emerson Hall in the notes taken by Bell and Hocking on those two dates.
3. This comment is written vertically in the upper right-hand corner of the page.
4. Up to this point the content has followed the sequence in lecture 43 in the notes taken by Bell in Emerson Hall on 10 February, but there seems to be a break here in Heath's notes and then from this point on the content does not seem to be a direct match for either lecture 44 or lecture 45 as given in Emerson Hall, although it does contain elements found in each plus material not found in either.

1.[1] If a is any ~~volum~~ ∨region∨, then there ∨exists∨ a region x which contains a. xKa

2. If a is any region, then there exists a region y which is contained in a. aKy.

It is to prevent clumsiness here that we've had K imply distinctness, else 1 would be a K a & therefore meaningless. It would, however, been more general to include K is identity.

Symbols for "there exists at least one value ∃" (first letter of exists to show it means one value)

∃x = there exists a ...

F'K = class of terms which are related by K (field)

∈ = is a.

a ∈ F'K = a either contains or is contained or both

|5|

1. a ∈ F'K . ⊃ . (∃x) . xKa

To say this of all values of a can either

$a \in F^cK . \supset_a . (\exists x) . xKa$

or

(a) : a ∈ F^cK . ⊃ . (∃x) . xKa

Two dots to show this (a) applies to whole.

2. (a) : a ∈ F^cK . ⊃ . (∃y) . aKy

3. (a, b) : aKb . ⊃ . (∃x) . aKx . xKb

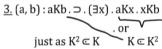

. or

just as $K^2 \subset K$ $K \subset K^2$

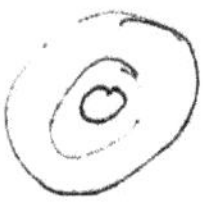

4. (a, b) : a,b ∈ F^cK . ⊃ . (∃x) . xKa . xKb

|6|

Points necessary to complete technique of Logic

p ⊃ q This is general form of deduction.

assert = ⊦ . p

∴ ⊦ . Q

Idea of complete generality

1. What does p ⊃ q mean.

Either not p or q i.e. mere comparison

~p ∨ q of 2 propositions as to their truth relation

---

1. There is also a circled 'a' above this '1.', the import of which is unclear.

To have implication important the relation must hold because of structure of p & independent of actual truth of either.

p q

All men are mortal & Socrates is a man . ⊃ . Socrates is mortal

x x

Formal implication

$(x) \quad . \phi(x) \quad . \supset \quad . \psi(x)$

A proposition of one variable.

$\phi(x)$ = a propositional function.

x is idea of some undetermined but particular function. Essential to remember the "individuality" of the variable.

Deductive reasoning consists in replacing some of your constants (like Socrates) by variables & seeing

|7|

that the implications still follow.

or getting propositional function $\phi(x)$ instead of proposition

[All logical structure out of and, or ∨, not ~ .]

chi

$(x) . \phi(x) . \supset . \psi(x) \quad = \quad (x)\, \chi(x)$

2 Another element of generalization which deals with existential idea.

When I've got hold of $\chi(x)$, question arises if there is any value of x for which it is true.

$(\exists x) . \chi(x)$ There is at least a value of x for which $\chi(x)$ is true.

Now in science you're dealing with a set of entities a,b,c ... → ∨class∨ α

There are the observational facts we believe to be true & then inductively generalize this & believe this to be true about all[1] the entities.

You feel shaky about the generalizations & want to tie them together; want to be able ∨to∨ see that if some are true all the others follow.

Furthermore you don't want it to depend any one set but to have it thrown into as many different deductive systems as possible, each of which involves

|8|

~~any~~ everything. In doing that you've enormously increased your probabilities.

In doing that you will have got beyond your particular science dealing with α for all propositions will have been there in the form

$(x,y,z)\ x, y, z \ldots \in \alpha . \supset . \phi(x,y,z)$

If it is merely a question of logical structures α becomes a variable & also often ϕ, and have merely an examination of deductive relation.

Then you start with bundle of concepts, i.e. class, sub-classes etc. – certain numbers of primitive ideas.

Also you get derived ideas definable in terms of primitive ideas.

1. While Heath has put 'of', we think Whitehead must have meant 'all'.

Logically definable in terms of primitive ideas. Question as to important ideas is necessary to whole go of science. Then primitive propositions, you assume. Then follows the whole deductive system, expressible as and, or, not.

|9|

2/17/25[1]

Get back to question of how we get an accurate geometry starting with idea of volume.

Saw need of grasp of logical method & ideal.

aKb

F'K – field of extending over.

(i) $aKb . \supset . a \neq b$ Arbitrary assumption.

(ii) $a \in F'K . \supset . (\exists x) . aKx$ } Divisible — Can have a sub-volume
$a \in F'K . \supset . (\exists y) . yKa$ } unbounded — or super-volume.

(iii) $a \neq b : bKx . \supset_x . aKx : \supset . aKb$

(iv) $aKb . bKc . \supset . aKc$ or
$K^2 \subset K$

(v) $a,b \in F'K . \supset . (\exists x) . xKa . xKb$

(vi) a intersects b . =Df . $(\exists x) . aKx . bKx$
K_{in} stands for or (a,bKx)
$aK_{in}b \equiv bK_{in}a$
is equivalent to

(vii) $bK_{in}x . \supset_x . aKb : \supset . aKb . \vee . a = b$
iii is particular case of vii

viii Things are separated if they don't intersect.

ix A dissection of a is class of F'K, all contained in a, all separated from each other, ∨no part∨ of a ∨which is not∨ intersector of at least one.

δ is a dissection of a . =Df . $x \in \delta . \supset_x .$ ~~x F'K~~. aKx :
$x,y \in \delta . x \neq y . \supset_{x,y} . \sim xK_{in}y$:
$aKx . \supset_x . (\exists y) . y \in \delta . xK_{in}y$.
(Look up Boutroux on Socratic method).[2]

|10|[3]

2/19/25

Two types of harmony –

1. due to keying down.
2. in spite of keying up.

Way to make everything definite is to take up each point-for-itself.

Rome at 9:30 is higher degree of abstraction than Rome during 9:–9:30. It represents division between 9:–9:30 and 9:30–10:00

When you have a surface junction of two ~~solids~~ ∨events∨ the two together make one event.

1. This date is added in the left margin, and seems to mark the transition to lecture 46 as delivered in Emerson Hall on this same day.
2. See footnote 2 on page 211, lecture 47, in Bell's notes from 19 February.
3. In Emerson Hall on 19 February, lecture 47 begins with the reference to Boutroux immediately above, and then continues as below.

Two events are joined when the two make one.
This can be expressed in terms of sheer extension.

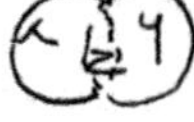

Is a dissection of z so that every point is either x, y or both.

Will not include  or 

|11|
another entity of same species.
4/ Things which coincide with one another are equal with one another.

[Curiosity about use of Greek word congruence. εφαμοιν.[1] Things which fit. Liddel & Scott ∨1860∨ refer to its use in Xenophon & Aristotle but not to Euclid.]

Euclid's famous method of superposition. Idea of rigid body. How do you know it's rigid? Can't get idea of equal in unless you have some structure & are comparing homologous elements in a structure.
5. Whole is > ∨any of∨ its parts.

Three ideas.
1. Equivalence. { Reflexive / Transitive / Symmetry }
2. Junction
3. Congruence.

|12|
Routes of Approximation

Let's talk of volumes (events are hard to think about.)
May approximate to a surface, line, point, ~~(circle?)~~, two points.

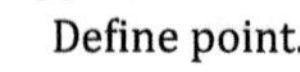

Define point.
1. α therefore a class converging to a point if

$$\alpha K_c \beta \,.\supset.\, \beta K_c \alpha$$

All right as long as don't have tangency – or as long as no α are converging to a point on boundary. Can't cut out that case without presupposing points.

Can't be certain of getting a class which covers & is covered by another. Must bring in some condition.

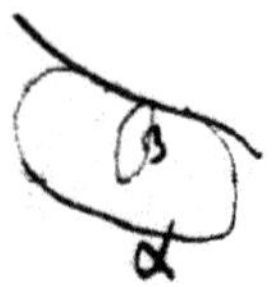

1920 Look up art in *Revue of Metaphysique & Morals*.[2]
Ingenious but a fallacy.
Are obviously on the correct track.

|13|
Adjunction – two events that are separated & yet joined.
Injunction – one inside another & yet have common external boundary.
y & x are injoined if there is a third w event which is adjoined to both.

1. While this is the Greek term Heath carefully transcribed, this coincides with the problem noted when Whitehead makes the same point in Emerson Hall. See footnote 1, page 223, lecture 46.
2. See similar comment made in Emerson Hall and identified with J. Nicod; see footnote 1, page 208, lecture 46.

Abstractive Classes. (How do we get at exact surface, line, point.)
Compare with theory of real numbers.

1	2	3	4	5
1/1…	5/4…	3/2	8/5	2/1

$\longleftarrow\!\!\longrightarrow$ a $\quad f_1\, f_2 \quad$ $\longleftarrow\!\!\longrightarrow$ b

A <u>dense</u> series because an infinite number of fractions between any two.

a./ The thing of class of numbers whose square is ∨greater than 1 &∨ less than 2.[1]

$1 < f^2 < 2$

b. The of class ∨greater than 2 &∨ less than 4.

$2 < f^2 < 4$

a. Can get as near to 1/1 as you like.

b. Can get as near to 2/4 as you like.

Can get $f_2 - f_1 < \epsilon$ (any number) but no fraction whose square is 2.

|14|

i.e. No jump from one class to another.

Then

Math ask what is $\sqrt{2}$?

(Kant – activity of intellect produces $\sqrt{2}$ – nonsense).

~~As~~ You have produce a whole set of entities – having general properties of numbers & yet square less than 2. You have it in whole class of fractions whose squares are less than 2. Then take class of fractions whose squares are less than 4. & more than 2.

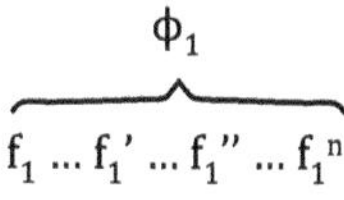

Then $\phi_1 + \phi_2$ (i.e. add all pairs of $f_1^{\underline{n}}\, f_2^{\underline{m}}$). Have to knock out end point.

Multiplication is therefore $f_1^n \times f_2^m$ i.e. Multiplication & addition are way of getting new classes out of $\phi_1\ \phi_2$. Thus you've got entities having all formal relations of numbers (addition & multiplication) & thus have got rid of gaps like $\sqrt{2}$.

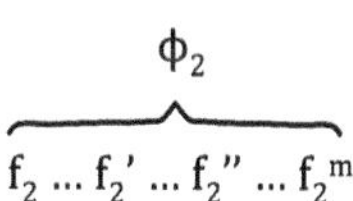

With real numbers have set in which same formal things hold commutative, associative & no exceptions.

All A – B[2]

Always give a good "second class" to anyone here is clearly wrong.

~~B~~. Part III of *Principles of Natural Knowledge*

---

1. In the expression that follows (1<f²<2), Heath definitely writes f² but, given the diagram, it surely should be f¹, and more likely should have a subscript number rather than superscript.
2. These notes were found, here, written on a loose, partial sheet of Heath's ring-binder paper, but with little relevance to the material recorded on nearby pages; they are written horizontally on the back of the sheet.

|15|
$^1/_1$, $^2/_1$ while not 1 , 2 ~~are~~ have same rules.

Point is that where ever you have a boundary ⟨exactness⟩[1] you can have an entity (a class) which defines that exactness. Are also alternative classes of entities which define this exactness but one must make a choice for convenience.

Returning to extension, –

Want entity to define surface of intersecting volumes (a volume of intersecting events).

Start with abstractive classes.

2/21/25[2]

(1) No event z which contains x & y. The two together do not make an event but are numbers of a class which is a more abstract term.

(2) Do for one event z. If our dealing with point idea had to make this distinction. Points of x & y in either can form class z but in

|16|
(1) class does not make up an event & in (2) it does.

Impartial relation of events (foundation of geometry.)

[Taking seriously the doctrine of internal relations. The entity is in itself the grouping together in essential unity of its relations to other entities.]

The Abstractive Class. i.e. How are we going to attain exactness? Must have a route of approximation, in which we have to stop but in which we consider that our stopping arises due to our limitation not to limitation in thing itself.

Approximating to a definite point P. But as we're dealing with volumes there isn't a class to approximate to. We're approximating to nothing. Ideal of exactness does not depend on there being a non-voluminous entity to which we approximate, but ~~that~~ another fact that we can get route of approximation, giving increasing exactness.

Also are alternative routes of approximation to same

|17|
facts. Not converging to any definite entity other than themselves. I.e. However exact you are, the route is equal to that degree of approximation.

$e_1, e_2, e_3, \ldots e_n, e_{n+1}, \rightarrow$ converges to nonentity.

What then are we doing. Well there is bundle of quantities relative to e'. Call it $q(e_1)$. Have process of getting smaller & smaller entities. in order that one may say something definite about them.

1. The term 'exactness' was added, directly above the word 'boundary'.
2. While Heath marks this as the beginning of the next lecture, lecture 48 as recorded by Bell and Hocking at Emerson Hall, on 21 February, seems also to contain material from the upper half of Heath's |15|.

$q(e_1), q(e_2), \ldots q(e_n), q(e_{n+1}), \rightarrow q(l)$

Converges to a class of limits. Physical constants q limitlessness

Tables are of this sort.

$e'_1, e'_2, e'_3, \ldots e'_n, e'_{n+1}, \rightarrow$ nonentity.

Alternative route to e_1.

Convergence of the associated quantities is

$q(e'_1), q(e'_2), \ldots\ldots q(e'_n), q(e'_{n+1}), \rightarrow q(l')$

To same limit as above.

|18|

Now we want to produce an entity to specify all routes of approximation which have same ~~physical quantities~~ convergence. [Cf. $\sqrt{2}$ in theory of numbers.]

Mathematics has also showed us that it doesn't matter how complex & abstract its definition be provided it have certain relations to other entities of same sort.

Call classes of entities which converge to nonentity. $\underbrace{\varepsilon', \varepsilon'', \varepsilon''' \text{ etc.}}$[1] therefore the class of the whole lot of them stares you in the face & $\tilde{\omega} \equiv$ p. i.e. point p is class $\tilde{\omega}$

All relations between volumes can be translated into relations between points & hence geometry. Now whole problem is before us.

1. How are we going to define $e_1, e_2, e_3, \ldots \rightarrow$ something in in terms of volumes.

2. How relation of two routes having same convergence.

First, let us face the problem of how man gets idea of <u>point</u>

|19|

if really it is so highly abstract. Well it's simple – first have volume so small that one does not need to consider its dimensions. Then seeing that it still has dimensions, one simply says – oh well, there <u>is</u> such an entity, & fact that there is such an entity – though abstract – prevents seeing the faults in this vaguely conceived <u>point.</u>

Business of philosophy is to clear this up.

Definition of an abstractive class –

An abstractive class psy ψ is a

$\hat{\Psi}[x,y \in \psi \,.\, \supset_{xy} : xKy \,.\, \vee \,.\, yKx :.$

$x \in \psi \,.\, \supset_x . (\exists y) \,.\, y \in \psi \,.\, xKy :. \sim (\exists w) : x \in F\text{‘}K \,.\, \supset_x . xKw\,]$

A class of ψ such that x & y member of ψ implies that for every value of x & y, x contains y or y contains x any x being y, implies for every value of x that there is a y ~~such~~ also member y ψ & x contains y, <u>and</u> there is not a w which they all contain.

1. It is not clear if these are meant to be epsilons or $\in$.

|20|

2/24/25[1]

Euclid in Greek Thomas Heath. Cambridge 1920.

Define idea of a route of ~~convergence~~ VapproximationV

Ab'K = the class of abstractive classes.

$\alpha \in \mathrm{Ab'K} . =\mathrm{Df} :: x,y \in \alpha . x \neq y . \supset_{xy} : xKy . \vee . yKx :.$

$\sim (\exists v) : x \in \alpha . \supset_x . xKv$

{ All men are mortal / x is a man . $\supset_x$. x is mortal } Peano.

When one abstractive class <u>covers</u> another

α covers β :. =Df . $\alpha K_c \beta$:.

=Df :. $\alpha,\beta \in$ Ab'K : – we consider only abstraction classes.[2]

$x \in \alpha . \supset_x . (\exists y) . y \in \beta . xKy$ – However small you take a there is always a β inside it[3]

Two abstractions classes are equal.

αK equal β . x =Df . $\alpha K_{eq} \beta$.

=Df . $\alpha K_c \beta . \beta K_c \alpha$

Thus have defined idea of having convergence to same quantity of physical

|21|

entites & have done it in terms of space. (or better space-time)

Two problems arise

1. Must know what sort you're dealing with.
2. Might converge to same thing & yet not be equal.

All these are cases either of tangency, or of common boundary.

Review of members

K, $K_{in,}$ $K_{c,}$ $K_{eq.}$ Ab'K.

(x), $\supset$, $\supset_x$, $\vee$, ($\exists$x)

Idea of equivalence is always turning up.

~~Involves three relations any. two can be got from the third.~~

There are three relations, which are properties of equality.

(1) Reflexive. aRa

(2) Symmetric aRb . $\supset$. bRa

(3) Transitive aRb . bRc . $\supset$. aRc

Verify[4] Notice that K is (3), not 1, & 2.

K_{in} is 2, &1, not 3

$K_{c,}$ is 1, 3, not 2

K_{eq} is 1, 3, 2.

From 2 & 3 you can get 1.

---

1. Heath marks the following as occurring at Radcliffe on 24 February, and it matches material beginning in the middle of lecture 49 from notes taken in Emerson Hall by Bell and Hocking on 24 February; but the Emerson Hall lecture began with the defining of abstractive classes, as appears above, presumably at the end of Whitehead's Radcliffe lecture on 21 February.
2. This is added in, lightly, at an angle.
3. As above, this is added in, at an angle.
4. This word seems to have been added, at an angle, in the left margin.

|22|
2/26/25[1]
Euclid's Axioms.

1. Things equal to same thing are equal to one another.

People went on happily thinking equal was a simple thing. Really belongs to a class of relations having properties of reflexiveness, symmetry & transitivity. Different notions of equality in different subject-matters.

What Euclid was really doing was pointing out main property (transitiveness ∧& symmetry∧) of equality.

1. aRb . cRb . ⊃ . aRc

Euclid assumed reflexiveness.

2. If and aRa

Then from 1 & 2 you get regular definition on previous page.

comes down to All members of a species, so far as that speciality is concerned are ~~members~~ equivalent.

2. If equals be added to equals, wholes are equal.

3. If equals be subtracted from equals remainders are equal.

Has idea of a junction & dissolution of junction. Two entities have some relation to each other such that these two entities uniquely define
|23|
Idea of conditions – ∨i.e.∨ class is of a certain species.

Call species of class σ

Consider idea of sharpest convergence of that species – call σ-Prime.

Call α or Prime σ

$$\alpha \in \text{Prime}(\sigma) = \text{Df} :. \alpha \in \sigma \cap \text{Ab}'K : y \in \sigma \cap \text{Ab}'K : \alpha K_c y . \supset_y . y K_c \alpha$$

Antiprime – fullest convergence

$$\alpha \in \text{Antiprime}(\sigma) = \text{Df} :. \alpha \in \sigma \cap \text{Ab}'K : y \in \sigma \cap \text{Ab}'K : y K_c \alpha . \supset_y . \alpha K_c y .$$

2/28/25[2]
Various auxiliary methods in philosophy.

All thought is abstraction. Concentrating on a few partial facts, which are true, but (holding doctrine of internal relations) these partial facts are only true because the omitted relationships which they require, can be conceived for some purposes under systematic conceptions (as for example space-time.)

Every science is taking a set of partial relationships & studying their relationships, & how that set (ignoring everything else) realize themselves in the actual world. Then that definitely means that certain ultimate relations which are supposed ultimately
|24|
held occurs in those sciences as simply unexplained laws.

1. Heath indicates the following notes were taken on 26 February, but this review of Euclid's axioms is part of lecture 49 in the notes taken by Bell and Hocking on 24 February, whereas the analysis of primes and antiprimes was taken up in lecture 50 at Emerson Hall, on 26 February.
2. Once again, Heath seems to be recording material from lectures Whitehead delivered in Emerson Hall on 26 February, and then carrying over into 28 February.

Philosophy is not another science with its own limitations, but is the intellect standing back & criticizing & harmonizing – asking reason of it all. For example – bringing into adjustment laws of physics (mechanical) & social responsibility of ethics. A criticism is wanted on both sides. Must study various sciences & see what their philosophical presuppositions are.

Besides attending to sciences, must attend to what repeatedly occurs in human thought. To try to find a certain unity of apprehension in all literature & expressed that. Get beyond formulation of a single epoch.

Here the exact methods of logic come in. See exactly what comes out of assumptions of a science.

|25|

Defined prime & anti-prime. Idea necessary to define point. An added condition to point is abstractive class that is covered by any other abstractive class which they covered.

Prime ∨class∨ was to show an abstractive class which satisfied certain conditions & which was covered & covers any class which also satisfied these conditions.

$$\alpha \in \mathrm{Prime}(\sigma) = \mathrm{Df} :. \alpha \in \sigma : \alpha K_c \beta . b \in \sigma . \supset_b . \beta K_c \alpha \supset_b . \alpha K_{eq} \beta$$

$$\alpha \in \mathrm{AntiPrime}(\sigma) = \mathrm{Df} :. \alpha \in \sigma : \beta K_c \alpha . b \in \sigma . \supset_b . \alpha K_c \beta \supset_b . \alpha K_{eq} \beta$$

These conditions aren't of very much value when σ gives regular primes. σ gives regular primes

if (this is the case:) –

$\alpha \in \mathrm{Prime}(\sigma)$ ~~. É : β∈σ . βKeqa~~ $. \alpha K_{eq} \beta . \supset_{\alpha b} . \beta \in \mathrm{Prime}(\sigma)$

Same for regular anti primes.

Unless this condition is satisfied can't deal with whole class of equal classes on same level.

|26|

A Geometric element is point, line etc.

Do not want to hook members[1] up to merely one type of convergence for point. To define point merely by route of convergence to it & as any route will do – must define point as class of equal routes

G_σ'K = all geometric elements which $\alpha \in \sigma : \alpha K_c \beta . b \in \sigma . \supset_\beta . \beta K_c \alpha$ have to do with condition σ.

Point =

$$P \in G_\sigma\text{'}K = \mathrm{Df} :: (\exists \alpha) :. \alpha \in \mathrm{Prime}(\sigma) : \beta \in P . \supset_\beta .$$

$$(1) \quad \overbrace{\beta \in \mathrm{Prime}(\sigma) . \alpha K_{eq} \beta}^{\text{necessary}} : \underbrace{\beta \in \mathrm{Prime}(\sigma).}_{\text{sufficient}}$$

$$\beta K_{eq} \alpha . \supset_b . \beta \in P$$

Gives idea of geometric element which is governed by condition σ.

$p \supset q . q \supset p$

$p \equiv q$ – equivalent – (as between propositions)

---

1. It is not clear whether this is 'members' or 'numbers'.

Equivalence is not one idea, but a class of things.

Therefore (1) might become

$P \in G_\sigma$'K =Df :: $(\exists x)$:. $\alpha \in$ Prime(σ) : $\beta \in P$. $\equiv_\beta$.

$\beta \in$ Prime(σ) . $\alpha K_{eq}\beta$ }necessary & sufficient

|27|

Similarly $\bar{G}_\sigma$'K = geometric elements of anti prime

$P \in \bar{G}_\alpha$'K =Df :: $(\exists\alpha)$:. $\alpha \in$ Antiprime(σ) : $\beta \in P$. $\equiv_\beta$. $\beta \in$ Antiprime(σ) . $\alpha K_{eq}\beta$.

When σ gives regular primes or antiprime when you have α is Prime(σ) $\alpha K_{eq}\beta$ then $\beta \in$ Prime(σ) & this[1] can be knocked out ---

3/3/25[2]

How time helps us get over our difficulties in analyzing the continuum of events so as to deduce from it geometrical elements, i.e. instantaneous spaces.

Summary. Time & Space come in together. Every event involves a transition to something beyond itself. May think of it as spatial transition. Time arises from universe considered under aspect of succession. (*sub specie successiones*) Time succession is <u>most</u> obvious fact. Descartes distinguishes between duration (in external world) & time. (way we think of world. [Look up in *Principles*]). Derives time from motion.

|28|

Brings him rather near to Bergson's distinction between <u>durée</u> (not divisible) & <u>time</u>.

Whitehead takes line of extreme realism. Cognition, qua cognition, creates nothing; though everything affects everything else. Cognition is of thing as it is for itself, but this is as it expresses universe. Maintains as argument[3] Bradley (illusion) that what universe is in its aspects, is what universe is in itself. Threefold aspects – past, present – future. Epicurus present contains within itself a past & future. Is <u>immediate</u> in a unique sense. Present is advancing into a future which is qualified by it but not fully determined so far as immediate apprehension is concerned. Present is simultaneous with our immediate self. Represents the immediacy of achieved value. Present is immediate definite achievement of the universe. What is ∨the∨ immediately given present is aspect of spatiality. Succession of time is aspect of unit as succession of spatialities. Succession of something, grasping as though it were <u>halted</u>.

|29|

This succession exhausts all that there is, & yields an ordered framework, in terms of which all relations of patterned aspects can be expressed. It is the systematic character of this framework which makes apprehension & thought on part of finite organisms possible. Space &

1. In the Bell notes for this lecture, what can be knocked out is $\beta \in$ Prime(σ).
2. This lecture starts off with material Whitehead presented in Emerson Hall, according to Bell's and Hocking's notes, through the last half of lecture 51, on 28 February. Heath's |28| and |29| then move on to material dealt with in Emerson Hall in lecture 52, on 3 March – but only roughly.
3. This word could also be read as 'agreement'.

time represent the systematic relation of things. Abstraction is possible because you can arrive at truths which only require the rest of universe systematically & therefore you are justified in neglecting the rest of universe.

Therefore Loathe to admit that space-time are changed casually by the casual presences of things in them. In space & time the way the totality expresses is its "systematic patience". Can express truths via systematic patience for partial entity.

Simultaneity – is the relation to the given event of all those events which are disclosed as exhibiting in given event aspects of immediate

|30|

achievement of things. This is a spatiality. Cognition does not keep step with the passage of things.

Simultaneity, – that grasped together is one spatiality.

Can you define an instant as exact boundary of duration.

σ_d = this class of duration.

Doubt as to whether every rule of succession reveals same simultaneity. Let us make no exemption here. Let be different rules of succession,

$(\exists x) :. \alpha \in \text{Prime}$

(if there are different rules.)

Moment $M_\in G_{\sigma_{d_1}}$ 'K =Df. ~~M∈ antiprime(σ)~~ $(\exists \alpha) \in \text{Antiprime}$

$(\sigma_{d_1}) . \alpha \in M : \beta \in M . \supset_b . \alpha K_{eq} \beta$

[Burnet –

|31|

3/5/25[1]

Use of hypothesis in search of truth.

1. Everyone does it.

2. Involves fact that there are grounds of knowledge beyond what is immediately known.

3. If make a general hypothesis, any ~~other~~ experential verification, is invalid as verification of complete hypothesis.

Chief use is not question of truth of hypothesis, but that it is an hypothesis which will lead to direct apprehension of immediate experiences that would otherwise remain unnoticed.

It is then a hypothesis which relates to various details which are connected. A wrong hypothesis may do good if it directs attention to right sort of details.

Vague apprehension comes much before precise formulation. Hypothesis is hypothetical formulation, put ∨hypothetically∨ exactly & grounded in direct though vague knowledge. & after a time leads to direct apprehension. Then if the deductions are verified, the hypothesis rises in importance.

1. This seems to match the material presented at the start of lecture 53 in the notes Bell and Hocking taken at Emerson Hall on 5 March.

Chief use of hypothesis is as a discipline of cognitive apprehension.
|32|
Whatever fits into a scheme is more important than casual experience.

Also mere self-conscious cognition, heightens & increases the elements which unconsciously you grasp together. Cognition, while it does not create entities, influences my immediate prehension.

Therefore Hypothesis→ cognition→ prehension,

i.e. Hypothesis actually influences what you are to be.

Dangers of hypothesis

1. Subjective prehension (i.e. way your environment is clothed to you partly depends on fusion of aspects of environment in itself & aspects of your own body.) → Bodily life may take command of whole situation.

Thus hypothesis may make you see whatever you wish to see. While what you want to observe is relation of things as divested of personal aspects. [Danger of pragmatism][1]

One of enormous advantages of logic is its imperious objectivity. The impersonal test

|33|
Begin with anecdote, Philip Wicksteed[2]
3/7/25.
Another attitude which impedes ~~active~~ accurate apprehension is the lazy hypothesis – making the assumption do instead of apprehensiveness. This lazy assumption is perfectly obvious ~~way in~~ in religion.

Also the dominant scientific theory appears to be always absolutely supported by all recorded facts. A theory does not get upset until man with bull-dog mind gets grip on some apparently silly little set of facts ~~th~~ & insists that it won't fit into the theory. Here comes value of new & more sensitive instruments. Illustration from Relativity.

A moral order among hypotheses. Are told you should always act on the nobler. This seems dangerous. We think we should take truer.

Whitehead says. Are layers in us of wider & wider apprehension. Whole pursuit of Philosophy is founded on widest &
|34|
vaguest (Descartes would use word *inspectio*) i.e. that reason can see way things dovetail into each other. Not a series of arbitrary starting-points but there is a harmony that can be approximated to. We feel that the broader hypothesis is the finer.

On the other hand an hypothesis that is simply made up & used in the form so that you can make an arbitrary jump, i.e. allows machinery of which you have no direct apprehension, is not a justifiable hypothesis (Cf. God & final causes.)

1. Closing bracket provided.
2. It is difficult to read the surname noted by Heath, but both Bell and Hocking have Whitehead telling this anecdote about Philip Wicksteed; see footnote 2, page 244, lecture 53.

When you come to immediate action & have to act on some hypothesis or other, again the idea of nobler hypothesis comes in again.

Function of science is not to bother about action.

|35|

Then there comes another point. At beginning of session read from Broad *Scientific Thought*, on various types of philosophy as critical (excellent) & speculative (condemning with faint praise.) Whitehead makes claim for utmost freedom of thought etc.

Must always allow for ragged-edges.

3/10/25[1]

Plunge into hypothesis derived from modern theory of relativity.

Is <u>time-series</u> of moments unique? In order to answer that we've got to consider whole question of <u>time</u> more narrowly.

Durations – either completely separate or one lies within other, or intersect in another ~~moment~~VdurationV of same system, & common part is also ~~moment~~ ΛdurationΛ of same system.

Then we have idea of these ~~moments~~VdurationsV as forming a series.

[Can define boundary V(moment)V as abstractive system such that ~~every~~ each duration of this class intersects (but does not lie inside or outside) duration D.] That there are two alternative (begin & end) moments bounding a duration is fact of immediate

|36|

perception. VFurtherV Any two moments of same system bound a duration (also from immediate apprehension.)

Define <u>order</u> in ideas we already have

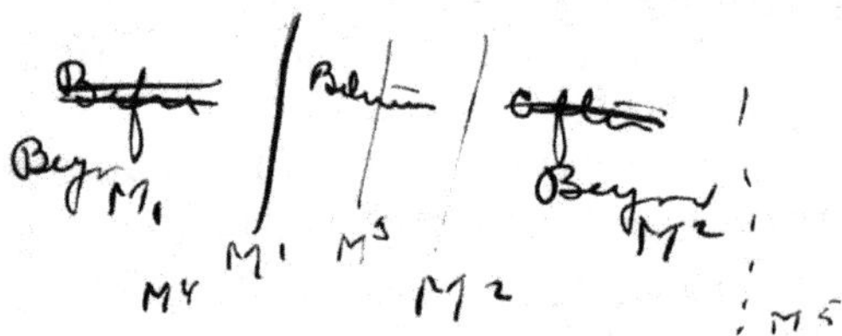

M^3 is <u>between</u> M^1 & M^2 if it is defined by an abstractive class having numbers contained in M^1 M^2

M^4 – abstractive class converging to[2] M_4 of which is at least one duration such that M_1 ~~& M_4~~ lies between its boundaries & M_2 does not & also at small end are durations which are completely separate from duration bounded by M^1, M^2.

M^5 – same except M_1 & M_2 are interchanged.

Later on will consider nature of series.

---

1. The notes taken on this date roughly match the content of lecture 55 in the notes by Bell and Hocking at Emerson Hall on 10 March
2. We have retained the superscripts and subscripts as Heath recorded them, but it is not clear from either the context or the diagram whether they should be differentiated.

Besides mere idea of series, in time we have another idea i.e. time sense, (of
|37|
going forward)

Get limiting relation – things at a moment. i.e. of an instantaneous three-dimensional space. Then we've got to ask what is it which gives us flatness ⟨planes⟩, straightness ⟨lines⟩,[1] etc; & you can't ⟨get⟩ all that out of mere idea of extension.

3/10/25[2]

Must get idea of what it is that differentiates a moment into ~~space~~ lines etc etc?

Founded on fairly obvious intuition – i.e. I see wall – loci of color etc. etc. i.e. One could easily show this differentiation in terms of casual relations, like color. But what one really wants is the geometrical aspect which colors have, expressed in terms of spatio-temporal facts like events without calling in casual elements. i.e. to show that differentiation is part of what is ~~casual in~~ systematic, not what is merely contingent.

As for our other problem – i.e. alternative time series – that obviously is not given in experience. On other hand the discovery of "simultaneity" depends on endurance of pattern.
|38|

With that root conception it is a rational question, as to whether the systems of simultaneity, discovered with respect to different enduring patterns, are all reducible to one? General evidence is very heavily against it.

In the first place it is luminously obvious that in our own individual experience there are not alternative time series at hand. Not necessary that it be some objective time system, but the succession of his organism is unified.

Owing to there being alternatives, we talk as though what we adopt was purely conventional. Not true. – not a convention that I judge this book same size as that – I have a direct perception of it. i.e. Idea that what we perceive is act of choice is nonsense.

Our delicate experiments give baffling phenomena explainable in terms of
|39|
alternative time systems.

Another point – is there any general theoretical reason for adopting ~~pre~~ alternative time systems, besides these definite experiences. There is, you then can explain the breaking up of your instantaneous present into

1. The two preceding terms in angled brackets appear directly above the word they follow, and in the next phrase it is simply a term the context calls for.
2. Despite the date Heath has given – she used 10 March previously – this looks, roughly, like the outset of lecture 56 in the Bell and Hocking notes taken in Emerson Hall on 12 March.

straight lines, etc. etc.

Therefore Whole differentiation of space-time continuum into a pattern of geometrical loci, can be looked on as being simply an exhibition of its properties arising out of ∨alternative∨ time series.

Basis of Whitehead's explanation is that looks on Euclid's property of "flatness" as expressive of the properties of a moment.

(There are enormous advantages in way of simplicity if think of this as giving you Euclidean space. It can be equally well (though more complicated) be worked out with uniform elliptic, or hyperbolic space. But that introduces an extra constant, not required in physics & therefore there seems no particular reason for it.)

|40|

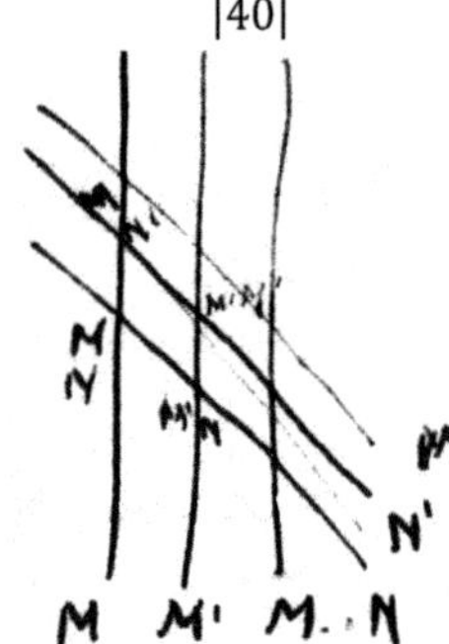

If have alternative time systems, diagram shows intersections. Must then define locus of the intersection of two moments.

Consider aggregate of abstractive classes covered by moment M. Also classes covered by moment N.

The locus of MN is locus of classes covered by both M & N.

A plane in an instantaneous space has its planeness because of its association with another moment.

Relations of moments of same system to each other[1] is basis of parallelism.

MN' & MN are two planes which can never intersect and can always get another between. Therefore A system of || planes in M is system of loci where M is intersected by an alternative time system.

<u>Here</u> as distinct from t<u>he</u>re is reflection into instantaneous plane

|41|

if ∨temporal∨ separation ~~due to tempora~~ due to alternative time systems.

MN is || to M'N

M'N is || to M'N' \ M'N' is || to MN[2]

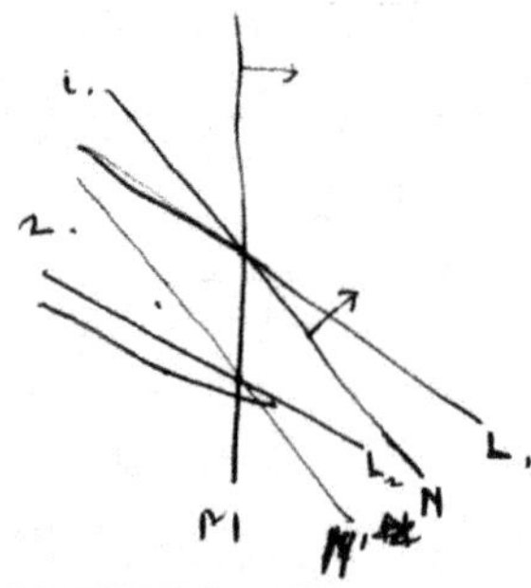

Now take a third time system.

L. might take many forms.

1. It might intersect M&N at these intersections LM=MN.

2. L intersects M in MN' where N' is M to N. LM&MN do not intersect ∴

L, M, & N have no common intersection.

3. LM is a plane which does not lie in any ~~plane~~ moment || to N.

1. These three words 'to each other is' were added in left margin.
2. Heath leaves out 'to' in this last formula, but it is surely intended.

Then [Prove or assume] That L, N, M have a common intersection.

Essence of 3-dimensional space is that if take time intersecting space they have to intersect all of them.[1]

|42|

Repetition

LM is either (i) a plane in each intersection if units are of different Time-Systems. General

or (ii) is null class special

L, M belong to same time system.

L, M, N is either (i) LMN = LM = MN = NL. is a plane VspecialV

or (ii) LMN is null class then L'MN is plane class L&L' are ||.

or (iii) LMN is not null (General case) & LMN is not = NN

then LMN is a straight line.

L, M, N, P. is either	i. Plane	Special
	ii Straight Line	Special
	iii Null Class	General Case
	iv[1.] Point	Special

[1.] The peculiar assumptions as to a point is this – if Q is another moment & LMNP is a point then either LMNP = LMNPQ or LMNPQ = null class.

μιρος[2] In other words a point is indivisible.

This theory has merit of giving exact meaning to divisibility. And all structure can be expressed in terms of alternative time systems.

|43|

Now we have idea of "without magnitude"

That comes in by abstractive classes – i.e.

Let x be any point.

– An abstractive class α be said to obey the condition ψ when α covers every abstractive class included in x.

A point, qua geometric element, is the ~~class~~ VsetV of primes which satisfy the condition ψ. i.e. set of ψ-primes. Now have a regular idea of how we will conceive of points.

---[3]

All our assumptions represent that an instantaneous point is ideally the final degree of simplicity.

Have to assume an identity of physical convergence. Our assumptions.

1. Indivisibility by a moment – i.e. cannot get any further simplicity of position.

2. Utmost simplicity from point of view of convergence of events – i.e. without magnitude.

1. The last three terms seem to have been added in the right margin.
2. While this appears to be a Greek term in the margin, it is not clear what it is.
3. Perhaps with this line Heath is signalling the beginning of Whitehead's next lecture. It does match, roughly, the material presented at the start of lecture 58 in the notes taken by Bell and Hocking in Emerson Hall on 17 March.

3. Utmost physical simplicity →

|44|

can't get convergence to different physical limits by i.e. all physical quantitative properties of K= classes converge to same set of limits.

We have hypothetically given a definite meaning to punctuality, straightness, planeness. All arise as properties of intersections of moments. Have also shown how ||[1] arises from properties of different moments. It is quite possible that assumption that moments of one time system never intersect & others always, may be wrong. [Einsteinians do talk of elliptical time.]

Descartes – in dealing with space separates it from time but says empty space is nonsensical idea & space is an abstraction, Similar to Whitehead, who says that

|45|

Space & Time are abstractions from events.

Cf. Descartes "moving vortices". Only a little step to say that true fundamental fact is something occurring. Whitehead replaces his vortices by electo-magnetic field.

Rather different from Leibniz & relational theory of space

In either Descartes or Leibniz the point ceases to be a primordial element.

Whitehead The extended occurrence is the fundamental element. (one thing a point has not got is extension.)

If an event is your concrete fact must then ask what an event is in itself. Here he runs theory of aspects. i.e. Are expressing in an impartial way the general pattern of aspects.

Now get to the structural relations between events as expressed in terms of relations of point.

Greeks founded geometry by discerning rather vaguely exact idea

|46|

of a point as utmost in simplicity & that all relations of geometry can be expressed in terms of relations of points. Can get rid of awkward relation of extending over.[2] Idea of a point lying in an event. P is a point, i.e. set of abstractive classes of some convergence.

Then an Event e can be formed such that e & tail end of α all are contained in larger event E. Holds for all $K_{eq.}$ classes.

An event is uniquely determined by the points which lie in it.

If event E contains E' it follows that any point Q which lies in E' also lies in E. All points of E' are some of points of E. So now may talk of event as set of points etc.

Then comes question of boundary of event. If β is any abstractive class belonging to B (on boundary) then every event E' which belongs to B intersects E but also ~~contains~~ is not wholly contained in E.

1. There seems to be a superscript following this symbol but we cannot make it out.
2. This 'awkward relation' Whitehead was himself introducing in earlier lectures.

|47|
Then probably as an assumption, have to state that an event is uniquely determined by points or its boundary. (3 dimensional)

No such thing thing[1] as motion at an instant. Instantaneous space gives configuration.

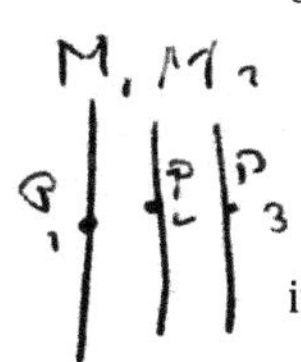

Therefore this question comes up.

Discern[2] P_1 at P_2 but P_1 & P_2 at P_3

If line only separates instants, motion does not come in at all.

According to Whitehead what you have is not configuration at an instant but general pattern with its endurance.

Rest in a time system is fact that it is the appropriate time system of that pattern.

Motion is that it is another time system.

Physics interprets every thing in a Common Space, enduring in a Timeless way.

So A geometrical point is a complex locus, an outcome of idea of endurance:- expresses the patience of universe of events for endurance.

|48|
Must not muddle instantaneous point (simple) & geometrical point (complex.)

|–|[3]
Essay.

Describe the general methods of symbolic logic. Discuss scope & limits of its futility as an auxiliary method in philosophy.

Give illustrations

Literature.

Introduction to Mathematical Philosophy – Bertrand Russell.

Principia Mathemtaica – Russell & Whitehead.

Vol I. Introduction & Chapter I.

Scientific Method in Philosophy.[4] Russell.

Logic – The Essence of Philosophy

Logic – Johnson[5]

Part II, Chapter on Symbolism & its functions.

Logic – Bosanquet.[6]

Lec. Ed. Volume II. Book II. Appendix to Chapter I. (pp.40ff.)

1. The doubling of the word 'thing' is likely Heath's oversight.
2. The second set of P_1 and P_2 were represented by what we take to be ditto marks, but they are not clear.
3. This page of notes appear on a separate, unnumbered sheet, was found at the end of the first semester notes. They carry a date, however, and so we have relocated them here. This is the only case where Heath retains notes on an essay assignment, whereas Bell records several such assignments for the men at Emerson Hall, including one on 19 March. None of those essays have been found, but at least three of Heath's own essays are still extant, including one entitled 'Elements of Geometry'.
4. The full title is *Our Knowledge of the External World as a Field for Scientific Method in Philosophy* (1914), in which chapter II is 'Logic as the essence of philosophy'.
5. William Ernest Johnson (1858–1931) was a Cambridge 'apostle', along with Whitehead.
6. Bernard Bosanquet's work was entitled *The Essentials of Logic: Being Ten Lectures on Judgment and Inference* (1895).

Tues. March 17.[1]

See *Principle of Relativity* page 37. ff.

Descartes draws a distinction between time ∨(as measure)∨ & duration ∨(in nature)∨

Couturat[2] ∨*La Logique de Leibniz*∨– apropos of symbolic Logic.

~~Proce~~ An event is locus of point-events.

It is therefore defined by points that lie in it or by its boundary.

Every geometrical ⟨spatial⟩[3] point intersects every moment in an instantaneous point (or point-event, or regular point).

∨Geometrical∨ Timeless points which intersect moment in a straight line are instantaneous straight line.

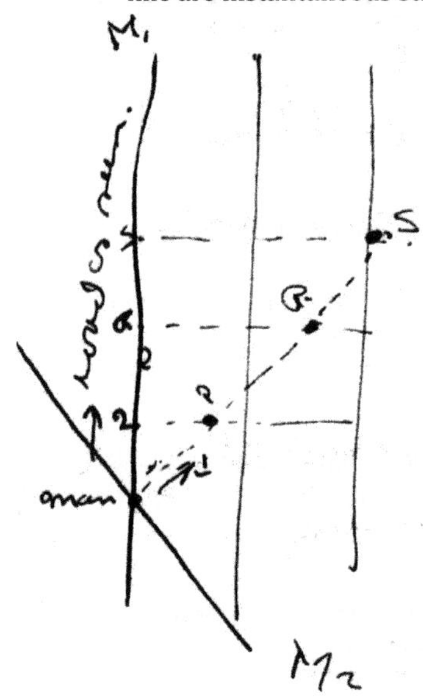

1 is Man's time system as he drives
He does not drive along road he sees but → to Q, R, S & may be entirely different due to explanation[4] of times.

|49|

When you are thinking of space as in your immediate apprehension, it is a close approximation to instantaneous space, but when you think of it as space in which events happen it is the complex idea.

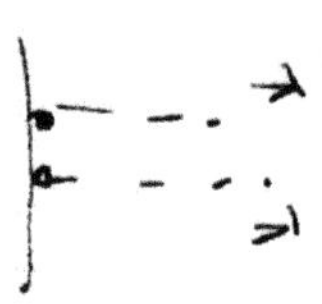

Space of physical science is conceptual, but has ~~same~~ objection to it when it is used as a cloak for refining phenomena to some unknown which relieves one of necessity of making further investigations. Word "conceptual" merely states problem. How is your concept realized in concrete fact from which you've abstracted it. Therefore Whitehead prefers to say it's a "logical construct."

1. Although Heath does not indicate as much, this is likely the outset of the next lecture, which roughly matches the material in lecture 59 in the notes taken by Bell and Hocking at Emerson Hall on 19 March.
2. Louis Couturat (1868–1914), mathematician and philosopher.
3. This term is added directly underneath 'geometrical'.
4. This word could as easily be 'experience'.

M_1 & M_2 intersect in ~~point~~ plane.

← l. is the straight line of road. projected back for Q, R, S.

Ought to be some relation between l & m_1m_2. Because pattern of the whole depends on the relations of symmetry – One we've already discussed is in parallelism, never intersecting.

This, where l & M_1M_2 intersect is ⊥ perpendicularity.

|50|

3/21/25[1]

Principles of Relativity III.

Congruence.

Euclid bases his treatment on idea of being able to transfer things – in doing that you really presuppose measurement.

Idea of matching.

You have a class of characters (c_1, c_2 ... c_n).

A matches B in relation & class (c_1, c_2 ... c_n)

$A = B \to (c_1, c_2 \ldots c_n)$ – This is class of mutually incompatible predicates.

quantities magnitude

$A = B \to \gamma$ (class of magnitudes)[2]

$B = C \to \gamma$

then $A = C \to \gamma$ } Transitive.

$A = B \to \gamma$

then $B = A \to \gamma$ } Symmetry.

$A = A \to \gamma$

Incompatibility gives can't have both $A=B \to \gamma$ = same predicate applies & $A \neq B \to \gamma$ different predicate applies

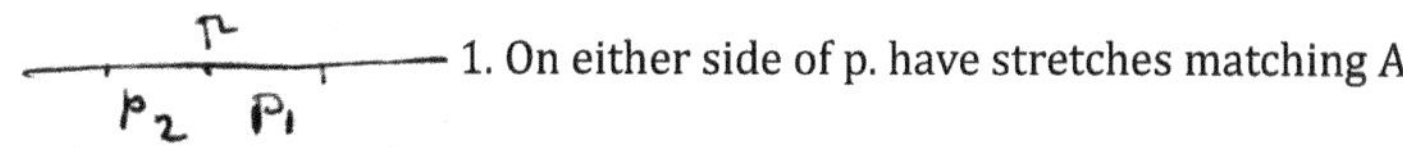

1. On either side of p. have stretches matching A.

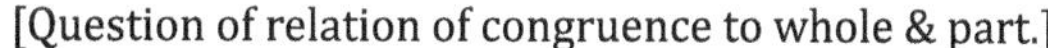

[Question of relation of congruence to whole & part.]

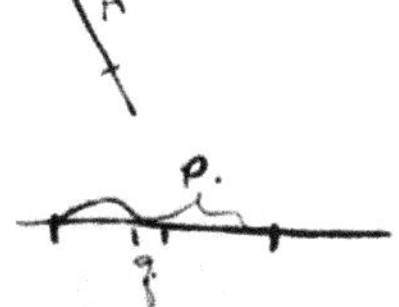

2. If p & q are time stretches & q is contained in p then $p \neq q \to \gamma$

1. The following material seems to match that presented at the start of lecture 60 in the notes taken by Bell and Hocking at Emerson Hall on 21 March.
2. From Heath's notes it is quite unclear what symbol is being used, and a few lines down she introduces 'y', but it is clear from the notes taken by Bell and Hocking that Whitehead is using 'γ'.

|51|
Question of addition coming in.

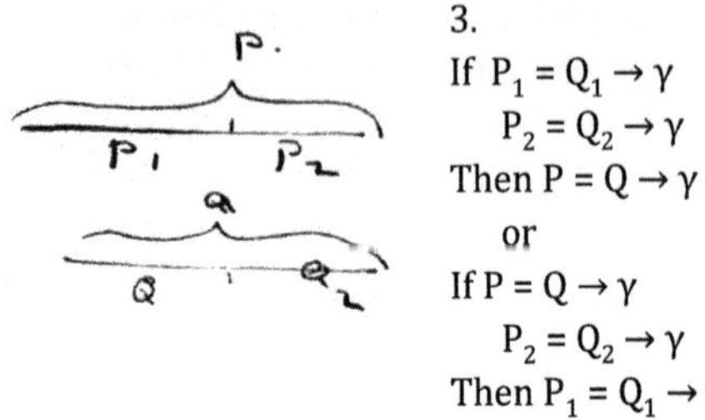

3.
If $P_1 = Q_1 \rightarrow \gamma$
$P_2 = Q_2 \rightarrow \gamma$
Then $P = Q \rightarrow \gamma$
or
If $P = Q \rightarrow \gamma$
$P_2 = Q_2 \rightarrow \gamma$
Then $P_1 = Q_1 \rightarrow \gamma$

I.e. Have structure in which having analogous positions means having same predicates.

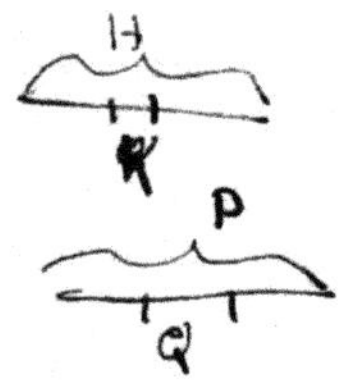

What do we mean by greater than?
We know what we mean by unequal.
Suppose you have whole H & part K & whole P & part Q
Cannot have P = K & Q = H owing to impossibility of getting that crisscross relation, that greater than, has a stable character.

Look up Archimedes

4. Idea of a sequence of equal lengths, all adjoining.

Name the length of total sequence as[1]
m e = c'
c'/m = c
If we have any stretch A
& m is any number.
it is possible to ~~find a sequence~~ subdivide A into a sequence of m equal parts
|52|
Then suppose we have any continuous sequence. A & B. Then we can always find a sequence, by taking one large enough, so that A to first segment & lengths in B = A can stretch beyond B.

That is axiom of Archimedes.
i.e. can find sequence on A of which A is first term & end stretches beyond B.
Comes to this
Whatever length VBV you get, & whatever length VAV you take, however small, & after a finite number, however large shall have landed into a bit whose end lies outside B.
When you've satisfied all those conditions you can run your theory of measurement.

1. This 'm' and the two below could as easily be 'n'.

Another interesting point turns up. When you have spotted one method of assigning predicate class γ, another class may turn up. γ'. i.e. Are an indefinite number of other methods which are inconsistent with each other.

One method will give you Euclidian geometry, another will give you Euclidian geometry with other measurements – other methods will give various sorts of non-Euclidean geometries.

|53|

3/24/25[1]

Addition means that have 2 quantities and they definitely indicate themselves as enter into a three-term relation with a third entity, the sum of the two. Depends on relation of whole & part among ~~rel~~ events. Intensive quantity viewed as measurable forms some correlation either with extensive quantity or counting.

Congruence in space & time.

Summary – Congruence involves analogous functions in spatial structure. – must depend on parallelism. Relations between two time systems are to be same whatever aspect they are revealed in due to uniformity of Space-Time.

Parallelism – moments of same time system, etc. reproduce themselves. with only difference of position. i.e. – What you mean by congruence is that only difference is one of position. VOneV Whole system, || lines in an instantaneous space has a definite method of congruence.

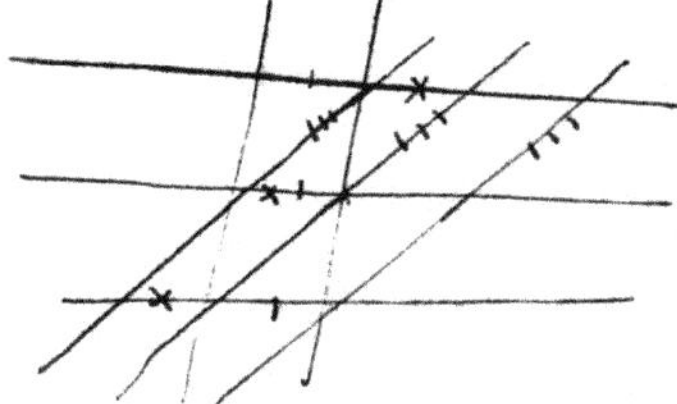

Congruence of one instantaneous space is correlation with another instantaneous space by consideration of points at rest.

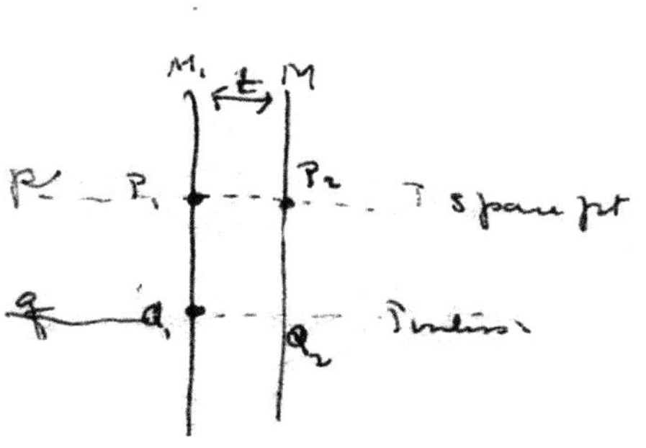

Distance between P_2 & Q_2 if p & q have been at rest is ~~same~~ congruent with the distance between P_1 & Q_1

1. The following material matches that in lecture 61 in the notes taken by Bell and Hocking at Emerson Hall on 24 March.

|54|

This is explanation in terms of Euclidean space. Great complications would follow from elliptical space but would be possible.

We've been trying by considering relativity very conservatively, to link it up with a metaphysical explanation. So far we've only got comparison for ∥ moments, lines, etc. If we're going to express the meaning of congruence, must provide something in the structure which is basis of our apprehension [we're opposed to mind adding anything.} Columbus discovered America did not create it.]

At present we've no basis for comparison indicated before o to o.

Orthodox relativists get structure out of measurements whereas really must get measurements out of structure.

Symmetry of lines which are not ∥ is symmetry which comes from ⊥ only

Basis for ⊥ is that of alternative time systems.

x between ⊥ to plane p. has same relation to all lines in .p.

i.e. is essential symmetry connected with right angles.

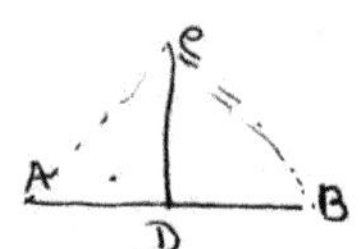

CD ⊥ to AB. AD = DB

The AC = CB. This gives us a real comparison of lines that are ∥.

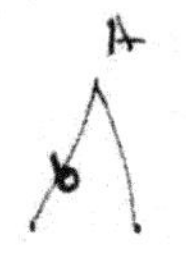

Another difficulty. Start with two lines & want to lay off distance = Ab. Another assumption is

|55|

needed, which may be made in several ways. Elaborate way in *Principles of Natural Knowledge*.

Below is simpler way. Must somehow appeal to continuity.

Have

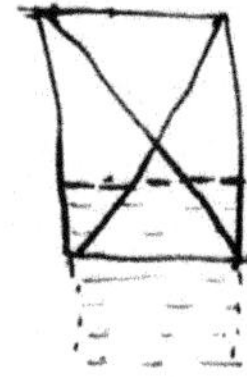

1. Will first say, if you've got a square, its diagonals bisect each other & are equal.

2. If start from rectangles & shorten it sooner or later reach a square. We're simply assuming the existence of a figure (square) which exhibits this particular symmetry of time systems.

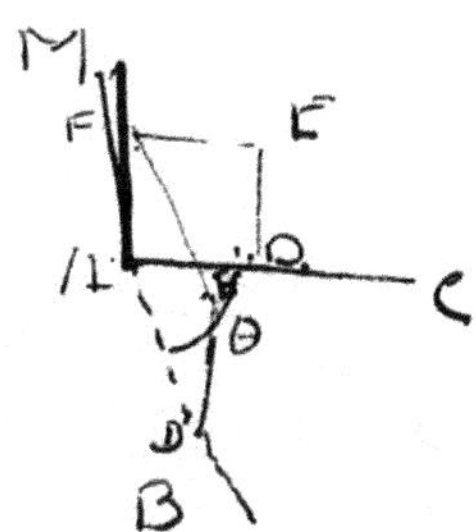

Have three dimensions
AM, AC, AB.
Describe squares FEDA, & FE'θ'A, both having one side AF, therefore everything is symmetrical.
Now question of Time. Time can be worked in same way.
Difficulty as to different times series.

3/26/25[1]
Can express congruence entirely in terms of moments.

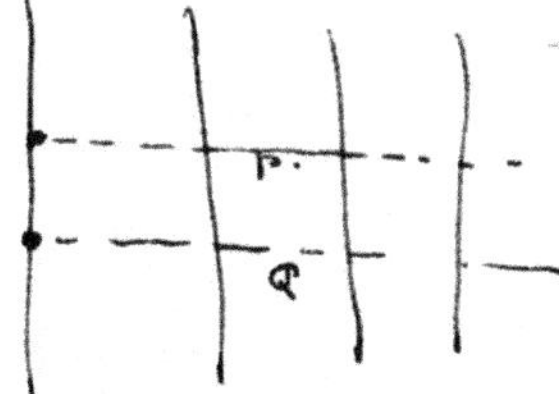

Get idea of point p. from notion of rest

If two particles remain at rest in same time system then the stretches between them in different moments are congruent.

|56|
Might Take another time system

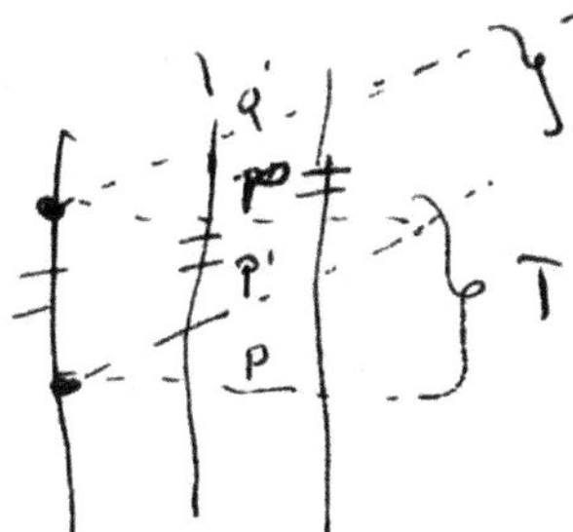

Q' & P' at rest = T^0

If two particles are mutually at rest with respect to each other, their spatial distances in any time system are congruent.

If two particles are at rest in their space their distance in another space is not necessarily congruent to their distance in their own space.

This explains rel<u>ativit</u>y.

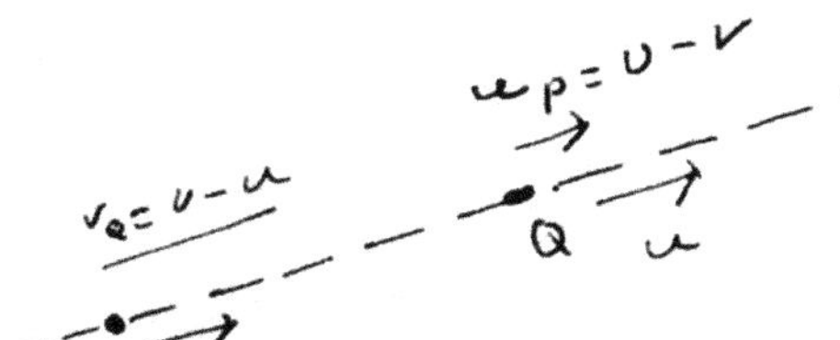

← Obvious formula
But does not hold in modern relativity.

1. This material matches that at the start of lecture 62 in the notes taken by Bell and Hocking at Emerson Hall on 26 March.

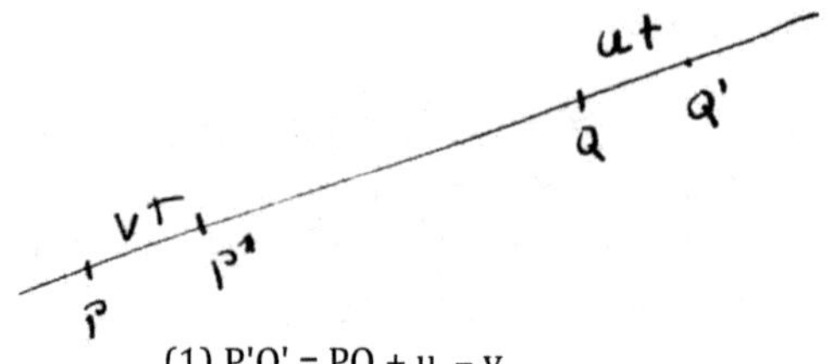

(1) $P'Q' - PQ + u_t - v_t$

(2) $\frac{P'Q' - PQ}{T}$ (rate at which distance is diminishing) $= u - v =$ differential velocity

This is in observer's space.

This does hold. [Lucky for us it does hold or else all our ideas of space are wrong]

|57|

But there's a different idea of relative velocity. Suppose you abandon observer's space-time system & get on p.

Here very meaning of distance & time has altered. And not so obvious that velocity of Q is u – v.

or if you're on Q that velocity of P is v – u.

Three spans

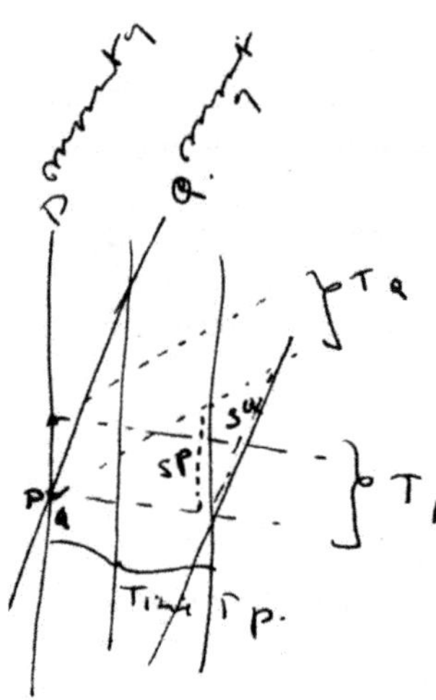

T_0 of observer

T_P of p

T_Q – q.

Cf. P & Q together at 0 time.

~~In Time~~ T_p, Q goes distance S_P/T_P (rest in Q)

p in Time ~~p~~ has gone the distance P_Q/T_Q (rest in p.).

These relative velocities are same in magnitude.

This presupposes that you have congruent ways of measuring your time.

May also say that the relative velocities are certain methods of expressing same physical fact & that choice of units is to be taken as giving congruence of time units.

i.e. can get your units of space congruent (see above) & then your time units are congruent if

|58|

magnitude of relative velocity in the two systems are equal i.e. if

$\frac{Q_P'}{T_P} = \frac{P_Q}{T_Q}$ Otherwise they would be simply incomparable.

---

x ⊥ plane yz

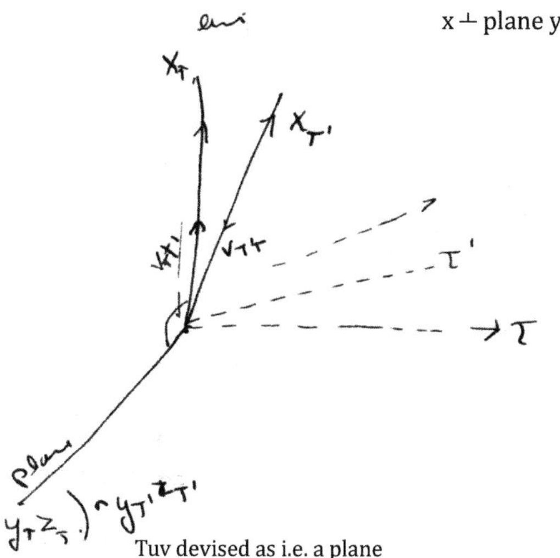

Tuv devised as i.e. a plane

T = time system.

Take another time system T' whose moments intersect T in plane

Rest in T' means velocity $V_{TT'}$ & rest in T means velocity $V_{T'T}$

And what we've said is the $\underline{\underline{V_{TT'} = V_{T'T}}}$ when units of space & of time are congruent.

Any other particle ~~moving will have~~ at rest in another time system will have

|59|

some relative velocity V in ~~T and u in~~ along X_r and u in plane $Y_T 2_T$ and also velocity r' in X_r' & V' in $Y_{T'}Z_{T'}$ & $\underline{\underline{u = u'}}$ i.e. the velocity reflected in that common plane is equal.

[1]

Different cars going along same road at different velocities – all intersect road at 0 time in planes at ⊥ to road.

Car at certain speed will be at rest in time system

T. & its

Faster car is T_2

Limiting case – when T— & T--- have approached each other. which corresponds to a definite velocity – = $\underline{c}$. biggest velocity. Has nothing to do with light but is inherent in structure of your spatio-temporal systems

1. This seems to be the juncture, although we have no indication from Heath where the next lecture begins; it would match lecture 63 in the notes by Bell and Hocking taken in Emerson Hall on 28 March.

An alternative representation is

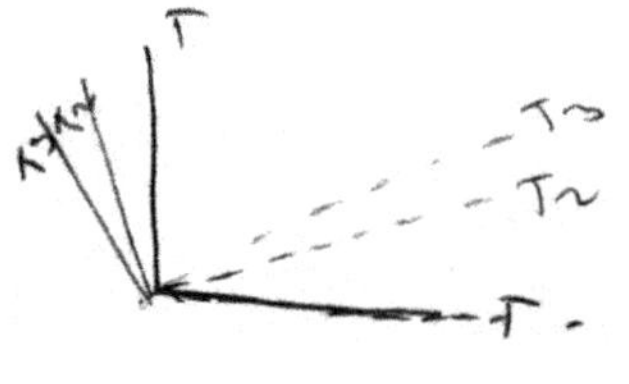

But here there is no distinction between time & space.

|60|

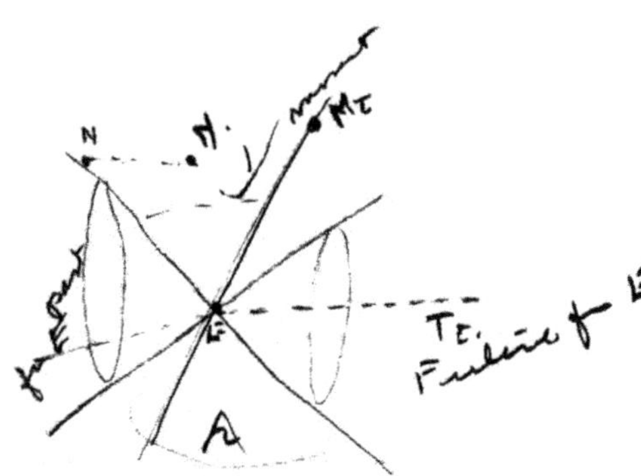

E = an event particle
All other event particles related to it.
1. Any thing enduring in 1 or can never be at E. Can meet in future, or been together in past, provided they are not in same time system. M & E are simultaneous with respect to one time system – M & E are with respect to another. But M & N need not be simultaneous – with respect to any time system.

What we at E perceive ∨as∨ at N is as it is revealed to us from N'. at limiting case coming to us with velocity e. i.e. velocity of light.

That's whole transmission theory, which has led to secondary qualities.

|61|

This is an abstract pattern in which you're endeavoring to express certain abstract relationships. Concrete fact is the grasping together by events of ~~the~~ its relationships with all other entities.

This abstract scheme is an attempt to be impartial. There is no impartial scheme except as logical abstraction. ~~There is no one~~ The concrete fact is e<u>ven</u>t.

You grasp events of vague present with potentialities derived from past.

|62|

3/30/25.[1]

Going back to Extensive Quantity

Kant on extensive quantity M. Müller. Vol 2. p. 143[2]

Part is necessary to whole.

Whole & Part as fundamental.

Quantity is continuous.

Look up in Heath's *Euclid* | Plato & Kant ∨& Whitehead∨ agree on nature of point & moment as limit. Aristotle differs.

---

1. This material matches a shift from lecture 63 to lecture 64, on 30 March or possibly 31 March. Unless Whitehead lectured at Radcliffe on a Monday this one time, it looks like Heath simply put down the wrong date, and it was actually Tuesday, 31 March.
2. On Max Müller, see footnote 3, page 300, lecture 64.

What Zeno should have said was that temporal existence is impossible. Each whole requires its parts before itself, etc. ad infinitum – this is irrefutable.

If try to run time as series of atomic instants – you get no satisfaction, for when you're in an instant, there's no next.

Descartes & Bergson both distinguished between *durée* – not divisible & time which is divisible. Whitehead calls it "passage" "expansion."

Modern relativity makes progress a less important thing in time, for instead of one there are indefinite bundle, each of which
|63|
is partial fact exhibiting some aspect of total expansiveness of Nature. & therefore each time progress in itself is not to be looked on as so fundamentally producing Nature. It hits at our old conceptions just at the point where the difficulty arises.

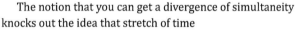

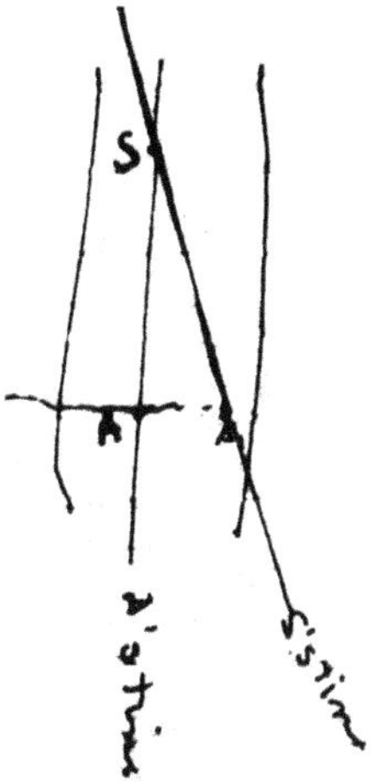

The notion that you can get a divergence of simultaneity knocks out the idea that stretch of time

$A–A_1$ is going to presuppose its parts, because A_1 is already there for man in motion by its item[1] S. i.e. Have to drop idea that $A–A_1$ is dependent on temporal succession, which is one of factors in our dilemma stated above.

Brings us to the point that Whitehead has made that event does not depend on parts, i.e. can know it as one entity, the parts are other entities, in very important relation to it.

Can you then abolish the parts? no because spatial & temporal relations are internal & therefore event A's relations to its parts are of the essence of A.

That is why he has always protested against "simple location" because it presupposed space & time as external relations.

Thing which puzzled Pythagoras was how unbounded got bounded. Also
|64|
puzzles Whitehead.

4/2/25[2]

By expressing this as paradox of motion Zeno put it in form easier to apprehend but also put in the background the area of his difficulty because he seemed to assume endurance.

Real crux is how is transmission possible, if by it you are generating extensive qualities because everything presupposes something of same sort. Not possible if look on part as prior to whole, must have whole on equivalent plane with parts. The whole is exhibiting its relationship

1. This term not at all clear and it could be an abbreviation for 'intersection'.
2. The following material is roughly the same as that of lecture 65 in the notes taken by Bell and Hocking at Emerson Hall on 2 April.

to its parts & that relationship as thus exhibited is thereby an internal relationship. i.e. relation of whole to parts are of essence of whole. it is a two-sided internal relationship – whole is necessary to parts & parts are necessary to whole. This is all involved in our "doctrine of beyondness."

Interesting to note that this is involved in difficulties put by Hume & Zeno.

Thesis problem In opposition to almost everyone we will consider Space & Time as internal relations – not external.

(William James showed us that we do perceive relationships)

|65|

Relation between moments is external but here we have a high degree of logical abstraction. It isn't really ~~intern~~ external when see what it is founded on – but internal relations may be disposed of as systematic.

Have external relations when talking space as points but many odd things arise.

1. Points do not exist outside space. –
2. all space of points is relative to something going on.
3. Have to have something located there. i.e. We're in a high degree of abstraction.

Its all relative & therefore point can't be fundamental entity. i.e. point is merely expressing relation to something else & therefore ought to start with these relations.

Thesis It's not realizing that space & time take these relations from internal relations – that these turn around difficulties for time.

|66|

If whole is given concurrently with its parts, time is not something produced from nonentity, but is in a sense already there – a character of the given. We have this in our perception – i.e. specious present which has endurance within itself.

Got somehow or other to have a theory that 1. gives all Space-Time on same level, & 2. to differentiate into past & future via some Atomic theory.

4/4/25[1]

According to modern view of relativity, where dating goes all way, sharp distinction between future & past does not hold.

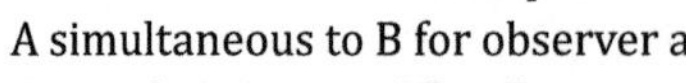

A simultaneous to B for observer at B

C simultaneous to A for observer at A

but C in future for A & B for observer at B.

Must be some meaning for All Space-Time, i.e. being in some sense already there. An ontology

|67|

which gives a meaning to them as element. That is Spinoza's ontology in making space-time attributes of substance. When you have that not only do space & time assimilate each other, but time loses its special direction.

1. The following is roughly the same as the material presented in lecture 66 in the notes taken by Bell and Hocking at Emerson Hall on 4 April.

i.e. the whole is viewed *sub-specie aeternitatis*. Then must separate idea of time & space qua its potentiality (~~as events~~) ~~& divisible.~~ ~~fr its eternal aspect.~~ from the actual things in it.

The divisibility of it must be given to it as potential. This division is really what is turning up everywhere, in our ways of apprehending there is thought for which thing thought of stands outside qua concept is outside generation. Then there is immediate sense experience which is apprehending within generation. Hard to separate the two. What ever you are directly experiencing can also be thought of.

Pythagoras propounds exactly that question of how the unbounded becomes bounded. Exactly the same
|68|
difficulty in Plato repeating & polishing the Pythagorean tradition. The world of Ideas somehow manage to adventure in world of generation of things. This point of view for a mathematician strikes one.

On other hand Aristotle was chiefly interested in biology. Therefore starts from discussion of generation. [His zoology of shores of Black sea is still best work on the fauna of that region.] But he has to bring in potentiality. This must be something outside & behind generation. Generation is within potentiality. Have got to have a positive view of potentiality.

Spinoza starts with One Substance, Space & Time as attributes of the substance. i.e. is trying to give a positive content to our idea of possibility. VJump toV Modes. Whitehead never quite emphasizes how modes get into space & time.

Then with subjects or people the idea gets very attenuated. Of course Berkeley saves himself here by the mind of God.
|69|
Hume a little observes this point & lands in absolute skepticism.

Kant's Things-in-themselves are out of Space & Time. Generation is made purely subjective. i.e. is procedure of organizing experience.

Hegel – back again. Have the Idea unfolding itself. i.e. the dialectic.

When try to fit it on to modern physical ideas, we have just another rendering of something that occurs in all metaphysics & which stares us in the face in all our immediate experience. Particular shape that it takes in metaphysics is always relevant to the stage of scientific & general knowledge.

Our immediate experience according to any theory [even if drop alternative time series] curiously enough even if generation reproduces something like that

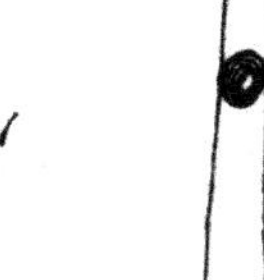

M. grasping together the pattern quā aspect of a totality. In that knowledge the generation falls to the background. i.e. Transition within the already given. [specious present]

Experience is self-knowledge of immediately enduring pattern which one is. Two sides –
1. Steady endurance & permanence.
2. More variable side which comes from aspects of environment

|70|
Furthermore, whenever you have – i.e. Space as over against Time, comes from the exhibition of this duration as given. i.e. it is doing in a limited way what is (according to metaphysical ideas of possibility) really unlimited. i.e. Even sense-experience gives us generation within the given.

In M. (see last page) you have the actual potentialities, selected from mere potentialities.

If you make space-time to be system of external relations you make whole thing quite meaningless.

Still got Zeno on our hands.

At beginning was emphatic that every eternal object pervades qua potentiality the Whole space-time & is a selection as it becomes actual.

Only one was out of Zeno infinite regress & this is atomic theory of generation. If take theory of togetherness of a pattern being realized in an event as unit of realization that is impossible to get hold of unless the event in some way is given as atom. And yet must have Space-Time as infinitely divisible. Whitehead suggests
|71|

Thesis | Where Newton got idea of fluxions. i.e. the flowing quality. [Turns up later in Kant.]

---

that the divisibility belongs to Eternal side & atomic side belongs to process & therefore the divisibility lies within the given. The complete divisibility is inherently potential but cannot be realized for you. [Fact of psychological experience]

4/7/25[1]
In what sense can there be an atomic theory of time. Cf. the Quantum Theory.

Does it naturally grow out of the metaphysical point of view of this course.

Seems to require it, for we've been saying, 1. that future must be there as well as past & future & 2. that get a blur between past, present & future due to relativity. As soon as have idea of "potentially there" you run into danger of putting time of future into another time. In making the "future there" some distinction must be made between "future qua there" & "future as subsequently brought on the stage." That subsequently must be explained not as something of future but as something left out of future.

Zeno's argument makes great difficulties over how extensive magnitude is possible.
|72|
Time & space differentiate themselves when come to endurance. Endurance is pattern of togetherness. Talked of it as spatialized & in

---

1. The following material matches that presented at the start of lecture 67 in the notes taken by Bell and Hocking at Emerson Hall on 7 April.

a moment, but a moment is a mere abstraction. Actually endurance must be spatialized in an event, although it may be a minimum event. Therefore distinguish event qua home of possibility & there qua scheme of extensiveness, its particular, not a super-occasional universal but an occasional universal. i.e. has definite relationships with actual already.

But qua mere possibility, qua unrealized, ~~event mere~~ it merely gives the extensional qualities, which thereby do not differentiate time from space.

Another point, the direction of time. This is generally merely assumed. If we slip in the obvious we ought to give it a fundamental character. This essential character of the "forwardness" of time should be
|73|
more emphasized than we've been doing. And a community of possible (actual or not) occasions gives no direction at all. Temporalization of continuum of events is its realization, which must be realization of patterns, situations. Fundamental patterns are totalities – all or none. Therefore you will get a ~~duration~~ pattern as within a duration grasped into realization atomically, i.e. as a whole. What is realized is divisible because it is extensive. Progression, continuous, in time is within the already realized. ~~That is what~~

Hardheaded opinion is that future is nothing. Then transition must be in given, i.e. can't have transition to nonentity, i.e. transition is relation[1] within already real. i.e. Temporalization may well be atomic.

Also must be prepared to allow degrees of realization. Realization is presentation of durations as a succession. Then the duration as real is the achievement of a real togetherness of a pattern, & this pattern as achieved in itself contains continuous transition, & therefore the directional character entirely comes from realization & is not in the extension.

But each atomic duration is only realized in respect to a certain pattern, & are different realizations
|74|
in respect to different patterns.

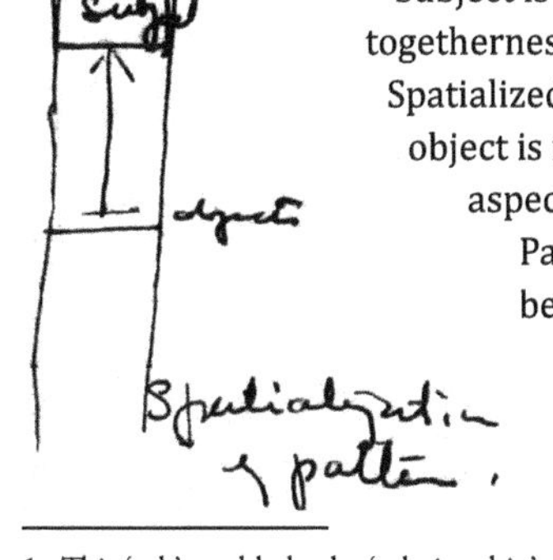

Question is how the subjectivist point of view can be adapted so as to exhibit the subject as an element in a total universe necessary to it.

Subject is realization of a pattern. The pattern is a togetherness of eternal objects under a spatialized aspect. Spatialized aspects mean exhibiting that in which the eternal object is merely an element as under the limitation of those aspects essential for the very nature of the subject. i.e. Pattern is exhibiting the subject as being what it is because of that which is beyond it.

Spatialization of pattern is realization of duration under limited aspects, i.e. as now in the subject, i.e. is simultaneity; i.e. we're dealing with lapses

1. This 'rel.' could also be 'relationship'.

of time, i.e. time for light to get from sun to me, but I see sun now not 8 minutes ago. i.e. We've got a complex relationship which can't get into subject predicate form.

|75|

4/9/25[1]

Where ever are dealing with extension you have continuity. Generation is distinct from relationship of extension, is being realized, in a way being temporalized, the coming of a pattern into real togetherness. That ~~can~~ requires to be conceived as being atomic. A mere succession of things in themselves extended. It was that successiveness which gave distinction between space & time. A partitioning via successive generations.

Remarks – That is obviously highly speculative.

Importance of ideas we inherit – chiefly by language which represents the popular philosophy – words, grammatical structure especially.

Thus it comes about that Science is pretty apt to asVcertainV the philosophy of the past.

The scientific dislike of speculative philosophy Whitehead recognizes this as well as critical philosophy. Speculative philosophy must make new adjustments which should give a more obvious harmony of all immediate apprehension of things.

After all this speculative method is necessary for science to progress.

|76|

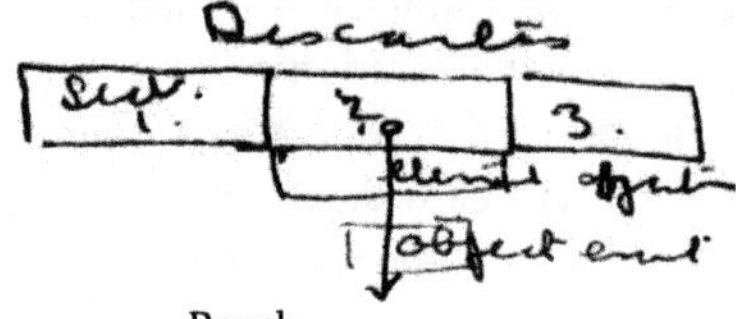

Psych.

Subject is event for itself. [cf. Descartes *Meditations*] is a present reality. The realization does not come, but is pure succession. An event being generated into a real becomingness, within which is succession.

Disagreement with Descartes is in grammatical form, talks of subject under guise of subject & predicate. I am cogitating. i.e. talks of Descartes cogitations as predicates of Descartes.

[Due to language] Whitehead – ~~subject is self-knowledge~~ our knowledge doesn't take the form.

Our self realization is the object event, via eternal objects. My mind is the grasping together of that which is "other" than it.

Subject is grasping together of the totality under limitation.

|77|

Notes go in here.[2]

1. The following material matches that at the start of lecture 68 in the notes taken by Bell and Hocking at Emerson Hall on 9 April.
2. It looks like these words are in Heath's hand, but it is not clear what to make of it.

4/11/25[1]

All the continuity should be looked on as more primordial. That is existent & as existent is in future; i.e. has that existent which the future & the possible have. That is why the future is already disclosed for us in present.

Then we have time as what you have in the becoming real.

Realization clinches the atomicity of things. That it is the achievement of or the emergence of substantial independence, or of a certain individuality. The separate events as in the continuous ~~have~~ are simply exhibiting the patience of the realm of existents. But the becoming, the temporalization, is the attaining of realization.

We're sharpening up two ideas.

1. Atomic time &

2. Future possibilities.

– Real togetherness involves Selection.

~~Pass~~ Future is not value yet, because it has all the possibilities there are.

Temporalization exhibits the relationships between the events as taking |78| on the aspects of internal relations.

Everything that is real has internal relations.

The subject is the whole, the one substance becoming a particular matter of fact.

There are two aspects.

1. The individualizing of universal activity is of the same person in a stream of time. Substantial identity of finite.

2. What it is in itself in any individualization is a spatialization of the whole.

4/14/25.[2]

Mathematical symbols are extraordinarily valuable & significant within its very narrow field.

1. The following material seems to correspond to lecture 69 in the notes taken by Bell and Hocking at Emerson Hall on 11 April – but only roughly.
2. The following material seems to correspond to lecture 70 in the notes taken by Bell and Hocking at Emerson Hall on 14 April – but only roughly.

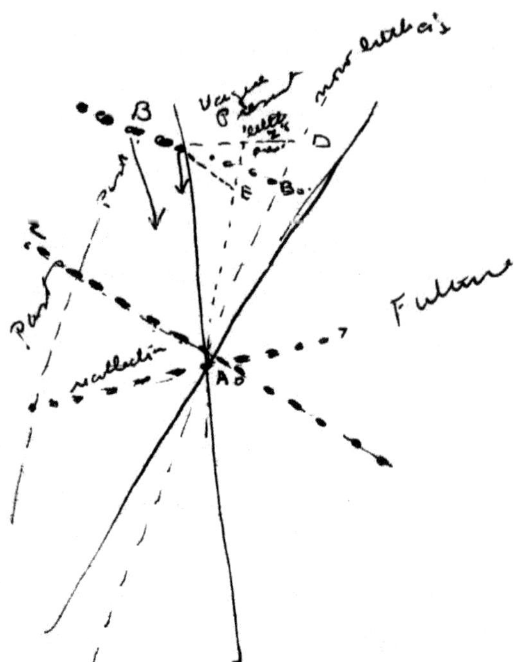

A_0 is embodiment of whole under two limitations.
1. As Now.
2. As Past. – recollection
[There is also The anticipation.]

Grasping the <u>now</u> is really B with its aspect at A represented as at D [at this moment it's really at B_0. Sees B as at E]

|79|
Z is aware of a [3] termed relation between
Z_0 B, E & Eternal objects
a is aware of a [3] termed relation between
A_0, B, D & Eternal object
A_0 almost = Z_0

Necessary to note that <u>B</u> is same [content] for both.

Why should you identify E or D with anything in external world? Question of the subjectivist.

But E is what you immediately apprehend & every thing else is defined via its relations to the present. If E is merely subjective then all the else is subjective too, for it's only apprehended through E. [Cf. Hume & Berkeley on secondary qualities] Our point is that it's E or D which is fundamental.

The way B comes in is as D & E exhibit a harmony of geometrical relations. Not identical geometrical relations but they fit into one harmonious scheme.

Also eternal objects turn up. We cannot compare your & my yellows but for each of us it is an in<u>dividual</u> which maintains its identity.

Thus the perceptual object is the apprehension of a <u>strain of control</u> i.e. our present perception is of object as conditioned by past.

|80|

Wyckenstein[1] on Symbolism. Requires reproduction of same picture

Thus we see enormous importance of abstract or mathematical reasoning because its completely abstractive. Not specify which eternal object – just a eternal object.

4/16/25.[2]

Spatial relationships are grasping the universe as now.

By saying that the present is our psychological field you get into ~~psy~~ subjectivist ~~difficulties~~ point which seems at first to be clear & definite, a simplification. This is because in subjectivism you are expressing the truth that you are an entity enjoying experiences. But this is an insufficient explanation of what "enjoying" really is.

This raises whole status of "abstraction."

The things which are habitual etc. are hard to express. Such things as space, Otherness, etc.

Reasons against subjective position.

1. The immediately present world is not only beyond itself by my resolution, but also by fact that it is not an instantaneously spatialized now, devoid of transition – transition within itself

|81|

but it essentially has embodied within itself, transition from beyond itself & to beyond itself.

2. A further reason – in looking at the details of the experience, do they bear out that idea of subjectivism. Stars, inside of earth, far side of moon – remote ages – are unknown in detail, & yet we say they existed.

3. Instinct for action. ~~which seem~~ Action seems to issue in instinct for self-transcendence.

4. Idea of common conceptual world seems to be implicit in essence of common world of sense.

5. Distinction between realism & idealism does not coincide with objectivism & subjectivism.

|82|[3]

For the individual concrete fact there must be a particular selecting energy which exhibits itself as the embodiment of the totality of selecting energy.

The essence of being real is to be a particular & concrete fact. This is where so much metaphysics goes astray in not accounting for the "irreducible & stubborn concrete fact."

---

1. Clearly, a note added at the top of the page, and quite likely refers to Wittgenstein. Thus Heath was guessing at the spelling. His *Tractatus Logico-Philosophicus* had been published in translation and with an introduction by B. Russell in 1922.
2. The following material seems to correspond to that presented in lecture 71 in the notes taken by Bell and Hocking at Emerson Hall on 16 April – but only roughly.
3. Heath does not mark the start of an 18 April lecture at Radcliffe, but somewhere in the following content we find a rough correspondence to lecture 72 in the notes taken by Bell and Hocking at Emerson Hall on 18 April – but only roughly.

The particular occasion exhibits the world-in-itself under grades of aspects. 1. The world as now for myself as now. The world as having ingredients in it – the eternal objects now. Every other aspect of world is always relative to the world as now. That is why if you construe the world as now as "private world" you can never get out of solipsism, because everything else exhibits itself as aspect of world now.

In science – all measurements are of the world as appearing now for me & therefore would lie within the private psychological world.

Therefore The spatio-temporal relationships arise from the aspects of the world now.

The world as now is not world as controlling but world as achieved as issue of control.

|83|[1]

We know ourselves as Ego-objects amid objects.

This is our immediate experience. The subject-object is a secondary stage of considering experience qua modifying one's essence. The perceptual object is the individualization of strains of control (cause).

Can't see how Hume escaped this except because he had the "old" idea of matter.

One's primary experience is of relationships as entering into the Ego.

It's an abstraction when we use subject-predicate terminology.

The ego-object is in double individualizations, i.e. 1. qua underlying the process, & 2. qua part of the process.

Now come to the future, to thought & to question of potentiality – the realm of ideas or concepts.

|84|

4/28/25[2]

Cognitive Experience

To be analyzed as the cognitive apprehension of how the essence of an immediate ~~apprehension~~ Voccasion of realizationV is modified by its relationships with other entities.

Emphasizes the subjective aspect of cognitive experience. And then cognitive apprehension is the ~~rela~~realization of the cognitive relation between an entity in its own essence qua composite.

Thus cognizance is a relation between an emergent entity & the composite entity from which it emerges as being unified realization of this composite potentiality.

At this point a paradox arises – not a contradiction – i.e. cognitive apprehension presupposes the cognizant entity (or the immediate occasion) & its essence qua composite. But it is obvious that the cognizance itself enters into the entity & therefore affects its modifications.

1. The words 'Miss Aven' were added, in pencil, at the top right of this page. It seems to have no relevance, so we have not included it.
2. After 'break week', the following notes seem to correspond to lecture 73 in the notes taken by Bell and Hocking at Emerson Hall on 28 April – but only roughly.

Quite obvious therefore that the cognizance presupposes the entity as less than itself & as emerging from a potentiality of essence, antecedent to the fact of cognizance. Thus the potentiality is permissive of
|85|
cognizance, but not inclusive of it.

It follows that the entity qua cognizant ~~follows from the~~ is other than the entity with respect to whose entity cognizance is relative. We distinguish therefore between the occasion as inclusive & as exclusive of cognizance. Exclusive is "the standpoint" inclusive is "the cognizant." It stands in the nature of cognitive apprehension that these two emerge into realization. The translucence of cognizance arises from this fact. i.e. you don't make VOr breakV a thing by knowing it.

A parallel distinction arises in respect to the "enduring object" which is partial realization through the flux. From cognizant point of view the enduring entity is the self, (which is something for its own sake). It's quite obvious that we must discriminate the self of the flux of cognizant occasions, from the self of the standpoint occasion. i.e. the cognizant self from the experient self. is industrious or action.

It follows that cognizance is fitful & broken & very variable in its partial analysis of the essence of each standpoint occasion. Accordingly, in respect to endurance, the important self is the experient self & properly speaking
|86|
there are many cognizant selves. This multiplicity gains a derogate unity by reference to the stability of the experient self.

The standpoint occasion is the immediate total experience of the experient self. This is merely the converse of the experient self's identity which connects the series of standpoint occasions. The cognizant occasion is immediate image which constitutes the immediate total knowledge of the cognizant self. ~~Having~~

The standpoint occasion is within the cognizant occasion [if you're going to use that term] Also the experient self is within the standpoint occasion. Furthermore, having regard to fundamental status of experient self & usual neglect of distinctions drawn above, we must say that self is within each of images whose existence constitutes its knowledge & reject "images in self."

That which is permanent is not cognitively mental & cognitive mentality is primarily associated with ~~concrete image~~ transient (than concrete) image.
|87|
Cognizant mentality is "a universal" therefore the same for all VcognitiveV occasions. Therefore The self, enduring since birth, is not cognitive. Many cognitive selves successively identified with experient self. & not fully assured of total experience of experient self.

Also involved in Plato's theory of knowledge is reminiscence. At any rate it is certainly required by any theory that explains endurance as pattern & emphasizes the continuity of historical route.

Whitehead thinks it is also required by any theory that maintains the translucence of cognizant.

4/30/25.[1]

Summary of last lecture – "Instead of saying that an image is mental, we should say that a mind is imagined. But on the whole we'd rather say neither."[2]

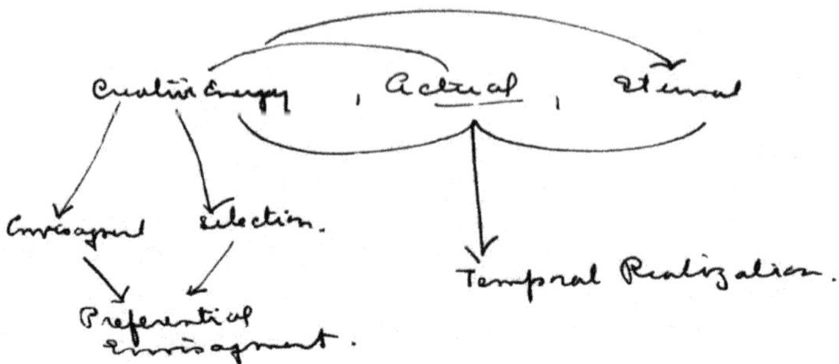

|88|

Experient self is not quite as definite as we might like. Can get different experient selves according to where you look for them.

This is recognized in ordinary common life – of religious conversion – shocks etc.

Goes badly with idea of mind as independent substance.

5/2/25.[3]

Cognitive apprehension is return of actuality into creating activity.

The taking account of the actual, emerges as a realized fact & thus the actual emerges into an additional fact within the realized process. i.e. instead of being a factor it has emerged into a fact, i.e. an occasion of knowing.

This knowledge also causes emergence into realization of values, i.e. emotion, pain, etc. etc. All of them are intimately connected with purpose, i.e. selectiveness.

The experient self is not quite as definite as one would like. We've said it is a pattern reiterating itself – but not as always the same. i.e. moods differ from day to day.

Two extremes for dating the experient self. 1. The immediate

|89|

experient self, i.e. that which remains permanent from an immediately experienced past. That, in a sense, has the fullest content. because it.

That immediate experient self embodies a great deal of rather transient fact.

1. The following material may correspond to lecture 74 in the notes taken by Bell and Hocking at Emerson Hall on 16 April, but only later on in that lecture does the role of 'creative energy' arise, and the diagrams do not at all match.
2. It is not clear why this passage is within quotes, nor what source is being quoted.
3. While Heath notes the date here, it is the line below which seems to mark the start of the lecture, where the material corresponds to the notes taken for that lecture 75 by Bell and Hocking at Emerson Hall on 2 May.

2. At other extreme get the self – i.e. organic experient self. There is an organism which maintains itself throughout a life history. It may be a trivial organism, in that sense the organic self will be a trivial thing. Or it may represent some fundamental individualization of creative character realizing itself throughout a life history. It has less content than the immediate self, but may be more important. It is only when that has risen in importance that you get knowledge.

So the immediate experient self is complete matter of fact identity & the organic experient self is schematic identity maintaining itself through a life history. The embodiment of limited unity of a creative purpose.

|90|

The image of cognitive experience is always a selection from the complete experience which is synthesized in stand-point occasion. The complete experience merely enters as systematic background of cognized elements.

[Knowledge is possible only on account of systematic character of relations.]

In this way partial knowledge is possible within the complex of systematic internal relations, because partial knowledge only needs the apprehension of whole as systematic.

We conclude therefore that cognitive experience always involves abstraction.

German.

5/5/1925.[1]

|91|

5/7/1925[2]

A puzzle.

Abstraction in cognitive experience.

1. Division of cognitive experience into thought and awareness of fact.

2. Cognitive experience is analytical. i.e. the abstractive character enters into its very texture, it is selection. The complete image, i.e. occasion of cognition, presents itself as analyzable into component images & each component presents itself as a possibly complete image. We are there carrying the principle that simultaneous images are independent. (see earlier lectures.) Immediate correlation of intrinsic character of past & intrinsic character of present, is of a sort which one's direct apprehension leaves one vague & sometimes wrong, (illusion) but whole point of science is to trace back this definite complex relation of apprehension & ingression. We grasp this relation under a rough & ready rule which cloaks the complexity & here error comes in.

External relations from "eternal objects" & ~~standpoint occasion.~~ & ~~involv All~~ ^any^ occasion ~~only involves system~~ is separable from others (except stand point) but does involve their systematic

1. Might Heath have had to miss this class for another class, or test, in German?
2. The following material seems to correspond to lecture 77 in the notes taken by Bell and Hocking at Emerson Hall on 7 May.

|92|

relations.

Points in dealing with puzzle. 1. Relationship of ingression is only internal for standpoint & systematic but external for all others. It is for that reason that measurement is possible.

More difficult. <u>Error</u> comes in. 2. Relationship to past takes form of a hypothetical entity in the present which is independent (hypothetically) of everything in ~~past~~ present, but as derivative from past is systematically related to everything in the present.

The hypothetical entity in present is derivative from past & our standpoint occasion of present, i.e. state of our body.

3. Involve ingression of eternal objects in 2 ways.

a. Ingression of sense objects.
b. Ingression of idea of a particular character.
a. Shape of moon etc.
b. Moon as an occasion-in-itself. The "thisness" of an occasion.

All connected with the sense of the past as issuing in the present.

|93|

5/9/25[1]

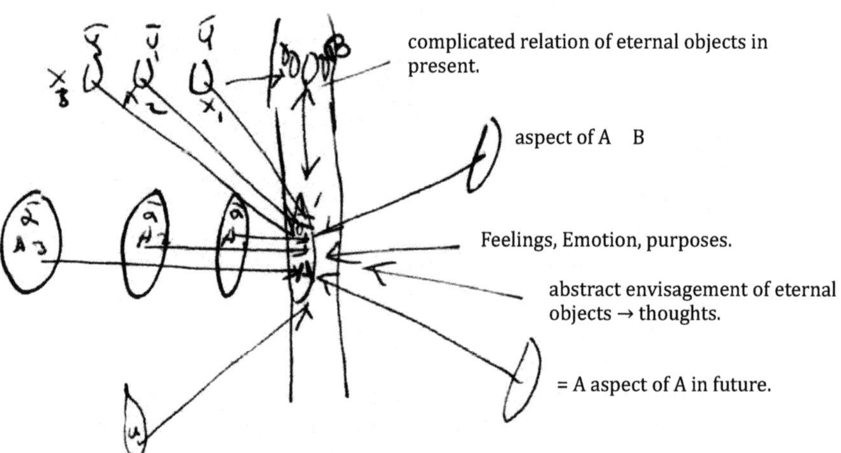

Denies private worlds. your thoughts $\bar{\psi}$ enter my experience[2] just as (qualitatively) mine $\bar{\alpha}$ do, though <u>quantitatively</u> less. Denies argument for existence of other VmindsV from an<u>alogy</u>. Only has to be said to see how ludicrous it is.

Past & Future as systematic. As a scheme. The essence of A is synthetic unity of experience which is A, as analyzed is

(i) { Aspect of A itself as conditionally $\bar{\alpha}$ realized in the experiences of $\widehat{A_3 A_2 A_1}$ }

1. While the diagram at the top of heath's page 93 does not appear in the notes taken by Bell at Emerson Hall on 9 May – at least not in this form – the content below corresponds with lecture 78.
2. It is not clear whether Heath has recorded Whitehead using a bar-over-y (both here and in the diagram above) or bar-over-psi, but clearly it is meant to correspond with what is in the diagram.

+ (ii) {Aspect of itself as conditionally ψ realized in experiences of $\overline{X_3 X_2 X_1}$ + - - - - etc.}

+ (iii) {A as standpoint for ingression of eternal object β in sequence of occasions $(B_1 B_2 \text{---})_\beta$ in the present. (Present is given as involving its transition.) & of y in sequence $C\,)_\gamma$ + etc.[1]}

over.

|94|[2]

Eternal objects {idea, meaning, existence, being} in one. Enduring self-hood is an eternal object.

+ iv {A as standpoint for ingression of enduring self-hood in sequence $(B_1 B_2 B_3)_\beta$}

+ v {A as conditionally realizing the future via system.}

+ vi {A as conditionally realizing the past as system.}

+ vii {A as standpoint for preferential abstract envisagement}

+ viii {A as standpoint for feeling emotion, purpose.}

5/12/25

exams[3]

|95|

5/14/25[4]

Considers details of system <u>very</u> alterable.

But have been running one or two principles rather ruthlessly & without connection of details.

As in

Lloyd Morgan *Emergent Evolution.*
Alexander's *Space-Time-Deity.*[5]

Namely –

Start with idea of process & seeing everything emerge in the process.

Whitehead – Alexander is in a muddle over Space-Time. Doesn't provide the niches for his emergents. i.e. fails to bring out relations which, commonsense tells us are there.

---

1. The subscripts here look to be the Greek letters provided, although they are unclear, and in (iv) below the subscript looks more like 'B', although it should be another beta, as it is in the Emerson Hall lectures.
2. The diagram with brackets has been added in the top margin of the page.
3. The single word explains the lack of notes for this day's lecture.
4. What follows corresponds to lecture 80 in the notes taken by Bell and Hocking at Emerson Hall on 14 May.
5. Alexander and Morgan are first mentioned, together, in lecture 13, 23 October, and then again in lecture 80; see footnote 3 on page 379, lecture 80.

Whitehead – procedure analogous to that of Husserl.[1] But there is a fundamental difference due to difference of empirical vs. *a priori* traditions.

Whitehead – we've got to base all we know in immediate experience of concrete fact.

The universals are at home in their own way throughout space & time. Actual is particular relation or selection

|96|

from universals. Not mere arbitrary selection from outside but in some sense intrinsic in totality itself. A actualize in form of real relevance – r<u>eal</u> achievement of synthetic unity.

A fresh actuality i.e. not this & that, but this <u>with</u> that.

As soon as you've laid down the principle of synthetic unity you can see that there's a niche for each thing but cannot define what that niche is-in-itself. There are universals that require synthetic unity of other universals for their entrance.

Relation of standpoint occasion to its essence qua analysed in an abstraction emerging as a real fact.

Must first make your knowledge translucent & then make it altering the fact. So your crude experience includes knowledge (as aspect) of the antecedent occasion. But qua simultaneous, emergents are independent except via common past.

|97|

"Consider, my brother, the gospel that is preached at my imperfect hands."[2]

|98|

5/19/25[3]

Abstraction

Mathematics as instance of extreme of mathematics.

If anything out of relation – then complete ignorance of it.

5/23/25[4]

1. How are actual events implicated in relation to each other & to eternal objects.
2. How are self-emergent objects implicated in relationships.
3. How are eternal objects implicated in relationships.

How this bears on our concept of what eternal object is, as an abstract from actuality.

---

1. Edmund Gustav Albrecht Husserl (1859–1938). It is not clear what Whitehead knew of Husserl or phenomenology, but Hocking (one of those taking notes for Whitehead's lectures in Emerson Hall) had spent time with Husserl in the early 1900s and Bell (the other note taker) went to Göttingen a few years later to complete his PhD under Husserl; see the Introduction to this volume.
2. This quoted passage is the only material on this page. It is not at all clear what the passage represents, or at which lecture it is noted. Lectures 81 and 82 appear to be, from Bell's notes, Whitehead delivering the paper on the history of mathematics he had prepared for presentation at Brown University. Heath leaves no notes for these lectures.
3. What follows seems to correspond, in brief, to lecture 83 in the notes taken by Bell and Hocking at Emerson Hall on 19 May.
4. What follows seems to correspond, if only roughly, to lecture 83 in the notes taken by Bell and Hocking at Emerson Hall on 21 May.

As we proceed from 1–3 the relationships progressively drop their internality. Can't have same actual event under different circumstances.

But same externality to relations of self.

Eternal object is exactly the thing
|99|
that does remain same. Are we then to say that eternal object is merely implicated in relations as external to itself.

Plato: Yes & Aristotle: No. Both right.

Consider eternal object A. If we ask how A finds itself in universe of things (∨this∨ metaphysical situation) we must assume that this status of A enters into the essence of A. i.e. what A is in itself involves A's status in the universe.

But in the essence of A there is on the one hand a perfect determinateness as to relation of A to all other eternal objects

There stands in essence of A an indeterminateness as to relationship of A to actualities. This indeterminateness in very essence of A & is A's patience for external relations.

Two principles.

1. An entity cannot stand in external relations unless in its essence there stands an indeterminateness which is its patience for such external relations.

2. An entity in internal relations has no being as an entity not in these relations. i.e. once with internal relations, always with internal relations. or Internal relations are
|100|
constitutive relations.

Thus when in essence of an entity there stands a determinateness as to certain relations, then relations are internal & are the character of entity.

Also ∨when∨ in the essence of these is an indeterminateness, these relations are external. But in no case can any entity stand in any relation not in its entity either as determined or undetermined. (a possibility that has not come off is negative fact. Cf. Aristotle on unmusical man becoming musical.)

Foundation for Theory of Negative Fact.[1]

or An essence is what it is

1. in itself

2 in its determinate relationship

3. in its "patiences"

Thus an eternal object A is 1 ∨(in itself)∨ & it is 2 (in all determinate relationships to every other eternal object) & 3 (as patient of relationships to actuality). These three aspects of A cannot be apart & are in fact eternal objects of higher grade which presuppose A and each other. The determinate

1. This is written vertically in the left margin, next to the paragraph that begins 'Also when in the essence'.

|101|
relationship of A to every other eternal object is how A is systematically related to every other eternal object. This Vdeterminate V relationship is A & all other eternal objects as pervading Space-Time in ways appropriate to ~~each~~ such objects.

Thus Space-Time is a general system of relationships for all eternal objects & it stands in the nature of all eternal objects that they should be thus related. Thus the complete[1] spatio-temporal relationship of A stands in the essence of A & is eternal to A. i.e. the complete possibilities for A. But in respect to actualities an indeterminateness stands in the essence of A. An actual occasion synthesizes in itself every eternal object, including its complete determinate relatedness to other eternal objects. But this synthesis involves limitation, not of content, but of actuality; so that grades of entry into this synthesis are inherent in each actual occasion. These grades can only be expressed as irrelevance of value. (Possibly only by contradicting oneself.) This irrelevance of value varies in grade from the inclusion of individual essence of A as element of aesthetic synthesis to
|102|
lowest grade which is inclusion of essence of A as element in synthesis. Lowest grade – unfulfilled possibility not contributing any aesthetic value. Thus A conceived as merely in its determinate relation to other eternal objects is A as not-being (cf. Hegel) ~~which does~~ in usual phraseology which does not fit Whitehead's. A in its patience of ingression into actual occasions is A as a possibility & A as in a synthesis of actual occasion is A as included or excluded as an effective element in synthesized value. In this sense every occasion is a synthesis of being & not-being.

Furthermore though some eternal objects are synthesized merely qua not-being (cf Aladdin) On other hand each eternal object which is synthesized qua being is also synthesized in same situation qua not-being. Follows because in every eternal object has more possibilities than synthesis in any one occasion.

Also there are different meanings of being corresponding to different grades of ingression.
|103|[2]
Being means effective in experient or self-creative synthesis.

Every object is in every occasion in different grades in indetermination.

The analysis of the determinate relatedness.

Consider set of eternal objects.

A, B, C. set ABC.

Total relatedness of A to all eternal objects & of all eternal objects to each other allow themselves to be analyzed into:

1. This term could also be read as 'complex'.
2. Heath does not indicate any lecture breaks here, but these topics when presented at Emerson Hall and recorded in the notes taken by Bell stretch over two lectures, and this page break may reflect a shift from lecture 83 to 84.

(1.) some one definite specific determinate relatedness of A, B, C among themselves exclusively.

(2.) into all other relatedness any eternal object including other relations if any of A, B, C. exclusively.

R (ABC) is specific relatedness of A, B C. in (1).

Also may get higher degrees of relationships between eternal objects.

1st grade ABC
2nd grade R_1 (ABC)
3rd grade S_1 (R_1 ... 2)
etc.

|104|
Each successive grade is in a certain sense a higher abstraction than the preceding grade. What you're abstracting from is not from actuality but from the realm of possibility. In first grade have A simply, in second have A under certain limitations, of the possibilities. And so the eternal objects as yet more & more complex are retreating further & further from the simplicity of A in itself.

Each grade retains the simplicity of the eternal object of that grade & restricts the simplicity of all lower grades. One never gets to the end of this possibility so as to get the class of it. i.e. danger of taking "the realm of possibility" as itself possibility, i.e. an entity.

General principle connecting the actual to the possible. Principle of the Translucency of actuality.

[Translucency of cognition is particular case of this.]

R(ABC) as in an actual occasion whether it is in it as excluded or in some grade of inclusion has no other content than just being R(A,B C). Variation of grade
|105|
is not a variation of content but of actuality. Actualization cannot be made a predicate in same sense that eternal object can. Thus if R(ABC) is synthesized in α under lowest grade it is synthesized as not-being. Also A as standing in R(ABC) will be synthesized in α as not being, even if A in some other R is synthesized in α as being. This synthesis as being & not-being is reason why every actualization is a limitation. (Old as Aristotle.)

So R(ABC) in its determinate realization in actual occasion α is not part of essence of R(ABC.)[1] (otherwise frightful difficulties as to finite knowledge.) Thus what is + actual is exclusive of the –. or actualization is a process of exclusion, or of abstraction from the full simplicity of possibility. i.e. abstraction consists of graded entry into actuality. Thus the complete entry into actuality ~~will depend on~~ is most concrete when have most complete determination in respect to actual occasion of the indeterminateness inherent in each eternal object simply. Thus abstraction from the actual is a

1. There could be marks over the B and C, but it is difficult to be certain.

|106|
return from the concrete determinateness to the indeterminate simplicity in the possible. So abstraction from the possible is an ascent toward concreteness in actual & vice versa.

The grades of inclusion.

A thought is something included in the actual occasion. (I'm thinking) But its status is not full realization. Grade of realization of thought is in occasion but not determinately. Cf. Hume on "faintness."

You cannot separate the form (eidos) from the matter. All forms are in every occasion.[1]

1. Bell and Hocking record one further lecture at Emerson Hall, lecture 85, given on 26 May, commenting on Miller's announcement of recent measurements on ether drift. There is nothing like this in Heath's notes.

Whitehead Seminary, October 1924–1925

Notes taken by W. E. Hocking on Phil 20h, 'Seminary in Metaphysics', delivered by Alfred North Whitehead

Within the folder containing the notes he had taken of Whitehead's first course of lectures throughout the academic year 1925–5 were found these notes W. E. Hocking had taken for at least some of the meetings of Whitehead's first 'seminary' – Phil 20h, 'Seminary in Metaphysics'. Whitehead himself explained in a letter to his son, North, that these were typically held on Friday evenings, and included primarily graduate students in philosophy as well as some colleagues. For reasons we have not uncovered, these notes include only four sessions, in October 1924.

|1|[1]
Whitehead, Seminary, October 10.

"An object is an existent in a particular mode of realization"
"Reality requires particularity and limitation"

Entities as existences are isolated.
Fusion requires limitation.
Reality is the achievement of reactive significance of these entities one for another
Space-time is the locus of realization and the machinery of limitation.

Envisagement.
"Knowing is a highly complex example of envisagement"
Cognition, envisaging its own envisagement
Is the term corresponding to Bergson's intuition
Images are not "in the mind" ; A knows B only as B disturbs A.
"Green is there for the molecules of my brain"

Reactive significance. of A and B
Means (1) A taking account of B (envisagement)
(2) being causally influenced by B
(3) enjoying achievement.

Relational essence (of green)
How it can achieve relations within space time

Question:
Is all realization in space-time? If so, is mind real?
Is there any true process from the existent universal to the particular? How can an eternal and omnipresent essence 'take up relationship'? It neither changes nor ceases to exist, – there is no 'from' in the 'to'
And there are no betweens in this alleged transition
In what sense do all the existents pre-exist?
They seem, in experience to EMERGE from the concrete and to exist only in our minds (Alexander, from nowhere)
Why suppose that all possibilities exist?

1. The first two pages of notes are less detailed and have been typed out. Where this occurred for a few Emerson Hall lectures, Hocking had arranged for someone else to take notes, but no comment to that effect has been found here. On the second page of notes, for 17 October, some handwritten notes overlap with the typewritten notes.

|2|

Whitehead. Seminary Oct. 17

1. Papers by Van der Walle and Shimer:[1] restatements.
Questions relating to the fundamental position of the physical field, the function of the primates, and how 'vibration of the primates' is to be conceived.

2. The physical field is not fundamental but an abstraction
(Essentially in Hegel's sense?)[2]
"The finiteness of the event essentially clamors for its relation to the beyond"

3. The meaning of the theory of "primates" – a warning
Quantum theory need not "imply discontinuity of space-time"
Conceive electrons as vibrant rather than as passive charges
Space-time as "uniform objects", whereas molecules are not

4. Not all objects are complex: a sense-datum, black

5. Space time may be abstractions, but
All realization is in space-time
Thinking of a name is not a realization?
A thought process is a weaker degree of reality
(Can space-time then be realized?)

6. Brain molecules may take account of a thought, subconscious
(Is there any sense in this?)

7. Envisagement has "the secret of what Lloyd Morgan and Alexander call emergent entities"
It is realization plus its conditions.

|3|

Whitehead seminary   Oct. 17, 1924

Van der Walle.   [Restatement]
Aliotta[3] vs Science. Problem of Induction in Lecture
Solution by absorbing one problem in a more general

1. Shimer and van der Walle were two of the graduate students taking this Seminary for credit. Whitehead had explained in letters to his son, North, that there were 17–20 men attending, including some staff. 'Someone reads a paper on a topic which I want elucidated, and then we discuss': Victor Lowe, *Alfred North Whitehead: The Man and His Work, Volume 2* (Johns Hopkins University Press, 1990), pp. 291–7. Later, McGill and Farber are also named as readers.
2. Closing parenthesis provided, here, and in all three instances on this page.
3. Antonio Aliotta (1884–1964) was an Italian philosopher who gained prominence for his 1912 work *La reazione idealistica contro la scienza*, translated in 1914 by Agnes McCaskill as *The Idealistic Reaction Against Science*.

And revision of assumptions – Bring in Change and Connectedness into the assumption

[This makes induction valid according to lecturer]

The "primate" shows that atomicity is not the negation of continuity.

All entities are "social", i.e., have a reactive significance for other. A value for each other?

[Are we to know how these entities control the field of display?]

Shimer [Restatement]

Is the field finite? Vibrations. Differentiation by vibratory motion engendering space-time.

What is meant by the frequency of vibration of the primates? Absolute motion or rest?

All reality is knowable & all Taken-account-of?

Comment by Whitehead

1. Repudiates the fundamental character of the physical field: it is an abstraction. "Here we are".

"Finiteness of the event essentially clamors for its relation to the beyond"

"Certain relata are peculiarly associated with the event.

This patch of yellow is there for me."

2. "My whole theory of primates" a warning –

Electron via its tubes of force extends throughout all space. As old as Faraday, 1847. Electron has its focus here.

Atom – a structural character –

Space-time the abstract idea of process as continuous

Then somebody says Quantum Theory implies Essential discontinuity of space-time –

Then show that the Clerk Maxwell equations --- ?

Can be made to elicit the identical

Can be got by vibrant theory as well as passive charge theory of elections –

And furthermore that other way has the advantage of supposing that the Clerk Maxwell equations hold within the nuclei. providing an odd field with ridges within[1] the nuclei.

So you are not bound down to an essential discontinuity of space-time. Cf. *Principles of Natural Knowledge*. 1918

[Uniform objects those that stay no matter how small the period of time. A time can't be a uniform object. Can a molecule? Not if it is a motion. Elections & Protons not uniform.]

Quantum ~~theory~~ relation actually deduced

There may be temporal minimum for physical field – not for space-time

---

1. These two terms especially difficult to decipher, but it is a speculation mentioned by C. Maxwell (attributed to Descartes).

---

3. Not all objects are complex. Sense-data, as black. doubt if it is complex.

Realization in space-time. A thought process is a weaker degree of reality. Thinking of a name. I affect object – object not me.

|4|
Whitehead. 10/17 – 2

[If all realization is in space-time – can space-time be realized?]

Brain molecules take account of a Thought (subconsciously) –
Reenvisagement makes this a thought
[Won't say there is any ignorance on the whole]
[Realizing King Arthur & Julius Caesar
King Arthur realized in sense-data – if not in space-time –
Mrs. X says the image of Arthur takes account of her as much as J.C.][1]

"Envisagement" has "the secret of what Lloyd Morgan & Alexander call emergent entities"

= Realization plus its conditions.

"The book is black" means just that, but it is an error. Its truth is that the book is black for <u>me</u> at this time – a "very complex" relation.

Refers to theory of secondary qualities – "only justifiable at end of a long evening" to refer to authority in this way.

To make your statement unconditionally is "very odd" – though you might not be displaced from position –

|5|
Whitehead. Saturday. Oct. 17[2]

1. Question raised in Seminary – would be put by Russell

Grant that to justify induction – beyond Hume

But we all do believe in induction – VForeseeV future. What is the least you can add to Hume to justify induction? That will be justified.

Assume the relevance of past to future – Van der Walle's question – Assume the whole thing? (add nothing?)

Tackles it as an anti-Kantian in the Kantian spirit.

We do not first observe a fully-determinate fact then plaster it over with assumptions from without

---

1. Closing bracket provided.
2. This date cannot be correct – Saturday was 18 October. These notes capture a discussion held on Saturday, 18 October in Whitehead's regular Emerson Hall lectures. There was some overlap in who attended his Emerson Hall lectures and these Friday evening Seminary lectures, and there seems to have been some cross-over in discussions as well. It happens again in Emerson Hall on 15 November. Happening at least twice underscores that Whitehead was considering similar issues in both courses. And just why this particular set of notes was found within the sequence of notes Hocking kept for the Seminary is not clear. Perhaps everyone was invited to join the Saturday discussion on these occasions.

Not a bloodless dance either, of sense-data or Hegelian categories.

Reality as experienced is the realization of essences fusing together. "It is the fusion which is realized."

"Its ingression into realization is its limited fusion with other factors." The existences qua existences are without value. Process has value – "Throwing forward" is the "qualification of the future by the past". Idea of "beyond" is essential.

Conjecture arises because of our partial knowledge of the present fact. If we knew more the throwing-forward would be complete.

Not a blank assumption brought in from without to qualify present fact complete before us but as part of the analysis of the present fact itself.

---

2. Is the proposition "All reality is process" the result of induction? Answer: Observation. But we can't observe it all.

"The idea of fitting-on to the present fact in the sense of being indicated by the present fact"

Dreams refuse to fit on. The only criterion: not vividness, nor the taking of notes. Dream of hovering & taking notes & resolving not to be deceived by the illusion that it was a dream.

Fitting-on is a spatio-temporal fitting-on. Can't be sure everything will fit on. That is what makes the unity of experience. There are doubtful memories. But we know that there was a past fitting on to the present.

|6|

3. Shimer. Is immediate experience only of process?

Don't we experience the static?

Answer. With Bergson – No instantaneous present is experienced. Always one in between any two "instants" of time. We don't experience a dense series of static configurations. We don't experience jumps.

What is immediately before me is the transition – or the "Endurance" of the walls, etc.

– Why call endurance process q.⟨?⟩ static?

Answer. The permanents are the universals that characterize the door, etc. The "world-path" of endurance is the individuality of the door.

– I experience the table as static & have to think to find it characterized by process.

Answer. You experience an enduring table. "You are thinking of your experience of tableness which is not in time at all but has its spatial realization in the present" Protest against this spatialization of experience, which he attributes to 'intellect' "Fallacy of misplaced concreteness"

– It has taken advanced physical theory to convince common sense that experience is process? Bell[1] brings common language to witness.

---

1. This likely refers to Winthrop Pickard Bell, whose notes have been transcribed for the Emerson Hall lectures. Bell's personal diaries suggest he did not regularly attend the Friday evening Seminary, but obviously he was present for this Saturday discussion.

Answer. Thinks he can interpret what common sense perceives
Refers to *Principles of Nature Knowledge* & *Concept of Nature*
Shelly vs. Wordsworth in attitude to science asadding to apprehension of reality. Drop of water –

|7|
Whitehead Seminary. Oct 24. 1924

What is value? Perry. Miscellaneous Essays.
Definition of Value –
Journal of Philosophy XI. #6.
Also. X. 167-8. Formulation of question

M^c^Gill. Objects in General. 1. Mathematical objects
Points & lines as limits of converging series of volumes – "Extensive abstraction" –
A moment of time – limit of converging faculty –

Holds with Montague that we see points & lines –

[Whitehead: we do see edge, but we are talking about a little blob that can be made smaller if we want. Mathematics gives the ideal of exactness. Not a line but a process of abrupt transition.][1]

Intersections –
Can't stimulate retina
Doesn't have to! We see it just the same

---

2. Objects in General – Russell's *Analysis of Mind*
1. Broad's criticism: Multiplication of entities –
2. Wouldn't it be "better to endure" 'substance'
3. Sum is a class & doubtful whether classes exist!

[Whitehead comments: careful to give meaning to classes.][2]

4. Assumes each aspect indubitable – Dogmatic
Broad's sensa are of objects – allow correction
Hicks, G.D.:[3] We are aware of table, not of patches of color. Sense datum are artificial abstraction

---

"There can be no such thing as a judgment about a mere sensum" Penny = round brown & round hard.

1. This comment is inserted in the left-hand margin.
2. This comment is inserted in the left-hand margin.
3. George Dawes Hicks (1862–1941) was a professor of philosophy at University College London while Whitehead was there. It is not clear what source is being referenced here, but Russell did publish a response to Hicks in *Mind* (1913), who had written an article critical of Russell, entitled 'The Nature of Sense-Data', which had appeared in *Mind* in 1912.

An object as "The sum of perceived aspects" Russell

Whitehead is phenomenalism accepted[1]

Broad's view that the real penny furnishes the most elegant explanation of its appearance

Moore's view that we know the real penny immediately. (Though it may be doubted)

What is the nature of the entity immediately before us

Accepts multiple-relation view as worth developing – G.D. Hicks & Broad

|8|

Objections to the class theory of objects –

1. Class can't be defined by enumeration
2. How define the class? Only as the class that clusters around the object

It isn't the definition we work by.

Doubts whether we can nail sense data to object.

Strange noises in the night & other stray sense datum

Close eyes & see greenness – not belonging to objects.

But the phenomenalist gets into a strange position unless you can somewhere point to the real(?) physical field.

[We see the surfaces & we understand the line etc., but what we understand with sense, that we experience.]

Recognizing a hat – or Venus when seen (a) as planet, (b) as inhabitant.

Secondary qualities – prevent you from taking simply the view that the color simply is there! Rainbow, etc.

Crimson & smoky cloud.

Whitehead objects to recognizing an object by its relations, because it is so easy –

Bushmen & dogs do it. Ms. Langer[2] needs Newton

Whitehead: how about seeing double? Rats in delirium

Ms Langer By relational structure.

---

1. We read this as a rhetorical question.
2. Susanne Katherina Langer (1895–1985) earned her undergraduate degree from Radcliffe in 1920 and her PhD in 1926. It is interesting to note that someone referred to as 'one of the first women to achieve an academic career in philosophy' participated in Whitehead's first graduate Seminary at Harvard.

Whitehead thinks the ground of recognition is denser than the relational Complex. Half of a sheep is mutton – not sheep. Organic unities must have some other ground of recognition.

Must we not assume density[1] somewhere –
May we not assume it anywhere?
"We apprehend the sense-data as controlled"

|9|
Whitehead Seminary. October 31, 1924

Farber . Perceptual object

That all conceivable objects are objects of perception. Perceptual Essential of General objects.

Acts have an empirical foundation vs. Russell.

Object is immediate & mediate also - "Alteration".

Whitehead's Impressions

1. A half & half compromise between the Kantian phenomenal object & a postulated object. Because we would get into such a muddle if we didn't postulate it.

Farber answers. he limits himself to experience but must regard perception as consummated in region of ignorance. – Determinable.

Phenomenology = extreme rationalism.

2. In laying such emphasis on the process of perception the real thing slips away. Be naive "I am very naive individual". Must be separate from self, or we can never disengage it. Can't get it at the end of a long procedure of perception. Immediately the object up against us.

There something of a heretic – Differ from Hume – impression a good word but missed the point – looks on impression as an act of cognizance on mental side.

What you are aware of is an impression, but the awareness is not the impression – the impression is a moment of realization, a moment of being real, i.e., your being real together with something else.

What becomes is that togetherness. That being real is a relatedness & that you are one of the *relata*.

Impression has nothing to do with mentality or cognizance at all – only arises – more complex cases. Essence of impression is a certain mutuality which we are Cognizant of. This is what we mean by the reality of the table – parts of the table are something for each other – & this the same mutuality of which we are primarily aware.

1. It is difficult to decipher this term but it looks like a related term four lines above.

|10|
Oct. 31 – 2

In observing the table we have an immediate awareness of this togetherness of the partial impression.

3. Thought of scene in Tolstoy's *War & Peace* battle of Austerlitz – night before – Austrian chief of staff reading out the battle orders & the old general going to sleep. Except that Napoleon turned up in quite another way. Napoleon was reality. The matter can't be quite as complex as that: a deep sort of skepticism I feel when I ~~feel a~~ hear that sort of thing. Bring a bigger unity into it.

Besides our different shots at describing it, what is there that gives us the knowledge that there is something to describe – naively given. [right!]
In the cognitive process is there some assurance that there is something definite for us to Know?

Farber tries to get away from common sense into the theoretical sphere, which has no presuppositions. How do we come to speak of objects –

Whitehead says we have to answer that by pointing.[1] Naiveté means state of the metaphysician

But if he can't discern an element that is an object no discussion can bring it in. Deprecates the division into the naive & the theoretical, because the theoretical side gets into a muddle with no particular objects. You don't know the process of knowing nearly as well as you know the watch: & again can't get away from a naive statement about something held up before your for knowledge.

|11|
Oct. 31 – 3

Intentional objects.
Whitehead. "Can't get hold of it" – What is it? With regard to this table, where am I to get hold of the intentional object?
Farber. It is all that is within knowledge.
Whitehead. In case of illusion or mirror, is there an intentional object?
Farber. Thinks this use especially good – Real would be a limiting case.
Whitehead. Doesn't see that the idea of the intentional object helps you to determine the status of the physical object.

No use for the transcendent object. Reduced to a "rather interesting habit of your mind to refer things to an x." –

1. It is difficult to decide between 'pointing' and 'positing'.

Either one has to take that account or else have got to look to see whether one hasn't omitted something in one's account.

[Whitehead wants to get the not-me[1] as a concrete *Cognitum*][2]

The issue he thinks is that Farber's object is passive, has nothing to do, like an Epicurean god. The object Whitehead sees is playing an active part. Science gets atoms & electrons – all active – doesn't go to an unknown X.

Sameness is an emergent value.

Maintains itself as being that kind of value. during a space-time continuity. The aggregate or shape or integral *eidos* of the table. You must have the emergent unity or you would never get the event to be one thing. The impressions have a form which endures – there is no such thing as an instant of time. A general value.

1. It is difficult to determine whether this unusual hyphenated term is perhaps 'not-me' or 'not-one'.
2. Closing bracket provided.

Appendix
Sample scans of original handwritten notes

Three pages of handwritten notes have been selected, one each from the notes kept by Winthrop Bell, W. E. Hocking and Louise Heath. They have each been selected from the notes taken during Whitehead's lecture on 14 April 1925, at roughly the same portion of the lectures in which a particularly important diagram was drawn by Whitehead on the blackboard. This should provide for a direct comparison. Each is fairly representative.

Technically, digital scans were taken of these pages by each of the archives holding the originals, and these were adjusted in contrast, only, to make each clear (to minimise the amount of text coming through from the backside, for example) but without introducing distortion, and saved as 300 dpi TIFF files.

Page |135| of Bell's notes (see pages 331–3)

6501

Whitehead : 14. April 1925

Associated Symbols vs Substitute Symbols

Essence of Language & its enorm. power from fact that it roams over all field & is at home there.
Like human being among animals. They can do everything better than men. General inefficiency as animal
But enormous adaptability. But in math. symbn spec. adaptn to a partic. region of thot.
As soon as you move one step beyd that field – no signce or
Wittgenstein put out well – Also Stout already in Psychol. of subst. symbs.
In the symbol, viewed merely as physl fact before you – you have reln wh. embody in genl
charn same chars as embodied in subj. matter. You see in the symbols same charcs as in
obj. Then, if Symbols easier to reason about than the more complex objs symbolized – – – &c.
Connectn of our Expce ("Percn" when we're aware of it) w. the Sp-Templ Continuum.
Advtg of a diagram = that it is itself spatial. Some of the charcs you're speaking of are
preserved in the diagram. Diagram must be used & kept in its specl sphere. Philost has
lost there a valble help.
How Expce involves Sp-templ Relatn: – Assuming altern Sp-templ reln (mod. Relativity).
Take event A_0 (has defte past & defte future) & then there's the "vague present." (That wh.
will fall within some durat of simultaneity accordg to some Sp-T system.

(Really a 4-dimensl diagram wanted, not of 2)
Now consider an enduring Entity [a Man, e.g.]
..... not as discrete but as atomic.
Same pattern reiteratg itself in the life-history.
Present Duratn for
[The "one" Substce sounds like one Thing among others but: –
underlying Activity whose essence is individualisn of what is all
in the different things. "Aspects of things in A_0". – Then have
a dual form – essentl related to one of 2 types. (1) Aspect of the
whole – the other things – as now – under aspect of being Now – Limitn of Whole into the Now.
Aspect of what the Whole other than a is for a at A_0 now. Unity of World
depends on there being a unity of Scheme of Limitn. Take aspect of another entity
(Green Entity b) – How is its aspect in A_0
(life-history course) Take the more direct aspect that comes w. vel. of light. [Anything
coming with less is to be looked on as indirect – to be looked on as imptn of this on this
on this ... & then there].
The main obstructn comes in your body. Bergson, rightly; – Body rather as a
method for keeping out – it's only what it selects that really is for one the aspect of
things. Finally the effective part of one's body – – – ?? — "Direct aspects" – Where
we have what we call normal or true percn. Owing to obstrn of body it's often
very vaguely located Sp-Temply. — It's at D for you
Under aspect of "Now" you get B at D.
Then aspect of the Past. Grasping whole essence of a together in F. So "this ["Recol"]
will be of the type of Recollection & this "["Perc"] of type of direct percn.
Now he puts on a purple entity just whizzing by. Z enduring pattern
or entity at Z_0 (that individualisn of the undl activity
Z's now What's happening at Z_0 for z
What's for z at Z_0 under aspect of being at E
& " a " A_0 " " D is entirely a representn of essence
of a, or Z_0 as outcome of same event; b at B
You've got a 3-term reln for each percd expce [Knowl. of a [3]-term reln
betn A_0 B & D. for a at $\{A_0, Z_0\}$
For z at Z_0 — 3 term reln betn $\{Z_0, A_0\}$, B, E
You're getting asst. of same fact in both. So this differce of Sp-T. relatedness

Page |April 14 – 1| of Hocking's notes (see page 334)

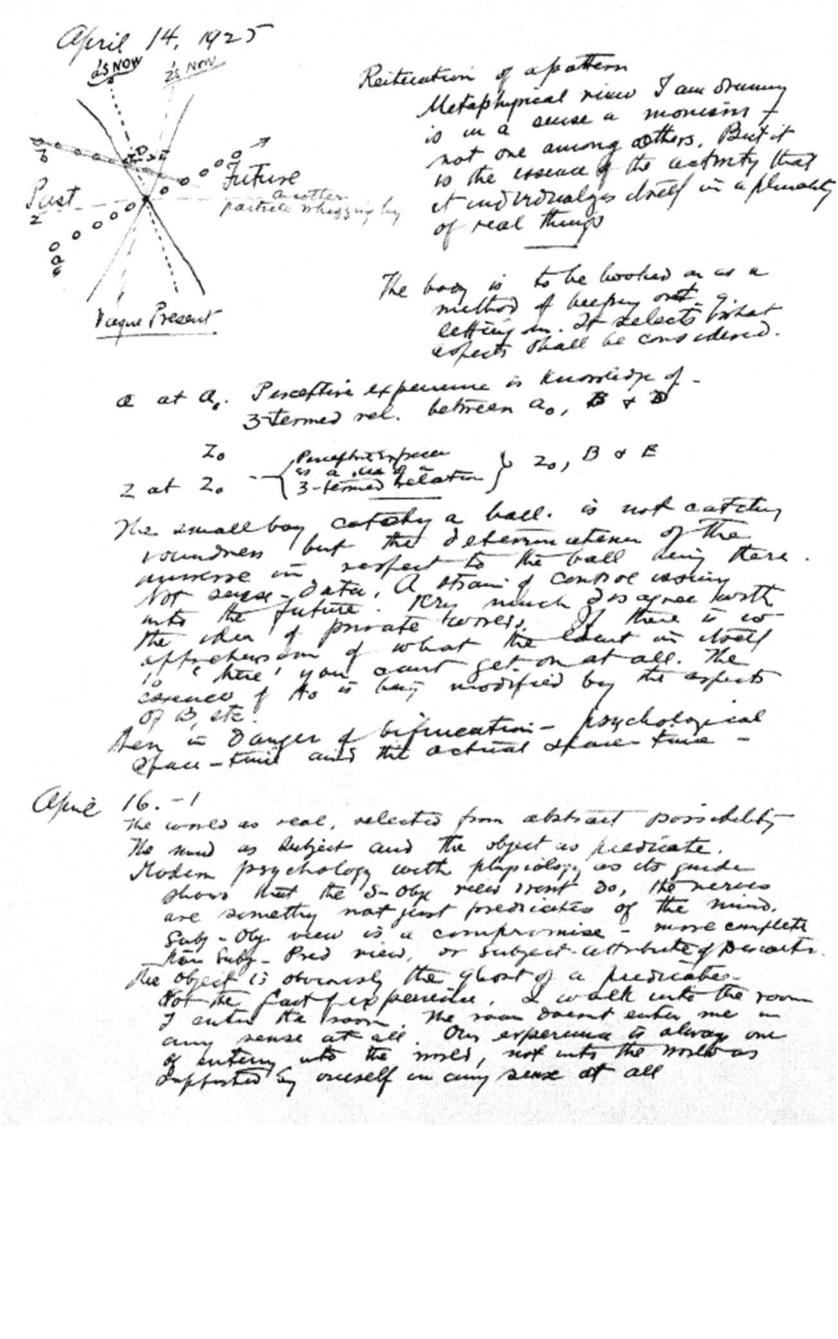

April 14, 1925

2s NOW

2s NOW

Future

Past

another particle whizzing by

Vague Present

Reiteration of a pattern

Metaphysical view I am drawing is in a sense a monism, it is not one among others. But it is the essence of the activity that it individualizes itself in a plurality of real things

The body is to be looked on as a method of keeping out — what letting in. It selects what aspects shall be considered.

a at a_0. Perceptive experience is knowledge of — 3-termed rel. between a_0, B & D

z_0

z at z_0 — { Perceptive space is a idea of 3-termed relation } z_0, B & E

The small boy catching a ball. is not catching roundness but the determination of the universe in respect to the ball being there. Not sense-data, a strain of control working into the future. Very much disagree with the idea of private worlds. If there is no apprehension of what the latent is, itself 'here' you can't get on at all. The essence of A_0 is being modified by the aspects of B, etc!

Then in danger of bifurcation — psychological space-time and the actual space-time —

April 16. – 1

The world as real, selected from abstract possibility —

The mind as subject and the object as predicate.

Modern psychology with physiology as its guide shows that the S-Obj view won't do, the nerves are something not just predicates of the mind.

Subj-Obj view is a compromise — more complete than Subj-Pred view, or subject-attribute of Descartes.

The object is obviously the ghost of a predicate —

Not the fact of experience. I walk into the room I enter the room. The room doesn't enter me in any sense at all. Our experience is always one of entering into the world, not into the world as supported by oneself in any sense at all

Page 78 of Heath's notes for the second semester (see pages 509–10)

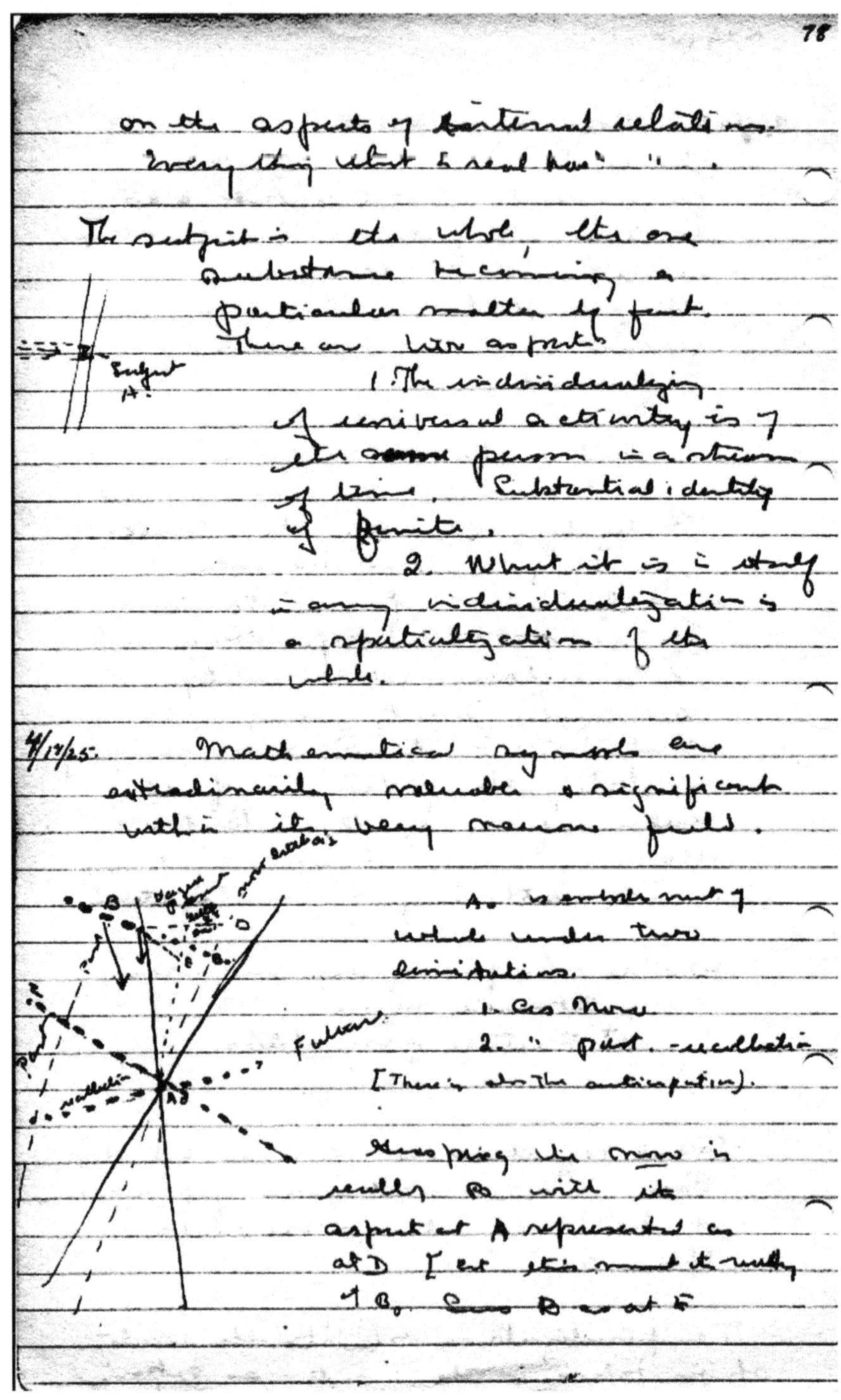

78

on the aspects of internal relations.
Everything that is real has' " .

The subject is the whole, the one substance becoming a particular matter of fact. There are two aspects.

1. The individualizing of universal activity is of the same person in a stream of time. Substantial identity of finite.

2. What it is in itself in any individualization is a specialization of the whole.

4/17/25. Mathematical symbols are extraordinarily valuable & significant within its very narrow field.

A is an entire unit of whole under two limitations.
1. As now
2. " past - recollection
[There is also the anticipation].

Assuming the now is really B with its aspect at A represented as at D [at this moment its reality of B, [illegible] B as at F

Index

Page numbers in *italics* refer to illustrations, 'n' refers to notes.